OXFORD MONOGRAPHS ON GEOLOGY AND GEOPHYSICS NO. 3

Series editors

P. Allen
E. R. Oxburgh
B. J. Skinner

OXFORD MONOGRAPHS ON GEOLOGY AND GEOPHYSICS

1. De Verle P. Harris: *Mineral resources appraisal: Mineral endowment, resources, and potential supply: concepts, methods, and cases*
2. J. J. Veevers (ed.): *Phanerozoic earth history of Australia*
3. Yang Zunyi, Cheng Yuqi, and Wang Hongzhen: *The geology of China*

The Geology of China

YANG ZUNYI
Beijing Graduate School, Wuhan College of Geology

CHENG YUQI
Ministry of Geology and Mineral Resources, People's Republic of China

WANG HONGZHEN
Beijing Graduate School, Wuhan College of Geology

CLARENDON PRESS · OXFORD · 1986

Oxford University Press, Walton Street, Oxford OX2 6DP
Oxford New York Toronto
Delhi Bombay Calcutta Madras Karachi
Kuala Lumpur Singapore Hong Kong Tokyo
Nairobi Dar es Salaam Cape Town
Melbourne Auckland
and associated companies in
Beirut Berlin Ibadan Nicosia

Oxford is a trade mark of Oxford University Press

Published in the United States
by Oxford University Press, New York

© *Yang Zunyi, Cheng Yuqi, and Wang Hongzhen, 1986*

All rights reserved. No part of this publication may be reproduced,
stored in a retrieval system, or transmitted, in any form or by any means,
electronic, mechanical, photocopying, recording, or otherwise, without
the prior permission of Oxford University Press

British Library Cataloguing in Publication Data
Yang, Zunyi
The geology of China—(Oxford geological sciences series; 3)
1. Geology—China
I. Title II. Cheng, Yuqi III. Wang. Hongzhen
555.1   QE294

ISBN 0-19-854460-X

Library of Congress Cataloging in Publications Data
Yang, Tsun-i.
The geology of China.
(Oxford geological sciences series)
Bibliography: p.
Includes index.
I. Geology—China. 2. Mines and mineral resources—China.
I. Ch'eng, Yü-ch'i. II. Wang, Hongzhen.
III. Title. IV. Series.
QE294.Y36 1985   555.1   85-7283
ISBN 0-19-854460-X

Typeset by Latimer Trend & Company Ltd, Plymouth
Printed in Great Britain by
Butler and Tanner Ltd, Frome, Somerset

PREFACE

This volume is written with the object of providing an up-to-date survey of the geology of China on the basis of numerous publications and unpublished papers that have appeared since the founding of the People's Republic of China. Four and a half decades have elapsed since *The geology of China* by the late Professor Li Siguang (J. S. Lee) was published in 1939. Although a comprehensive survey of the geology of China is now in preparation elsewhere, the present work, the second attempt to introduce the subject in a foreign language, should prove to be useful to readers.

In this volume Chinese names are spelt throughout according to the Pinyin system, although some authors would retain the old Wade spelling for stratigraphical names; for instance, Chihsia Formation instead of Qixia Formation, and Shihhotse Formation instead of Shihezi Formation. The stratigraphical names in both the Pinyin and Wade systems are given in the index for the benefit of readers familiar with the old spelling. Also, place names of minority nationalities are spelt in accordance with the usage of the Cartographic Publishing House (Beijing), e.g. Alxa instead of Alashan, Qamdo instead of Changdu; and, when they are not available, they are transliterated in Hanyu Pinyin.

Regional stratigraphical tables of various regions and provinces and simplified stratigraphical correlation tables published independently by the Geological Publishing House (Beijing) and Science Press (Beijing) form the indispensable basic information for us. They are listed in the selected references, though they are not specifically mentioned in the text. All the sketch maps showing rock distribution and stratigraphical regions except the Sinian and the earlier ones are reproduced with minor changes by permission of the authors concerned and the Geological Publishing House.

The literature on the geology of China is voluminous, but, as space is limited, only selected references are given. Names and abbreviations of isotopic age-dating institutions are given in the Appendix; in the text the isotopic datings are usually followed by abbreviations in parentheses.

The authors acknowledge the help of their colleagues in the preparation of this work. In particular we must mention Wu Shunbao and Li Fenglin, who helped Yang in preparing columnar sections as well as fossil plates, Song Shuhe, Xu Huifang, Zhang Shouguang, and Wu Jiashan for assisting Cheng in many ways; and Jiang Yinchang, Cui Xinxing, and Jia Weimin, who helped Wang in various ways. Thanks are also due to Tang Yuanqing, Zhao Yudong, Li Hong, and Song Yinnian for the excellent figures and maps they produced. Also, we would like to express our appreciation for Weng Fa's service throughout the preparation of the manuscript. Finally, our special thanks are due to the editors of the Oxford Geological Sciences Series for their encouragement and support.

Beijing
July 1985

Y.Z., C.Y., W.H.

CONTENTS

Plates appear between pages 278 and 279

Part I: Background — 1

1. Review of the development of Chinese geology *Yang Zunyi* — 3
    Ancient mining and metallurgy — 3
    Chinese literature concerned with geological observations — 3
    Modern geology—The period 1912–49—A new era: 1949 to the present — 3

2. The physical features of China *Yang Zunyi* — 6
    Relief features — 6
    Drainage systems — 8

Part II: Stratigraphy — 9

3. Introduction to the stratigraphy of China *Wang Hongzhen* — 11
    Stratigraphical provinces of China — 11
    Stratigraphical classification and terminology — 11
    Brief summary of the stratigraphical development of China—The Precambrian—The Palaeozoic—The Mesozoic—The Cainozoic — 13

4. The Archaean *Cheng Yuqi* — 16
    Introduction — 16
    Southern part of north-eastern China (Jilin and Liaoning)—The Anshan Group of Liaoning—The Anshan Group of Jilin — 16
    Eastern part of Shandong Peninsula and south-eastern Shandong Province — 18
    Central and western Shandong — 19
    Huaiyang region — 20
    Yinshan–Yanshan regions—Yinshan Region—Yanshan Region — 20
    Taihang–Wutai–Luliang region — 22
    The northern slope of the eastern Qinling Mountains and the neighbouring regions — 23
    North-western China — 25
    Correlation of the Archaean formations — 25
    Certain evolutional characters of the Archaean of China — 25
    Some chemical features of certain important types of Archaean metamorphic rocks of the North China Platform — 28

5. The Proterozoic *Wang Hongzhen* — 31
    The Lower Proterozoic—North China—North-west China—South China — 31
    The Middle and Upper Proterozoic—North China—North-west China—South China—Other parts of China — 35

6. The Sinian System *Wang Hongzhen* — 50
    Introduction — 50
    The Yangzi Platform — 51
    South-east China — 54
    North-west China — 55

CONTENTS vii

    North China—Western Henan—Southern Liaodong—The Xuzhou–Huainan region   57
    Other parts of China—North-east China—The Kunlun–Qinling region—Western and southern Yunnan   60
    The main palaeogeographic features of China in the Sinian Period—South China—North China and North-west China   62

7. **The Cambrian System**   *Yang Zunyi*   64
    Introduction   64
    Stable or platform type of sedimentation (North China type)—The Yangzi region—Eastern Yunnan—North China region—Tarim region   65
    Mobile or basinal type of sedimentation—Northernmost part of China—Qilian region—South-east region   68
    Relatively active or transitional type (Jiangnan type) of sedimentation—Qinling Mountains—Jiangnan region   69
    Boundary problems—The Sinian–Cambrian boundary—The Lower–Middle Cambrian boundary—The Middle–Upper Cambrian boundary   69

8. **The Ordovician System**   *Yang Zunyi*   73
    Introduction   73
    Stable sedimentation type—Yangzi region—North China region—Tarim region   76
    Mobile or basinal type of sedimentation—South-east region—Tianshan–Hingan region   78
    Transitional sedimentation type (Jiangnan type)   78
    Boundary problems   81
    Correlation of Ordovician subdivisions   81

9. **The Silurian System**   *Yang Zunyi*   82
    Introduction   82
    Stable sedimentation type—Yangzi region—Kalpin, Tarim region—Himalayan region   83
    Mobile sedimentation type—Hingan and Tianshan—Tianshan–Nei Mongol region—Qilian region—West Qinling region—South-east region   86
    Transitional sedimentation type—South Anhui—Baoshan, West Yunnan   87
    Boundary problems—Lower limit—Lower–Middle and Middle–Upper Silurian—Upper limit   88
    Correlation of Silurian subdivisions   88

10. **The Devonian System**   *Yang Zunyi*   91
    Introduction   91
    Stable sedimentation type—South China region—South-east region—Qomolangma area—Qilian region   92
    Mobile sedimentation type—Saerburte Mountain, West Junggar—Da Hingan Mountains—Longmenshan–Qinling region—Têwo, West Qinling—West Sichuan–North Xizang region   97
    Boundary problems—Silurian–Devonian boundary of the marine regime—Lower–Middle Devonian boundary—Continental Middle–Upper Devonian boundary   99
    International correlation   99

11. **The Carboniferous System**   *Yang Zunyi*   102
    Introduction   102

Stable sedimentation type—South China—Qinling (Têwo, Gansu Province)—North China region—Tarim region—South Qilian region—Qilian–Helan region—Qamdo, West Sichuan–North Xizang region—Qomolangma area ... 103
  Mobile sedimentation type—Kunlun–Qaidam—Beishan, Gansu—Burnhan Buda, Kunlun–Qaidam region ... 110
  Biogeographic provinces in Carboniferous times—South China province—Northern basin province—West China province ... 111
  Boundary problems—Devonian–Carboniferous boundary—Lower–Upper Carboniferous boundary and Carboniferous–Permian boundary ... 111
  International correlation of Carboniferous rocks ... 112

12. The Permian System   *Yang Zunyi* ... 113
  Introduction ... 113
  Permian chronostratigraphic units of China—Lower Permian Qixian—Lower Permian Maokouan—Upper Permian Longtanian—Upper Permian Changxingian ... 114
  Platformal sedimentation type—Qinglong, Guizhou (South-west region)—Jiangnan region and South China—South-east region (Yongding, Fujian)—Taiyuan, North China region—Kalpin, Tarim region—Himalayan region ... 117
  Basin sedimentation type—Northern Basin (eastern part)—Northern Basin (western part) ... 119
  Palaeobiogeographical provinces—Southern (warm water) or Tethys province—Northern (Boreal) province—Mixed faunal type ... 120
  Boundary problems—Lower–Upper Permian boundary—Permian–Triassic boundary ... 121

13. The Triassic System   *Yang Zunyi* ... 126
  Introduction ... 126
  Stable sedimentation type—Marine: Yangzi region—Longmenshan, Sichuan—South China region—Southern zone of the Himalaya region—North China region—Junggar (Karamay area), North Xinjiang–Beishan region ... 127
  Mobile sedimentation type—Garze–Yajiang area, Western Sichuan—Xizang–West Yunnan region—Northern Himalaya region ... 131
  Transitional sedimentation type—Youjiang region, West Guangxi ... 133
  Boundary problems—Lower limit of the Triassic—Lower–Middle Triassic boundary—Anisian–Ladinian boundary—Upper boundary ... 133
  Correlation ... 135

14. The Jurassic System   *Yang Zunyi* ... 140
  Introduction ... 140
  Marine Jurassic—Qinghai–Xizang region—Northern Guangdong region—Natan Hada (Wanda Hills) region ... 140
  Non-marine Jurassic—North-west region—North-east region—South-west region—Central–South region—South-east region ... 142
  Boundary problems—Triassic–Jurassic boundary in the marine regime—Upper limit of the Jurassic: the Jurassic–Cretaceous boundary ... 146
  Correlation of Jurassic sequences ... 147

15. The Cretaceous System   *Yang Zunyi* ... 153
  Introduction ... 153

CONTENTS ix

	Non-marine life—Early Cretaceous—Middle Cretaceous—Late Cretaceous	153
	Marine Cretaceous—Yarlung Zangbo area—Kashi, Tarim Basin—North Xizang–East Karakorum	157
	Non-marine Cretaceous—South-east region—Hengyang, Yangzi region—East Shandong, North China—Songliao, North-east region—Hailar, Gobi region—East Gansu, North-west region—Lanping–Simao area, West Yunnan	158
	Boundary problems—Marine sequences—Non-marine sequences	160
	Correlation of Cretaceous subdivisions	161
16.	The Cainozoic *Yang Zunyi*	168
	The Tertiary System—Introduction—Marine deposits—Non-marine deposits—Tertiary boundary problems	168
	The Quaternary System—Introduction—Quaternary stratigraphy and palaeontology—Lower limit of the Quaternary—The problem of Quaternary glaciation in China—Correlation	173

Part III: Magmatic and metamorphic rocks of China — **187**

17.	Magmatic rocks and magmatism in China *Cheng Yuqi*	189
	Introduction	189
	Chronological sequence of igneous activity—Basic and ultrabasic intrusives—Granitic rocks and granitoids—Volcanic rocks	189
	Major regions of igneous rocks—Tianshan–Yinshan–Da Hing'an Mountains region—Tarim Alxa–North China (Platform) region—Kunlun–Qilian–Qinling region—Qinghai–Xizang Plateau and Transverse Ranges—South China region (South-central and South-east China region)	193
	The Yarlung Zangbo ophiolite thrust belt—an example of the ophiolite suite	200
	Mesozoic continental volcanic rocks of the eastern part of China	202
	Granitic rocks of the eastern part of South China—Geochronological groups—Structural control of the granitic activity—Petrochemical and evolutionary features—Granite series	203
	Magmatic series of the ultrabasic intrusions	209
18.	Metamorphic series and metamorphic belts of China *Cheng Yuqi*	210
	Introduction	210
	Metamorphic series and metamorphic belts—their distribution and characteristics—Metamorphic series chiefly of Archaean metamorphic age—Metamorphic series of dominant Early Proterozoic metamorphic age—Metamorphic series and belts of dominant Middle to Late Proterozoic metamorphic age—Metamorphic series and belts chiefly of Early Palaeozoic (Caledonian) metamorphic age—Metamorphic series and belts chiefly of Late Palaeozoic (Hercynian) metamorphic age—Metamorphic series and belts chiefly of Early Mesozoic (Indosinian) metamorphic age—Metamorphic series and belts chiefly of Late Mesozoic (Yanshanian) metamorphic age—Metamorphic series and belts chiefly of Cainozoic (Himalayan) metamorphic age—Summary	210
	Evolutional features of metamorphism	219
	Polymetamorphism and polystage metamorphism	221
	Examples of Archaean and Early Proterozoic metamorphic terrains—The Wutai	

    Mountain and part of the Taihang Mountains—Metamorphic and migmatitic rocks of the Taishan Group ... 223
    Localities of glaucophane-schist and related rocks and paired metamorphic belts ... 230
    Types of migmatization and examples ... 231

Part IV: Geotectonic development of China ... 235

19. The tectonic framework and the geotectonic units   *Wang Hongzhen* ... 237
    Introduction ... 237
    Geotectonic units of China—Terminology—The principal tectonic domains of China ... 237
    The North (Siberian–Mongolian) Continental Margin Domain—The North-west region—The North-east region ... 240
    The North China Continent and Continental Margin Domain—The North China Platform and adjoining continental margins—The Tarim Platform and adjoining continental margins—The South-western continental margin tract ... 244
    The South China Continent and Continental Margin Domain—The Yangzi Platform—The western elements of the domain ... 248
    The Circum-Pacific Continental Margin Domain—The Cathaysian Caledonides—Hercynian and Indosinian Fold Zones—Yanshanian and Himalayan Fold Zones ... 253

20. Geotectonic development   *Wang Hongzhen* ... 256
    Introduction ... 256
    Megastages and stages in the crustal development of China ... 256
    Megastage of formation of continental nuclei: crustal development of China before the Proterozoic ... 260
    Megastage of formation of the Platform: crustal development of China in the Middle and Upper Proterozoic (pre-Sinian) ... 261
    Megastage of formation of Pangaea: crustal development of China from Sinian to Triassic—The Caledonian Stage—The Hercynian (Variscan) and Indosinian Stages ... 263
    Megastage of disintegration of Pangaea: post-Indosinian crustal development of China—The new tectonic framework of East China: basin development, volcanic activity, and formation of marginal seas—Basins and ranges of North-west China—Northward movement of the northern Gondwanan massifs and formation of the Qinghai–Xizang Plateau ... 270
    Concluding remarks ... 273

Appendix: Abbreviations of isotopic dating institutions ... 277

Selected references ... 279

Stratigraphic index ... 291

Subject index ... 297

PART I
BACKGROUND

1. REVIEW OF THE DEVELOPMENT OF CHINESE GEOLOGY

Yang Zunyi

ANCIENT MINING AND METALLURGY

China has long been known as one of the ancient cultural centres in the world. Passing over crude stone implements made by Palaeolithic man (Plate I.1, I.2) and fine stone and bone implements and ornaments used by Neolithic man (Plate I.3), the use of copper for implements was discovered in China about 4000 years ago (2000 BC), bronze ware was made in 1600 BC (Plate I.4) and iron implements were commonly used not later than 450 BC. A few years ago people were amazed to learn of the discovery of an old mining site estimated to date from 700 ± BC in Tonglu Mountain, Daye, Hubei Province, where metallurgical equipment with a melting capacity of 1–1.5 tons a day was unearthed. The ancient Chinese people had evidently acquired a substantial knowledge of useful copper and iron minerals, as well as of metallurgy, and an embryonic stage of mineral prospecting must have begun at least 2000 years ± BC. In the subsequent several thousand years sufficient mineral resources such as metallic ores, coal, salt deposits, clay, and precious stones were excavated. (See Xia *et al.* 1980 for further details.)

CHINESE LITERATURE CONCERNED WITH GEOLOGICAL OBSERVATIONS

In ancient Chinese literature there are many acute observations on minerals, rocks, geological processes, and fossil remains, etc., much of which was brilliantly described and is worth mentioning. A few examples will suffice. *The classic of the mountains and rivers* (Shan Hai Jing) written 500–300 BC contains chapters which describe five mountain systems, 347 ranges, 258 rivers and lakes, and give details of 73 mineral species, including metallic ores such as gold, silver, copper, iron, and tin, as well as non-metallic ones (jades, realgar, chalk, etc.).

In about the same period (the fourth century BC) in *The book of master Guan* (Guan Zi) there is reference to stream erosion and the formation of meanders. The famous classic *Commentary on the waterways* (Sui Jing Zhu), written about 512–518 AD, describes volcanoes, earthquakes, hot springs, karst, and, most interesting of all, fossil fish found in Xiangxiang, Hunan Province.

In his *Notes on the altars to the immortals on Ma-Gu Mountain,* Yan Chenqing, a famous Tang scholar, wrote: '... on a high cliff are snails and clams [molluscan shells], which might have resulted from the change of sea into land ...'. *The dream pool essays* (Meng Qi Bi Tan), written in 1086, by Shen Kua in the Song dynasty, contain an explanation of the sculpturing of land-forms by stream erosion in the Yandang Mountain, Zhejiang Province. In 1077 Shen made observations on fossil shells on the Taihang Mountains and correctly explained that they were formed near a seashore. Needham (1959) gives detailed and systematic descriptions of the physical geology and palaeontology of this period, although some criticisms of the book have been made by Li and Wang (1981).

MODERN GEOLOGY

It was not until the end of the nineteenth century, however, that modern geology developed in China, as evidenced by the translation into Chinese (by Hua Hengfang in 1872) of J. S. Dana's *Textbook of mineralogy* and C. Lyell's *Principles of geology* by the same scholar in 1873. In the latter part of the nineteenth century and the early part of the twentieth century foreign travellers and geologists came to China with various motives. Among them were F. von Richthofen (1860–72), L. von Loczy (1877–80), V. A. Obrutschew (1892–1909), Sven Hedin (1893–1908, 1929–33), and B. Willis and E. Blackwelder (1903–4). Their geological reports and papers dealt with the stratigraphy, palaeontology, structure, and mineral resources and provide basic information on the geology of China.

The period 1912–49

Modern Chinese geological studies started in 1912 when under the Ministry of Agriculture and Commerce of the Peking Government two organizations were set up: the section of geology, which later became the Geological Survey of China headed by V. K. Ting and the Geological Institute, practically a geological training school under H. T. Chang, who together with Ting and W. H. Wong acted as chief instructors. It was here that in the spring of 1916 the first group of about twenty brilliant Chinese geologists graduated. Ten of them, including L. F. Yih and C. Y. Hsieh, immediately joined the National Geological Survey. The geological school was discontinued in 1916, but the department of geology in Peking University was restored in 1918 to take the school's place and two important figures, J. S. Lee and A. W. Grabau, joined the teaching staff in 1920, thus continuing the training of geologists in China. Geological or geoscience departments were set up in other universities, such as Shanxi University, South-east or Central University, Sun Yat-Sen University, and Tsinghua (Qinghua) University in the middle and late 1920s and Chongqing University in the middle 1930s. They too trained qualified geologists in the 1930s and 1940s, though the number was not great.

The National Research Institute of Geology, Academia Sinica, was founded in 1928 with J. S. Lee as its director. Its publications include memoirs, contributions, and monographs.

The National Geological Survey set up in Beijing in 1916 was moved to Nanjing in 1935 and continued to operate there until it was forced to move to Beipei, Chongqing, Sichuan Province before the fall of Nanjing to the Japanese invaders. Its publications include bulletins started in 1920, a series of memoirs, *Palaeontologia Sinica*, and some geological maps. A *Bulletin* and the *Geological Review of the Geological Society of China* were published regularly.

Among provincial geological surveys may be mentioned the Hunan Survey set up in 1927, the Guangdong and Guangxi Survey, also set up in 1927, the Jiangxi Survey (1937), and the Sichuan Survey (1938), all of which carried out geological explorations.

During the Sino–Japanese War the National Geological Survey was operating under difficult conditions in Sichuan with its regional offices at Kunming in Yunnan Province, Lanzhou in Gansu Province, and Urumchi in Xinjiang. The Bureau of Exploration for Mineral Resources headed by C. Y. Hsieh carried on investigations chiefly on metallic and non-metallic ores. In general, the Chinese geological profession experienced its worst times during this period. The total number of geologists actually working before the liberation was only about 300.

During the period 1912–49 a number of foreign geologists also took part, either singly or in groups, in the study of the geology of China. These included J. G. Andersson, Teilhard de Chardin, and A. W. Grabau.

A new era: 1949 to the present

Since the founding of the People's Republic of China development of the national economy has become one of the chief concerns of the government, and geology, like the other branches of science, has played an important part in national construction. In 1952 the Ministry of Geology was established with the urgent task of promoting geological exploration and mineral prospecting. This task has been shared by provincial bureaus of geology, which were progressively set up. In the mean time the Fuel Ministry was also organized to control exploration for petroleum and coal. Geological exploration teams were formed in other industrial ministries, and later the Fuel Ministry was replaced by the Ministry of Petroleum Industry and the Ministry of Coal Mining Industry.

Under the Chinese Academy of Science (Academia Sinica) the Institute of Geology, Institute of Geology and Palaeontology, Institute of Geochemistry, and other institutes were established, all of which have made great contributions to geological science.

Reorganization of the then-existing universities and colleges resulted in the establishment of three geological colleges in Beijing, Changchun, and Chengdu—now increased to five, including the Hebei and Xian Geological Colleges—with the purpose of training many new geologists. Also, nearly twenty colleges and universities under various ministries (Education, Mining, Petroleum, Metallurgy, etc.) set up geological departments or specialist courses for the same training purpose. Geological schools have now been opened to train field and laboratory workers.

Under the Ministry of Geology and Mineral Resources (formerly Ministry of Geology) is the Chinese Academy of Geological Sciences, which includes more than a dozen institutes. Consequently, there are now over 70 000 geological personnel, mostly college and technical school graduates.

Today geological research workers in China cover a wide range of subjects, theoretical as well as practical, and their results are published regularly in a total of about fifty periodicals and monograph series, mostly with summaries or abstracts in English.

It is worth mentioning that over 130 kinds of mineral resources have so far been explored in detail and their reserves made known. Many discoveries are concerned with metallic and non-metallic ores. Of special significance is the discovery of several great oilfields in East China, including the famous Daqing and Shengli fields.

Noteworthy also is the vast amount of work done on various types of surveying and mapping. These embody:

(1) maps compiled from comprehensive geological surveys at a scale of 1:1 000 000 for 98 per cent of

the land area, and at a scale of 1:200 000 for nearly 60 per cent of the land surface;
(2) areal geophysical surveying by various methods on different scales, both from the air and on land;
(3) offshore marine geological mapping, chiefly geophysical, for extensive oil exploration; and
(4) geochemical survey and geological mapping for special purposes.

The Geological Society of China, founded in 1922, has now become one of the important academic bodies. So far it has enrolled about 40 000 members who take an active part in academic exchanges within 23 professional sections. Its publications include the well-known *Acta Geologica Sinica* (being the continuation of the *Bulletin*) and the *Geological Review*.

At the meeting held in August 1982 commemorating the Society's Sixtieth Anniversary, the main achievements in geological sciences in China were fully reviewed by its president (Huang, T. K. 1982) and the following aspects were discussed at length by various professional sections (*Geological Review* 1982): structural geology, technology of comprehensive utilization of mineral resources, exploration geophysics, and exploration geochemistry.

2. THE PHYSICAL FEATURES OF CHINA

Within the limited space available here it is inappropriate to give a full description of the physical geography of China. Two salient features will be treated briefly: relief features and drainage systems.

RELIEF FEATURES

China is a country with a great expanse of hills, mountains, and plateaux. It is marked by great changes in relief: highest in the west, becoming lower and lower to the east, like a giant inclined staircase. Over this staircase flow eastward great rivers, such as the Yangzi, the Yellow River, and the Heilongjiang, which empty into the Pacific (Fig. 2.1). The general relief embraces three giant steps of the staircase (Fig. 2.2). The highest step is the Qinghai–Xizang (Tibetan) Plateau, which is composed of exceedingly high mountains (including the Himalayas, the Gandise, Tanggula, Kunlun, the Transverse Ranges) and great plateaux with an average elevation of 4000–5000 m above sea-level, forming the well-known 'roof of the world'. Between the outer skirt of the first step and the Da Hingan Mountains, the Taihang Mountains, Wushan Mountain, and Xuefeng Mountain form the second step, which encloses chiefly broad plateaux and great basins. Eastwards are the Nei Mongol Plateau, the loess plateau, Sichuan Basin, Yunnan–Guizhou Plateau; northwards are great basins enclosed by great mountain ranges, such as the Tarim Basin, studded between the Kunlun and Tianshan Mountains, and the Junggar Basin enclosed between

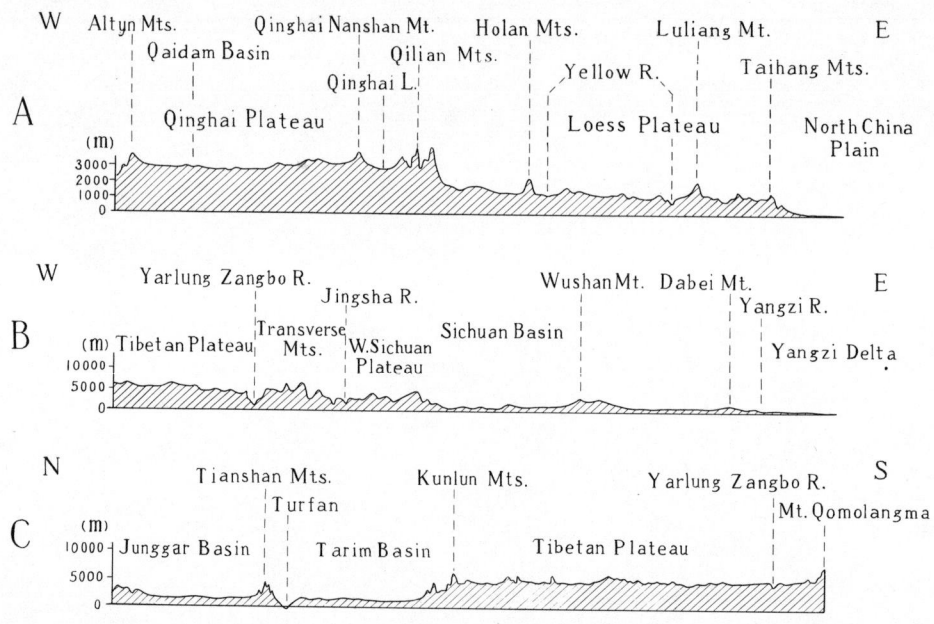

Fig. 2.2. Topographic profiles of China: A, E–W profile from the Qinghai Plateau to the North China Plain; B, E–W profile from the Tibetan (Xizang) Plateau to the Yangzi delta. C, N–S profile from the Junggar Basin to the Tibetan (Xizang) Plateau. (Modified, after Ren, Yang, and Bao (1980).)

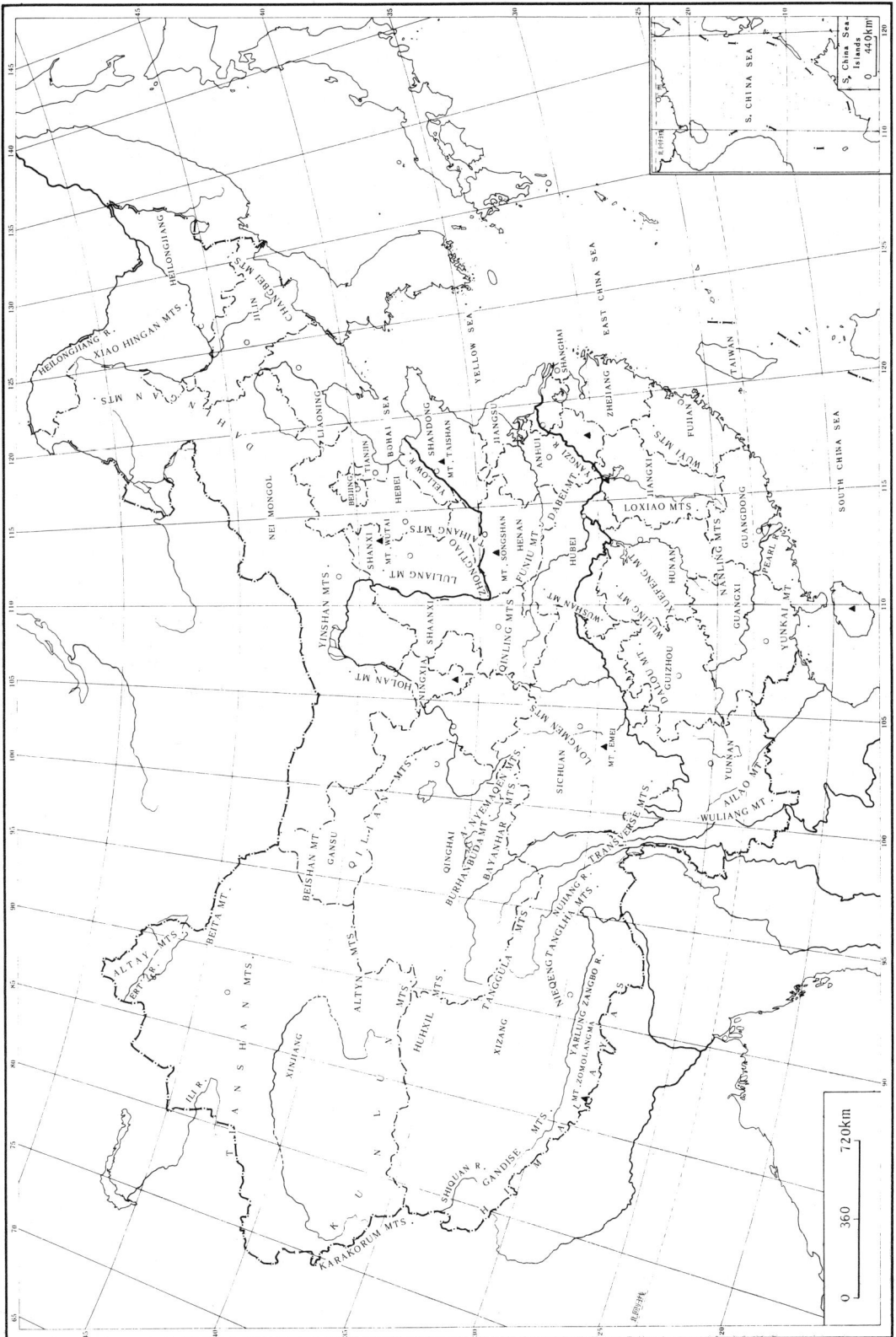

Fig. 2.1. Outline map of China showing the main mountain systems, drainage systems, and the administrative divisions. Abbreviations for most of the administrative divisions are given in parentheses. Anhuei (Wan), Fujian (Min), Gansu (Gan), Guandong (Yue), Guanxi (Guei), Guizhou (Qian), Heilongjiang (He), Hebei (Ji), Henan (Yu), Hubei (E), Hunan (Xiang), Jiangsu (Su), Liaoning (Liao), Ningxia (Ning), Qinghai (Qing), Shandong (Lu), Shanxi (Jin), Shaanxi (Shan), Sichuan (Chuan), Taiwan (Tai), Xizang (Zang), Yunnan (Dian), Zhejiang (Zhe); e.g. Shan–Gan–Ning Basin stands for Shaanxi–Gansu–Ningxia Basin; Qing–Zang, for Qinghai and Xizang, etc.

the Tianshan and the Altay Mountains. The lowest step, dropping 1000–1500 m from the second step, forms the vast eastern plains and hills, including (from north to south) the North-east China Plain, North China Plain, Weihe Plain, and the plain covering the middle and lower reaches of the Yangzi, all of which join in an almost continuous strip.

DRAINAGE SYSTEMS

Streams in China may be divided into those having external and those having internal drainage. To the external ones belong the Pacific, Indian, and Arctic drainage systems, distributed respectively in the eastern, southern, and north-western parts of China, covering 63.8 per cent of the whole domain of the country. The internal drainage system lies within the eastern part of the great Eurasian inland drainage system, covering the arid region in the western part of China (Nei Mongol and Xinjiang) and the interior of the Qinghai–Tibetan Plateau, occupying 36.2 per cent of the whole area of China. The line dividing these two drainage systems starts from the western foot of the Da Hingan Mountains in the North and extends NE–SW through the southern foot of the Nei Mongolian Plateau, Yinshan Mountains, Holan Mountain, Qilian Mountains, Riyue Mountain, Bayanhar Mountains, Nieqeng Tanglha Mountains, and Gangdise Mountain, and ends at the western border of China. This clear-cut line generally follows mountain ridges and foothills. To the east of this line external drainage dominates, apart from the Ordos Plateau and the Songhuajiang–Nenjiang Plain where there is some internal drainage; everything to the west of this line, except the Ertixhe River in the north-western corner of Xinjiang, belongs to the internal drainage system. The Ili River in West Xinjiang empties into the inland Lake Balkhash in the USSR.

The Pacific drainage area covers 56.8 per cent of the country's total area, equivalent to 88.9 per cent of the external drainage area. It includes China's important rivers, such as the Yangzi, the Yellow River in Central China, the Heilongjiang in North-east China, and the Pearl River in South China. The Indian drainage area is small, only 6.5 per cent of the total area; it includes the streams distributed in the southern part of the Qinghai–Xizang Plateau, namely the Nujiang (the Upper Salween), the Yarlung Zangbo River, the Shiquan River, and the Xiangquan River.

The Arctic drainage area is smallest in extent, with a single river, the Ertixhe River of north-west Xinjiang, covering 0.5 per cent of the country's territory.

The main mountain systems and the drainage systems most often mentioned in the book are shown in Fig. 2.1. This figure also shows the administrative divisions of China. There are altogether 27 provinces and autonomous regions, and three municipalities.

PART II
STRATIGRAPHY

3. INTRODUCTION TO THE STRATIGRAPHY OF CHINA

Wang Hongzhen

Systematic study on the stratigraphy of China began in the early 1920s, and the main stratigraphic successions of most regions in the eastern part of China were established toward the end of the 1930s, when the late Professor Lee Siguang (Lee, J. S. 1939) prepared the first comprehensive treatise on the geology of China. Since the foundation of New China in 1949, extensive investigations of all aspects of geology have been carried out in most parts of the country, including the remote frontier regions. It is now possible to give a comprehensive general review of the stratigraphical development of China.

STRATIGRAPHICAL PROVINCES OF CHINA

The aim of synthetic stratigraphical provincialization or regionalization is to reflect the main features and the overall characteristics of the entire stratigraphic succession in different regions. In the 1930s Li Siguang set up fifty-six stratigraphic tables representing different regions in his *Geology of China*. At the first All-China Stratigraphic Congress held in 1959, Huang Jiqing (1962) reviewed the principles of stratigraphical regionalization and proposed a classification of stratigraphic provinces of China, mainly on a tectonic basis. Wang, H. (1978) indicated that sedimentation types and sedimentary associations provide the basic criteria for the recognition of palaeogeographic and palaeotectonic regimes under which they are formed, and the contact relations between the successive stratigraphic units or sequences serve to mark palaeogeographic and palaeotectonic changes in the different periods. A third important factor is the palaeobiogeography, which may indicate the kinship of faunas and floras between different continents at different times, and the differences caused by isolation of continents by oceanic basins that have long since disappeared (Wang, H. 1978). Thus, through analyses of all these aspects, we are able to recognize the tectono-palaeogeographic frameworks and to trace their changes through geological history. These provide the basis for stratigraphical regionalization.

In general, stratigraphic types and sedimentary associations may be classified according to their sites of deposition, which have an environmental as well as a tectonic significance. Three stratigraphic types, stable, intermediate, and mobile, may be recognized in both the continental and the marine category, thus making up six types altogether, under which sedimentary associations may be further distinguished according to their main facies and general composition.

Applying the disciplines cited above, we can recognize three types of stratigraphic super-regions, which are further subdivided into stratigraphic regions. The first type may be called continental. It includes most regions consolidated before the Sinian Period (*c.* 850 Ma), and is represented by a stable type of Phanerozoic cover sequence on the platforms. The second type is intercontinental: i.e. regions situated between the continental super-regions. It includes both the mobile fold zones and the semistable median massifs. The third type comprises the continental margin tracts, which are mainly composed of mobile zones that for most of geological time have faced the oceanic basins. Continental margins bordering the long-existent Pacific oceanic basin, probably since the Middle Proterozoic, may have been unique in geological history.

Subdivision of a stratigraphic super-region into regions is based on spatial relations and on the changes of nature and mode of stratigraphical development in different tectonic stages. A preliminary categorization of the stratigraphic provinces of China is shown in Fig. 3.1.

STRATIGRAPHICAL CLASSIFICATION AND TERMINOLOGY

Although the publication of the *International Stratigraphic Guide* has met with the general approval of many stratigraphers, the principles of stratigraphical classification remain a much-discussed problem, and the stratigraphic terminology used at present is by no means unanimous. Diverse opinions seem to centre on the mutual relations between the different categories of stratigraphic units, and the use of chronostratigraphic units, especially in the Precambrian.

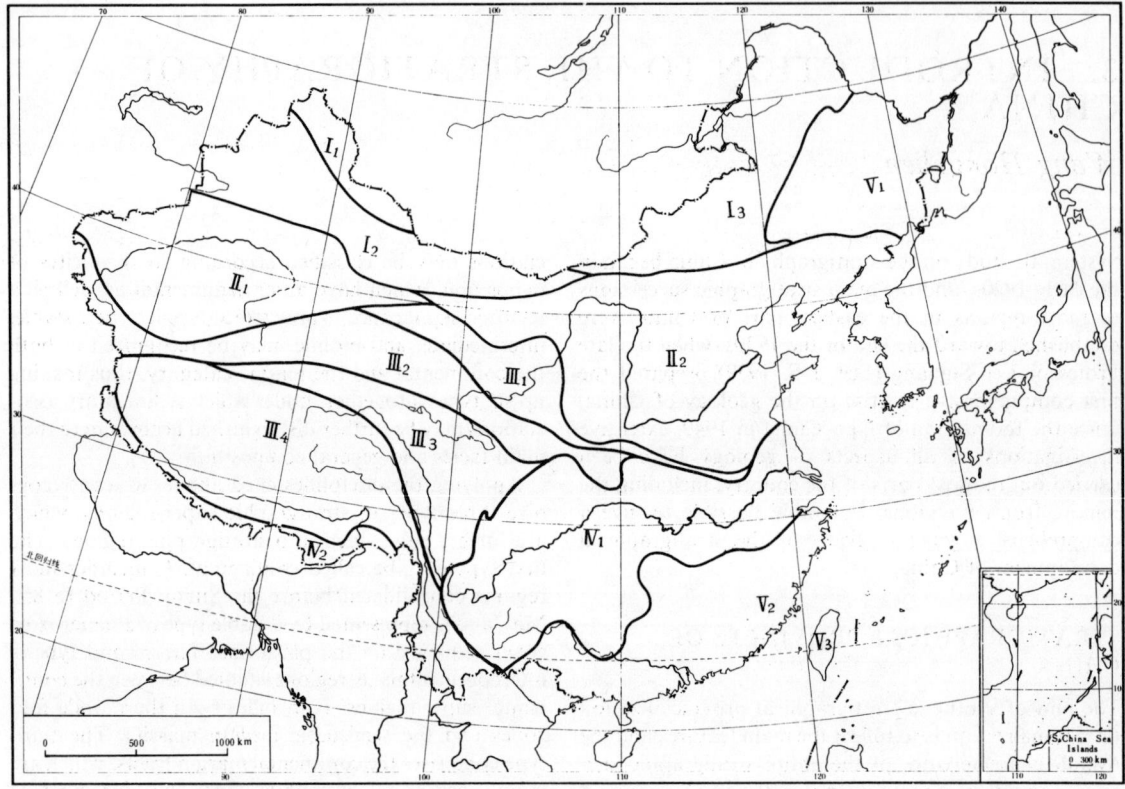

Fig. 3.1. Stratigraphic Provinces of China (simplified after Wang, H. (1978)).
I. Northern Intercontinental Super-region: I_1, Altai–Bayitik Region; I_2, North Tianshan–Beishan Region; I_3, Hingan––Nei Mongol Region. II. Northern Continental Super-region: II_1, Tarim Region; II_2, North China Region. III. Central Intercontinental Super-region: III_1, Qilian–North Qinling Region; III_2, Kunlun–South Qinling Region; III_3, Bayan Har–West Sichuan Region; III_4, Xizang–West Yunnan Region. IV. Southern Continental Super-region: IV_1, Yangzi Region; IV_2, Himalaya Region. V. Eastern Continental Margin Super-region: V_1, Songhuajiang–Yanbian Region; V_2, South-east Region; V_3, Taiwan Region.

There seem to be two opposite views on the classification of stratigraphic units. The German school represented by O. H. Schindewolf (1970) would consider the chronostratigraphic category as unique and essential and regard all other categories as preliminary or preparatory, and as having no independent and permanent position. Schindewolf and his co-workers insist on a unified stratigraphical classification. More popular is the viewpoint of H. D. Hedberg (1976), who holds that, although chronostratigraphic units are of special importance, we may have as many stratigraphic categories as the properties and methods used in stratigraphic studies. Hedberg advocates a manifold stratigraphical classification. Two kinds of stratigraphic classification should be specially emphasized (Wang 1982b). The first, the chronostratigraphic, aims at setting up a world-wide standard in terms of chronological subdivisions based on temporal criteria. The second, the lithostratigraphic, seeks to establish local standards of stratigraphic subdivision, based on physical criteria of the strata. These two categories are distinct from all other categories in that they provide an overall and complete subdivision of the strata concerned without overlaps or gaps, and afford a strict hierarchy of different ranks of stratigraphic units.

In practice, 'group' and 'formation' are the most commonly used lithostratigraphic units. For the Precambrian, 'supergroup' and 'subgroup' may also be useful, but 'supergroup' is seldom used in Chinese stratigraphic terminology. As a subdivision of a formation, the term 'member' is often used in China, more or less in the sense of 'subformation' in Soviet usage rather than in the original meaning used in American literature.

There has been much discussion about the relationship between 'stage' and 'chronozone'. A 'chronozone' is regarded here as the basic unit in chronostratigraphy and is a constituent part of a 'stage'. Various kinds of biostratigraphic zones may probably be regarded as preparatory for the establishment of chronozones, although biostratigraphic units themselves have their own significance and are independent. Sometimes the term 'assemblage' is used for faunal zones of regional value, possibly separated by barren intervals. 'stage' and 'chronozone' may be of world-wide application, but they are usually regional and conform to the main biological realms of the time. Special difficulties arise in the chronostratigraphic classification of the Precambrian. There is recently a tendency to establish chronostratigraphic units, mostly of System rank, of regional nature, e.g. in the Proterozoic of Australia, which may be called regional chronostratigraphic units. We have used, below the Sinian System in the Middle and Upper Proterozoic of China, preliminary System names, but they are only regional chronostratigraphic units and are not of the same status as the Sinian and Palaeozoic Systems.

BRIEF SUMMARY OF THE STRATIGRAPHICAL DEVELOPMENT OF CHINA

The Precambrian

The classification of the Precambrian has been much discussed in recent years, and no scheme of chronostratigraphic units within the Archaean or the Proterozoic has found general acceptance. It is probably appropriate first to set up regional standards within the scope of a continent, and then to attempt to establish international chronostratigraphic units of 'system' rank. In order to avoid complications, we are inclined for the time being to confine the use of 'aeon' and 'aeonthem' to the traditional Cryptozoic and Phanerozoic, the boundary of which may be drawn between the Sinian and the Cambrian. Thus the Archaean and the Proterozoic are here considered as eras, the boundary between which is tentatively drawn at 2500–2600 Ma.

The Precambrian is widely distributed in the Northern Continental Super-region (Fig. 3.1, I), notably on the North China Platform, or the Sino–Korean Platform of Huang. Within the Platform the Archaean is well developed, especially on the Shanxi Plateau and in the Jiaodong–Liaodong region. The Lower Proterozoic is well represented and forms the basement complex together with the Archaean, and is therefore sometimes inseparable from the latter. The Lower Proterozoic includes two parts, the lower Wutaian covering the time-span 2600–2200 Ma, and the upper Hutuoan the time-span 2200–1850 Ma, both considered as regional chronostratigraphic units that can be correlated throughout most parts of China. In many cases the Lower Proterozoic forms a folded basement, in contrast to the Archaean, which is more widespread and constitutes the crystalline basement of the platforms. The Middle and Upper Proterozoic display an extremely clear and continuous succession in the Yanshan region of North China. The well known Jixian section amounts almost to 10 000 m in thickness and covers a time-span of over 1000 Ma, approximately from 1850 to 850 Ma. It is on this excellent sequence that the three Middle and Upper Proterozoic 'Systems', the Changchengian (1850–1400 Ma), the Jixianian (1400–1050 Ma), and the Qingbaikouan (1050–850 Ma), are established. But apart from the Yanshan region, the Proterozoic becomes much thinner and is usually incomplete. The uppermost part of the Upper Proterozoic, the Sinian System proper, is confined to the southern and eastern border parts, and is generally lacking in the interior of the platform.

In the western part of the Northern Continental Super-region, along the northern border of the Tarim basin, in Quruktagh, the whole Proterozoic, including the Sinian, and possibly part of the Archaean, is well developed and contains several unconformities within the whole sequence. The Archaean is also inferred to be present underneath, mainly on the evidence of aeromagnetic surveys, under the thick Mesozoic and Cainozoic to the south of latitude 38 °N, in the southern part of the basin. The Proterozoic is also met with in the central belt of Palaeozoic folded mountains, notably in the Central Tianshan and Central Qilian of North-west China.

The Yangzi Platform of South China, which constitutes an integral part of the Southern Continental Super-region (Fig. 3.1, IV), has a basement composed mostly of the Proterozoic, but here the Sinian System is widespread and forms the first genuine cover sequence on the platform, overlying the basement rocks with a flagrant unconformity. The nuclear part and some uplifted border parts of the Yangzi Platform are of the Lower Proterozoic, and are usually of diverse sedimentary types in different parts of the platform.

To the south and east of the Yangzi Platform in the inner parts of the circum-Pacific Eastern Continental Margin Super-region (Fig. 3.1, V), only the Upper Proterozoic is definitely known. It is of a geosynclinal facies and is not readily separable from the overlying

Lower Palaeozoic. Also in the extensive Qinghai–Xizang Plateau, indubitable Precambrian strata, probably of Middle and Late Proterozoic age, are known in the Himalayas and also in the Nyainqentanglha, both belonging to the northern border of the ancient Gondwana continent.

The Palaeozoic

Palaeozoic strata are extensively distributed on the stable continental super-regions in the eastern parts of China. In South China, the Palaeozoic and the Triassic are almost entirely marine and abundantly fossiliferous throughout, on the Yangzi Platform as well as in the surrounding geosynclinal regions. In the South Qinling and the Guangxi–Hunan region, both of Caledonian age, a nearly complete Palaeozoic sequence of great thickness is usually met with. The Lower Palaeozoic shallow-sea facies is especially well preserved on the Yangzi Platform, and passes eastwards into the geosynclinal facies of South-east China, the continental margin of that time. The Cambrian and Ordovician on the platform are abundant in carbonates, but the Silurian contains more argillaceous and clastic rocks. In the continental margin tract of South-east China, the main part of the Lower Palaeozoic belongs to the Cambrian and the Ordovician, while the Silurian is generally lacking.

As compared with the Lower Palaeozoic, the Upper Palaeozoic is less widespread on the Yangzi Platform. The Devonian and the Carboniferous are usually confined to the border areas, and are in general lacking in the interior of the platform. It was not until Permian times that transgressive shallow seas once more covered the whole Yangzi region and brought about the extensive deposition of carbonates. In South-east China, the Upper Palaeozoic, unlike the Lower, is as a rule not metamorphosed, although facies change is much more pronounced than on the platform.

In the Northern Continental Super-region, the Palaeozoic shows a considerable difference between the North China and the Tarim platforms. In North China, the Lower Palaeozoic is represented by the Cambrian and the Lower to Middle Ordovician, consisting mainly of carbonates of littoral to shallow-sea origin, except for the Middle Ordovician, which on the whole approximates to open-sea type. A large depositional gap representing the whole of the Silurian and Devonian is almost universal on the platform. The subsiding belt to the west of the Ordos Massif is peculiar in that Ordovician argillaceous strata containing graptolites, similar to a geosynclinal facies, are found: there was evidently a connection with the Qilian geosyncline at that time.

The Upper Palaeozoic on the North China Platform begins with the Middle Carboniferous. It is usually paralic and partly continental in facies, and passes upwards into terrestrial deposits of Late Permian to Triassic age. The subsiding belt on the western side of Ordos is characterized by a much thicker sequence of strata.

Great progress in stratigraphical research has been achieved in North-west China in the last two decades as a result of extensive geological mapping on a scale of 1:200 000. The Palaeozoic succession of a stable marine type has been largely established on the northern border region of the Tarim Platform, in juxtaposition with the mobile type developed in South Tianshan. The Cambrian and the Ordovician of these regions are similar to each other, both belonging to an intermediate sedimentary type. From the Silurian upwards the mobile facies prevailed, until in the mid-Permian when the Hercynian orogeny brought an end to the geosyncline. The Palaeozoic of Tarim and South Tianshan is characterized by Tethyan faunas, but the sea-ways seem to have been via Central Asia in the west rather than through the Qilian geosyncline in the east.

In the extensive Northern Intercontinental Super-region (Fig. 3.1, I), important work has been done in Palaeozoic stratigraphy. In Altai and Junggar, to the north of Tianshan, the Palaeozoic faunas are of Boreal type, with the Silurian as the most noticeable. The Carboniferous and the Permian are partly terrestrial and yield an Angaran flora, which is usually complicated by the presence of European forms such as *Callipteris zeilleri*. The Palaeozoic succession has also been established and correlated in North-east China, where Boreal faunas and Angaran floras are predominant, but the stratigraphic sequence is less clear and often discontinuous owing to intensive diastrophism.

The wide and complicated terrain lying between the Alxa (Alashan) region of the North China Platform in the north and the Himalaya Region (Fig. 3.1, IV$_2$) in the south, including several median massifs and fold zones of various ages, may be conveniently called the Central Intercontinental Stratigraphic Super-region (Fig. 3.1, III). The Palaeozoic is well developed in the Qilian Mountains and on the Qaidam Massif. Typical geosynclinal sedimentary associations of Early Palaeozoic age are well developed in North Qilian and South Qilian, and were folded by the Caledonian orogeny. In Late Palaeozoic time, a marine regime prevailed only to the south of the Qilian Mountains. It was only in late Early Carboniferous and early Middle Carboniferous times that extensive marine transgressions again occurred and once more flooded the Qilian region.

In the Himalayas, an almost complete marine

Palaeozoic sequence of stable type is found on the northern slope. The Lower Palaeozoic contains marine faunas partly akin to those of the Yangzi Platform. The Upper Palaeozoic, however, shows faunas and floras of more clearly Gondwanan type, which extend even to the north of the Bangong Lake region.

The Mesozoic

At the end of the Palaeozoic, marine basins had entirely disappeared from the Northern Intercontinental and the Northern Continental Super-regions. In the Triassic, the vast regions to the north of a line joining West Kunlun in the west via Altyn and Qilian to Qinling in the east, were dominated by a continental regime with various kinds of basin deposits. To the south of this line, the Triassic is mainly marine, especially in the extensive belt of western Sichuan and Bayan Har, where thick Triassic flysch and paraflysch are the chief component rocks of the Indosinides.

The Indosinian orogeny brought about a basic reform of the geotectonic frame and a consequent change in the stratigraphic development of China. Except for southern Qinghai, Xizang, and part of western Yunnan, where marine conditions still prevailed, the main part of China was incorporated in the Eurasian continent. Three kinds of inland basin deposits may be recognized. North-west China was dominated by large intermontane basins, such as Junggar and Qaidam. A large cratonic basin was developed in the western part of Tarim, which was partly transgressed by Late Cretaceous seas through the Kashi region. In the western part of East China, large cratonic basins, the Shaanganning in the north and the Sichuan–Yunnan in the south, reached their widest extent in Late Triassic and Early Jurassic times. After the Middle Jurassic, the basins diminished in size and had almost entirely disappeared before the end of the Cretaceous. The eastern part of East China, i.e. the regions east of the Hingan, Taihang, and Wuling Ranges, is characterized by Jurassic volcanic and sedimentary rocks and by Cretaceous downwarped- and faulted-basin deposits of various types. Volcanic eruptions began in the Early Jurassic and became prevalent in Late Jurassic times, when volcanic activity became general throughout the circum-Pacific belt of Eastern Asia. Further to the east, marine Jurassic rocks occur along the Ussuri River in the north, and marine Cretaceous rocks are known in Taiwan province. Both regions represent the continental margin at that time.

It is mainly in southern Qinghai and Xizang that normal marine facies of the Jurassic and Cretaceous are extensively developed. Sedimentation of geosynclinal type is common, but a comparatively stable facies is known in the Qiangtang, the Gangtise, and the Karakhorum. Late Cretaceous marine intercalations are also found along the southern border of the Tarim basin. The marine Mesozoic sequence may be followed from eastern Xizang to western Yunnan, where the Cretaceous becomes mainly terrestrial. Volcanic and volcano-sedimentary rocks are abundant in the Mesozoic of the Qinghai–Xizang Plateau, especially in the Cretaceous of the Gangdise region.

The Cainozoic

The distribution and development of the Cainozoic of China closely resemble those of the Mesozoic. In East China Palaeogene continental deposits usually occupy rifted basins in the huge subsiding belt that extends from the Bohai Bay in the north to the Leizhou Peninsula in the south, which contains important oil-producing horizons in the middle and upper part of the sequence, especially in North China. The Palaeogene usually rests on the Cretaceous, usually the Upper Cretaceous, which forms the basal beds of the basin. Small faulted red basins of graben or semi-graben type are common in South-east China. They are mainly Palaeogene in age, but some date from the Late Cretaceous. The Neogene rocks usually form a universal cover on the Palaeogene in the subsiding belt, and may reach a huge thickness to the east of the Taihang Range. On the Shanxi Plateau they occur in the Fen-Wei rift valleys and pass upwards into the Quaternary. In Taiwan, Palaeogene miogeosynclinal deposits occur extensively in the Central Range, while Neogene and Quaternary beds, including typical ophiolites, are distributed along the eastern coast.

In the large intermontane basins of North-west China, the Cainozoic follows the Mesozoic and commonly overlaps the latter towards the interior of the basins. The greatest known thickness of the Cainozoic is recorded in the Qaidam basin and in the southern part of the Junggar basin. Palaeogene and Miocene marine incursions are known in the western part of the Tarim basin. Pliocene molasse deposits, which pass upwards into the even coarser Quaternary molasse, are distributed all along the piedmont belts of the mountains. On the Qinghai–Xizang Plateau, Tertiary basin deposits, including the open lacustrine type in the central part and the intermontane type in the surrounding mountains, are also widely distributed. The typical marine nummulitic facies up to Middle Eocene in age is mostly restricted to the south of the Yarlung Zangbo and south of the Karakhorum. Pliocene and Quaternary molasse is also reported recently from the northern piedmont belt of the Himalayas.

4. THE ARCHAEAN
Cheng Yuqi

INTRODUCTION

Archaean formations of China are found chiefly in North China and the southern part of North-east China, and also subordinately in adjacent regions in East and North-west China (Fig. 4.1). These exposures, together with the overlying lower Proterozoic metamorphics, form the ancient crystalline basement of the North China Platform, or the 'Sino–Korean Platform' of Huang Jiqing. There are also rocks of disputable Archaean age in the far north-west. These formations consist of metamorphic complexes, migmatitic complexes, and various igneous rocks, most of which have undergone complicated changes throughout their long geological history, especially during the Archaean. Almost all the rocks have undergone medium- to high-grade metamorphism, commonly in the amphibolite facies and partly in the granulite facies, almost to the exclusion of the greenschist facies. The discrimination between the two types of Archaean terrain in China, i.e. the high-grade region and the greenstone belt, therefore calls for comprehensive investigations of the geological environment for the formation of protolithic rocks and the tectonic evolution they have undergone, as well as their composition and petrological characteristics.

In terms of stratigraphy and geological evolution, the Archaean may in some districts be subdivided into two major volcano-sedimentary cycles. These are well developed in part of the Yanshan region, Hebei Province. So far as is known, the Archaean of China has probably witnessed two periods of regional metamorphism and associated migmatization ending at about 2900–3000 Ma and 2500–2600 Ma respectively. Chemically the metamorphic rocks in many regions are generally rich in potassium.

The metallic mineral deposits or mineralization horizons and localities found in these ancient formations are those of iron, gold, nickel, and chrome. The non-metallic minerals are apatite and mica. But for some of the known occurrences the exact age of mineralization is still uncertain.

Very little geological work was done on the Chinese Archaean metamorphic rocks until the late 1940s, and before that time geological ages were often assigned without a sound scientific basis. Geological studies of these ancient formations have, however, made rapid progress since 1949 as a result of the unprecedented development of geological work throughout China. The manuscript of the monograph on the *Precambrian of China* (Institute of Geology, Ministry of Geology, China 1962), submitted to the First All-China Stratigraphic Congress convened in 1959, made the first overall though very brief description of the Archaean. The evidence used there included the first isotopic age data in China. This monograph was followed by the publication of two papers summarizing the Archaean stratigraphy and related problems. The first of these, dealing with northern and north-eastern China, appeared in 1973; the second, dealing with the whole country, in 1982 (Cheng *et al.* 1973, 1982*a*). This chapter presents a further revision of the previous publications. It is essentially in the form of a general stratigraphic account of eight separate Archaean regions in China, each characterized by certain salient geological features. A brief summary of stratigraphic correlations and certain evolutional characters and a brief description of some petrochemical features of the main rock-types are also presented.

SOUTHERN PART OF NORTH-EASTERN CHINA (LIAONING AND JILIN)

The Archaean metamorphics are exposed in eastern Liaoning and south-eastern Jilin, to the south of 43 °N. They are distributed mostly in a northern belt stretching from the Anshan–Fushun–Tieling region in Liaoning east-north-eastward into Jilin, mainly along the Longgang Mountain range, and in a southern belt extending for some distance near the coast of the Liaodong Peninsula. The regional trend of the foliation of the rocks is mostly ENE–WSW, with the eastern part of the northern belt swinging to NE–SW and also locally deviating to other directions. It is probable that the present trend was influenced both by Archaean and by Proterozoic tectonic activity.

Archaean formations in the Liaoning province have been named the Anshan Group, and those in Jilin the Longgang Group, but the latter term is falling into disuse. These formations are seen to be unconformably

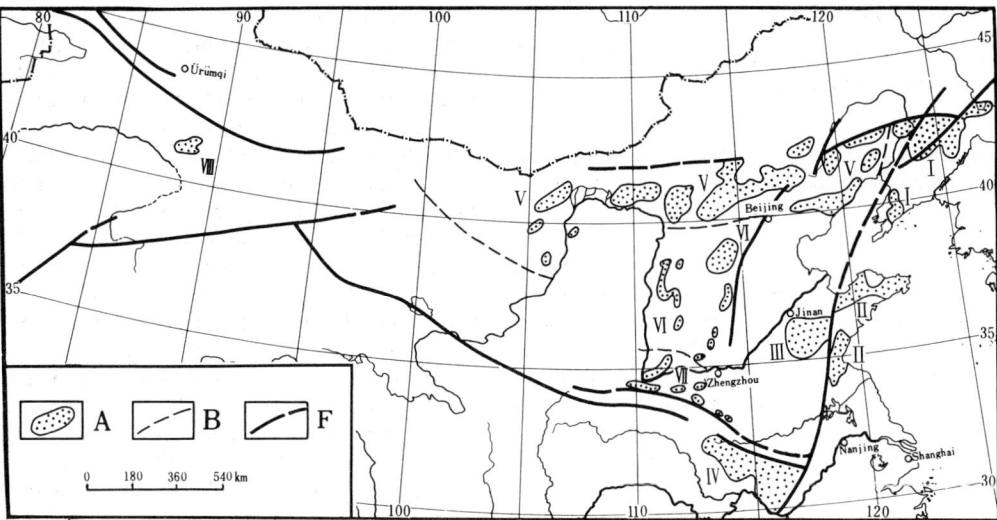

Fig. 4.1. Sketch map showing the distribution of the Archaean of China. A, Archaean terrain; B, boundary between different regions; F, fracture; I, Liaoning and Jilin; II, Eastern and South-eastern Shandong; III, Central and Western Shandong; IV, Huaiyang region; V, Yinshan and Yanshan region; VI, Wutai–Taihan–Luliang region; VII, Northern slope of eastern Qinling Range; VIII, North-west China.

overlain by the less metamorphosed Proterozoic formations at certain localities, as in the vicinity of Anshan and southern Jilin. But there are also places where the upper Anshan Group passes upward into the Proterozoic Lower Liaohe without marked hiatus. The stratigraphic boundary between the two is then somewhat ambiguous after the overprinting of Proterozoic migmatization and there is no remarkable difference in the nature and degree of metamorphism. K–Ar ages of both mica and whole-rock samples of both groups have been determined.

The Anshan Group of Liaoning

This is subdivided into five formations in ascending order as follows:

(1) Chengzitan Formation. The lower part consists of hornblende-plagioclase-gneiss, biotite-hornblende-gneiss with plagioclase-amphibolite, biotite-granulitite,* biotite-gneiss and even pyroxene-amphibolite and hornblende-granulite intercalations; the upper part, biotite-hornblende-plagioclase-gneiss, hornblende-plagioclase-gneiss, biotite-feldspar-gneiss with intercalations of plagioclase-amphibolite, biotite-granulitite, leucogranulitite and banded iron beds† and/or ores, locally with marble lenses; about 3800 m.

(2) Tongshicun Formation. Biotite-granulitite and biotite-plagioclase-gneiss with plagioclase-amphibolite and banded iron beds and/or iron ores; about 2000 m.

(3) Cigou Formation. The lower part consists of thick-bedded plagioclase-amphibolite with banded iron ores (Plate XIX, 8); the upper part consists of biotite-granulitite with plagioclase-amphibolite (Plate XVIII, 8); about 2500 m.

(4) Tayugou Formation. Biotite-granulitite, leucogranulitite, mica-granulitite with layers of mica-quartz-schist and plagioclase-amphibolite, and occasionally also some marble and banded iron beds; about 1000 m.

(5) Yindaoyuan Formation. Muscovite-quartz-schist, sericite-chlorite-schist, mica-granulitite, biotite-granulitite, and a thick sequence of banded iron ores; 200–400 m.

The tectogenesis responsible for the unconformity between the Anshan Group and the overlying Proterozoic Liaohe Group near Anshan is usually known as the Anshan Movement.

---

*Corresponding to the 'pepper-and-salt Moine' type of quartz-feldspar granulites of Scotland. In order to distinguish the Moine-type 'granulite' from the granulite-facies 'granulite' of higher grade of metamorphism, a new name 'granulitite' was provisionally proposed for the former and other rocks showing similar structure, texture, and degree of metamorphism by the author (Cheng et al. 1973).

†Iron formations of no economic importance are here denoted as 'iron beds'.

The Anshan Group of Jilin

This is generally subdivided in ascending order into the following three formations:

(1) Sidaolazi Formation. Mainly biotite-plagioclase-gneiss, biotite-quartz-schist, and plagioclase-amphibolite, with local intercalations of lenticular banded iron beds; 3200 m.

(2) Yangjiadian Formation. The lower part is chiefly various types of mica-gneisses and mica-granulitites, with some plagioclase-amphibolite and biotite-quartzite; upper part, mainly biotite-hornblende-plagioclase-gneiss and plagioclase-amphibolite; about 3500 m.

(3) Sandaogou Formation. Chiefly plagioclase-amphibolite and chlorite-schist, containing two formations of banded iron ores; about 3000 m.

So far as present knowledge goes, the generalized stratigraphic successions of Liaoning are better represented. The lower part is characterized by the abundance of biotitic and/or hornblendic plagioclase-rich gneisses, and by the common presence of granulitite of similar composition, both belonging to the amphibolite facies, with accessory rocks of the lower grade of the granulite facies, such as those exposed to the north of the Anshan–Benxi region and Jinxian. The original rocks were mostly semipelitic, probably of greywacke composition, partly mingled with intermediate to acid tuffaceous materials, with intercalations of basic types. These basic types include basic lavas, pyroclastics and volcano-sedimentary rocks with or without sodic affinities, with associated intrusions, and sometimes including iron-rich marly sediments. The middle part is marked by the dominance of biotite- or mica-granulitites and plagioclase-amphibolite and other hornblendic types, all of the amphibolite facies; and the original rocks are chiefly of basic types and subordinately of semipelitic rocks similar to those of the lower part, with accompanying banded iron formations that are probably partly of the Algoma type. The upper part consists of granulitites and schists, partly of amphibolite facies and partly of greenschist facies, chiefly derived from silty rocks with associated banded iron formations of the Lake Superior type. The absence of the protolithic pelitic and quartzitic types, and the scarcity of marble in the sequence, are noteworthy.

In addition to the intercalated iron ores or iron beds in this group in the Anshan and Benxi mining districts, Liaoning, there are associated copper and nickel sulphide and even minor gold mineralizations lower in the sequence at certain localities. It is also noteworthy that the Anshan rocks have been migmatized and granitized both in Archaean and later times. The resulting migmatites and migmatitic granite are of various types, accompanied by pegmatitic patches or dykes. In many districts they were probably formed by remelting as well as by alkaline metasomatism.

It will be evident from the above description that the rocks of the Anshan Group show some resemblance to those of the Archaean greenstone belt elsewhere in the world both in their protolithic rock-types, and also in the geotectonic background of their formation. However, they differ greatly in their distinctly higher degree of metamorphism. Other Archaean terrains of greenstone belt affinity in other regions in China are similar.

Isotopic age determinations have been carried out by the U–Pb, K–Ar, Pb–Pb, and Rb–Sr isochron methods (Geoch. Inst., Shenyang Inst., Guilin Inst., Geol. Inst.-A.S., etc.). Those made up to 1979 (Cheng *et al.* 1982*a*), have shown that the rocks of the Anshan Group underwent at least two periods of regional metamorphism, probably of the medium-pressure type, and migmatization, at about 3000 and 2500 Ma respectively. Further studies by Wu Jiahong *et al.* (Shenyang Inst. 1981) on zircon from the granite gneiss invading the ancient metamorphics give an additional Pb–Pb isochron age of 2635 Ma and a U–Pb age of 2632 Ma on a concordia diagram. So far no unconformity has been found within the Anshan metamorphic sequence. Both geochronological and field investigations suggest the superposition of early Proterozoic metamorphism (1850 ± ? Ma), migmatization, and related granitic and/or pegmatitic activities on these rocks.

No isotopic age data have been obtained for the ultrabasic rocks intruded into the Anshan Group.

EASTERN PART OF SHANDONG PENINSULA AND SOUTH-EASTERN SHANDONG PROVINCE

The Archaean formations of this region crop out to the east of the Tancheng–Lujiang Fault Zone and are characterized by the frequent occurrence of eclogite lenses and bodies. Those of the eastern part of the Shandong Peninsula (the Jiaodong region), known as the Jiaodong Group, show some resemblance to the rocks of the Anshan Group of Liaoning. They consist of a thick sequence of gneiss, granulitite, and amphibolite, with marble and graphite-bearing rocks, mostly of amphibolite facies and partly migmatized and locally of greenschist facies. The protolithic types were mainly fine clastic sediments, including pelitic rocks, with calcareous beds, and probably also include some basic volcanics and volcano-sediments.

The Jiaodong Group in the western part of the

Jiaodong region, i.e. the Penglai–Laiyang area, is subdivided into three formations in ascending order as follows:

(1) Pengkuang Formation. Chiefly plagioclase-amphibolite, biotite-granulitite, and biotite-schist; about 4300 m.
(2) Minshan Formation. Biotite-granulitite, biotite-plagioclase-gneiss, plagioclase-amphibolite, biotite-schist, and marble; about 2600 m.
(3) Fuyang Formation. The lower and middle parts are chiefly intercalations of biotite-plagioclase-gneiss and hornblende-biotite-plagioclase-gneiss with accessory biotite-granulitite; the upper part consists of nodular garnetiferous biotite-schist; about 1800 m.

In the eastern part of the Jiaodong region the Group consists of three formations with dominant hornblende-biotite-plagioclase-gneiss and intercalations of leuco-granulitite and marble, with a total thickness that is probably more than 2500 m. Graphite-schist and graphite-gneiss are found in certain horizons in both districts. There are also exploitable apatite-rich metamorphic rocks.

The Archaean rocks of the south-eastern part of Shandong and the adjacent northernmost region of Jiangsu are usually ascribed to the Jiaodong Group and are subdivided into three formations. They are chiefly mica-(potash) feldspar-gneiss, biotite-granulitite, and leuco-granulitite with lenses of marble, and with an aggregated thickness that is probably more than 15 000 m.

Exposures of the Jiaodong rocks at many localities have been shown to be influenced by later igneous activity and migmatization. The isotopic ages for their regional metamorphism have thus been modified to a greater or lesser degree. For instance, among the small group of isotopic age values so far obtained, an amphibolite in the top part of the group shows a K–Ar age of 1721 Ma, which is coincident with the K–Ar age of 1774 Ma for a pegmatite (Cheng et al. 1973; Cheng et al. 1982a). It is also possible that a part of the Jiaodong Group is of Early Proterozoic age.

CENTRAL AND WESTERN SHANDONG

The Archaean Taishan Group, formerly known as the Taishan Complex, is also of fairly wide distribution to the west of the Tancheng–Lujiang Fault Zone. It appears as a southward-tapering triangular area. The foliation direction of the rocks is mostly NW–SE, with a WNW trend at its extreme south-western part, changing to NNW towards the eastern portion, and then to NNE within the fault zone, thus exhibiting a north-facing, fan-shaped distribution of the foliation directions, as well as the rock exposures. This is in contrast to the dominant NE–SW foliation trend with local swinging to E–W or even WNW–ESE to the east of the fault zone. These tectonic patterns, which are similar in age to that of the initial stage of the development of the Tancheng–Lujiang Fault Zone, must be of great antiquity, though their age is an intricate problem which needs further investigation.

This group is a massive succession of biotite-plagioclase-gneiss and hornblende-plagioclase-gneiss and also plagioclase-amphibolite and biotite-granulitite. This has been metamorphosed to amphibolite facies, probably of the moderate-temperature, moderate- to low-pressure facies series, and partly further migmatized at two different periods to various degrees, or locally retrograded to greenschist facies by late dynamic metamorphism. The migmatites are genetically related to the presence of migmatitic granitoids along certain anticlinal axes or other tectonic elements; they are probably also related to some para-autochthonous or rheomorphic granitoidal bodies such as are found in the Taishan–Xintai district (see below). The rocks are generally subdivided into the following four formations in ascending order:

(1) Wanshanzhuang Formation. Mainly biotite-plagioclase-gneiss with intercalations of hornblende-biotite-plagioclase-gneiss, biotite-granulitite, biotite-hornblende-schist, hornblende-schist, and plagioclase-amphibolite; about 2000 m.
(2) Taipingding Formation. Chiefly biotite-plagioclase-gneiss with intercalations of hornblende-biotite-plagioclase-gneiss, mica-plagioclase-gneiss, hornblende-plagioclase-gneiss, muscovite–quartz-schist, biotite-quartz-schist, biotite-granulitite and also scanty hornblende-quartz-schist, and plagioclase-amphibolite; about 4600 m.
(3) Yanlingguan Formation. Mainly plagioclase-amphibolite and hornblende-granulitite, intercalated by biotite-granulitite, talc-schist and actinolite-tremolite-schist, with scanty graphite-phyllite near the top. It also contains workable banded iron ores, especially in the upper part; about 1000–1400 m.
(4) Shancaoyu Formation. Mainly biotite-granulitite, mica-quartz-schist, and muscovite-granulitite, locally with hornblende-biotite-gneiss and also scanty plagioclase-amphibolite near the base; more than 2600 m.

It is, however, important to note that the field evidence for the stratigraphic relationship between the Taipingding Formation and the Yanlingguan as shown above is still rather weak and needs further detailed investigation.

The protolithic rocks of the two lower formations are probably mainly fine siltstones partly of greywacke type intercalated with basic volcano-sedimentary or even volcanic rocks. The Yanlingguan Formation, at the type locality Yanlingguan, Xintai (Cheng et al. 1982b), for part of the succession at least, was a neritic volcanic to volcano-sedimentary series of back-arc affinity composed mainly of basic lavas, fine pyroclastics, and various tuffaceous rocks, intercalated with thin silty layers and locally by marly beds. They were probably partly formed in a transitional belt between the island arc area and a deep marine basin, with associated minor basic and ultrabasic intrusions, or even probable komatiitic lavas. Many lavas in the upper part are evidently of subaerial to subaqueous character (Plate XVII.3) and exhibit variolitic and complex amygdaloidal structures (Plate XIX.6). It is noteworthy that there are peculiar metamorphosed concordant tuffaceous conglomeratic to arkosic beds up to 105 m thick, which probably derived part of their materials (for instance granitic and gneissic pebbles and alkali-feldspar debris) from a now-concealed ancient land area. Also noteworthy is a thin stratiform pyrrhotite-sulphide layer associated with graphitic rocks indicative of a lagoonal environment near the top. This is seen to pass gradually upward to the Shancaoyu Formation, composed mainly of a thick protolithic silty sequence (often of greywacke and probably even turbiditic in nature) that was originally deposited in a more stable environment, mainly in deeper water (Chapter 18).

Judging by the rock types and their petrological characters, as well as the geological background of their formation, the rocks of these two upper formations show some resemblance to those of the greenstone belt sequence of certain Archaean regions in other countries. They differ from the latter by their higher degree of metamorphism and the absence of part of the generally accepted succession (Chapter 18).

The average K–Ar age of biotite samples from granulitites of the Yanlingguan Formation, practically free from the influence of migmatization, is 2450 Ma; that of the muscovite from pegmatites in the same formation is about 2500 Ma (Cheng et al. 1964); and the Rb–Sr isochron age of the first stage Na- and K-migmatization is 2586 Ma (Geol. Inst.-C.A.G.S. 1976). Hence it is safe to infer that the age of the conclusion of the regional metamorphism, which should have begun earlier than the earlier stage of migmatization, is definitely Archaean. The age of the second migmatization and granitization with prominent potash effects is c. 2230 Ma.

HUAIYANG REGION

Within the rugged terrain of the Dabie, Dahong, and Tongbai Mountains in the Huaiyang region, the Dabie Group, probably of Archaean age, together with the Proterozoic metamorphic formations, extends roughly in a NW–SE direction up to the Tancheng–Lujiang rupture belt for about 400 km in the borderland of Anhui, Henan, and Hubei provinces. The Dabie group is unconformably overlain by the Proterozoic Hongan Group in Hubei. It is an immense succession, probably over 15 000 m thick, composed mainly of hornblende-plagioclase-gneiss and biotite-plagioclase-gneiss with subordinate plagioclase-amphibolite, biotite-granulitite, leucogranulitite, and marble; and contains also intercalations of banded iron (magnetite) beds and magnetite-bearing amphibolite. The rocks belong chiefly to amphibolite facies and are derived mainly from basic volcanics, tuffaceous siltstone, siltstone, and greywacke with carbonate rocks, arkose, and cherty iron beds.

The Archaean age of the rocks was formerly assigned mainly on the basis of regional stratigraphic correlation. It has recently been supported by two U–Pb isotopic determinations for zircon from two localities within the Dabie Group terrain. One is 3120 Ma (Ur. Geol. Inst. 1981) and the other, about 2500 Ma (Yichang Inst. 1981). While the 2500 Ma age probably indicates the age of the metamorphism the rocks underwent at the close of Archaean, it is, however, uncertain whether the 3120 Ma age represents the age of formation or that of an earlier metamorphism. Another U–Pb isotopic date of 2080 Ma, also for zircon from the Dabie Group, probably indicates a Proterozoic event (Cheng et al. 1982a). All the K–Ar isotopic age data available lie within the time span from Middle Palaeozoic to Late Mesozoic and probably represent modified values of the original metamorphic ages. It is also possible that the age of formation of this group extends from the Archaean to the Proterozoic.

YINSHAN–YANSHAN REGION

In southern Nei Mongol (Inner Mongolia), the northern parts of Shanxi, Hebei, and Beijing, and western Liaoning, the Archaean rocks, formerly generally known as the Sanggan Group, are distributed mostly in a belt which includes the main granulite zone in China within the limit of 40–42 °N. They extend from 105 °E eastward to 122 °E for a distance of over 1200 km, including the Langshan and Yinshan Ranges in the west and the Yanshan and neighbouring mountains in the east. The regional trend of the foliation of the rocks is in general approximately E–W and to a lesser extent

NE–SW, swinging locally to other directions and distinctly NE–SW and even NNE–SSW towards the eastern end of the belt.

Yinshan region

The Archaean Sanggan Group was subdivided in the middle 1970s (Cheng et al. 1982a) by the Nei Mongol geologists, into a lower Jining Group and upper Wulashan Group using lithological differences, but this was without sound field evidence. The lower Jining Group is of great thickness, probably over 10 000 m, and is exposed mainly to the east of Huhhot, Nei Mongol. Its lower part consists chiefly of hypersthene-bearing or hypersthene-free granulites and gneisses, containing either plagioclase or potash feldspar, of the granulite facies derived from basic types and semipelitic rocks, with metamorphosed cherty iron beds characterized by a diopside–magnetite–quartz association. The middle part consists mainly of garnetiferous sillimanite-gneisses which are partly graphite-bearing, with subordinate sillimanite–garnet–quartzite and garnetiferous biotite-granulitite or leucogranulitite and also some amphibolite and serpentine marble. These are chiefly amphibolite facies rocks of pelitic origin and are probably derived from acid to intermediate and even basic volcanic (volcano-sedimentary) and psammitic rocks. The upper part is characterized by the dominance of different types of marbles of the amphibolite facies originating from a carbonate formation with intercalated semipelitic and sandy types metamorphosed to plagioclase-gneiss, feldspar-quartzite, leucogranulitite, etc. Most of the rocks have been further migmatized, resulting in the formation of different types of migmatites and migmatitic granites. Old basic dykes of hypersthene-plagioclase-granulitic composition are also present. The middle pelitic formation has been invaded at certain localities by numerous pegmatite dykes containing workable muscovite of Proterozoic age.

The Wulashan Group occurs mostly in the western part of this region and probably has a total thickness even greater than that of the Jining Group. It consists of hornblende-bearing or hornblende-free biotite–plagioclase(feldspar)-gneisses and various granulitites of pelitic and intermediate to basic volcanic (volcano-sedimentary) origin with some marble occasionally containing asbestos veins and quartzite. Most of the rocks are of the amphibolite facies. They are often migmatized, and are further invaded by Proterozoic pegmatite in some places. Hypersthene-hornblende-plagioclase-granulite also occurs in the lower part.

Recent work by Shen Qihan, Sun Dazhong, and others[*] since 1980 seems to indicate that the two groups mentioned above are actually parts of an Archaean sequence which shows the following general upward succession: (1) dominant hypersthene-bearing or hypersthene-free pyroxene-granulites, (2) dominant garnet-(biotite-)sillimanite-gneisses, (3) dominant plagioclase-amphibolite and hornblende-gneisses, (4) (graphite-)marbles with sillimanite-gneisses, and (5) dominant marbles, marked by amphibolite facies rocks with the exception of the granulitic types in the lower part. This sequence as an entity shows great resemblance to the Archaean formations of the Yanshan Region (see below) in the degree of metamorphism, lithological characters, and geotectonic style.

Among the scores of U–Pb and K–Ar age determinations of various minerals from the invading pegmatite and the metamorphic rocks, a U–Pb date of 2359 Ma of anorthite from a pegmatite gives the highest value. The rest fall into four age groups for pegmatites, i.e. 2111 Ma, 1900 Ma, 1800 Ma, and 1700 Ma (Cheng et al. 1973). It seems that the K–Ar ages of the metamorphic rocks have been modified by pegmatitic activity as well as by Proterozoic metamorphism.

Yanshan region

The Archaean formations of this region, formerly known collectively as the Sanggan Group, have been subdivided recently into the following two groups.

The lower Qianxi Group, probably over 30 000 m thick, is composed mainly of hypersthene-granulite and/or diopside-granulites, gneisses, plagioclase-amphibolites, and pyroxene-bearing banded iron formations. They were derived from basic to intermediate volcanic and volcano-sedimentary rocks and also semipelitic types, the lower to middle parts being of the granulite facies and the rest of amphibolite facies, and were mostly further migmatized. It is subdivided into three formations. Isotopic ages of over 3000 Ma have been obtained by K–Ar, Rb–Sr isochron, and Pb–Pb isochron determinations (Geol. Inst.-A.S., Geomech. Inst., Geol. Inst.-C.A.G.S.) in 1975–9.

The upper Dantazi Group is divided into three formations and is composed mainly of biotite-granulitite, schists, and gneisses, with intercalated banded iron formations, mostly of amphibolite facies and partly of greenschist facies. Most of the original rocks were

---

[*] Oral communication and preliminary reports of the Geological Institute and Tianjin Institute of Geology and Mineral Resources, Chinese Acad. Geol. Sci. and the Regional Geological Surveying Party, Nei Mongol Geological Bureau.

semipelitic and pelitic types, with basic volcanic as well as intermediate-acid tuffaceous sediments, and local calcareous rocks. The K–Ar age determinations for both hornblende and whole-rock samples range from 2435 to 2660 Ma (Cheng et al. 1982a). There is also an Rb–Sr isochron age of 2523 ± 139 Ma (Geol. Inst.-C.A.G.S. 1981).

However, the above two groups have never been found in direct contact.

Geologists of the Tianjin Institute of Geology and Mineral Resources recommended another twofold subdivision of the Archaean of this region at the Second All-China Stratigraphic Congress, 1979. In ascending order the sequence is as follows:

(1) Qianxi Group in the limited sense. This contains the lower part of the Qianxi Group of the above stratigraphic division. Its thickness is over 2700 m and it consists of granulites and gneisses of the granulite facies, being derived from basic to intermediate-acid volcanic and volcano-sedimentary rocks. The rocks have undergone strong migmatization, including anatectic transformation, of charnockitic affinity, often associated with tonalitic gneiss. It is from this portion of the metamorphic formations that the specimens with isotopic ages over 3000 Ma have been collected. More recent data for specimens from approximately the same horizon are the U–Pb isochron ages of 2480 Ma (R. T. Pidgeon 1980) for zircon, and 2590 Ma (Sun Jiashu and Cui Chengyu: Geomech. Inst. 1982) for zircon and apatite, and whole-rock Rb–Sr and Sm–Nd isochron ages, and a U–Pb zircon age of c. 2500 Ma (Zhang Zongqing and Jahn Borming 1982). K–Ar whole-rock and mineral determinations give ages of 2860–2923 Ma (Sun Jiashu and Cui Chengyu: Geomech. Inst. 1982). Such contradictory data may be explained by the fact that the metamorphic series has suffered two epochs of metamorphism and related migmatism in the Archaean, as was pointed out by Gao Jifeng in 1981. It is quite probable that the age of the earlier granulite facies metamorphisms associated with the formation and evolution of anatectic migmatitic granite relates to the older age group of over 3000 Ma, and that the later amphibolite-facies metamorphism, accompanied by the generation of migmatitic granite of metasomatic type, relates to the younger age group around 2500 Ma. The superposition of the latter metamorphism on the former is evidenced by pseudomorphs of aggregates of later diopside granules after earlier larger diopside (Plate XVII.1), by the presence of relic diopside in a later hornblende (Plate XVII.2), and by a series of optical and geochemical features of minerals such as biotite, hornblende, garnet, and plagioclase of the two epochs of metamorphism. These have been worked out in some detail by Gao. It seems that the K–Ar ages of about 2900 Ma may represent modified isotopic values of the earlier epoch.

(2) Badaohe Group. This contains the upper part of the former Qianxi Group and the lower part of the Dantazi Group. It is composed mainly of amphibolites and granulitites of amphibolite facies, with some pyroxenite in the lower part and amphibole-bearing banded iron formations in the upper. Some rock types are geochemically quite high in gold content. While most of the original rocks are intermediate-acid tuffaceous sedimentary and semipelitic types, there are also basic lavas and related fine pyroclastic and cherty iron beds. They often show evidence of being migmatized. As well as the K–Ar and Rb–Sr isochron ages listed above for the Dantazi Group, there is also a U–Pb concordia age of 2494 ± 24 Ma (Ur. Geol. Inst. 1981) for zircon from the migmatitic granite occurring in this group.

It is evident that on the whole the metamorphic rocks just described exhibit features similar to those of Archaean high-grade terrains of other countries.

TAIHANG–WUTAI–LULIANG REGION

The Archaean rocks, often accompanied by Lower Proterozoic metamorphics, are found in two belts in this region. One stretches from 40 °N south-westward along the borderland of Shanxi and Hebei Provinces to 35 °30 ′N, constituting the major part of the Taihang Mountains. The second branches off from the northern end of the first discontinuously in a south-westward and westward direction to a point near 111 °E, 37 °N, including the mountainous country of Wutai, Luliang, and other mountains in Shanxi.

Uncomformably overlain by lower Proterozoic formations, such as the Wutai Group of the Wutai region, the Archaean metamorphics are known as the Fuping Group and the unconformably overlying Longquanguan Group in the north Taihang–Wutai region, the lower part of the Zanhuang Group in the south Taihang Mountains, and the Jiehekou Group in the northwestern Luliang district. The trend of the foliation varies from place to place. It is, for example, mostly N–S in south Taihang, NNE–SSW in the Luliang Mountains, and roughly E–W at Fuping in northern Taihang, where it shows a conspicuous dome structure. The rocks are mainly different types of gneisses with variable proportions of magnesian marbles, granulitites, plagioclase-amphibolite, and leucogranulitite of the amphibolite facies, containing also granulite and garnetiferous hypersthene-amphibolite and hypersthene-bearing iron beds and lenses of the granulite facies in the lower part. In many places, the rocks have been quite intensively migmatized with the production of corundum in quartz-sillimanite nodules. The parent rocks were mostly of semipelitic composition, containing frequent pelitic and also arkosic, carbonate and basic types. It is apparent that the Archaean regions described are of the high-grade type.

The Fuping and Longquanguan Groups have been more thoroughly investigated in recent years than others. The former is subdivided into seven formations in ascending order as follows:

(1) Sujiazhuang Formation. Biotite- or hornblende-oligoclase-gneiss and plagioclase-amphibolite with intercalations of magnetite-hypersthene-quartzite, leucogranulite, and magnesian marble; about 2000 m.

(2) Tuanpokou Formation. Hornblende-biotite-plagioclase-gneiss, potash feldspar-leucogranulite and magnetite-garnet-quartzite, with frequent intercalations of gneiss containing quartz-sillimanite-nodules in the middle part and marble and graphite-schist in the upper; about 2000 m.

(3) Nanying Formation. Biotite-plagioclase (feldspar)-gneiss, plagioclase-amphibolite, and leucogranulite with marble and banded iron beds, containing sillimanite-quartz nodules in the middle to lower parts; about 1300 m.

(4) Manshan Formation. Chiefly thick-bedded feldspar-leucogranulitite, containing biotite-plagioclase-gneiss, plagioclase-amphibolite, diopside-granulitite beds and marble lenses and locally also corundum-bearing sillimanite–quartz nodules. At places there are repeated upward leucogranulitite–biotite-gneiss–diopside-marble successions which reflect the original rhythmic sandy pelitic-(semipelitic)–carbonate deposits; about 1200 m.

(5) Muchang Formation. Chiefly marbles and biotite- or hornblende-plagioclase-gneiss with interbedded plagioclase-amphibolite, diopside-granulitite, and leuco-granulitite; about 3000 m.

(6) Sidaohe Formation. The lower part is chiefly feldspar-leucogranulitite with local sillimanite-quartz nodules; its upper part, various marbles with gneiss, frequently biotite-bearing, granulitite, schist, and plagioclase-amphibolite; about 2300 m.

(7) Hongtupo Formation. Chiefly biotite- and magnetite-bearing leucogranulitite with thin intercalations of biotite-plagioclase-gneiss, and biotite-granulitite, and also tremolite-schist, diopside rock, and serpentine marble towards the top; about 1000 m.

Some plagioclase-amphibolite beds in the first three formations still preserve the amygdaloidal structure of the protolithic basic lava.

The Longquanguan Group, being less metamorphosed and migmatized than the Fuping Group, and containing rocks of both amphibolite and greenschist facies, is subdivided into two formations in the following ascending order:

(1) Paoquanzhang Formation. Chiefly biotite-granulitite and occasionally sillimanite-bearing biotite-plagioclase-gneiss with impersistent tremolite rock, marble, and plagioclase-amphibolite beds; up to 1100 m.

(2) Yushuwan Formation. Biotite (hornblende)-plagioclase-gneiss with interbedded plagioclase-amphibolite, biotite-granulitite, muscovite-quartz-schist and tremolite-chlorite-schist, and local magnetite-bearing leucogranulitite; up to 4000 m.

The Longquanguan Group was invaded by a pre-Wutaiian granitic body with a U–Pb isochron age of 2560 Ma* (determined on zircon). An older age of 2800–2830 Ma* has been given by the same method on detrital zircon from the lower horizon of the Fuping Group. This is referred to by some geologists as the possible lower limit for the age of its formation. Also of interest, some metamorphic zircon samples give isotopic dates of about 2600 Ma, which is probably the approximate age of the metamorphism. A biotite from the upper part of the Fuping Group shows a K–Ar date of 2310 Ma, which is probably a modified value of the original metamorphic age. The Fuping rocks have also been intruded by at least three periods of pegmatite dykes with K–Ar ages of 2000–2100 Ma, 1900 Ma, and 1700 Ma respectively (Cheng et al. 1973).

THE NORTHERN SLOPE OF THE EASTERN QINLING MOUNTAINS AND THE NEIGHBOURING REGIONS

The Archaean formations of this region are found in a belt more than 500 km long, stretching from 109°E to 114°E, within a N–S boundary between 33°N and 35°30′N. Their outcrop thus includes the northern slope of the eastern Qinling Mountains in Shaanxi in the west, and the mountains along the eastward and

---

*Determinations made by Liu Dungi of the Institute of Geology, Chinese Academy of Geological Sciences.

east-south-eastward extension in Henan, such as a part of the Xionger and Funiu Mountains, and also in the north the Zhongtiao Mountain in Shanxi. This may further extend south-eastward for over 100 km to the buried metamorphics under the Quaternary in the district of Huoqiu, Anhui (32°20′N, 116°20′E).

The Archaean metamorphics, mostly of amphibolite facies metamorphism but locally retrograded to greenschist facies and showing various degrees of migmatization, have been named the Dengfeng Group, the Taihua Group, and the Linshan Group in different parts of this region. Of these, the Taihua Group has the widest distribution and is well developed in Lushan, Henan Province, where a type section consisting of the following five formations in ascending order has been observed by geologists of the Yichang Institute of Geology and Mineral Resources:

(1) Zhanggou Formation. Chiefly biotite-plagioclase-gneiss with subordinate hornblende-plagioclase-gneiss and plagioclase-amphibolite, with some other types of gneisses; about 1800 m.

(2) Dangzehe Formation. Mainly plagioclase-hornblende-gneiss and to a smaller extent, biotite-plagioclase-gneiss, with some other intermediate types of gneisses; about 2000 m.

(3) Tieshanling Formation. Dominantly biotite- or hornblende-bearing plagioclase-gneiss and subordinately plagioclase-hornblende-gneiss; the middle and lower parts contain workable banded iron formations composed of quartz-type magnetite ore, marble, plagioclase-amphibolite, and scanty quartz-schist; about 3000 m.

(4) Shuidigou Formation. Chiefly various types of graphite-bearing or graphite-free gneisses and marbles with scanty plagioclase-amphibolite; about 700 m.

(5) Xuehuagou Formation. Composed mainly of plagioclase-hornblende-gneiss and hornblende- or biotite-bearing plagioclase-gneiss, the middle part usually graphite-bearing and intercalated with albite- or oligoclase-bearing leucogranulitite, magnetite-bearing granulitite, and marble; about 1600 m.

The lower part of the Taihua Group exposed in Shaanxi further west consists mostly of biotite-plagioclase-gneiss. This is occasionally hypersthene-bearing, and it also contains hornblende-pyroxene-gneiss and hornblende-pyroxene-granulite of the granulite facies, thus exhibiting higher metamorphism than the average of the group.

The buried metamorphics of Huoqiu, Anhui, known as the Huoqiu Group or Zhouji Formation, have been correlated with the middle to upper part of the Taihua Group.

It is very probable that most of the protolithic rocks of the Taihua were in general members of volcano-sedimentary rocks relatively rich in iron and semipelitic types, with the volcanic rocks changing from basic to intermediate-acid composition, the semipelitic rocks increasing in amount, and some graphite-carbonates and banded cherty iron beds appearing from the lower part upward.

The Dengfeng Group occurs in a comparatively limited area in the Dengfeng–Linru–Xuchang district of Henan. The regional trend of the foliation and the fold axes of the metamorphic rocks is mainly NNW–SSE to N–S, in great contrast to the general E–W or WNW–ENE direction of the Taihua Group. The rocks are mostly of the amphibolite facies but partly, especially in the upper part of the Group, of greenschist facies. Those of the lower part are mostly biotite- or hornblende-bearing migmatites, including homogeneous migmatite, or gneisses. Those in the middle part are mainly (garnetiferous) plagioclase-amphibolite, plagioclase-hornblende-gneiss, biotite-(mica)-granulitite, and biotite-schist (biotite-plagioclase-gneiss). The original rocks are chiefly intermediate to acid volcanics, with basic types as well as semipelitics and occasional cherty iron beds and thin carbonate beds. The upper part is composed of mica-schist, mica-gneiss and plagioclase-amphibolite, derived from pelitic and also from basic and semipelitic types.

Metamorphic rocks of the Linshan Group are exposed in the Jiyuan district in North Henan and are mostly of amphibolite facies. The lower part of the group consists chiefly of hornblendic rocks, mica-gneiss, and migmatites; the upper part consists mainly of granulitite and leucogranulitite, with scanty marble.

It is probable that part of the Archaean formations of this region are of high-grade type.

The U–Pb, K–Ar, Pb–Pb isochron, and Rb–Sr isochron determinations (Cheng et al. 1982a; Yichang Inst. 1980) so far obtained, including the Rb–Sr isochron of 2796 ± 69 Ma for whole-rock specimens from the buried Huoqiu Group, Anhui, reported recently (Ur. Geol. Inst. and Anhui Geol. Inst. 1982), have shown that the Archaean metamorphics of this region exhibit at several localities three age groups, i.e. about 2600, 2200–2300, and 1800–2000 Ma. These probably represent three stages of metamorphism and related migmatization. The first indicates the isotopic date for the metamorphic epoch towards the end of the Archaean; the second and third may reflect the imprints of the two successive periods of Lower Proterozoic metamorphism undergone by the Wutai and the Huto Groups respectively. It is interesting to note that there is a whole-rock Rb–Sr isochron age of 2986 ± 180 Ma for the rocks from the lowermost part of the Dengfeng Group exposed in Linru and Dengfeng (Ur. Geol. Inst. 1978), but it is as yet uncertain whether this determination denotes the depositional age or that of an earlier metamorphic period.

NORTH-WESTERN CHINA

Old metamorphic formations are scattered in many districts in the north-west, but are mainly found in the western part of the Kunlun, Tianshan, Altyn, and Qilian Mountains, Alxa, etc. Their pre-Middle Proterozoic age and subdivision into Archaean and Early Proterozoic (where possible) are based chiefly on geological reasoning and correlation, such as unconformable contacts within the metamorphics and those with the overlying Middle Proterozoic or younger rocks. All the available isotopic data for the related rocks are K–Ar values and have probably been modified by later tectonic or igneous activity. The oldest date is 1682 Ma for muscovite from a pegmatite in the Longshoushan Group in northern central Kansu Province, which is probably lower Proterozoic rather than Archaean in age.

There is a well-defined metamorphic succession developed in the Quruktagh Mountain, eastern Tianshan region. The Archaean Daklakbulak Group consists mainly of garnetiferous biotite-schist, biotite-quartz-schist, and hornblende-schist of semipelitic or even pelitic parentage and some marbles, often migmatized to various degrees. This Group is unconformably overlain by the Early Proterozoic Xingditag Group.

There is as yet not enough evidence for the presumed Archaean age for the Muzhart Group of the western Tianshan region and the Karakax Group of the western Kunlun region. Hence they are not shown in Fig. 4.1.

CORRELATION OF THE ARCHAEAN FORMATIONS

A tentative scheme of correlation of the Archaean stratigraphical units of China is given in Table 4.1. In many regions the Archaean rocks are separated from the overlying Proterozoic by an unconformity, which was formed during the Fuping (Tiepu) Movement, an orogenic movement which gave rise to the unconformity between the Fuping or Longquanguan Group and the Wutai Group. The type locality is at Tiepu, Wutai in Shanxi Province. Hence it is also known as the Tiepu Movement. The Fuping movement roughly corresponds to the Kenoran Orogeny of Canada (Stockwell 1968) and its equivalents elsewhere in the world. The chronological demarcation has been fixed at 2500–2600 Ma, as evidenced by the isotopic age data from five major regions from the eight listed. In two regions the Archaean is further divided into two groups. For the Anshan Group of Liaoning, the Qianxi Group in Hebei, the Dabie Group in the Huaiyang region, and the Dengfeng Group in Henan, there are isotopic age data for samples from their lower parts that approximate to, or are over 3000 Ma. This suggests an earlier Archaean metamorphic period. The geological interpretation of these dates is still uncertain.

CERTAIN EVOLUTIONAL CHARACTERS OF THE ARCHAEAN OF CHINA

Archaean rocks are found chiefly in the broad North China Platform and are composed mainly of biotite–plagioclase-gneiss or hornblende–plagioclase-gneiss, biotite-granulitite, and plagioclase-amphibolite of the amphibolite facies. The lower part is exposed in several regions containing biotite–pyroxene(hypersthene)-plagioclase-gneiss and pyroxene(hypersthene)-granulite of the granulite facies. The parent rocks of plagioclase-amphibolite and pyroxene(hypersthene)-granulite are mostly basic volcanics and volcano-sedimentaries; those of the biotite–plagioclase-gneiss and biotite–pyroxene(hypersthene)-gneiss, generally corresponding to greywacke and tuffaceous siltstones and possibly partly to intermediate to acid volcanics and volcano-sedimentary or even pelitic rocks. Those of biotite(mica)-granulitite correspond largely to semipelitic siltstones and/or tuffaceous siltstones partly of greywacke composition—and probably partly to intermediate to acid volcanic and volcano-sedimentary rocks. The Archaean was thus essentially a volcano-sedimentary complex prior to metamorphism. This constitutes the main part of the Archaean of the North China Platform and probably also a portion of the initial continental crust, composed mainly of rather immature, not well-differentiated sedimentary and volcano-sedimentary rocks, in addition to the frequent occurrence of volcanic rocks of basic to intermediate–acid composition. These were probably the prevailing conditions of sedimentation and volcanism in most districts of this major geotectonic region, which shows features both similar to and also dissimilar from greenstone belts elsewhere. There are also districts where to various extents the rocks are intercalated with normal sedimentary types, such as highly magnesian marble, mica-schist or sillimanite-gneiss, feldspar-quartzite, banded iron beds and/or graphite-schist or graphite-gneiss, derived from carbonate rock, pelite–semipelite, arkose, cherty iron beds, and carbonaceous rocks.

Table 4.1. Correlation of the Archaean of China.

	I	II	III	IV	V	V	VI	VII	VII	VIII
	Liaoning and Jilin	Eastern Shandong	Central and Western Shandong	Huaiyang	Yanshan	Yinshan	Wutai–Taihang–Lüliang	Northern slope of E. Qingling		North-west China
Lower Proterozoic 2.5–2.6 b.y.	Liaohe Gr	Fenzishan Gr	Taishan Gr	Hongan Gr	Shuangshanzi Gr	Sanheming Gr	Wutai Gr	Songshan Gr		Xingditag Gr
					(Tiepu movement) Fuping movement					
Archaean 2.9–3.0 b.y.	Anshan Gr (M)	Jiodong Gr		Dabie Gr	Badaohe Gr	Wulashan Gr (M)	Longquan-guan Gr	Dengfeng Gr	Taihua Gr (M)	Daklakbulak Group
					Qianxi Gr (M)	Jining Gr (M)	Fuping Gr (M)			

(M): With granulite facies rocks   Gr: Group.

From the stratigraphic and evolutional points of view, the Archaean in some districts of this area of original crust may be subdivided into two major volcano-sedimentary cycles which are well developed in part of the Yanshan region. There the lower one, i.e. the Qianxi Group in the limited sense, exhibits an upward transition from basic volcanics to intermediate–acid volcanics and related pyroclastics, and the upper Badaolhe Group shows an upward gradation from basic volcanic rocks to intermediate–acid tuffaceous sedimentary and also sedimentary rocks. In most of the other regions only formations of the second cycle are present as an independent entity, and the first is either absent or incompletely developed or preserved. The basic volcanic rocks of the second cycle are probably represented by the Cigou Formation of the Anshan Group, the Pengkuang Formation of the Jiaodong Group, the Yanlingguan Formation of the Taishan Group, the Wangchang Formation and the Wanzhangzi Formation of the Badaohe Group, the Muchang Formation of the Fuping Group, the upper part of the Jiehekou Group, the Dangzehe Formation of the Taihua Group, and the lower middle part of the Dabie Group. It is still uncertain whether these two major depositional cycles could be used separately for stratigraphic correlation between various regions. It is to be noted that banded iron formations or beds are often or occasionally present in the upper part of the first cycle and in the middle–lower and/or upper part of the second. This would serve as one of the arguments to support a similar evolutionary tendency for the Archaean volcano-sedimentary rocks in various regions of this platform.

The dissimilar nature of the intercalated beds or formations in the major cycles and some related minor volcano-sedimentary cycles of certain regions is often apparent. For instance, in the Archaean along the borders of the Platform, there are normal sediments and quite well-differentiated carbonate and sandy rocks (quartzose) and even pelitic types, such as the Jianping Group and Dantazi Group along the northern flank of the Yanshan and Jining and Wulashan Groups of the Yinshan on the north side; the Fuping Group of the Taihang Mountains and the Jiehekou Group of the Lulian Mountain on the west; the Taihua and Dabie Groups on the south; and the Jiaodong Group of eastern Shandong on the east. Together with various volcano-sedimentary rocks, these sediments often constitute repeated minor depositional cycles, especially along the western and southern borders. Such sediments are rare in the central part of the Platform, including the regions of eastern Liaoning and southeastern Jilin (Anshan Group), the southern flank of the Yanshan Mountain (Qianxi Group), central and western Shandong (Taishan Group), central north Henan (Dengfeng Group), and probably also the buried metamorphic basement beneath the Hebei Plain. (Drilling by various organizations during recent years has revealed that the buried ancient metamorphic rocks below the surface of the Hebei Plain are mainly of the Taishan Group.) The sequences are, however, characterized by the frequent presence of poorly differentiated sedimentary rocks, such as the greywacke–turbidite association metamorphosed to biotite-(mica)-granulitites. This reflects the rather mobile nature of the central region and the conditions of differentiation of the border zone for a particular period of time. It is also inferred that the regions on both the north and south flanks probably represent a deeper marine environment and that certain districts of the central part of the Platform, such as western and central Shandong and probably also the Hepei Plain, became elevated, at least in the later Archaean. This is evidenced by the neritic and even partly subaqueous and subaerial nature of basic volcanic activity shown by the rocks of the Yanlingguan Formation of the Taishan Group, and also by the transition from the continental basic volcanics of central–north Henan southward to the basic oceanic types. It is, however, still difficult to fit this evidence into the ancient plate tectonic pattern as usually proposed. This must have undergone successive complicated changes throughout ensuing geological time, of which our knowledge is as yet very meagre.

As we have seen, there are indications from some isotopic age determinations that there are three, and probably four, regions (described above) affected by an early stage of regional metamorphism that often reached the granulite facies and was accompanied by migmatization at about 3000 Ma or a little earlier. It is also clearly evident that towards the end of the Archaean at about $2500 \pm$ Ma a prolonged, probably second, period of medium- to high-grade regional metamorphism came to an end. This was probably mostly characterized by moderate pressure, often associated with rather intensive migmatization. As a result, almost the entire old continental crust of North China was transformed into a rather intensively or repeatedly metamorphosed, structurally complicated, and regionally migmatized mass, possibly a 'mini-continent'. A long northern granulite belt is, however, still preserved, which now extends from the eastern part of Liaoning westward to Yanshan, thence southward to the northern section of the Taihang Mountains, and further westward through Yinshan to end in the mountainous area south-west of Langshan. The relics of a southern granulite zone are now found only in eastern Shaanxi along the northern slope of the Qinling and a few other localities. The formation of the granulites in these two belts seems to be related to the depth-control effect exerted by the huge pile of Archaean formations lying above. However, a better explanation may be a cumulative metamorphic effect under appropriate temperature, pressure, and geological conditions during a certain time interval in the Archaean.

There is no well-marked linear metamorphic zonation in the Archaean terrain, and rocks metamorphosed at the same time and distributed within particular geographical limits in different areas have usually undergone metamorphism of about the same degree. Wherever more than one metamorphic facies does exist, the boundaries of the stratigraphical units are usually concordant with those of the metamorphic grades.

The Archaean strata were extensively folded. It is clear that there has been the superposition of at least two late Archaean major fold systems generated by different tectonic events and with different orientations. The boundaries of metamorphic facies as well as those of intact stratigraphical units have been involved in the folding.

All the striking features pointed out above indicate that, during Archaean times, the geological environment over an extensive area was quite uniform yet tectonically fairly active. No rigid tectonic boundaries have been found to occur in the terrain.

The Lower Proterozoic rocks such as the Wutai Group and the corresponding formations were formed in the marine basins (some of them trough-shaped) that developed on the possible Archaean sialic mini-continent mentioned above. They are often characterized by the rather widespread occurrence of partially sodic volcanic rocks and the abundance of semipelitic and pelitic types and turbidites, with the frequent association of coarser terrestrial sediments in the lower part.

The Archaean, together with the unconformably overlying lower Proterozoic formations, suffered further metamorphism, partly of a retrogressive nature, and other geological transformations during Proterozoic and even later times, at least around 2200–2300 Ma and 1800–2000 Ma.

SOME CHEMICAL FEATURES OF CERTAIN IMPORTANT TYPES OF ARCHAEAN METAMORPHIC ROCKS OF THE NORTH CHINA PLATFORM

As a result of investigations carried out by Sun Dazhong (Sun and Wu 1981) and others, certain important Archaean rock types of the North China Platform are known to be characterized by particular petrochemical features. For instance, the most widespread biotite-plagioclase-gneiss and biotite-granulitite, exhibit textural dissimilarities and certain compositional differences (Table 4.2). There are also some chemical differences between rocks of the same type from different regions, but they are, in the main, marked by fairly high silica and K_2O percentage, medium iron oxides and CaO content (2.28–4.89 per cent), a rather low mean value for MgO and a K_2O/Na_2O ratio close to 1. Compared with average greywacke (Pettijohn 1949),

Table 4.2. *Mean chemical composition of Archaean gneisses and granulitites from different regions of the North China Platform (after Sun et al. 1981, except III*).*

	Biotite-granulitites					Biotite-plagioclase-gneisses		
	V	VII	I	VI	III*	V	I	IV
SiO_2	65.41	62.89	64.93	56.75	64.70	60.57	62.94	70.15
TiO_2	0.39	0.50	0.44	0.66	0.44	0.73	0.48	0.46
Al_2O_3	15.48	15.41	14.89	16.40	15.93	14.98	16.61	13.93
Fe_2O_3	1.78	2.02	1.65	3.35	2.05	3.23	3.29	1.29
FeO	2.65	3.39	4.21	4.41	3.81	3.72	2.81	2.50
MnO	0.06	0.09	0.09	0.12	0.08	0.09	0.09	0.08
MgO	1.90	2.12	2.40	3.78	2.05	3.05	2.52	1.24
CaO	3.43	3.41	2.42	4.89	3.10	3.85	2.28	2.74
Na_2O	4.19	4.49	4.04	3.80	3.70	3.95	3.40	3.98
K_2O	2.92	2.24	2.79	2.27	2.65	3.29	2.84	2.67
P_2O_5	0.14	0.21	0.09	0.28	0.22	0.23	0.17	0.11
Samples	8	23	51	13	7*	14	2	19

For Roman numerals, refer to explanation of Fig. 4.1. III*, Xintai, Shandong (after Cheng *et al.* 1977).

they show slightly lower MgO and CaO contents, and higher K_2O and Na_2O. In fact, sedimentary features, both macroscopic and microscopic, are often clearly observed in quite a number of districts. In addition, specimens of the same type but from different localities show variations of some chemical constituents, thus indicating a mixed origin. This does not of course exclude the presence of some volcanic and volcano-sedimentary rocks of intermediate composition.

The plagioclase-amphibolites, which are also common in the Archaean metamorphic formations, are mostly lavas and related volcano-sedimentary types of tholeiitic affinity and subordinately of sedimentary origin. Their compositional ranges (Table 4.3) are: MgO, 6.46–7.58 per cent; Na_2O, 2.50–2.90 per cent; K_2O, 0.77–0.94 per cent. The average value of K_2O for the six regions listed in the table is 0.89 per cent, which is twice that of oceanic tholeiite and is close to the mean K_2O content (1 per cent) of continental tholeiite (Sun and Wu 1981). On Cr–FeO/MgO diagrams and Ni–FeO/MgO diagrams, using data from the southern part of North-east China, the Yanshan Mountains, the northern slope of the Qinling Mountains, and the west–central part of Shandong, most of the rocks are in the field of basalts from stable continental and oceanic areas.

The composition of the hornblende-plagioclase-gneiss (Table 4.4), which is less common than the amphibolite, lies between the amphibolite and the biotite-plagioclase-gneiss. The protolithic rock may include rather basic lava and tuffaceous types.

Migmatite and granite are often present in the Archaean terrains, as previously stated. Their chemical composition as shown in Table 4.5 is characterized by a rather high K_2O content.

Table 4.3. *Mean chemical composition of Archaean amphibolites from different regions of the North China Platform (after Sun et al. 1981).*

	V	VII	I	VI	III*	IV
SiO_2	49.94	49.27	50.20	49.38	49.91	48.68
TiO_2	0.84	1.11	0.95	1.00	0.68	1.48
Al_2O_3	14.03	14.44	14.59	14.49	14.41	15.21
Fe_2O_3	4.39	3.65	3.31	4.00	2.86	3.84
FeO	7.81	9.05	8.69	8.64	10.15	8.53
MnO	0.18	0.26	0.18	0.17	0.21	0.19
MgO	7.50	6.46	7.04	7.29	7.58	6.46
CaO	9.23	9.31	8.68	8.87	10.45	9.31
Na_2O	2.90	2.50	2.73	2.56	2.72	2.70
K_2O	0.94	0.92	0.94	0.94	0.77	0.88
P_2O_5	0.17	0.17	0.18	0.20	0.08	0.22
Samples	10	29	33	36	14	6

For Roman numerals, refer to Fig. 4.1. III*, protolithic basic lava and tuff of the Yanlingguan Formation, west-central Shandong (Cheng *et al.* 1977, 1982*b*).

Table 4.4. *Mean chemical composition of Archaean hornblende-plagioclase-gneiss from different regions of the North China Platform (after Sun et al. 1980).*

	VII	I	VI	III	IV
SiO_2	55.46	49.63	56.24	55.71	59.58
TiO_2	0.67	0.78	0.66	0.58	0.85
Al_2O_3	16.52	14.51	15.54	16.46	15.35
Fe_2O_3	3.17	5.08	3.39	2.97	1.98
FeO	4.62	5.97	5.23	5.71	4.71
MnO	0.17	0.19	0.12	0.19	0.14
MgO	4.28	6.74	3.97	2.73	3.29
CaO	6.62	7.77	5.70	6.90	5.72
Na_2O	4.13	3.05	3.86	3.60	3.85
K_2O	1.56	1.58	2.40	1.50	2.76
P_2O_5	0.33	0.11	0.23	0.21	0.29
Samples	18	17	26	1	12

For Roman numerals, refer to Fig. 4.1.

Table 4.5. *Mean chemical composition of migmatitic granites from different regions of the Archaean of the North China Platform (after Sun et al. 1980).*

	V	VII	I	VI	III	IV
SiO_2	67.78	71.50	72.66	72.49	71.59	73.71
TiO_2	0.50	0.26	0.18	0.16	0.19	0.18
Al_2O_3	14.64	14.15	14.17	13.42	14.36	13.71
Fe_2O_3	2.41	2.15	1.48	1.04	0.46	0.83
FeO	3.29	0.55	1.41	1.23	1.69	1.01
MnO	0.12	0.06	0.06	0.08	0.04	0.03
MgO	1.72	0.61	0.52	0.65	0.71	0.35
CaO	1.86	2.04	0.73	0.95	1.68	1.00
Na_2O	3.35	3.42	3.77	3.49	4.19	3.85
K_2O	4.45	4.36	4.36	4.19	4.25	4.85
P_2O_5	0.20	0.06	0.05	0.10	0.10	0.03
Samples	21	3	30	23	11	8

For Roman numerals, refer to Fig. 4.1.

Postscript. (1) According to isotopic determinations made by Huang Chengyi *et al.* of the Tianjin Institute of Geology and Mineral Resources, CAGS, in 1983, the Rb–Sr isochron age (whole rock) of the Jining and Wulashan Groups of the Yinshan region is close to 2600 and 2450 Ma, respectively.

(2) Jahn and Zhang reported in 1984 a whole rock Sm–Nd isochron age of 3515 ± 115 Ma for the Tsaozhuang metabasic rocks of granulite and amphibolite facies metamorphism of the Qian'an area in the Yanshan region, E. Hebei Province (Jahn and Zhang, 1984). [Radiometric ages (Rb–Sr, Sm–Nd, U–Pb) and REE geochemistry of Archaean granulite gneisses from eastern Hebei Province, China, given in Kröner *et al.* 1984.] The said rocks have been ascribed to the Qianxi Group by most Chinese geologists.

5. THE PROTEROZOIC
Wang Hongzhen

The Proterozoic as used here covers the time-span from 2500–2600 to 850 Ma. The Sinian System (850–600 Ma) will be dealt with separately in Chapter 6.

A threefold classification of the Proterozoic with 1600 Ma and 900 Ma as boundary ages of the three subdivisions has been recommended by the Precambrian Subcommission of the IUGS Stratigraphic Commission (1982), on the principle of synchronous geological events. We have set the boundaries at 1850 Ma and 1050 Ma (Table 20.1), because in China the Luliangian (1850 Ma) is the most pronounced orogenic episode that has produced widespread unconformities in the related stratigraphic sequences on the stable platforms, and profound changes in palaeogeography and in the organic world occurred at about 1050 Ma. The Lower Proterozoic rocks are metamorphosed and constitute an integral part of the ancient foundation of the platforms. The Middle and Upper Proterozoic are, however, unmetamorphosed or only slightly metamorphosed, yet they nevertheless show a clear difference from the Sinian and the Palaeozoic, which form a true cover sequence on the platforms. The Middle and Upper Proterozoic may therefore be called paracover strata.

Apart from the platforms, the Proterozoic is also well developed on the median massifs and along the anticlinal axial belts in the folded regions. As the Lower Proterozoic is a part of the basement complex, it will be treated separately in the following account.

THE LOWER PROTEROZOIC

The Lower Proterozoic of China covers the time-span from 2500–2600 to 1850 Ma. It may be further subdivided into the lower Wutaiian and the upper Hutuoan, which are separated by a pronounced unconformity, dated at approximately 2200–2300 Ma. The terms Wutaiian and Hutuoan are used here as regional chronostratigraphic units; correlation beyond the scope of the North China Platform still remains obscure. We shall first give a brief description of the Lower Proterozoic of North China, and then attempt to correlate it with contemporaneous rocks of the remaining regions.

North China

In North China, the ancient foundation began to form at the close of the Early Proterozoic. It is composed of the Archaean crystalline rocks and the Lower Proterozoic folded terrains, including the Wutaiian and the Hutuoan.

The Lower Proterozoic is widely distributed on the North China Platform, and is best developed in four regions: the interior of the platform, the northern border, the eastern border, and the southern border regions. The interior part includes the Shanxi Plateau and part of western Henan. The northern border region is best represented in the Yanshan and Yinshan Mountains, which extend from 108 °E to 120 °E, approximately along 41 °N. The eastern and southern border regions comprise respectively the Jiaodong–Liaodong region and the narrow belt to the north of the eastern Qinling Mountains.

The interior of the North China Platform

In the interior of the platform, the Lower Proterozoic is well exposed in the mountainous area on the Shanxi Plateau. The Wutaiian is typically developed in the Wutai region, situated about 150 km north-west of Taiyuan. The type locality of Hutuoan is also within the Wutai region. It is named after the Hutuo River which flows around the mountains from the north to the west and then eastward, to join the Ziyahe before entering Bohai Bay.

The Wutai Group of the type region consists of some 5000–6000 m of granulitites, plagioclase-amphibolites and amphibole-schists. Bimica-schists, garnetiferous biotite-schists, and chlorite-schists are also common. Taconites of the BIF type occur in several horizons within the sequence. In a paper submitted to the second All-China Stratigraphic Congress of 1979, Li, Ji, et al. divided the Wutai Group into three formations in descending order as follows (Table 5.1, 1):

(1) Pushang Formation. This is the most widespread of the three formations. The main iron ore layers are in the lower part, locally called the Wenbiyan Member. The main rock types comprise albite-chlorite-schist and

albite-epidote-schist, with intercalated albite-granulitite, chlorite-quartz-schist, marble, and taconite. Metaconglomerates also occur occasionally. The intercalated relationship of iron ore layers and amphibolites and amphibole-granulitites indicates that the Si–Fe deposits are related to contemporaneous volcanic activity.

(2) Taihuai Formation. This formation is divisible into two members. The lower member is chiefly composed of biotite-gneiss and hornblende-plagioclase-gneiss, the original rocks being probably clastics and intermediate to basic eruptives. The upper member consists mainly of mica-schist and chlorite-schist, and also amphibole-schist and plagioclase-gneiss. The original rocks were argillaceous to silty sediments with a minor proportion of basic volcanics.

(3) Shizui Formation. This is the lowest formation of the Wutai Group. It contains as main rock types arkose quartzite, biotite, muscovite-quartz-schist, diaspore-marble, and taconite. A thin plagioclase-amphibolite member with intercalated taconites occurs near the base, and contains an important iron ore horizon.

Each of the lower two formations corresponds to a volcano-sedimentary cycle, beginning with clastic deposits and passing upwards into intermediate to basic volcanic and volcano-sedimentary rocks. The Pushang Formation contains, however, intermediate to acidic volcanics. From a tectono-lithological viewpoint, the Wutai Group belongs to the ophiolitic and spilitic sequence. Tholeiitic basalt makes up the majority of the original rocks of the Taihuai Formation. The Pushang Formation includes abundant flysch deposits, some of which are typical turbidites.

In a paper submitted to the Archaean and Earlier Proterozoic Geologic Evolution and Metallogenesis International Symposium held at Salvador, Brazil, in September 1982, Cheng, Bai and Sun recommended a threefold subdivision of the Wutai into Shizui, Taihuai, and Gaofan subgroups. Their scheme was based on a fresh study of the field relations of the various stratigraphic units by Bai Jin, Xu Chaolei, *et al.*

The Gaofan Subgroup lies unconformably on the Taihuai Formation and contains only 800 m of epimetamorphic sedimentary rocks, mainly turbiditic in character. An isochron U–Pb age of 2521 Ma was obtained from a quartz-keratophyre in the Taihuai Formation or Subgroup. The Lanzhi granite, with a U–Pb age of 2561 Ma, is in all probability in sedimentary contact with the Wutai Group according to Cheng and Wu.

There are various opinions concerning the subdivision and the chronological position of the Wutai Group. Before the 1950s it was attributed to the Upper Archaean. As K–Ar isotopic age data obtained until recently did not exceed 2300 Ma it has been included in the Lower Proterozoic. As we have put the upper boundary of the Archaean at 2500–2600 Ma, the Wutai Group is here retained in the Lower Proterozoic.

Within the Shanxi Plateau, equivalents of the Wutai Group are found in the Luliang Mountains. The Luliang Group here is over 12 000 m thick, and consists in the main of biotite-granulitite, plagioclase-gneiss, hornblende-schist, and chlorite-schist, with some phyllites. Magnesian and siliceous marbles occur in the upper part, and taconites and phyllites are known at more than one horizon. The Luliang Group is unconformable with both the underlying and the overlying formations, and is comparable with the Jiangxian Group in the Zhongtiao mountains.

The Hutuoan is also well exposed in the Wutai area. F. von Richthofen first gave it the name 'Huto Schiefer' in 1870. Bailey Willis used the name 'Huto System' and divided it into the Doucun (Toutsun) Series and the Dongye (Tungyeh) Series. The terms 'System' and 'Series' are, of course, not used here in the modern sense of chronostratigraphic units. The Hutuo Group of the type area, as recently reinvestigated and classified, consists of three subgroups, the Doucun, Dongye, and Guojiazhai (Table 5.1), and twelve formations.

The Doucun Subgroup is coarsely clastic in the lower part and is composed of metaconglomerates and arkose quartzites, with occasional phyllites and marble intercalations in the upper part, containing *Kussiella* and other stromatolites. Above a local unconformity comes the much thicker Dongye Subgroup, over 5500 m in thickness. The lower part of the Dongye Subgroup is largely arenaceous and argillaceous, and contains metabasites. The middle part is mainly dolomitic, and the Yaochicun Formation is especially rich in stromatolites, with *Paraboxonia* (Plate II.11), *Pilbaria*, and *Gymnosolen* as common forms. The upper part of the Subgroup is again argillaceous. Stromatolites collected from the middle part include such forms as *Straticonophyton, Grunneria,* and *Pseudogymnosolen*. The whole assemblage closely resembles forms reported from the Lower Proterozoic (2500–1800 Ma) Great Slave Supergroup and the Epworth Group of western Canada; it is certainly different from those found in the Middle Proterozoic of Jixian.

Still higher in the sequence, a clear unconformity separates the overlying Guojiazhai coarse clastics, which amount to about 1000 m in thickness and are peculiar in their reversed, upward coarsening.

In the middle segment of the Taihang Mountains, some 150 km to the south, there is developed a much thinner epimetamorphic sequence referable to the

Hutuoan, locally called the Gantaohe Group. Apart from being less thick, it is characterized by a generally more clastic lithology with more frequent volcanic eruptives ranging from basic to intermediate to acidic in composition. On the evidence of stromatolites, the Gantaohe may be correlated with the Dongye Subgroup of the Wutai region. It evidently represents a more mobile type of sedimentation.

The time-span represented by the Hutuoan, especially its relationship with the Changchengian, has been the subject of considerable debate. The greatest age so far obtained from the Hutuoan in the Wutai region, from rocks that are mostly metamorphic and partly intrusive, is a K–Ar date of 1862 Ma (Yichang Inst. 1975), obtained from a hornblende in the Dongye Subgroup. The Dongjiao Group in the Taihang Mountains, believed to be equivalent to the Guojiazhai, has yielded a metamorphic K–Ar age of 1870 Ma. Thus the upper boundary of the Hutuoan may probably be placed at 1850 Ma, which is quite consistent with the inferred lower age limit of the Changchengian (see below). The lower boundary of the Hutuoan is however more uncertain. As no direct data are available, we may use the upper boundary of the Wutaiian, which yields a group of metamorphic K–Ar ages around 2300 Ma (see above) and is certainly older than the Hutuoan. Thus 2200–2300 Ma may be taken as the age of the thermal event that occurred after the initial metamorphism of the Wutaiian and prior to the deposition of the Hutuoan.

In the south-western part of the Shanxi Plateau, the Hutuoan is represented by the Lanhe and Yejishan Groups in the lower part and by the Heichashan Group in the upper part in the Luliang Mountains. The whole sequence is in unconformable contact with the underlying Lower Proterozoic Luliang Group and the overlying Middle Proterozoic. The two lower groups comprise three sedimentary cycles, each beginning with conglomerates and quartzites of the parent rocks and ending in dolomites, and followed at the top by sodium-rich metabasites.

In Zhongtiao Mountain, the lower part of the Zhongtiao Group is renowned for its copper deposits, and yields stromatolites much the same as those found in the Hutuo Group. The upper part is diverse in lithology and shows a weak metamorphism. The whole sequence seems to be characterized by alternating schists and marbles, and is generally more deeply metamorphosed than in the north. On the top rests the coarse clastic Danshanshi Group, probably comparable with the Guojiazhai, which is in turn overlain by the Middle Proterozoic Xiyanghe basalt.

The Songshan Group of Songshan, northern Henan, differs from the Hutuoan of Shanxi in its more stable type of sedimentation and smaller thickness. It was probably deposited near the western margin of the sea that flanked the old land to the east in Middle Proterozoic times.

The northern border region

In the northern border of the platform, the Lower Proterozoic is extensively exposed in the Yinshan and Yanshan Mountains, but the succession and correlation are not very clear. In the Yanshan Mountain of eastern Hebei, above the Archaean Badaohe Group is the Shuangshanzi Group. This is composed of plagioclase-granulitite, biotite-schist and hornblende–chlorite-schist, with frequent taconite intercalations, and is referred to the Wutaiian on the basis of a metamorphic Rb–Sr isochron age of 2400 Ma (Sun Dazhong et al. 1979). Unconformably overlying the Shuangshanzi is the Qinglonghe Group, also composed of granulitites and schists with parent rocks of pelitic and turbiditic types, and a Rb–Sr isochron age of 2390 Ma (Geol. Inst. C.A.G.S. 1982). As both these groups are comparatively thin, they are all assigned to the Wutaiian, and it is uncertain whether the Hutuoan is actually represented in this region.

Along the Yinshan Mountain of Nei Mongol, above the Archaean crystalline basement complex and below the Shinagan Limestone of Middle Proterozoic paracover sequence, there occur several stratigraphic units generally attributed to the Lower Proterozoic. The representative strata are the Sanheming Group, Erdaowa Group, and Majiadian Group, which are demarcated by clear unconformities. The Sanheming metamorphics are mainly of basic volcanic origin and locally migmatized; on the whole they are comparable with the Wutaiian. The Erdaowa Group is composed of quartz–mica-schist and plagioclase–amphibole-gneiss in the lower part, and marble and amphibole-schist in the upper part. The original rocks are arenaceous and pelitic, but carbonates and keratophyres are also common. A low radiometric age of 2363 Ma has been obtained from zircon in the basal conglomerate (Geoch. Inst. 1975). This probably represents the age of deposition and seems to suggest a Hutuoan age for the group. The Majiadian Group is restricted in distribution and is covered by the Shinagan Limestone with a pronounced unconformity. The U–Pb zircon age from the basal conglomerate gives a maximum value of 2230 Ma, which also fits in well with the time-span of the Hutuoan. The lower part of the Majiadian Group comprises epimetamorphic conglomerates and quartzites, slates, and phyllites, and marble lenses. Intercalated intermediate to acidic volcanics occur frequently.

The upper part is dominated by various kinds of marbles containing stromatolites.

There are two more Proterozoic epimetamorphic groups in the Yinshan region, the Chaltai and the Bayanobo, which are distributed on the southern and northern slopes respectively of the Yinshan Range. No direct contact relation has been observed between them and the Majiadian Group. The Chaltai Group has yielded a syndepositional isotopic age of 1600 Ma, obtained by the Pb–Pb method; the Baiyanobo has given an average value of 1514 Ma (Geoch. Inst. 1975). In view of their tectonic setting, it would seem that the two groups may have ranged from Early Proterozoic upwards into the Middle Proterozoic and represent a continuous sequence formed in the passive northern continental margin of the North China Platform.

The eastern and southern border region

In the eastern border region of the platform, to the east of the Tancheng–Lujiang fracture zone, the Lower Proterozoic is well developed on the Liaodong and Jiaodong Peninsulas and in the Huainan region. In the Liaodong Peninsula, the Liaohe Group forms the major part of the Lower Proterozoic and rests on Archaean basement with a clear unconformity. It is directly overlain by the Upper Proterozoic Xihe Group. Near the Fuzhou District, the Liaohe Group may be divided into five formations. The lower two, the Langzishan Formation and the Lieryu Formation, consist of granulitites and schists and various kinds of metavolcanics, and are renowned for their phosphate- and boron-bearing horizons. In the upper part, the lowest Gaojiayu Formation is characterized by carbonaceous and calcareous slates rich in acritarch remains such as *Asperatopsophosphaera* and *Trachysphaeridium*. The succeeding Dashiqiao Formation is chiefly composed of magnesian marbles and contains important magnesite deposits. Stromatolite assemblages comparable with those found in the Hutuoan in Shanxi have been reported. The uppermost Gaixian Formation is among the widest in distribution. It includes various kinds of schists and quartzites, and attains a total thickness of over 5000 m in some places. K–Ar ages for the metamorphism and migmatization of the Liaohe Group range from 1900 to 1750 Ma (Cheng *et al.* 1973). An Rb–Sr isochron age of 2040 Ma and a U–Pb age of 1960 Ma were recorded by Tao Quan in 1981. It is generally held that the Liaohe is comparable with the Hutuoan, although it has been suggested that part of the Middle Proterozoic may have been included in the upper part.

In southern Jilin, the Lower Proterozoic includes a lower Ji-an Group and an upper Laoling Group. The Ji-an Group, some 4000 m in thickness, consists mainly of granulitites, gneiss, and serpentine-marbles and is boron-bearing. The Laoling Group amounts to more than 15 000 m and contains metaconglomerate, carbonaceous slate, and various kinds of marble. The two groups are therefore comparable with the lower and the upper part of the Liaohe Group respectively. Some geologists correlate the lower part of the Liaohe Group and the Ji-an Group with the Wutaiian, but it is more likely that they belong to the Hutuoan. Another point worth noting is the presence of the Yushulazi Group above the Gaixian Formation, which is coarse clastic and metamorphic, thus resembling the Guojiazhai Subgroup of Shanxi. This is probably further evidence that the Liaohe Group is entirely Early Proterozoic in age.

In the southern part of the eastern border region, the Lower Proterozoic of the Huainan area is represented by the Fengyan Group, consisting of muscovite–quartz-schist in the lower part, quartzite and quartz-schist in the middle part, and dolomite and amphibolite in the upper part. Along the northern slope of eastern Qinling, the Lower Proterozoic is found in a belt stretching from 110° to 114°E and within 34° to 35°N, probably representing the Hutuoan. The representative strata are the Tiedonggou Formation of the Xiaoqinling area, which directly overlies the Archaean Taihua Group and is followed by the Middle Proterozoic Xionger volcanics with an unconformity.

North-west China

Within the scope of North-west China are included the Tarim Platform, the Qaidam Massif, and the Central Qilian, where the Lower Proterozoic is exposed as a part of the basement complex. All these ancient massifs belong to one stratigraphic super-region (Fig. 3.1).

On the Tarim Platform, ancient rocks are found only in the border uplifts, i.e. the Quruktagh Mountains and the Kalpin region in the north, and the Tieklik Range in the south.

The oldest rocks in the border regions are Lower Proterozoic, possibly including part of the Archaean. In Quruktagh, the lowest unit is the Daklakbulak Group of gneiss and schist, metamorphosed to a granulitite facies and probably Early Proterozoic to Archaean in age. Unconformably overlying the Daglakbulak is the Xinditag Group, which is mainly composed of biotite-schist, quartz-schist, and marble, with intercalated taconite, about 5000 m thick in total. The Xinditag is overlain by Middle Proterozoic carbonates with a clear unconformity, and is therefore Early Proterozoic in age. In the Kalpin region and the adjoining South Tianshan, the basement rocks include the lower Muzat gneiss and the upper Aksu quartz-schists and marbles, corresponding respectively to the Daklakbulak Group and the Xingditag Group. At the south-eastern end of the Central Tianshan, the Xingxingjia Group or Keyaobliubulak Group, locally strongly metamorphosed and migmatized, may also belong to the Lower Proterozoic.

The southern border range of the Tarim basin is the Tieklik, situated at the northern slope of West Kunlun. Here the oldest strata are the Ailiankate Group, which is over 10 000 m thick and is divided into two parts. The lower part consists of quartz-schist, chlorite-schist and marble; the upper part of sericite-quartz-schist and phyllite, from chlorite to amphibolite metamorphic facies. The parent rocks are characterized by basic volcanics and silicolites. As no chronometric data are available, these rocks are tentatively referred to the Lower Proterozoic, probably the Wutaiian.

The Qaidam Massif and the Central Qilian might have been a unified block before the opening of the South Qilian geosyncline in Early Palaeozoic times. The oldest strata of the basement form the Hualong Group, exposed around the Hualong basin at the south-eastern end of the Central Qilian. The Hualong Group has a total thickness of over 5500 m and is divisible into three formations. It consists mostly of biotite-schist, amphibole-schist, plagioclase-gneiss, and quartzite, with lenticular marbles in the upper part. The ancient parts of the massif proper, such as the Jinshuikou Group in the south and the Dakendaban Group in the north, are also Early Proterozoic in age. On the evidence of stromatolites reported in recent years, both groups are comparable with the Hutuoan.

South China

In South China, the Lower Proterozoic is known only on the Yangzi platform. Three types of basement strata may be recognized. Along the northern border of the Upper Yangzi are the Wudang Group and the Dabie Group, developed in the mountains bearing the same names. The Wudang Group includes four formations with a total thickness of over 15 000 m. The dominant component rocks are quartzite and mica- and quartz-schist, with a minor amount of carbonaceous slate, alternating with various kinds of metavolcanics, chiefly metabasites and meta-andesites. An isotopic age of more than 2000 Ma has been reported from the Wudang Group, but part of the Middle Proterozoic may also be included in it. The Dabie Group is on the other hand more deeply metamorphosed and migmatized, consisting mostly of amphibolites, plagioclase-gneiss with occasional marble layers, and belonging partly to the Archaean (see above). Unconformably overlying the Dabie Group is the Susong Group, containing more marbles and phosphatic rocks. The U–Pb ages of the Susong Group range from 1850 (apatite) to 2343 (zircon) Ma according to a preliminary report by Z. Yao and Sh. Zhang of the Anhui Provincial Bureau of Geology. The higher figures probably represent the initial metamorphic age and the upper age limit of deposition. Thus the Susong Group may be safely assigned to the Lower Proterozoic.

In the interior of the platform, in the Yangzi gorges of Hubei, the Sandouping Group (Kongling Group) is well exposed along the Huangling anticlinorium, which is directly overlain by the Sinian System. The lower part of Sandouping comprises biotite–plagioclase-gneiss and amphibolites; the middle part contains quartz-schist, plagioclase-gneiss, and marble; the upper part is again dominated by gneiss and amphibolite. The total thickness is about 3000 m. A metamorphic age of 1688 Ma has been recorded in the Sandouping Group (Xu 1982), which is evidently older than the nearby unmetamorphosed Middle Proterozoic Shennongjia Group and forms the basement complex underneath the Sichuan basin.

The second type of the basement is distributed on the Kham–Yunnan axis, and is represented by the Dahongshan Group which lies beneath the Middle Proterozoic Kunyang Group, probably with an unconformity. The Dahongshan Group, exposed in Xinping District near the Red River fault zone, is characterized by an alternating succession of metamorphosed sediments and alkaline volcanic rocks. The equivalent of the Dahongshan, the Hekou Formation in north-eastern Yunnan and south-western Sichuan, occupies a similar stratigraphic position and is of similar lithology; it is overlain by the Middle Proterozoic Huili Group. A U–Pb age of 1750 Ma of zircon has been obtained from the middle part of the Dahongshan Group, and isotopic K–Ar ages around 1900 Ma have been recorded from the intrusive rocks, according to J. Duan of the Yunnan Institute of Geology (1981). In addition, ultrabasic rocks, mostly augite-peridotites found near Huili and Xichang, which are correlated with the Hekou Formation and the Dahongshan Group, have yielded isotopic ages of 1704–1958 Ma, from the same source.

THE MIDDLE AND UPPER PROTEROZOIC

The Middle and Upper Proterozoic are widely distributed as paracover sequence or folded basement of the platforms and median massifs; they are also found in folded regions (Fig. 5.1). As the Middle and the Upper

Proterozoic are not always clearly divisible, they are here treated together, covering the time-span 1850–850 Ma, and not including the Sinian System (850–600 Ma), as stated at the beginning of this chapter. It is probably appropriate here to give a brief review of the Middle and Upper Precambrian stratigraphic terminology of China, especially the term 'Sinian System'. In 1922 A.W. Grabau defined the 'Sinian System' as the non-metamorphosed or slightly metamorphosed sequence between the Cambrian and the basement complex. He referred to two standard sections, the Nankou section near Beijing and the Yangzi Gorge section of Hubei, and attempted a world-wide correlation. Subsequently J.S. Lee and Y.T. Chao (1924) made a fresh study of the Sinian section of the Yangzi Gorges, and Kao (Kao, Hsiung, and Kao 1934) worked out the detailed succession of the Jixian. The Sinian has since been used as a formal system name in China, and both the Jixian section and the Yangzi Gorge section have been cited as standards for the System.

Since the late 1950s it has been realized that the Jixian section occupies a stratigraphic position lower than the Yangzi Gorge section. It was subsequently found to cover a time-span of at least 1000 Ma (approximately 1850–850 Ma). Thus the question arose as to whether the term 'Sinian System' should be applied to the northern or the southern section. As an attempted solution, the term Sinian Subera was introduced in the mid 1970s to cover the major part of the Proterozoic (1850–600 Ma), the term 'Sinian System' (850–600 Ma) being reserved for the Yangzi Gorge section. This is, however, inappropriate, for it is contrary to the regulations of stratigraphic nomenclature

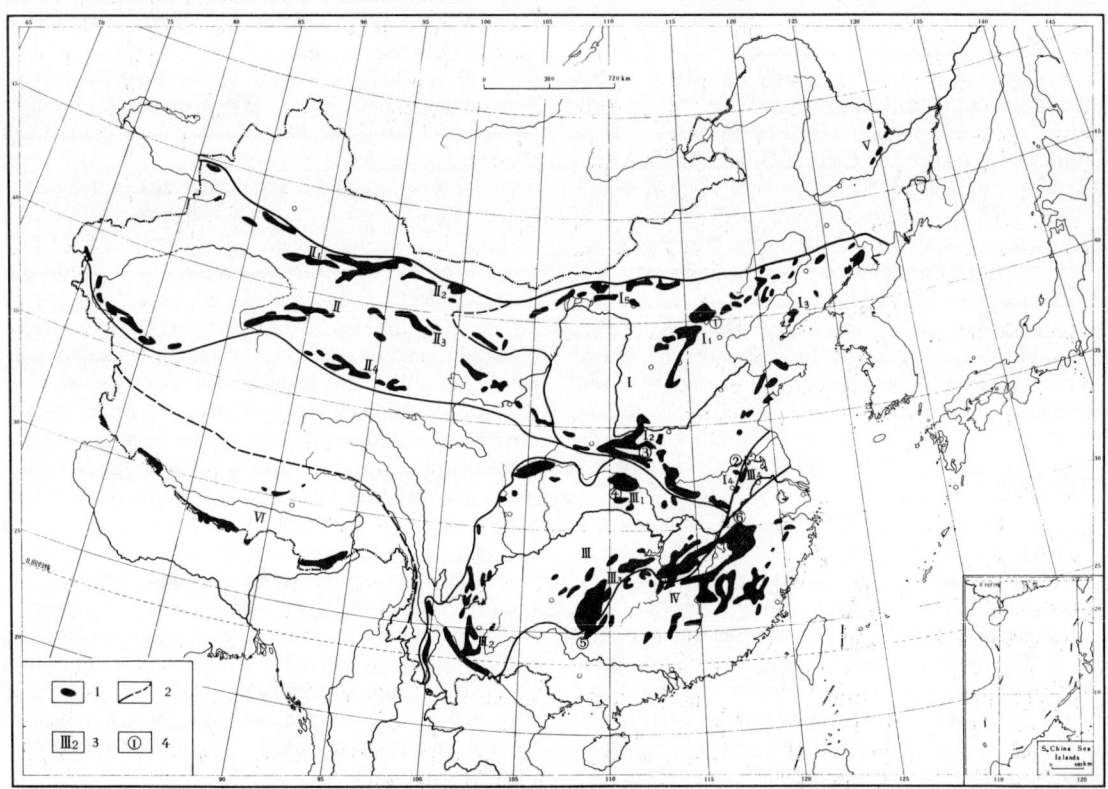

Fig. 5.1. Sketch map showing the distribution and the sedimentary regions of the Middle and Upper Proterozoic (pre-Sinian) of China (modified after Qiao Xiufu (1985)). 1, outcrop of Pt_{2-3} (pre-Sinian), Xizang region including Sinian; 2, boundary between sedimentary realms; 3, number of sedimentary regions; 4, localities of columnar sections (see Fig. 5.2). I. North China Realm: I_1, Yanshan–Taihang Region; I_2, West Henan Region; I_3, Liaodong Region; I_4, North Anhui Region; I_5, Yinshan Region. II. North-west China Realm: II_1, Quruktagh Region; II_2, Beishan Region; II_3, Qilian Region; II_4, Burhan Buda Region. III. Yangzi Realm: III_1, Shennongjia Region; III_2, East Yunnan Region; III_3, Jiangnan Region; III_4, Lower Yangzi Region. IV. South-east China Realm. V. North-east China Realm. VI. Xizang Realm.

in using the same name for the Subera and a constituent part of it. In this connection it is worth pointing out that the All-China Stratigraphic Commission has recommended that the Sinian System should be restricted to the Yangzi Gorge section, and that the remaining part of the Middle and Upper Proterozoic represented in the Jixian section (1850–850 Ma) should be subdivided into several units, presumably with the rank of systems. This scheme is followed here (Chapter 3, p. 13).

The distinction of sedimentation realms in China in the Middle and Late Proterozoic is now clear (Fig. 5.1). The Northern Continental Super-region was characterized by a stable or semistable type of sedimentation and is dominated by carbonate deposits in North China and in Tarim; an intermediate type is met with in the border parts of the platforms. As the Yangzi Platform had not yet grown to its present size in Middle and Late Proterozoic times, the corresponding sediments show an intermediate to mobile nature, much like those on the median massifs in the intercontinental regions. The major part of the Northern Intercontinental Super-region and the Eastern Continental Margin Super-region were probably occupied by oceanic crust, where little sedimentary record is available.

North China

The Yanshan–Taihang region (Fig. 5.1, I_1)

In North China, the most complete succession of the Middle and the Upper Proterozoic is found in Jixian, between Beijing and Tianjin on the southern slope of the Yanshan Mountains. During the last twenty years comprehensive research work has been carried out by members of the Tianjin Institute of Geology and Mineral Resources. Some 9500 m of strata are excellently exposed in a continuous and regular succession with its basal part directly overlying the Archaean at Malanyu, near the Great Wall (Fig. 5.2, 1) (Table 5.1, 1). Acritarch remains are abundant, and six stromatolite assemblages have been recognized in various sections in the Yanshan region (Zhu 1978).

Changchengian

The Changchengian is divisible into two parts, the lower Changcheng Group and the upper Nankou Group, each representing a sedimentary cycle from clastic to carbonate deposits and separated by a disconformity. The total thickness in the type section amounts to 2266 m.

In the lower group, the basal Changzhougou Formation contains conglomerates and coarse sandstones in the lower part and quartzitic sandstones with intercalated sandy shales in the upper part. The pebbles in the conglomerates are mainly vein quartz and quartzite averaging 3 cm in diameter, well rounded and sorted. The sedimentary facies ranges from fluvial to upper tidal zone from below upwards.

The Chuanlinggou Formation is composed of dark grey shales with minor amounts of thin-bedded siltstones. Carbonaceous dolomite intercalations occur in the upper part, which passes upwards into dark grey argillaceous and siliceous dolomites of the Tuanshanzi Formation. Mud cracks and ripple marks are found, and horizontal and crumple bedding are also reported in the lower part. The carbonaceous shales are chiefly littoral and intertidal, with the middle part probably subtidal. The Tuanshanzi dolomites may be partly lagoonal in facies; pyrite concretions and rock salt pseudomorphs are common.

Acritarch remains found in the Lower Changchengian comprise minute forms of *Leiominuscula* and *Margominuscula* (Plate III.22), which are quite distinct from those found in higher horizons. Characteristic stromatolites, including *Grunneria* and *Omachtenia*, belong to the *Grunneria–Xiayingella* Assemblage, which is peculiar to this group.

The upper part of the Changchengian, the Nankou Group, has a total thickness of 2000 m and is divisible into two formations. The Dahongyu Formation is clearly transgressive and consists mainly of littoral quartzitic sandstones and arkose sandstones, generally medium-bedded to massive, usually marked with mud cracks and ripple marks. Noteworthy is the occurrence of volcanic rocks, including agglomerates, tuffs, and lavas, erupted in three episodes. The principal rocks are highly potassic trachyte, with a K_2O content as high as 10 per cent. Tuffaceous potash shale of emerald green colour often occurs and may be of economic value. The eruptions were partly submarine and partly terrestrial: the amygdales are sometimes filled with carbonates. The environment was probably small islets newly emerged as a result of volcanic activity.

The Gaoyuzhuang Formation is Mn-bearing and contains a variety of dolomitic rocks. It attains a thickness of more than 1500 m and ranges in lithology from terrestrial detrital cherty dolomite, dolomitic siltstones, and sandstones to argillaceous and siliceous dolomites.

The Nankou Group is comparatively meagre in fossil remains; *Asperatopsophosphaera* is so far the only form recorded. Characteristic stromatolites are, however, abundant (Plate II.7–9). They include the well-known *Conophyton cylindricum, C. garganicum,* and *Tabuloconigera*. *Conophyton* and *Tabuloconigera* are the two representatives of the second stromatolite assemblage.

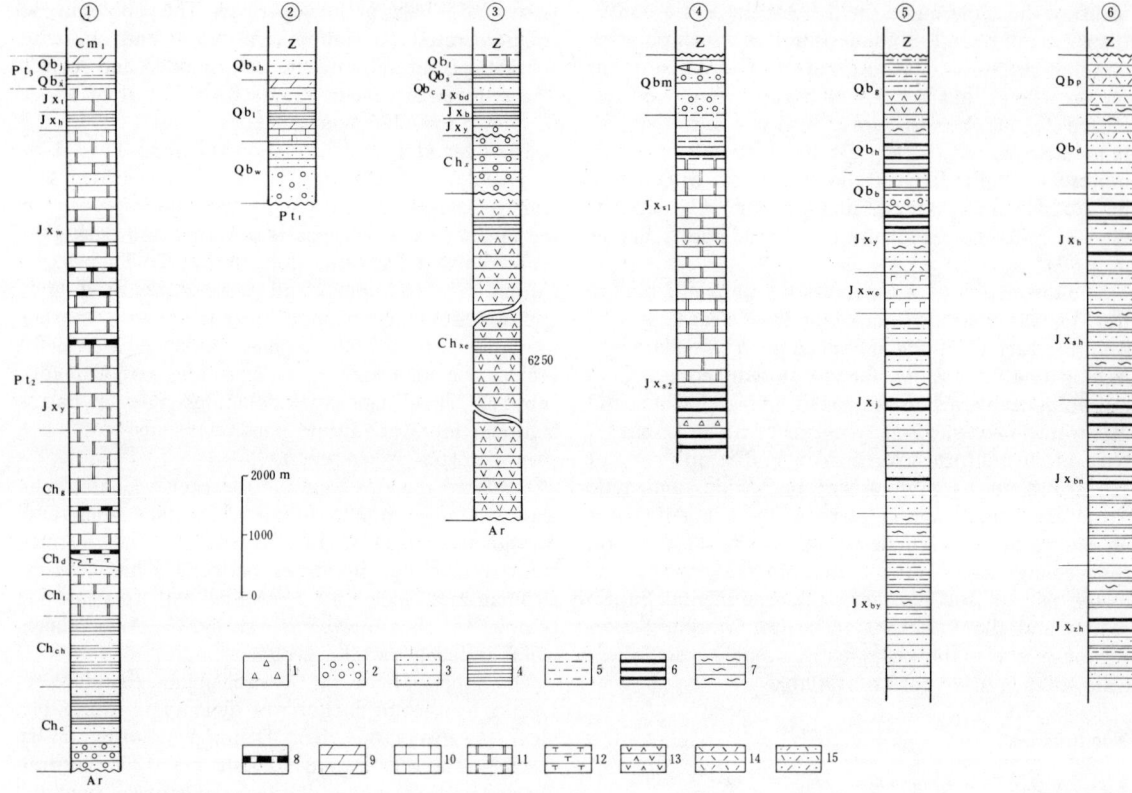

Fig. 5.2. Columnar sections of the Middle and Upper Proterozoic (pre-Sinian) of East China. 1, breccia; 2, conglomerate; 3, sandstone; 4, shale; 5, sandy shale; 6, slate; 7, phyllite; 8, silicolite; 9, marl; 10, limestone; 11, dolomite; 12, andesite; 13, andesite–porphyrite; 14, rhyolite and dacite; 15, tuff. 1 Jixian, Tianjin. Ch_c, Changzhougou Fm; Ch_{ch}, Chuanlinggou Fm; Ch_t, Tuanshanzi Fm; Ch_d, Dahongyu Fm; Ch_g, Gaoyuzhuang Fm; Jx_y, Yangzhuang Fm; Jx_w, Wumishan Fm; Jx_h, Hongshuizhuang Fm; Jx_t, Tieling Fm; Qb_x, Xiamaling Fm; Qb_j, Jingeryu Fm. 2 Fengyang, Anhui Province. Qb_w, Wushan Fm; Qb_l, Liulaobei Fm; Qb_{sh}, Shouxian Fm. 3 Ruyang, Henan Province. Ch_{xe}, Xionger Gr; Ch_x, Xiaogoubei Fm; Jx_y, Yunmengshan Fm; Jx_b, Baicaoping Fm; Jx_{bd}, Beidajian Fm; Qb_c, Cuizhuang Fm; Qb_s, Sanjiaotang Fm; Qb_l, Luoyukou Fm. 4 Shennongjia, Hubei Province. $Jx_{sh}1$, Lower Shennongjia Gr; $Jx_{sh}2$, Upper Shennongjia Gr; Qb_m, Macaoyuan Fm. 5 Luocheng, Guangxi Province. Jx_{by}, Baiyanding Fm; Jx_j, Jiuxiao Fm; Jx_{we}, Wentong Fm; Jx_y, Yuxi Fm; Qb_b, Baizhu Fm; Qb_h, Hetong Fm; Qb_g, Gongdong Fm. 6 Xiuning, Anhui Province. Jx_{zh}, Zhangqian Fm; Jx_{bn}, Banqiao Fm; Jx_m, Mukeng Fm; Jx_n, Niuwu Fm; Qb_d, Dengjia Fm; Qb_p, Puling Gr. For locations of 1 to 6 see Fig. 5.1.

There has been much discussion about the age of the lower boundary of the Changchengian System. The isotopic age from the Changchengian most frequently cited is 1922 Ma, reported from the Chuanlinggou Formation (Zhong 1977). But this figure seems too high: a number of age determinations made on granites and pegmatites below the unconformity underlying the Changchengian approximate to 1800–1850 Ma, and a U–Pb isochron age of 1839 Ma has been obtained from apatite in the basal part of the Changzhougou Formation (Tianjin Inst. 1982). Thus the time-span covered by the Changchengian may be tentatively taken as 1850–1400 Ma.

Jixianian

The Jixianian includes four formations comprising two sedimentary cycles, but there seems to be no break in the sequence. The Yangzhuang Formation is marked by a peculiar lithology of brick-red to pinkish dolomite, often silty and argillaceous at many horizons. The Wumishan Formation is continuous with the Yangzhuang and is the thickest formation (3336 m) of the whole sequence. The principal constituent rocks are shelf and littoral deposits that are clearly rhythmic and rich in organic content. The lower part is composed of alternating thick white and dark grey to black bituminous dolomites; the mid–upper part is characterized by intercalated layers of purplish-red silty and argillaceous dolomites. Oolitic and pisolitic structures are common, and both horizontal and undulating laminations of bituminous dolomites are present. The presence of silts, fine to coarse sand grains, turbulent bedding, and fragmented, redeposited chert slabs in the dolomites indicates high energy condition in a shallow-water environment.

Above the Wumishan is the Hongshuizhuang Formation of limited thickness and composed of dark green to black shales. Sometimes the carbon content is so high that the rock is combustible. The highest unit of the Jixianian is the Tieling Formation, the outstanding rock-type being stromatolitic limestones. The Tieling Formation begins with a medium- to thick-bedded sandstone passing upwards into interbedded sandy and Mn-bearing dolomites, which are in turn followed by purplish- and greenish-grey shales. Upon the shales is a slightly denudated surface with fossil soil layers. The upper part of the formation consists mainly of alternating dolomites and dolomitic limestones, some horizons being entirely composed of stromatolite bioherms and reefs. Layered chert nodules are abundant.

Palaeontologically the Jixianian is characterized by the occurrence of some peculiar acritarch remains such as *Triangumorpha* and *Leiofusa* (Plate III.11), and the advanced alga *Taeniatum*. Two stromatolite assemblages, the lower of *Micristylus–Pseudogymnosolen*, and the upper of *Conophyton lituum–Jacutophyton*, may be recognized in the Wumishan, and a *Baicalia–Chihsienella* assemblage is found in the Tieling Formation (Plate II.5, 6).

The isotopic dates obtained from the upper part of the Tieling Formation range from 1010 to 1152 Ma. There is a remarkable erosion interval between the Tieling and the overlying Xiamaling Formation, and thus the upper limit of the Jixianian may be taken as 1050 Ma.

Qingbaikouan

In the Jixian section, as well as in Qingbaikou near Beijing (whence the name is derived), the Qingbaikouan is marked by a pronounced erosional surface both at the base and at the top. The Qingbaikouan of Jixian includes two formations, the Xiamaling below and the Jingeryu above. Isotopic K–Ar ages of approximately 850 Ma are recorded from glauconite in the Jingeryu Formation. The time-interval between the Qingeryu and the overlying Early Cambrian Fujunshan Formation thus amounts to at least 250 Ma, corresponding probably to the entire Sinian Period.

The Xiamaling Formation begins with a ferruginous residual deposit upon the eroded karst surface of the Tieling limestones. The main part is composed of illitic shales interbedded with siltstones and lenticular fine sandstones. Another erosional discontinuity separates the clearly transgressive Jingeryu Formation from the Xiamaling. The lower part of the Jingeryu Formation is dominated by conglomerates and sandstones passing upwards into dark purple and light green shales. The upper part contains indurated opal-green to purple thin-bedded dolomitic micrite. Owing to its obvious distinction in lithology, the lower part of the formation is sometimes separated under the name Longshan Formation.

The Qingbaikouan in the Jixian section is almost devoid of stromatolites, but its equivalent in the western Yanshan Mountains has yielded the *Inzeria–Linella* Assemblage (Plate II.1, 2), which is closely comparable with the Upper Rifean forms of the USSR.

In the foregoing we have described the standard development of the Middle and the Upper Proterozoic in the Yanshan region. This E–W trending subsiding belt was bounded by the Inner Mongolian Axis of Huang in the north and the Shanhaiguan Massif in the south-east, and was continuous with the more or less longitudinal Taihang trough, where the Changchengian is also well developed. In the Taihang region, the basal clastics rest on the Hutuoan with a clear unconformity. On the eastern slope of the Taihang Mountains at Yangshahou in the Quyang District, the Chuanlinggou and Tuanshanzi Formations at least are represented in a clear succession, and near Dongyetou in the Xiyang District on the western slope, very thick clastic rocks are found below the Chuanlinggou shales. Within the Yanshan region the overlap of the Dahongyu Formation is pronounced, but still more widely transgressive is the Gaoyuzhung Formation, which is found to the west and north of the Yanshan and the Taihang regions, as represented by the Shinagan limestone in Nei Mongol and the Chafangzi limestone in the Wutai region.

The western Henan region (Fig. 5.1, II$_2$)

Another Middle and Late Proterozoic marine realm was developed in the southern part of the platform, in south-western Shanxi and western Henan. This was probably a shelf sea opening southwards to the Qinling marine basin. Distinct sedimentary facies belts with a NW–SE trend may be discerned, which are partly terrestrial and clastic in the north and are dominated by marine carbonate deposits in the south. In the northern belt (Fig. 5.2, III, Table 5.1, 2), the lower part of the Xiyanghe or Xionger Group is a complicated volcanic series with minor sediments, mainly composed of dark purple to green andesite porphyrites with interbedded quartz-porphyries and volcanoclastics. Recent analyses show a particularly high K_2O content in the ultrapotassic trachyte, which is characteristic of rifting volcanism. Acritarch remains, *Leiominuscula* and *Trematosphaeridium*, indicate that the environment was at least littoral or partly marine.

Three formations may be recognized in the Ruyang Group, the middle part of which has a total thickness of 2000 m. The basal Yunmengshan Formation is a thick, coarse clastic sequence of fluvial to deltaic origin. The Baicaoping Formation comprises variegated shales and quartz sandstones; cross-bedding and ripple marks indicate a northern source of the detritus. The overlying Beidajian Formation is characterized by shallow-sea glauconitic and arkose quartz sandstones interbedded with green siltstones and shales. Sandy dolomites with syndepositional breccias occur in the topmost part and may amount to 100 m in thickness.

The upper part, the Luoyu Group, also includes three formations, the lower Cuizhuang, the middle Sanjiaotang, and the upper Luoyukou, composed of variegated shales, glauconitic quartz sandstones, and stromatolitic dolomites respectively. Simultaneous deposits in the Songshan region show essentially the same succession and lithology, but with more marginal facies. Carbonate deposits in the upper group, the Heyao Formation, are much thicker. The lower volcanic series contains primitive acritarch remains and yields an Rb–Sr isochron age of 1675 Ma (Guan *et al.* 1980). *Trematosphaeridium holtedahli, Taeniatum,* and *Leiopsophosphaera* have been found in the Baicaoping, and *Laminarites antiquissimus* from the Beidajian Formation, which also yields *Baicalia baicalica* and *Inzeria*. Similar fossils occur in the upper group, and doubtful red algae, *Multisiphonia* and *Praesolenopora*, have also been reported from the Luoyukou Formation. Glauconite K–Ar ages from the Sanjiaotang Formation range from 1012 to 1089 Ma (Guan *et al.* 1980). In the light of these data, the Xiyanghe or the Xionger Group may be correlated with the Dahongyu of the upper Changchengian, taking into account the geochemical similarity of the volcanic rocks, and the Ruyang and Luoyu Groups may be roughly correlated respectively with the Jixianian and the Qingbaikouan.

The southern sedimentary belt is exposed mainly in the Xiaoqinling region of western Henan and consists of the Xionger volcanics, a lower clastic sequence, and an upper carbonate sequence. The clastic Gaoshanhe Formation begins with pebbly coarse sandstones and purple-red mudstones and may attain a very great thickness. The upper calcareous formations are composed chiefly of various kinds of dolomites, sometimes with green calcareous shales and sandstones. *Tielingella* and *Chihsienella*, commonly met with in the Tieling Formation of Jixian, have been found in the middle part of the carbonate sequence. An isotopic age of 999 Ma has been recorded from a granite intruded into the topmost Fengjiawan Formation (Yichang Inst. 1975). The whole carbonate sequence may therefore be referred to the Jixianian, and the Gaoshanhe Formation may range downwards to the Upper Changchengian.

However, the correlation studies of the rock sequences between the northern Yanshan–Taihang and the south-western Henan realms have proved to be difficult. There seems to have been no adequate eastward outlet from the Yanshan–Taihang marine basin to the open sea, and there should therefore have been sea-ways connecting the northern realm to the Qinling marine basin, presumably through western Henan in Changchengian times. Geologists working in Shanxi claim that the Changzhougou sandstones could be 'traced over' to the Beidajian Formation of Henan. But this is contrary to palaeontological and chronometric evidence. Further study is necessary, and the fact that seemingly continuous formations are diachronous over wide areas calls for special attention in correlation.

In this connection, it is appropriate to comment on the Proterozoic of East Qinling. Further south from western Henan, a metamorphic sequence consisting of the lower Kuanping Group, mainly of metavolcanics, quartzites, and schists, and an upper Taowan Group, chiefly marbles and schists, is found all along the northern slope of the East Qinling Range. Acritarch remains of Late Proterozoic type have been found in the Taowan Group. It is obvious that this metamorphic sequence represents the time equivalent of the Middle and Upper Proterozoic of Henan, and marks the southern continental margin of the North China Platform.

The eastern border region (Fig. 5.1, I_{3-4})

The third sedimentary realm that deserves notice is the eastern border of North China, including the Jiaodong–Liaodong region (Fig. 5.1, I_3) in the north and the Central Anhui region in the south (Table 5.1, 3–4). A peculiar feature in these regions is that only the Qingbaikouan is developed, and this rests directly on Lower Proterozoic and older rocks. In southern Liaoning, the lower part of the Qinbaikouan is represented by the Yongning Group, which is composed of rapidly deposited clastic sediments, and is confined to the southern part of the Liaodong Peninsula. The constituent rocks are conglomerates, greywackes, and lithic sandstones, which are highly variable both in thickness and facies. The upper part is called the Xihe Group. It is a widespread and transgressive littoral to shallow-sea deposit, and contains two formations, the Diaoyutai glauconitic sandstones below and the Nanfen variegated shales above. *Chuaria* has been found in the Nanfen Formation, which is comparable with the Jingeryu of the Yanshan region. The Qingbaikouan is on the whole constant in thickness and facies and represents true cover sequence on the platform. But the contemporaneous Penglai Group of Jiaodong region is metamorphic and somewhat variable in sedimentary type.

In the Huaiyang region (Fig. 5.1, I$_4$; Fig. 5.2, 2), the Qingbaikouan includes three formations, the Bagongshan below, the Liulaobei in the middle, and the Shouxian above. The Bagongshan Formation contains conglomerates and sandstones bearing ferruginous deposits in the lower part, and quartzitic sandstones, sometimes glauconitic, in the upper part. It is continuous with the overlying Liulaobei Formation, which consists of variegated shales and marls, exactly like the Nanfeng Formation of the Liaodong region. Abundant fossils including *Chuaria circularia* and *Ellipsophysa* (Plate III.1, 3) are found in the shales. The next succeeding formation is the Shouxian, composed of calcareous siltstones and glauconitic sandstones, also rich in *Chuaria*. The chuarids are found in the Qingbaikouan over most of North China. They can be correlated with the Grand Canyon Group of North America. The Shouxian Formation is argillaceous and shows transitions both downwards and upwards. The Huainan area was probably in a deeper part of the marine basin. The boundary between the Qingbaikouan and the Sinian and the allocation of the thick calcareous formations above the Qingbaikouan in this region are discussed in Chapter 6.

North-west China

North-west China includes two separate extensive regions. The first region includes the Tarim Platform and its surrounding mountains, the northern slope of the West Kunlun in the south, and the Tianshan in the north. The Beishan Mountains to the north of Alxa (Alashan) actually belong to the northern border of the North China Platform, but are included here for convenience. The second region includes the Qaidam Massif and the Qilian Mountains. It also is discussed here, since the Middle and Upper Proterozoic are of the same sedimentary type as in Tianshan and Beishan.

Within the area of the Tarim Platform in the first region, the Middle and Upper Proterozoic are found only in the Quruktagh Mountains (Fig. 5.1, II$_1$, Table 5.1, 6), where they are overlain by the Sinian with a flagrant unconformity. The lowest member of the sequence, the Yangjibulak Group, consists of quartzites and phyllites, with intercalated marble layers. No fossils have ever been found, and some geologists regard the Group as a lateral equivalent of the Early Proterozoic Xinditag Group. The overlying Airjigan Group and the Pargontagh Group, both dominated by carbonates, are comparable with the Jixianian and the Qingbaikouan respectively. The Airjigan Group is composed mainly of dolomitic marbles yielding *Kussiella kussiensis* in the lower part and *Baicalia baicalica* in the upper part. Chert bands and silicolite intercalations are frequent. The Pargontagh Group is much thinner and is divisible into a lower clastic and an upper carbonate part. Chlorite-schists and calcareous phyllites are the main component rocks of the lower part; various kinds of dolomitic and stromatolitic marbles abound in the upper part. *Inzeria, Katavia, Patomia,* and *Kotuikania* are found in the upper horizons, all indicating an Upper Riphean Age.

The Middle and Upper Proterozoic are extensively exposed along the axial belt of the Central Tianshan, where they are known as the Kawabulak Group. The Kawabulak Group has an average thickness of over 5000 m, and in some places amounts to more than 9000 m. The main constituent rocks are siliceous and dolomitic marbles, but quartzite, biotite-schist and amphibolite-gneiss layers are also common. Chlorite-schists and metagreywackes and metaconglomerates are frequent in the upper part. *Tungussia, Conophyton,* and *Gymnosolen* are reported from the group.

Carbonate deposits of a more stable type, consisting mainly of dolomites and siliceous limestones, have been discovered on the northern slope of the Borohoro Mountains, and are called the Kuximchik Group. This Group is no more than 2600 m thick and yields *Grunneria* and *Omachtenia* in the lower part and *Jurusania* and *Jacutophyton* in the upper part. Ancient crystalline basement rocks are known in the Jinghe District, which probably represents the northern boundary of the original Tarim Platform.

In the Beishan region (Fig. 5.1, II$_2$), the Lower and Middle Proterozoic are developed along an E–W belt that is directly continuous with Central Tianshan (Table 5.1, 7). The lower division, the Baihu Group, is especially thick and strongly metamorphosed; it contains many metabasite horizons. The middle group, the

Pingtoushan Group, is characterized by cross-bedding and ripple marks; oolitic and wurmkalk limestones probably indicate a high-energy littoral sedimentary environment. Stromatolites found in these beds point to a Jixianian age. The upper group, the Dahuoloshan Group, is assigned to the Qingbaikouan on the evidence of the presence of *Kotuikania* and *Linella*.

On the northern slope of the western Kunlun, along the upper Yarkant and the Qipan Rivers, stromatolites including *Jurusania* and *Boxonia* are found in a carbonate succession. Similar finds have also been reported to the north of the Altun Mountains. They seem to represent the Jixianian, which is probably the widest-spread group in this region.

In the second region, in the North and South Qilian Mountains and in the region surrounding the Qaidam Basin (Fig. 5.1, II_{3-4}), the Middle and Upper Proterozoic show an obvious change of facies. Three facies belts may be recognized; they are represented respectively by North Qilian, Central Qilian, and Qaidam. In North Qilian, three groups may be recognized. The lower Zhulongguan Group is a submarine volcano-sedimentary series over 3000 m thick and divisible into three parts. The lower and upper parts are mainly sedimentary, composed of metamorphosed fine clastics with minor carbonates. The middle part contains in the main tholeiitic and andesitic basalts. The Zhulongguan Group is in fault contact with the two overlying groups, the Jingtieshan and the Daliugou, which are separated from each other by an unconformity. In general, the Jingtieshan is characterized by ferruginous clastic and argillaceous flysch in the lower part and by alternating clastics and carbonates in the upper part. The Daliugou Group is also rhythmic and contains argillaceous limestones, banded slates, and siltstones. Owing to the extremely complex structure, the succession within the groups is by no means well established, and the correlation with the Jixianian and the Qingbaikouan is only preliminary.

In the Central Qilian (Table 5.1, 8), a carbonate facies with intermediate to basic volcanic rocks is predominant. In the western part, two groups may be recognized. The lower is the Danghe Group, which overlies the Lower Proterozoic Yemashan Group unconformably, the outstanding rock-type being dark-coloured slates and purple pebbly sandstones. Thin intercalated limestones in the mid-upper part yield *Kussiella* and *Collenia*. The Danghe is unconformably overlain by the Tuolainanshan Group, which is over 6000 m thick and divisible into three parts. The dominant rocks in the lower and upper part comprise argillaceous and dolomitic limestones, often oolitic and brecciated. The middle part in contrast is made up mainly of phyllites and slates, with andesite porphyrites and tuffaceous lavas in the upper portion. Stromatolites so far found in the upper part of the Tuolainanshan Group are *Gymnosolen, Tungussia,* and *Baicalia*; the Group can thus in general be correlated with the Jixianian. Still higher is the Duoroner Group, composed of purple and brown clastic and argillaceous deposits. This is usually attributed to the Sinian, but the Qingbaikouan may also be represented.

In the eastern part of Central Qilian, near Huangyuan and Lanzhou, contemporaneous strata are the more clastic and much thinner Huashishan Group, and probably only the Jixianian is represented. The underlying Huangyuan Group is strongly metamorphosed and is of Early Proterozoic age.

The third facies belt covers the Qaidam massif, where the Proterozoic is exposed on the northern and the southern border regions (Table 5.1, 9). The basement rocks are called the Dakendaban Group in the north and the Jinshuikou Group in the south. Both groups are characterized by migmatized gneiss and schists, followed by thick-bedded dolomitic marbles. In the southern part, the Jinshuikou Group is followed by the Binggou siliceous limestones and dark-coloured slates, which are probably to be correlated with the Jixianian and the Qingbaikouan.

South China

The term 'South China' as used here encompasses the regions lying to the south of the Qinling and Dabie Mountains and to the east of the western boundary of the Yangzi Platform. In terms of stratigraphic provinces, it includes the major part of the Southern Continental Super-region and the Eastern Continental Margin Super-region, roughly corresponding to the Yangzi Platform and the Caledonides of South-east China respectively. Three different sedimentary types of the Middle and Upper Proterozoic may be discerned on the platform: a paracover type developed in the interior and the northern part; a miogeosynclinal type in the south-western part; and an eugeosynclinal or island-arc type in the south-eastern part along the Jiangnan Oldland. They are treated separately in the following account.

The western Hubei region

The first type is represented by the Shennonjia Group of western Hubei (Fig. 5.1, III_1), which is an unmetamorphosed, mainly carbonate sequence about 6000 m thick that is divisible into three subgroups and eight formations (Fig. 5.24), Table 5.1, 10). The lower subgroup consists mainly of sandstones and phyllites, with carbonates and spilitic basalts in the upper part. *Conophyton garganicus* and *Trematosphaeridium* are found in the carbonate horizons. The middle and upper subgroups are generally composed of dolomites, but slates and sandstone intercalations are also common. A whole-rock U–Pb age of 1332 Ma for the middle subgroup and a K–Ar age of 950 Ma for diabase intruded into it have been recorded (Yichang Inst. 1975), thus suggesting a mainly Jixianian age. Unconformably overlying the Shennonjia Group and below the Sinian System is a clastic series called the Macaoyuan Group, composed chiefly of conglomerates and coarse sandstones with calcareous cement. It is restricted in distribution and probably represents basin-filling deposits on a faulted basement. The Macaoyuan Group is usually referred to the Qinbaikouan, although it probably represents only a part of it.

As stated above, the Proterozoic is extensively exposed in the Wudang Mountains. Recent work shows that the lower part, the Yindonggou metavolcanic formation, may be assigned to the Lower Proterozoic on account of its high isotopic age (over 2000 Ma). The middle and the upper parts are characterized respectively by metavolcanics and carbonaceous schists, both of limited thickness but separated from each other by a disconformity. The whole sequence represents semi-stable deposits developed on an old massif.

The Kham–Yunnan region

The second sedimentary type of the Middle and Upper Proterozoic on the platform is well displayed on the Kham–Yunnan Oldland (Fig. 5.1, III$_2$). Here the epimetamorphic basement strata are mainly confined to the east of the Luzhijiang fault; they are known as the Kunyang Group in eastern Yunnan and as the Huili Group in southern Sichuan (Table 5.1, 11). All Chinese geologists agree that they are contemporaneous deposits, but the subdivision and correlation problems are by no means resolved. Eight formations have been recognized in the main part of the Kunyang Group, each of four formations making a subgroup, but there is a difference of opinion as to which subgroup is the higher, i.e. whether the two subgroups are in normal or reversed order. In the following account the sequence is taken to be reversed, for this seems to provide a more satisfactory correlation with the Huili Group in the north.

The general succession of the Kunyang Group is integrated from various sections and may be divided into three parts. The lower four formations have a total thickness of over 2500 m. The Yinmin and Luoxue formations are purple slates and grey siliceous dolomites and are constant throughout the region. The third formation, the Otouchang, is mainly argillaceous and carbonaceous, but pebbly sandstones occur in the upper horizons. The fourth, the Luzhijiang Formation, contains a thick sequence of alternating thick-bedded and laminated siliceous dolomites, with marly beds in the basal part. The middle part of Kunyang Group comprises the upper four formations. The basal Dayingpan Formation is peculiar in its dark-coloured slates and siltstones. The Heishantou Formation contains rhythmic sandstones and slates in the lower part, and andesitic to basaltic tuffs in the upper part. The overlying Dalongkou Formation (including the Meidang Formation) is the only completely carbonate deposit in the Kunyang Group and is especially rich in stromatolites. The upper part is unconformable upon different members of underlying strata and contains mainly volcano-sedimentary deposits known by various names. These include the Junshao Formation of Yimen, the Liubatang Formation of Jinning, the Niutouhan Formation of Luliang, and probably also the Zhegui Formation of Yuanmou. The Junshao is mainly arenaceous and has a basal conglomerate up to 100 m thick. The Liubatang and Niutoushan are characterized by carbonaceous and siliceous slates, siltstones, and sandstones with graded bedding. Intermediate to acidic volcanics and tuffaceous rocks are common in both formations. Geologists of the Yunnan Institute of Geology believe that the order given above for the three formations may represent the normal succession from below upwards, although no complete succession is met with at either locality.

Important progress has been achieved in recent years in studies of the palaeontology and geochronometry of the Kunyang Group. Stromatolites occur mainly in two formations. The lower, the Luoxue Formation, has yielded *Kussiella, Jacutophyton,* and *Scopulimorpha*; the upper, the Dalongkou Formation, including Meidang, contains *Baicalia baicalica, Jurusania,* and *Minjaria*. These formations can probably be compared respectively with the upper Changchengian and the Jixianian. The recent discovery of *Inzeria* and *Katavia* in the Liubatang Formation is significant, since a late Riphean or Qingbaikouan age is clearly indicated.

Isotopic age-determinations have been carried out using different methods by the Yunnan Institute of Geology. Serial data for the various horizons based on the Pb–Pb method provide a reasonably reliable record. Values obtained from the various formations are: Luoxue, 1708–1760; Otouchang, 1519–1598; Luzhijiang, 1300–1410; Dayingpan, 1296–1333; Heishantou, 910; Dalongkou and Meidang, mostly 780–850 Ma. No ages higher than 800 Ma are known from the Junshao

Formation. A number of K–Ar age values cluster around 850 Ma and 1050 Ma respectively, corresponding to the two pronounced orogenic epochs, the Yangfang Movement and the Jinning Movement. This threefold subdivision seems also to be applicable to the Huili Group, the three subdivisions corresponding to the Upper Changchengian, the Jixianian, and the Qingbaikouan respectively.

On the western and north-western borders of the Yangzi Platform, Middle to Late Proterozoic epimetamorphic and volcano-sedimentary rocks constitute the folded foundation of the platform and represent contemporaneous continental margin deposits. The Yanbian Group, developed on the western side of the Yalong River, to the north-west of Huili, is mainly arenaceous and argillaceous with spilitic metabasites in the lower part. In the southern part of Longmen Mountain, contemporaneous strata of similar type are known as the Huangshuihe Group. A little to the east, in the Liangshan region, equivalent deposits are called the Obian Group; this includes, in addition to sandstones and slates, more metabasites and marbles.

The Jiangnan region

The third sedimentary type of the Middle and Upper Proterozoic of South China is distributed along the Jiangnan Oldland, stretching in a north-easterly direction for a distance of about 1500 km. This long belt may be subdivided, from south-west to north-east, into three parts. The western part encloses the Fanjing Mountain area of eastern Guichou and the Yuanbao Mountain area of northern Guangxi, representing probably two simultaneous tectono-sedimentary zones of that time (Fig. 5.1, III$_3$). The median part of the Jiangnan Oldland encompasses the Jiuling Mountain and parts of the Huaiyu Mountain of northern Jiangxi; the eastern part covers the hilly regions of southern Anhui and western Zhejiang. In all the three regions it is possible to distinguish two metamorphic sequences separated by an unconformity, marking the boundary between the Middle and the Upper Proterozoic.

In the Fanjing Mountain area, the Middle Proterozoic Fanjingshan Group is subdivided into three parts and seven formations. The lower part includes an alternating sequence of metasandstones and slates with basic volcanic rocks. There are numerous layers of metamorphosed diabase which form the host rocks of polymetallic deposits. The sedimentary rocks are chiefly siltstones and slates, occasionally calcareous and argillaceous. The topmost Xiaojiahe Formation is characterized by the occurrence of ultrabasic rocks. The middle Huixiangping Formation comprises a complicated thick succession of volcano-sedimentary rocks ranging from agglomerates, spilites, and keratophyres to pyroxenites and peridotites. Typical pillow structures in amygdaloidal spilites are frequent, but a diagnostic ophiolitic suite has not been identified. The upper part is in general dominated by arenaceous sedimentary and tuffaceous rocks, especially in the uppermost part, in which metagreywackes and pebbly sandstones occur in more than one horizon. They presumably represent the coarse flysch that finally filled up the basins.

The overlying Banxi Group of Late Proterozoic age shows a clear differentiation of sedimentary facies. To the west of Fanjing Mountain, it displays a prevalent grey and green colour and has a thickness of more than 3500 m. Two sedimentary cycles may be recognized, each consisting of two formations; the Jialu and Wuye Formations in the lower cycle, and the Fanzhao and Qingshuijiang Formations in the upper cycle. The Fanzhao Formation often contains thick, coarse quartz sandstones and arkose sandstones, purplish in colour, especially in the northern area. The Qingshuijiang Formation consists mostly of grey and green arenaceous and argillaceous beds and is especially rich in volcanics. Sandstones, sometimes coarse sandstones, abound in the north, but give way to siltstones and slates in the south. Tuffs and ignimbrites are common volcanic rocks, and silicolitic rocks probably originated from volcanism. The upper members of the Qingshuijiang Formation are often lacking in the north, although it is the widest distributed among the formations of the Banxi Group.

To the east of Fanjing Mountain, the Banxi Group is characterized by a thick sequence of purplish-red and green sandstones in the lower part and by thin greenish-grey slates in the upper part. The Hongzixi Formation in the lower part begins with a basal conglomerate and is composed chiefly of alternating purple and green mudstones and slates that originated in a very shallow sea. The upper part, called the Qingshuijiang Formation, comprises several hundred metres of dark grey slates and tuffs. The type locality of the Banxi Group is situated at Banxi, Yuanling, where it includes two formations, the Madiyi below and the Wuqiangxi above (Table 5.1, 15). The basal member of the Madiyi Formation is usually a thick-bedded arkosic quartz-sandstone, which passes upwards into purplish-red marly dolomites. The higher layers are usually alternately red and green and become predominantly green upwards. The Wuqiangxi Formation begins with a pebbly coarse- to medium-grained sandstone, but consists mostly of green and grey banded slates, often tuffaceous and silicolitic. Graded bedding and flute marks are frequent in the slates, which may represent flysch beds of genuine turbiditic character, especially to the south of Yuanling and Chenxi. Further to the south, in south-eastern Guizhou and adjacent parts of Hunan, the Banxi Group becomes much thicker and entirely green in colour. Thus in eastern Guizhou and western Hunan the Banxi Group comprises two different facies, a red and a green, probably representing different water depths during deposition.

A more complete and much thicker sequence of the so-called green facies is developed in the Yongjiang region of south-eastern Guizhou. Here the Banxi Group is over 10 000 m thick and overlies the Middle Proterozoic with a clear unconformity. Above the Qingshuijiang Formation and below the Sinian there occurs the Longli Sandstone, which contains conglomerate lenticles in the lower part and greenish grey slates in the upper part. As the Banxi Group is almost devoid of carbonates, no stromatolites have been discovered, and acritarch remains are also very meagre.

It is in northern Guangxi that the chronology and correlation of the Middle and Upper Proterozoic are made relatively clear (Table 5.1, 14). In the Yuanbao Mountain area of the Yong'an District the oldest strata, the Sibao Group, are best exposed (Fig. 5.2, 5), and an obvious unconformity separating the Sibao from the overlying Banxi (Danzhou) Group indicates the well-known Sibaoan Movement.

The Baiyanding Formation, the lowest of the Sibao Group, has an exposed thickness of over 4000 m, and is mainly composed of grey and green quartz-schists, sericite-phyllites, and granulitites. Rhythmic deposition is evident, but cross-bedding is also known. In the Jiuwan Mountains, the thickness is much reduced and there is also an obvious increase in clastic material as the formation is traced eastwards. The main constituent rocks of the succeeding Jiuxiao Formation are arkose metasandstones, phyllites, and chlorite-schists. In general, the lower part is more arenaceous, and the upper more argillaceous, with a spilitic layer some 50 m thick occurring in the middle part. The third formation, the Wentong Formation, is a complex volcano-sedimentary series mainly composed of metasandstones and slates with intercalated volcanic rocks, among which spilites, keratophyres, and ignimbrites are prevalent. Volcanic breccia, tuffites, and jasperites are common in the lower layers; keratophyres and ignimbrites, including tuffaceous breccias, prevail in the upper layers. An upward change in composition from basic to intermediate is also obvious. The uppermost member, the Yuxi Formation, has a preserved thickness of 1500 m and consists chiefly of siltstones and slates of a rhythmic nature; it probably represents typical turbidites.

The Sibao Group is everywhere overlain by the Banxi (Danzhou) with a conspicuous unconformity. The Banxi is much thinner here than in eastern Guizhou, and is divisible into three formations. Lithologically the Baizhu Formation is clastic in the lower part and calcareous in the upper part. The lower part begins with a basal conglomerate that passes upwards into pebbly slates and phyllites, and further upwards into chlorite-schists, quartz-phyllites, and slates. The upper part consists of various kinds of schists with interbedded marbles and calc-phyllites. The topmost calcareous beds serve as a clear marker between the Baizhu and the overlying Hetong Formation. The latter is also divisible into two parts, and is characterized by green and light grey slates and phyllites in the lower part, and dark grey carbonaceous slates and phyllites in the upper part. In the eastern part of the region, in the Longsheng District, this formation is dominated by metasandstones and carbonaceous slates, with lenticular marble layers and eruptive rocks. Three volcanic sequences have been observed in Longsheng and Sanmen, where each sequence may contain several eruptive cycles. The Gongdong Formation, the highest unit of the Banxi Group, is mainly arenaceous and argillaceous in composition and is transitional with the underlying Hetong Formation. Arkose quartz-sandstones and unstable conglomerate beds occur in the middle part near Sanmen, and lenticular dolomites are reported from near Longsheng. The Gongdong Formation is also characterized by rhythmic successions formed of alternating arenaceous and argillaceous beds, often with cross-bedding, syndepositional breccias, and scoured surfaces. The sediments were evidently deposited in a high-energy environment.

In the regions discussed above, carbonates are extremely rare, stromatolites are practically lacking, and acritarch remains are also very meagre. Thus the main criteria for correlation with other regions are radiometric age data. Fortunately some critical age values have been obtained from both above and below the unconformity surface which resulted from the Sibaoan Movement. The Sibao Group is intruded by the Bendong granite, for which whole-rock isochron Rb–Sr ages of 1065 Ma and 1109 Ma have been obtained (Yichang Inst. 1975). Thus the upper age limit of the Sibao is about 1150 Ma, falling well within the scope of the Middle Proterozoic. Two significant figures may be cited for the Banxi Group. The andesite in the mid–lower part of the Madiyi Formation yields an Rb–Sr age of 950 Ma, and an intrusive diabase layer in the Hetong Formation gives an age of 837 Ma (Yichang Inst. Geol. 1975). It therefore seems reasonable to refer the Banxi Group to the Qingbaikouan (1050–850 Ma).

The median segment of the Jiangnan Oldland encloses the main part of northern Jiangxi, including the Jiuling Mountains in the west and the Huaiyu Mountains in the east (Table 5.1, 12). The Middle Proterozoic Shuangjiaoshan Group in the western part can be divided into two subgroups, but the detailed succession has not been worked out. The lower subgroup is mainly distributed in the southern slope of the Jiuling Mountains. It is characterized by the presence of ophilitic suites, comparable with the Sibao Group in lithology. The upper subgroup is developed in the Xiushui region and is represented by grey and green tuffaceous siltstones and slates, with a total thickness of 3000 m. The contact with the underlying subgroup is probably an unconformity. Rich acritarch remains were discovered in the turbidite flysch beds locally called the Anluolin and the Xinmin Formations. The main forms are *Protoleiosphaeridium infriatum, Leiopsophosphaera minor, Trachysphaeridium hyalinum, Tr. rugosum,* and *Margominuscula rugosa.* These probably belong to the Qingbaikouan.

Above the upper Shuangjiaoshan and below the Lower Sinian there occurs the peculiar Luokedong Formation, which is restricted in distribution and is so far known at only two localities. Near Luokedong in the Wuning District this formation has a thickness of 236 m, consisting of purplish-red and yellowish-green tuffaceous conglomerates, sandstones, and tuffites. Volcanic breccia and quartz-porphyries also occur. The Luokedong probably represents fault-basin deposits on the eroded surface formed after the folding of the upper Shuangjiaoshan, situated at the southern border of the Yangzi Platform. The only known fossils are some acritarch remains that indicate a Late Proterozoic age. It seems that both the upper Shuangjiaoshan and the Luokedong are attributable to the Upper Proterozoic. In the Huaiyu Mountains of north-eastern Jiangxi, the lower Shuangjiaoshan is known only near Jingdezhen, whence its outcrop continues to southern Anhui. The main constituent rocks are tuffaceous sandstones, sandy siltstones, and black slates. In some places these are metamorphosed to phyllites and chlorite-schists. A whole-rock Rb–Sr age of 1410 Ma was reported from the tuffaceous phyllites by the Geological Survey Team of Jiangxi Province in 1980. This indicates that the Lower Shuangjiaoshan may be correlated with the Sibao and the Fanjingshan of the western region. The upper Shuangjiaoshan Subgroup is more widespread and consists of arenaceous and argillaceous flysch up to 10 000 m in thickness. Three divisions may be recognized: the lower characterized by volcanic breccias and purplish-red banded slates; the middle by black slates and marble layers, all rich in tuffaceous material; and the upper by andesite to rhyolite lavas.

The eastern segment of the Jiangnan Oldland encompasses two belts trending roughly E–W, southern Anhui and western Zhejiang. In Xiuning and Qimen in southern Anhui, below the basal Sinian Xiuning sandstone, two sequences are developed (Fig. 5.2, 6), the lower being metamorphic and forming the basement, the upper containing clastics and volcanics of a paracover type. The lower metamorphic series is known as the Shangxi Group and contains four formations representing two sedimentary cycles (Table 5.1, 16).

The Zhangqian Formation is a flysch or paraflysch sequence composed of green and grey phyllites and phyllitic slates, occasionally metamorphosed to chlorite- and quartz-schists. It is continuous with the overlying Banqiao Formation, which contains dark-coloured sandy slates in the lower part, light-coloured calcareous phyllites and phyllitic slates in the middle part, and alternating slates and sandy siltstones in the upper part. Meta-andesites are also found in the upper horizons. The Mukeng and Niuwu Formations constitute the upper sedimentary cycle. The Mukeng Formation is composed of more than 3000 m of bluish-grey phyllitic sandstones, slates, and phyllitic lithic sandstones, which are incorporated into different ranks of rhythmic deposits. The Niuwu Formation is the highest unit of the Shangxi Group and has the character of a typical arenaceous turbidite. It is divisible into three parts, all with a prevalent green and yellow colour. Pebbly lithic sandstones and siltstones are characteristic of the lower part. The middle part contains sandy marls and calcareous phyllites. The upper part is entirely composed of siltstones and lithic sandstones.

Unconformably overlying the Shangxi Group, the Dengjia Formation contains whitish-grey arkosic sandstones and conglomeratic lithic sandstones in the lower part, and tuffites and sandy phyllites in the upper part. The topmost Puling Formation is chiefly volcanic, with smaller amounts of intercalated phyllites. The constituent volcanic rocks comprise rhyolite and dacite lavas and corresponding tuffites and porphyries. Andesite tuffs and lavas are prominent near Puling, Qimen District. The K–Ar ages of granites intruded in the Shangxi Group, and probably before the deposition of the Xiuning sandstone, are 908 and 913 Ma, and an age of 936 Ma was obtained from the Puling Formation (Liu *et al.* 1973). Thus the Dengjia and Puling may be reliably assigned to the Qingbaikouan.

In western Zhejiang, near Zhuxi and Shaoxing, pre-Sinian metamorphic strata are well developed (Table 5.1, 17). Three formations may be distinguished. The lower, strongly metamorphosed, sequence is called the Shuangxiwu Group. Within it, two volcano-sedimentary cycles may be discerned, each beginning with intermediate to basic lavas and volcano-clastics and ending in sandstones and mudstones with limestone lenticles. Spilites and andesite basalts, representing part of ophiolitic suites that are reminiscent of the Sibao Group of Guangxi, occur in the mid–lower part of the group. The overlying Luojiamen Formation is a marine flysch consisting of conglomerates and arkosic sandstones with intercalated acidic volcanic rocks in the lower part, and typical rhythmic siltstones and siliceous argillites in the upper part. Another unconformity separates the Luojiamen from the overlying Hongchicun Formation, made up of littoral to shallow-sea greywackes. The prominent rocks are purplish-grey to brown coarse arkosic sandstones interbedded with thin-bedded siltstones. The Hongchicun Formation is usually thin and passes upwards into the Shangshu Formation, a thick sequence containing up to six volcanic cycles. These begin with intermediate–basic eruptive rocks and end with acidic eruptives, thus showing an upward increase in acidic materials. Intercalation of siltstones and mudstones commonly occur between the successive cycles.

The volcanic sequence below the Sinian is not only widely distributed in the Jiangnan region, but is also found further to the north in the Lower Yangzi region (Fig. 5.1, III$_4$). Near Chuxian, above the Feidong Group which is the basement complex of the platform, a pre-Sinian succession is found, consisting of an epimetamorphic Beijiangjun Formation below, which is succeeded by a volcanic complex, the Zhangbaling Group. The latter consists mostly of rhyolitic and andesitic lavas and tuffs, including spilites and keratophyres. The Zhangbaling Group is comparable with the Puling and Shangshu, but the isotopic ages are a little higher: 894 to 1031 Ma, as reported by Lu Wuyun (1980).

In the region south of the Jiangnan Oldland (Fig. 5.1, IV), pre-Sinian stratigraphic records are generally meagre and uncertain. The pre-Sinian is probably exposed near Yingyangguan in northern Guangdong, near Shenshan in western Jiangxi, and in the Changting–Longyan region of south-west Fujian. These deposits will be discussed, for convenience, in the next chapter.

Other parts of China

The regions to be reviewed in this section include North-east China (Fig. 5.1, V), western Yunnan, and southern Xizang (Fig. 5.1, VI).

Proterozoic strata are mainly exposed in two belts in Heilongjiang Province, a western in the northern Hingan Mountains, and an eastern in the Laoyeling and Nadanhada Mountains. In the northern Hingan, in Huma District, a metamorphic sequence referred to as the Xinghuadukou Group contains four formations and attains a total thickness of over 7000 m. The main rocks include gneiss, crystalline schist, granulitite, and marble together with various types of migmatites. Amphibolites and taconites are also frequent. In general, plagioclase-gneiss, amphibolites, and magmatized granulitites abound in the lower part; carbonates, chiefly graphite-marble and dolomitic marble, are frequent in the mid–lower part; and various kinds of crystalline schists, predominantly biotite- and hornblende-schists, are characteristic of the upper part. The Xinghuadukou Group is overlain by the Cambro-Sinian Jiageda Group with a disconformity.

The Proterozoic in the eastern belt is usually subdivided into two sequences, the lower Heilongjiang Group and the upper Mashan Group. The Heilongjiang Group is primarily a volcano-sedimentary series dominated by greenschists, migmatization being generally weak. The principal constituent rocks are sodium-plagioclase-mica-schist, chlorite-schist, amphibole-schist, and marbles, amounting to more than 8000 m in thickness. The Mashan Group, on the other hand, is usually deeply metamorphosed and migmatized. Intercalation of graphite- and diopside-marbles occurs throughout the whole sequence, and the migmatized gneiss represents for the most part volcano-clastic and clastic rocks. In the easternmost part, in Baoqing District, a very thick metamorphic sequence comparable with the Heilongjiang and Mashan groups is exposed below the Upper Palaeozoic. The lower part comprises greenschists and metabasites, the middle part contains marbles, and the upper part includes metamorphic argillaceous deposits and intermediate to acidic volcanic tuffs and lavas. So far only metamorphic K–Ar ages of 600 to 900 Ma have been obtained from the Heilongjiang and the Mashan groups.

In western Yunnan, metamorphic rocks definitely proved to be of pre-Sinian or Precambrian age are the Ailaoshan Group in the Red River valley, the Lancang Group along the Lancang River valley, and the Cangshan Group in the Cangshan Mountain. The first crops out in a narrow belt along the Red River for a distance of about 300 km, and is composed of gneiss, schists, amphibolites, and marbles. Although the oldest overlying strata are Ordovician, the Ailaoshan is a direct continuation of the Red River Complex of Northern Vietnam, where a metamorphic age of 2200 Ma has been reported. Lei (1982) reported an acritarch assemblage of Jixianian age from the Lancang Group of the Huimin District. Thus the Middle Proterozoic at least is included. The Cangshan Group has a similar litho-

Table 5.1. Correlation of the Proterozoic (Pre-Sinian) of China.

Regions	North China					North-west China			
	1 Yanshan–Wutai	2 Zhongtiao–Songshan	3 Liaodong	4 Huanian	5 Nei Mongol	6 Quruktagh	7 Beishan–Alxa	8 Central Qilian	9 S. Qaidam
850 — Pt₃ Qingbaikouan	Cm₁ Fujunshan Fm; Qingeryu Fm; Longshan Fm; Xiamaling Fm	Z₂ Luoquan Fm; Luoyukou Fm; Sanjiaotang Fm; Cuizhuang Fm	Cm₁ Jianchang Fm; Nanfen Fm; Diaoyutai Fm; Yongning Gr (Xihe Group)	Z Jiayuan Fm; Shouxian Fm; Liulaobei Fm; Wushan Fm	Cm; Shinagan Fm	Z₁ Baiyisi Fm; Pargangtag Gr	Z Dahuoluoshan Gr; Pingtoushan Gr	Cm₂; Beimenjia Fm (Huashishan Group)	Cm₁; Binggou Gr
1050 — Jixianian	Tieling Fm; Hongshuizhuang Fm; Wumishan Fm; Yangzhuang Fm	Beidajian Fm; Baicaoping Fm				Airjigan Gr		Kesuer Fm	
1400 — Pt₂ Changchengian	Gaoyuzhuang Fm; Dahongyu Fm — 1650 —; Tuanshanzi Fm; Chuanlinggou Fm; Changzhougou Fm	Yunmengshan Fm; Xionger Gr				Yangjibulak Gr	Baihu Gr		
1850 — Pt₁ Hutuoan	Guojiazhai Subgr; Dongye Subgr; Doucun Subgr (Hutuo Gr)	Danshanshi Gr; Zhongtiao Gr	Gaixian Fm; Dashiqiao Fm; Gaojiayu Fm (Liaohe Group)	Songji Fm; Qingshi-shan Fm; Yinjiajian Fm (Fengyang Gr)	Majiadian Gr; Erdaowa Gr	Xingditag Gr	Longshoushan (Dunhuang) Gr	Qingshipo Fm; Moshigou Fm (Huangyuan Group)	Jinshuikou Gr
2300 — Wutaian	Pushang Fm; Taihuai Fm; Shizui Fm (Wutai Group)	Jiangxian Gr	Lieryu Fm; Langzishan Fm	Wuhe Gr.	Sanheming Gr	? Daklak-bulak Gr		Dongchagou Fm; Liujiatai Fm	
2600 — Ar	Ar₂ Longquanguan Gr		Ar Anshan Gr		Ar Wulashan Gr				

logy and is intruded by a granite with a K–Ar age of 660 Ma (Yichang Inst. Geol. 1972). It is thus definitely Precambrian and probably pre-Sinian in age.

In Xizang, Precambrian strata are definitely known to occur in the high Himalayas, known as the Nyalam Group or the Qomolungma Group. Near Mt. Qomolungma the metamorphic rocks are divisible into two parts: the lower part consists of sillimanite-gneiss and biotite-schist; the upper part is composed chiefly of banded migmatites intercalated with gneisses and marbles. The Nyalam Group has yielded a radiometric age of 640–660 Ma (Geoch. Inst. 1973), but the same metamorphic sequence in the high Himalayan belt beyond the Chinese border has yielded an age of 1400 to 1800 Ma. Strata of similar lithology are extensively exposed in a NE–SW belt along the Nyainqentanglha Range. As the stable type non-metamorphic Ordovician and Silurian were discovered in the Xainza region, some 159 km to the west, the Precambrian age of these metamorphic rocks is beyond doubt. An isotopic age of 1250 Ma was recently obtained by the Sino–French Expedition in Xizang. As is shown by the later stratigraphic record, both the Himalaya and the Nyainqentanghla were parts of the ancient Gondwanaland.

Table 5.1 (cont.)

South China								Regions
10 W. Hubei	11 C. Yunnan	12 N. Jiangxi	13 Central Anhui	14 N. Guangxi	15 Central Hunan	16 S. Anhui	17 W. Zhejiang	
Z_1 Liantuo Fm	Z_1 Chengjiang Fm	Z_1 Dongmen Fm	Z_1 Zhougang Fm	Z_1 Chang'an Fm	Z_1 Jiangkou Gr	Z_1 Xiuning Fm	Z_1 Zhitang Fm	—850— Pt_3 Qingbaikouan
Macaoyuan Fm	Liubatang Fm	Luokedong Fm ~?~ Xiushui Gr	Zhangbaling Gr	Gongdong Fm / Hetong Fm / Baizhu Fm (Danzhou Gr)	Wuqiangxi Fm / Madiyi Fm (Panxi Group)	Puling Gr / Dengjia Fm	Shangshu Gr / Hongchicun Fm / Luojiamen Fm	—1050— Jixianian
Shennongjia Group: Upper Subgr / Lower Subgr	Kunyang Group: Upper Subgr / Lower Subgr	Jiuling Group: Upper Subgr ~?~ Lower Subgr	Beijiangjun Fm	Yuxi Fm / Wentong Fm / Jiuxiao Fm / Baiyantang Fm (Sibao Group)	Lengjiaxi Gr	Niuwu Fm / Mukeng Fm (Shangxi Group)	Shuangxiwu Gr	—1400— Pt_2 Changchengian —1850—
	Dahongshan Gr		~?~ Susong Gr					Hutuoan Pt_1
Sandouping Gr			~ Dabie Gr					—2300— Wutaian —2600— Ar

6. THE SINIAN SYSTEM
Wang Hongzhen

INTRODUCTION

The Sinian System covers the time interval from 850 to 600 Ma, and is typically represented by the Yangzi Gorge section near Yichang, Hubei Province. Since A. W. Grabau first defined the Sinian System in its modern sense in 1922 (see Chapter 5), it has been used as a formal system name for over half a century both in China and abroad. It would therefore seem that the term Sinian System, being the earliest established, and having world-wide usage, has a good claim to meet the requirements of the recent international project to establish a new system for universal application in the uppermost part of the Precambrian.

There has been much discussion about the geochronological position of the Sinian Period. In fact, the Sinian System is invariably represented by a stable sequence on the ancient platforms and many of the larger old massifs, in China as well as in other parts of the world. Thus, from a tectonic point of view, the Sinian may be better regarded as the first system of the Palaeozoic Era, a view advanced by Grabau in his 1922 paper, and one that is followed by many Chinese geologists. The essential point in this context relates to the palaeontological characteristics of the Sinian and the concept of the Palaeozoic. If we take the abundance of fossils that can be used to establish chronostratigraphic stages and chronozones as the main criterion of the Palaeozoic, it would seem appropriate to begin the latter at about 600 Ma, when the first hard-shelled animals appeared in abundance, both as taxa and as individuals, and chronozones can be adequately established. As for the Sinian, although fossil remains abound in certain horizons, e.g. the *Ediacara* fauna and the sabelliditids in the upper part, there is no continuous succession of fossil assemblages or fossil zones that could be established for the whole System. Acritarch remains and stromatolites are abundant in many beds, yet they are not adequate for zoning or world-wide correlation. For these reasons we prefer to put the Sinian in the uppermost part of the Proterozoic.

Another important problem is the lower boundary of the Period. There should probably be no fundamental difference in the principles of chronostratigraphic classification between the Precambrian and the Phanerozoic. Two points need to be considered here. First, the actual time range of a period may inevitably be longer in the remote geological past of the Precambrian; secondly, more emphasis should probably be laid on crustal movements in the absence of palaeontological data. Some would use the widespread glacial deposits dated at about 700 Ma for the lower boundary of the System, but the opposing opinion is that glacial deposits cannot be contemporaneous everywhere, and are therefore inadequate for use as chronostratigraphic boundary marks. On the other hand, in China, and probably also in the northern continents as a whole, a very conspicuous orogenic movement, the well-known Jinningian Movement, took place at around 850–800 Ma, after which detrital deposits were formed at many places as the first platform cover sequence. Thus we may employ the pronounced unconformity surface resulting from the Jinningian Movement as the lower boundary (a step followed here). The general correlation is then convenient, although the palaeontological criteria may seem a little weak.

The Sinian System is generally divided into two series: the Lower—clastic and partly of glacial origin; and the Upper—mainly argillaceous and calcareous. The boundary between the two series has been the subject of much discussion in recent years. An apparent unconformity was early observed below the tillites in eastern Yunnan, on which basis the Chengjiang Movement was established. But subsequent studies have shown that this discontinuity usually dies away in other regions and a discontinuity also occurs above the tillite beds. In view of the fact that marked changes in the organic world, such as the appearance of advanced metazoans and acritarch remains, occurred after the glacial epoch, the boundary between the two series is now tentatively drawn above the tillite formation.

In the Sinian Period, the broad features of sedimentation realms and sedimentary types are distinct (Fig. 6.1), but conditions in the mobile regions remain obscure. To the north of the North China and Tarim platforms, in the Northern Intercontinental Superregion, the Sinian is not clearly separable from the metamorphic Precambrian and no definite succession

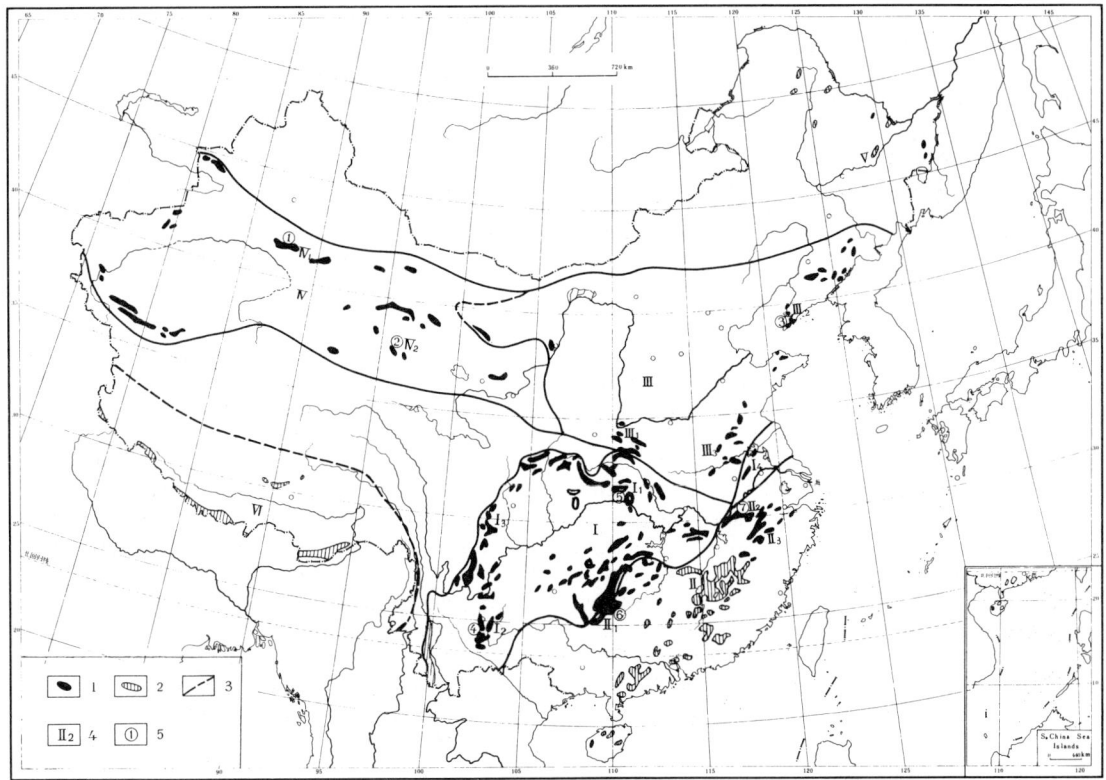

Fig. 6.1. Sketch map showing the distribution and sedimentary regions of the Sinian of China (modified after Ma Lifang (1984)). 1, Sinian outcrop; 2, outcrop of strata including Sinian; 3, boundary between sedimentary realms; 4, number of sedimentary regions; 5, number of columnar sections. I. Yangzi Realm: I_1, West Hubei Region; I_2, East Yunnan Region; I_3, South-western Sichuan Region; I_4, East Anhui Region. II. South-east China Realm: II_1, North Guangxi Region; II_2, South Anhui Region; II_3, West Zhejiang Region. III. North China Realm: III_1, West Henan Region; III_2, South Liaoning Region; III_3, North Anhui Region. IV. North-west China Realm: IV_1, Kuruktagh Region; IV_2, Quaidam Region. V. North-east China Realm. VI. Himalaya Realm. For locality names of 1 to 7 see Fig. 6.2.

has been observed. One of the most complete Sinian successions is found in the northern border region of Tarim. But it is in South China, which embraces the Yangzi Platform and its south-eastern continental margin, that the Sinian is best developed and many continuous Sinian–Cambrian sections are situated. The Yangzi region will form the basis of a discussion of Sinian stratigraphy.

THE YANGZI PLATFORM

The Yangzi Platform was finally consolidated as a result of the Jinningian Movement, and the Sinian forms the genuine cover strata on the folded basement. At the beginning of the Sinian Period, the Kham–Yunnan Oldland existed in the west, thus causing the whole Yangzi region to tilt slightly to the east. Complete successions of the Sinian are found in the Yangzi gorges where the stratotype section is situated (Zhao *et al.* 1980), and in the eastern Yunnan subsiding belt lying to the east of the southern segment of the Kham–Yunnan Oldland (Cao *et al.* 1980).

The Sinian sequence in the Yangzi Gorge (Fig. 6.1, I_1) amounts to a thickness of about 1000 m, and is divisible into two series and four formations as originally proposed by Li (Fig. 6.2, 5). The Lower Sinian contains two formations, the Liantuo below and the Nantuo above. The Liantuo Formation has an average thickness of about 160 m and begins with a very thin basal conglomerate consisting mostly of vein quartz pebbles. Purplish-red pebbly and coarse sandstones, sometimes arkosic, are predominant in the lower part. The upper part is more argillaceous, comprising purple and red tuffaceous siltstones and fine sandstones. A medium-grained thick quartz sandstone bed containing well-rounded black silicolite pebbles occurs in a low horizon, and greyish-green silty argillite and siltstone characterize the top part of the formation. The succeeding formation, the renowned Nantuo tillites, rests on the Liantuo with a clear disconformity and includes about 60 m of greyish-green to yellowish-green glacial deposits. The main rocks are various kinds of boulder clays, with unsorted boulders and pebbles ranging in size from a few millimetres to over half a metre. They are well striated and entirely without bedding in the lower part, but medium-grained sandstone lenses begin to appear in the middle part, and towards the top there is an evident decrease in pebble content and an appearance of stratification. Purplish-red to greyish-green, pebbly, silty mudstones are frequent in the upper part of the formation. Another disconformity separates the Nantuo from the overlying Doushantuo (Toushantuo) Formation. In the type section, the main bulk of the Doushantuo Formation is composed of some 200 m of siliceous and occasionally carbonaceous carbonates, including thin- to medium-bedded micritic and siliceous dolomites intercalated with numerous beds of black sandy shales and carbonaceous shales. Chert nodules are abundant in many horizons. The Doushantuo is continuous and transitional upwards with the Dengying (Tengying) Formation.

The Dengying Formation is the most conspicuous in the Sinian sequence. Along the Yangzi River from Nantuo to Shipai, the Dengying is excellently exposed and contains four members, the highest Tianzhushan Member being assigned to the Lower Cambrian on account of the appearance of a rich small shell fauna. The lowest member contains light-coloured sparitic to micritic dolomites, usually with oolitic structure, and includes in particular chert nodules and syndepositional breccias. The second member is characterized by thin- to medium-bedded dark grey to black asphaltic and micritic limestone. The third member, by far the thickest, is essentially composed of thick-bedded cherty dolomites. The relation between the third member and the Early Cambrian Tianzhushan Member is continuous and transitional, and provides an ideal case for the designation of the Sinian–Cambrian boundary.

Palaeontologically the Lower Sinian is characterized by the arcritarch *Trematosphaeridium hyalinum* (Plate IV.1) and *Laminarites antiquissimus*, but these two forms are also present in the Qingbaikouan below and in the Upper Sinian above. A remarkable change in the organic world took place in the Late Sinian, when the vendotaenids, the primitive worm *Micronemaites*, and the advanced algae *Polyedryxium* (Plate IV.3) appeared for the first time. Minute fossil remains attributed to primitive hyolithids are found in the upper part of the Dengying Formation, and an individual specimen of *Charnia* has also been reported from the same horizon. So far the radiometric data obtained from the Sinian rocks in the Gorge section are an Rb–Sr isochron age of 693 Ma of the Doushantuo Formation (Zhao *et al.* 1980) and a recent result of 740 Ma at the top of the Liantuo Formation (Yichang Inst. 1982). The Early Cambrian Shuijingtuo black shale has yielded 613 Ma by the same method, evidently somewhat higher for the horizon. It seems reasonable to set the upper limit of the Sinian at about 600 Ma and the boundary between the Lower and the Upper Sinian at about 700 Ma. The lower boundary of the System presents a difficult problem, for no direct data are available. Metamorphic and intrusive ages obtained by various methods for the Sandouping Group and intrusive bodies range from 805 to 880 Ma, indicating the obvious geothermal event associated with the Jinningian Movement. An approximate age of 800–850 Ma may therefore be taken as the lower limit of Sinian deposition in this area.

A distinguishing feature of the Sinian in the subsiding belt of eastern Yunnan (Fig. 6.1, I_2) and southern Sichuan (Fig. 6.1, I_3) is the abundance of volcanism and the absence of glacial deposits in the Lower Sinian. The Suxiong, Kaijianqiao, and Liegulu formations are mostly tuffaceous and without glacial deposits. The Upper Sinian Hongchunping dolomite is transitional with the overlying Early Cambrian Meidiping Formation, thus providing another favourable locality for boundary research. In eastern Yunnan, the Sinian begins with the Chengjiang Formation of clastic and paramolasse deposits. This is highly variable both in lithology and in thickness (Fig. 6.2, 4). Resting unconformably upon the Chengjiang Formation are the purplish-red tillites and laminated shales, probably varved clays, collectively known as the Chengjiang tillites. The Dengying Formation is divided into two parts by a disconformity, the lower part characterized by rock salt pseudomorphs and the upper part by intercalated purplish-red mudstones and yellow sandstones. The Dengying Formation is conformably overlain by the Early Cambrian Meishucun Formation. In the major part of central Sichuan and northern Guizhou the Lower Sinian is much reduced or entirely absent. The transgressive Upper Sinian was also partly eroded away before the advent of the Early Cambrian sea. The Dengying Formation contains very thick salt deposits in the local depressions in south-western Sichuan.

In the northern part of the Lower Yangzi, in eastern Anhui near Chuxian (Fig. 6.1, I_4), a semistable type of Sinian strata with slight metamorphism and great

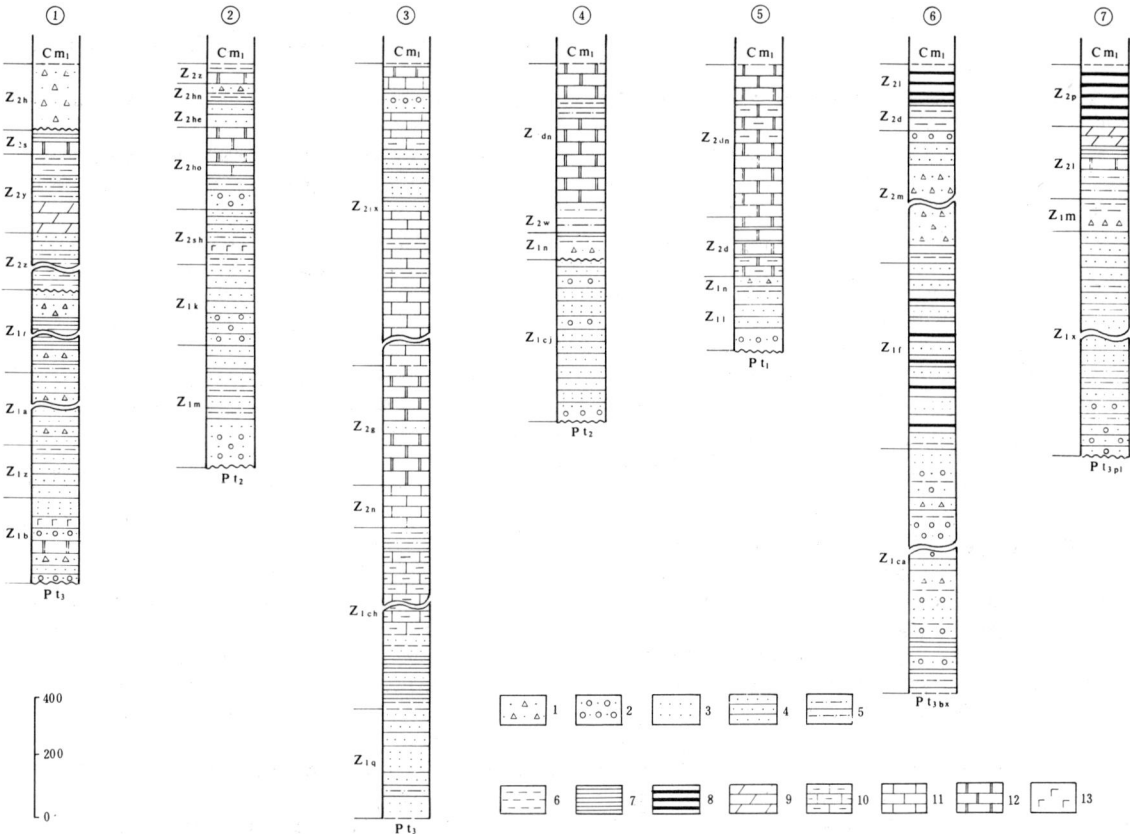

Fig. 6.2. Columnar sections of the Sinian of China 1, tillites and diamictites; 2, conglomerate; 3, massive sandstone; 4, bedded sandstone; 5, sandy shale; 6, mudstone; 7, shale; 8, slate; 9, marl; 10, argillaceous limestone; 11, limestone; 12, dolomitic limestone and dolomite; 13, intermediate to basic volcanics. 1 Quruktagh, Xinjiang Province. Z_{1b}, Baiyisi Fm; Z_{1ch}, Zhaobishan Fm; Z_{1a}, Altongol Fm; Z_{1t}, Tereeken Fm; Z_{2z}, Zhamokti Fm; Z_{2y}, Yukengol Fm; Z_{2hs}, Shuiquan Fm; Z_2, Hangelhaok Fm. 2 Quanji, Qinghai Province. Z_{1m}, Mahuanggou Fm; Z_{1k}, Kubomu Fm; Z_{2sh}, Shiyingliang Fm; Z_{2ho}, Hongzaoshan Fm; Z_{2he}, Heitupo Fm; Z_{2hn}, Hongtiegou Fm; Z_{2z}, Zhoujieshan Fm. 3 Fu Xian, Liaoning Province. Z_{1q}, Quiaotou Fm; Z_{1ch}, Changlingzi Fm; Z_{1n}, Nankuanling Fm; Z_{2g}, Ganjingzi Fm; Z_{2jx}, Jinxian Gr. 4 Chengjiang, Yunnan Province. Z_{1cj}, Chengjiang Fm; Z_{1n}, Nantuo Fm; Z_{2w}, Wangjiawan Fm; Z_{2dn}, Dengying Fm. 5 Yichang, Hubei Province. Z_{1l}, Liantuo Fm; Z_{1n}, Nantuo Fm; Z_{2d}, Doushantuo Fm; Z_{2dn}, Dengying Fm. 6 Sanjiang, Guangxi Province. Z_{1ca}, Cheng'an Fm; Z_{1f}, Fulu Fm; Z_{1n}, Nantuo Fm; Z_{2d}, Doushantuo Fm; Z_{2l}, Laobao Fm. 7 Xiuning, Anhui Province. Z_{1x}, Xiuning Fm; Z_{1m}, Majin Fm; Z_{2l}, Lantian Fm; Z_{2p}, Piyuancun Fm. For positions of 1 to 7 see Fig. 6.1.

thickness is developed. The Lower Sinian here has a total thickness of over 2000 m and the tillite beds are represented by pebbly phyllites intercalated with meta-andesites. The boulders are mostly volcanic, probably derived from the underlying Zhangbaling Group. The preserved thickness of the Upper Sinian amounts to about 1800 m. The lower formation, often phosphatic and occasionally manganiferous, is an alternating sequence of dolomitic carbonates and silty argillites. It probably marks the southern margin of the much narrowed platform in the Lower Yangzi. This transitional belt of intermediate sedimentary type may be traced throughout the south-eastern margin of the platform. It is especially rich in volcanic content in the Lower Series and in silicolites and argillites in the Upper Series (see over).

SOUTH-EAST CHINA

The regions lying to the south-east of the Yangzi Platform, which are customarily called the Jiangnan and the South China regions, roughly conform to the Eastern Continental Margin Super-region mentioned in Chapter 3. In the Jiangnan region, two representative areas may be mentioned: western and central Hunan (Fig. 6.1, II_1) and western Zhejiang (Fig. 6.1, II_3).

In western Hunan and northern Guangxi, the conspicuous lateral change of facies of the Sinian has caused difficulties in correlation and engendered diverse views regarding the lower boundary of the system. Here the Lower Sinian includes three formations, the Chang'an diamictites, the Fulu volcanics, and the Nantuo glacio-marine deposits, collectively known in Hunan as the Jiangkou Group (Fig. 6.2, 6). The Chang'an Formation may either be a glacial deposit or a solifluction product formed under complicated topography in the coastal belt. As stated above, the Chang'an Formation is underlain by the Gongdong Formation without pronounced disconformity, but the former does not have the flow cleavage which affected the latter. On this account the Chang'an Formation is assigned to the Sinian and the Gongdong to the pre-Sinian. The Jiangkou Group as a whole represents a thick pile of sediments which thins out rapidly towards east and north, and the Upper Sinian is composed entirely of black carbonaceous silicolites of limited thickness, obviously formed in a restricted and starved marginal sea. The silicolite sequence bears the name Laobao Formation in norther Guangxi and Liuchapo Formation in central Hunan. Among chitinozoan forms *Acanthochitina* is found in the Laobao Formation in northern Guangxi (Wang Yangeng *et al.* 1980).

In western Zhejiang the Sinian displays in general a volcanic and jasperitic facies with pronounced lateral changes. The Lower Sinian Zhitang Formation, as seen in Jiande, attains a thickness of about 910 m and is composed mostly of purplish-red and greenish-grey to brick-grey tuffaceous sandstones and sandy shales; it overlies the Shangshu Group with a weak disconformity. A basal conglomerate often occurs but is variable in thickness, ranging from 30 m in the uplifted areas to over 150 m in the subsiding troughs in the Kaihua and Changshan districts. The main part of the formation consists of greenish-grey greywackes and andesite tuffs, which may be very thick locally. Load casts and flute marks are common at many horizons, indicating a turbiditic origin. The upper part of the formation is characterized by silicolites and jasperites with intercalated tuffaceous sandstones and quartz sandstones. The succeeding Leigongwu Formation is formed primarily of volcanic clastics but may be partly glacio-marine in origin; its thickness seldom exceeds 130 m.

The Upper Sinian was originally known as the Xifengsi Formation, which is composed mainly of siliceous carbonates. Recently the lower argillaceous part was separated out to form the Sanliting Formation. The newly defined Xifengsi Formation contains at its thickest about 680 m of siliceous dolomites occasionally intercalated with argillaceous dolomites and siliceous argillites. The variation in thickness in the Upper Sinian is less obvious than in the Lower, but a differentiation of the sea bed in NE-trending uplifts and troughs may still be recognized. The Late Sinian sea seems to have deepened southwards and no old land seems to have existed in the neighbourhood. On the other hand, the topography was probably more complicated in the north. Crossing the Jiangnan Oldland, on which the Sinian is much reduced in thickness, we come to another subsiding faulted trough in southern Anhui (Fig. 6.1, II_2, Fig. 6.2, 7). Here above the Xiuning purple and green arenaceous and tuffaceous deposits are pebbly volcanic deposits locally known as the Majin Formation. The Upper Sinian includes the Lantian clastics in the lower part, and the Piyuancun black banded silicolites in the upper part. The southern Anhui trough is bounded to the north by another geanticline, and further to the north appears the Sinian sequence of Chuxian referred to above.

In the regions to the south and east of the Jiangnan Uplift described above, the Sinian is of a mobile sedimentary type, and the contact relations are transitional with both the underlying and the overlying strata. Three regions are remarkable: northern Guangdong, central and southern Jiangxi, and western Fujian. At Yingyangguan near He Xian on the Guangdong–Guangxi border, the Precambrian is divisible into three parts. The lower part contains profuse intermediate to acidic volcanic eruptives including spilites and keratophyres, apparently of calc-alkalic association; the middle part includes banded phyllites; the upper part is dominated by metamorphic clastics and argillites. A reasonable correlation is that the lower two parts belong to the Banxi Group and the upper part is equivalent to the Lower Sinian Chang'an Formation. It is to be noted that the whole sequence is continuous, probably characteristic of the inner part of the geosyncline.

In northern Guangdong, near Luochangjia, the Precambrian is represented by two groups, the Lower Luochangjia and the upper Liumei, which are followed by the Lower Cambrian Bacun Group. Both groups are characterized by arkose quartz sandstones, silicolites, and carbonaceous slates, with a total thickness of more than 5000 m. Most stratigraphers agree to refer them to the Sinian.

In central and southern Jiangsi and western Fujian, the Precambrian is widespread and is distributed in two belts. The western belt includes south-western Jiangxi and is represented by the Yongxin area to the south of Wugong Mountain. The oldest strata are the Shenshan Group, which consists of a lower part of carbonaceous phyllites and greyish-green metasandstones, and an upper part of yellowish-green phyllites and siltstones with conspicuous rhythmic structures, amounting to more than 5000 m in total. This is succeeded by the Shangshi Group, which is nearly 5000 m thick and is composed mainly of variegated phyllites and metasandstones, although carbonaceous beds are also known. These two groups are in all probability pre-Sinian in age. The succeeding formations represent the Sinian, for they are overlain by the Lower Cambrian Niujiaohe Group bearing *Protospongia*. The lower part of the Sinian is called the Xiafang Formation and contains at the base pebbly tuffaceous sandstones followed by magnetite layers and pyritic phyllites with interbedded calcareous beds. The upper part of the Sinian includes, in addition to green phyllites and thick-bedded metasandstones, carbonaceous phyllites with occasional lenticular white limestones. The coarse clastic beds with magnetite layers are probably equivalent to the Lower Sinian Jiangkou Group of Hunan (Wang *et al.* 1978, 1980) and the underlying groups can thus be referred to the Qingbaikouan. Another view is to put the whole sequence in the Sinian. In any case, this sequence of huge thickness represents a geosynclinal type of continuous sedimentation.

The eastern belt of the Precambrian is characterized by extensive exposures of basement rocks and absence of the Lower Palaeozoic. The age of the lowest rock group, the Jian'ou Group, is controversial, as the succession within the group and its relationship with the supposedly overlying strata have not been established. Near Mingxi the Precambrian is represented by a thick sequence divisible into three formations: the Dingwuling below and the Nanyan and Huanglian above. The Dingwuling Formation is composed of more than 3000 m of marine clastics including silicolite and phosphate beds with tuffaceous layers in the basal part. The Nanyan Formation is rich in volcanic materials, and includes interbedded pyritic iron layers. Coarse quartz sandstones with well-rounded quartzite pebbles occur at the base, but no discontinuity of deposition is indicated. The Huanglian Formation is mainly composed of fine sandstones and siltstones with intercalated silicolite and phosphate layers. If we take the pebbly sandstone as the base of Sinian, the Dingwuling Formation would be referred to the pre-Sinian.

In south-western Fujian, the oldest strata are the Louziba Group, which is well exposed in the Changting–Shanghang districts. It is divisible into three parts, the lower and the upper, silicolitic, and the middle, more calcareous, including phosphatic beds at many horizons. It is here that the Sinian succession is best displayed. The Dingwuling Formation at its type locality rests on the Louziba Group with a clear disconformity and begins with 5–30 m of basal conglomerate containing angular striated pebbles, possibly of glacial origin. The Dingwuling is generally a coarse clastic deposit rich in volcanic materials. The Nanyan Formation is variable in thickness and includes lenticular dolomites and breccia, and sometimes stone coal and graphite beds. The Huanglian Formation is on the other hand composed of silicolites and carbonates, especially of thick-bedded white and purplish-grey banded silicolites. The clastic deposits in the Sinian were evidently derived from nearby uplifts and strongly indicate the presence of pre-Sinian basement rocks in the neighbourhood. It may therefore be inferred that island groups or submarine uplifts composed of pre-Sinian rocks existed in the open seas to the east and south of the Jiangnan region in the Sinian Period.

NORTH-WEST CHINA

In the northern border part of the Tarim Platform the Sinian is well developed in two regions, in the Quruktagh (Fig. 6.1, IV$_1$) and in the Kalpin region. The Sinian System here is well known through the early work of Erik Norin, who reported the presence of Sinian tillites and correlated them with those of South China (Norin 1937). Research by Gao Zhenjia *et al.* has added greatly to our knowledge and a new classification of the Sinian has been presented (Gao, Z. *et al.* 1980). The complete section so far observed is near Zhaobishan in central Quruktagh (Fig. 6.2, 1). The Lower Sinian includes four formations (Table 6.1, 6) containing several glacio-marine horizons. The basal Bayisi Formation rests on folded pre-Sinian strata with an evident unconformity and is composed of more than 1500 m of greenish-grey tillites, tuffaceous conglomerates, dark grey sandstones and siltstones, andesites, and andesite porphyrites. The presence of acritarch remains in the tillites points to its marine origin. The Zhaobishan Formation is a clastic deposit, mainly quartz sandstones with occasional conglomerates and slates. The upper two formations, the Altungol and the Tereeken, are dominated by glacial and volcanic rocks amounting to

2500 m in total thickness. The Altungol Formation overlies the Zhaobishan unconformably and contains fine sandstones and banded slates, porphyries, and trachytes, with at least two tillite layers. The Tereeken Formation is on the other hand almost entirely composed of tillites with intercalated banded slates.

The Upper Sinian has a total thickness of about 2100 m and is divisible into four formations, which were originally collectively called the Yukengol by Norin. The basal unit, the Zamuktee Formation, lies partly unconformably on the Tereeken tillites and usually begins with a thin limestone. Its major part comprises variegated well-bedded sandstones and slates, basalt and agglomerate layers being found in the upper part. The succeeding formation is the Yukengol *sensu stricto*, which is composed mainly of greenish-grey mudstones and passes upwards into the Shuiquan Formation of stromatolitic limestones and black shales. Most conspicuous is the topmost unit, the Hangelhaok tillite which rests unconformably on the Shuiquan Formation. Grey to brown varve clays occur at the base and the top, and lenticular sandstones are occasionally found in the middle part. The whole sequence is constant in thickness and lithology, and is followed by the Early Cambrian Xishanbulak Formation with a slight disconformity.

The presence of three tillite formations and of several discontinuities within the sequence in Quruktagh is of particular importance, but unfortunately no radiometric age data are available. The lower tillites (Bayisi) occur in several layers and are intercalated with marine clastic sediments and volcanic clastics. The striated pebbles are even-sized and are mainly composed of siliceous dolomites, while the matrix is chiefly of volcanic materials. The middle tillites (Tereeken) are glacio-marine in origin and are the most widespread of the three. There are five to eight tillite beds intercalated with normal marine deposits, attaining a total thickness of 1640 m, whereas the underlying Altungol Formation contains only a few tillite layers. The glacial deposits are in general massive, the pebbles being variable both in size and in composition. The uppermost tillites, the Hangelhaok, are widespread and of even thickness.

Some doubt has been cast on the true origin of the Bayisi Formation. In stratigraphic position the Bayisi may be compared with the Chang'an Formation of northern Guangxi, which is also a diamictite of doubtful origin. The widespread Tereeken tillites are generally correlated with the Nantuo of South China, and the Hangelhaok is referred to the Upper Sinian and correlated with the Luoquan tillites of Henan (see below).

In the Kalpin region of north-western Tarim, the Sinian System has a thickness of about 2000 m and overlies the Early Proterozoic Aksu Group with a conspicuous unconformity. It begins with the Qiaoenbulak Formation, which consists of sandy shales and siltstones with intercalated tillites and attains 1380 m in thickness. The Upper Sinian includes two formations, the lower Sugaitbulak, which is chiefly argillaceous, and the upper Qigbulak, mainly stromatolitic dolomites. The Kalpin section differs from the Quruktagh succession being of stable sedimentary type and in the partial absence of the Lower Sinian. A stable type of Sinian carbonates and clastics is also found in the Guozigou area near Huocheng; this probably represents the northern extension of the Kalpin type, for the area of the Tarim Platform was then much broader than now.

Another succession deserving notice lies in the Oulongbruk region to the north of the Qaidam Basin (Fig. 6.1, IV$_2$), where the Sinian Quanji Group rests on the basement rocks, the Dakdaban Group, with a flagrant unconformity (Wang, Yunshan et al. 1980) (Fig. 6.2, 2). Two clastic formations, the Mahuanggou and Kubaimu, make up the Lower Series and attain a total thickness of 650 m. The Mahuanggou Formation is coarsely clastic and consists of conglomerates and purple to grey conglomeratic sandstones with frequent cross-bedding. The Kubaimu Formation is also clastic, but contains muddy layers with conspicuous mudcracks and ripple marks. Both formations show an evident westward thickening and range from fluvial conglomerate facies in the east to fluvial and possibly littoral facies in the west. Above a disconformity surface rests the Upper Sinian, which includes five formations with a total thickness of no more than 1000 m. The lowest Shiyingliang Formation contains an unstable basal conglomerate and is marked by mudcracks and ripple marks. The succeeding Hongzaoshan Formation includes in the upper part massive algal and stromatolitic dolomites. Rock salt pseudomorphs are found near the base. The dolomites pass upwards into dark grey to black carbonaceous shales, the Heitupo Formation, which yields problematic scolecodont remains in addition to *Taeniatum* and *Trematosphaeridium*. Still higher in the succession comes the yellowish-green to purplish-red Hongtiegou tillites, about 20 m thick, which are evidently comparable with the upper tillites of Quruktagh (see above). The uppermost unit of the Quanji Group, the Zhoujieshan Formation, is mainly composed of fine sandstones and siltstones and bears sabelliditids in the upper part. The Zhoujieshan Formation is disconformably overlain by the Early Cambrian Xiaogaolu Formation, which contains hyolithids and acritarch remains.

As stated above, the Dakdaban Group forming the basement of the Qaidam Massif may be attributed to the Lower Proterozoic; the Middle and Upper Proterozoic are absent from the Quanji section. It is noticeable that the Shiyingliang Formation yields K–Ar ages of 698 and 700 Ma (Wang, Yunshan et al. 1980), and *Micrystridium* begins to appear in the Hongzaogou Formation. Thus the boundary between the Lower and the Upper Sinian may be drawn below the Shiyingliang Formation, although the Nantuo tillite horizon does not seem to be represented. In view of the similar lithological character and stratigraphic position, the Hongtiegou tillites may be safely correlated with the Hangelhaok of Kuruktagh, and the Sinian–Cambrian boundary may be drawn immediately above the Zhoutieshan Formation. The occurrence of scolecodonts below the tillites is of special interest and might be an indication of the possible earlier appearance of advanced life in geological history than is generally thought.

NORTH CHINA

The Sinian System is generally absent in the interior of the North China Platform, and is distributed in the border parts only. Two regions may be defined, the south-western region of western Henan (Fig. 6.1, III_1) and the eastern region, which may be further divided into two parts, the northern Liaodong Peninsula (Fig. 6.1, III_2) and the southern Xuzhou–Huainan region (Fig. 6.1, III_3).

Western Henan

Before the discovery of the Luoquan tillites in western Henan in the late 1950s, no Sinian strata were definitely known in the extensive regions of North China. For a long time after the discovery of the tillites, opinions differed as to whether they should be referred to the Sinian or the basal Cambrian. Recent work has revealed that Sinian strata are developed both below and above the Luoquan tillites. In the Jiunudong section of Lushan, the Qingbaikouan Luoyukou dolomite is overlain by some 300 m of carbonates and clastic rocks with a clear disconformity, which are in turn followed disconformably by the Luoquan tillites. Two formations may be distinguished below the tillites. The Huanglianduo Formation, about 130 m thick and composed mainly of banded cherty dolomite with conglomerates and quartz sandstones, is transgressive on different members of the Qingbaikouan, and is mainly distributed in the southern part of Henan. The overlying Dongjia Formation begins with a grey pebbly sandy conglomerate followed first by glauconitic sandstones and subsequently by yellow argillaceous dolomitic limestones. The glauconite yields a K–Ar age of 650 Ma (Guan et al. 1980). The two formations are tentatively referred to the Lower Sinian, each representing a separate marine sedimentary cycle.

The Luoquan tillites are much more widespread and may rest on strata ranging from Sinian to Archaean. The classical section is in Linru, where the tillites are followed by the Dongpo Formation, which is in turn overlain disconformably by the Lower Cambrian Xinji Formation. The Luoquan tillites here amount to 180 m and consist mostly of stratified pebbly sandstones and mudstones. Acritarch remains occur at many horizons, with *Laminarites antiquissimus* and *Taeniatum* as leading forms. In the northern areas, in a belt to the north of Suiping, Linru, and Lingbao, scratched and deeply grooved rock floors indicating continental or mountain glacial erosion are frequently met with. It would seem that glaciofluvial deposits were formed in the piedmont zones of the northern glaciated mountains, while glaciomarine beds were deposited in the southern areas opening probably to the Qinling seas.

The Dongpo Formation overlying the Luoquan tillites, generally less than 100 m thick, consists of grey to green pebbly sandstones passing upwards into purplish-red calcareous shales and glauconitic sandstones. Its distribution is restricted to the southern belt and it was probably formed in a warmer climate. For this reason it is separated from the Luoquan tillites. Glacial deposits equivalent to the Luoquan are widely distributed to the west of the Helan Mountain; they are known as the Zhengmuguan Formation and are transgressive over older strata of various ages.

Southern Liaodong

In the eastern part of the North China Platform, on the Liaodong Peninsula, the Precambrian is peculiar in the absence of the Middle Proterozoic and in the special lithology of the Sinian (Fig. 6.2 3). As stated above, the Qingbaikouan is of a stable type of sedimentation. While the major part of North China underwent general uplifting in the Sinian, the Liaodong region continued to subside and received more than 4000 m of sediments. The Nanfen shales at the top of the Qingbaikouan are followed by the Qiaotou sandstone with a sudden lithological change. The Qiaotou Formation is succeeded by the Wuhangshan Group, which is divisible into three formations mainly composed of carbonates with subordinate slates in the lower part. The lowest is the Changlingzi Formation, made up of variegated shales and thin-bedded micritic limestones and attaining a thickness of over 1500 m. Problematic medusoid fossils are reported from its uppermost part, and indubitable sabelliditids were recently discovered at the same horizon. The Nanguanling Formation in the middle is a stromatolitic limestone. The Ganjingzi Formation, the uppermost of the three, is mainly composed of sandy dolomite with layered cherts. The succeeding Jinxian Group includes five formations with a total thickness of about 1700 m. The main constituent rocks are various kinds of micritic and oolitic limestones in the lower part, called the Yingchengzi Formation, and stromatolitic limestones and sandstones in the upper part, including from the bottom upwards the Shisanlitai, Majiatun, Cuijiatun, and Xingmincun Formations. Cross-bedding, scouring surfaces, mudcracks, and ripple-marks are common, and stromatolite reefs are also found in the upper horizons. These beds were obviously deposited in a high-energy littoral environment.

In southern Jilin, the Qingbaikouan Xihe Group is covered by the Hunjiang Group, in the middle part of which are found metazoan fossil remains. The Hunjiang Group is equivalent to the Wuhangshan Group of southern Liaoning, and the Jinxian Group is not represented.

Table 6.1. *Correlation of the Sinian System of China (for stratigraphic regions see Fig. 6.1).*

		South-west China		South-east China			
	I_1 W. Hubei	I_2 E. Yunnan	I_3 SW. Sichuan	II_1 N. Guangxi	II_2 E. Anhui–N. Jiangsu	II_3 S. Anhui	II_4 W. Zhejiang
600	Cm_1 Tianzhushan Fm	Cm_1 Meishucun Fm	Cm_1 Meidiping Fm	Cm_1 Qingxi Fm	Cm_1 Huanglishu Fm	Cm_1 Huangboling Fm	Cm_1 Hetang Fm
Upper Series	Dengying Fm *Vendotaenia* *Charnia dengyingensis* 250–670 m	Dengying Fm *Micrhystridium* *Lophominuscula* 745–1670 m	Hongchunping Fm *Acus Actinophycus Balios* 940 m	Laobao Fm 131 m	Dengying Fm 1100 m	Piyuancun Fm 80 m	Xifengsi Fm 170 m
	Doushantuo Fm *Micrhystridium* *Lophosphaeridium* 150–230 m	Wangjiawan Fm *Lophosphaeridium* *Margominuscula verrucosa* 180–360 m	Guanyinya Fm *Palaeomicrocystis* 47 m	Doushantuo Fm 49 m	Doushantuo Fm 796 m	Lantian Fm 65 m	Sanliting Fm 68 m
680 Lower Series	Nantuo Fm *Trachysphaerdium* *Laminarites antiquissimus* 90–150 m	Nantuo Fm 0–33 m	Lieguliu Fm 0–204 m	Nantuo Fm 967 m	Sujiawan Fm 1000 m	Majin Fm 88 m	Leigongwu Fm 127 m
	Liantuo Fm *Laminarites cf. antiquissimus* 50–260 m	Chengjiang Fm 300–1200 m	Kaijianqiao Fm 680 m	Fulu Fm 675 m Chan'gan Fm 962 m	Zhougang Fm 166 m	Xiuning Fm 484 m	Zhitang Fm 966–4200 m
850	Pt_1 Sandouping Gr	Pt_{2-3} Kunyang Gr	Pt_3 Suxiong Fm	Pt_3 Gongdong Fm	Pt_3 Zhangbaling Gr	Pt_3 Puling Fm	Pt_3 Shangshu Gr

Correlation of the Sinian in Liaoning with other regions is difficult, for the lithological succession is peculiar and radiometric data are not available. If the Qiaotou Formation is taken as the base of the Sinian, the boundary between the Lower and the Upper Series would lie somewhere below the top of Changlingzi Formation on account of the metazoan fossils found at that horizon. Although the metazoans seem to be more primitive and not directly comparable with the well-known *Ediacara* fauna, it is probably appropriate to regard the fossil horizon as Upper Sinian, for no fossils of this kind have been reported prior to the late Sinian throughout the world. Three stromatolite assemblages may be recognized in the sequence: one in the Wuhangshan and two in the Jinxian groups. Common forms include *Jurusania*, *Inzeria*, *Linella*, and *Gymnosolen*, all known in the Qingbaikouan and the Sinian. *Conophyton occularoides* (Plate IV.11) is a characteristic form, which is also found in corresponding beds in northern Anhui. Among the acritarch remains the presence of *Lophosphaeridium* in the Changlingzi Formation is significant, since it is well known in the Upper Sinian Doushantuo Formation of South China. *Trachysphaeridium hyalinum* is also a common form in the

Table 6.1 (cont.)

North China		North-west China		S. Australia	N. Europe	W. Africa	
III₁ Liaodong	III₂ Huainan	IV₁ Quruktagh	IV₂ N. Qaidam				
Cm₁ Jianchang Fm	Cm₁ Houjiashan Fm	Cm₁ Xishanbulak Fm	Cm₁ Xiaogaolu Fm				600
Jinxian Gr: Xinmincun Fm, *Boxonia* 292 m, Cuijiatun Fm 79 m, Majiatun Fm, *Praesolenopora* 199 m, Shisanlitai Fm 155 m, Yingchengzi Fm 894 m	Gouhou Fm 116 m, Jinshanzhai Fm, *Multisiphonia* 23 m, Wangshan Fm 377 m, Shijia Fm, *Multisiphonia* 902 m, Weiji Fm	Hangelhaok Fm 465 m, Shuiquan Fm *Taeniatum* 331 m, Yukengol Fm 583 m	Zhoujieshan Fm, Sabellititids 22 m, Hongtiegou Fm 110 m, Heitupo Fm 50 m, Hongzaoshan Fm *Micrhystridium* >100 m	Marinoan; Wilpena Group	Pound qtzt.; Breivik Fm; Stappogiedde Fm	Upper Kundelungu Group	Upper Series
Wuhangshan Gr: Ganjingzi Fm 400 m, Nanguanling Fm 400 m, Changlingzi Fm Sabellititids 1539 m	Xuhuai Gr: *Linella* 274 m, Zhangpu Fm 132 m	Zhamokti Fm 793 m	Shiyingliang Fm 150 m				680
	Jiutingshan Fm 243 m, Yiyuan Fm 242 m, Zhaowei Fm 144 m	Tereeken Fm 2685 m, Zhaobishan Fm 570 m	Kubaimu Fm 200 m, Mahuanggou Fm	Sturtian; Torrensian	Verrangian; Lispak	Small conglomerate tillite; Lower Kundelungu Group	Upper Series; Lower Series
Qiaotou Fm 565 m	Jiayuan Fm 445 m	Baiyisi Fm 1560 m	400 m				850
Pt₃ Xihe Gr	Pt₃ Huainan Gr	Pt₃ Pargangtag Gr	Pt₂ Dakdaban Gr	Willouran		Great cgt. tillite	

lower part of the sequence. It is significant that the red alga-like *Praesolenopora* and *Multisiphonia* occur in abundance in the Majiatun Formation. A tentative correlation of the various groups with other regions is given in Table 6.1 8).

The Xuzhou–Huainan region

In this region the Sinian is exposed in two areas, the southern near Huainan and Fengtai, and the northern to the south of Xuzhou (Yang *et al.* 1980) (Table 6.1 9). As described in Chapter 5, the Qingbaikouan is on the whole stable in lithology and thickness. In the southern area, the Sinian is represented by two formations, the Jiuliqiao below, and the Sidingshan above. The Jiuliqiao Formation is probably conformable upon the Qingbaikouan Shouxian Formation. It consists of glauconitic argillaceous limestones bearing *Chuaria* and stromatolites, with a thickness of no more than 40 m. The Sidingshan Formation is composed of carbonates about 230 m thick, mostly pink to grey thick-bedded and banded cherty dolomites. Calcareous siltstones and shales and wurmkalks occur in the upper part. The leading stromatolites are *Gymnosolen* and *Conophyton lijiadunensis*, both known from the Jinxian Group of Liaodong referred to above. The Sidingshan Formation is disconformably covered by the Fengyang conglomerate. This conglomerate was formerly assigned to the basal Cambrian, but recent work has revealed its glacio-marine origin. It is therefore comparable with the Luoquan tillites of Henan.

In the northern area, the lower three formations of the sequence, the Zhaowei, Yiyuan, and Jiudingshan (Table 6.1 9), comprise mostly dolomites and dolomitic limestones. They correspond to Jiuliqiao and Sidingshan formations in the south. The succeeding 1300 m of carbonates and clastics are divisible into four formations, which are lacking in the south. The lower two formations, the Zhangqu and Weiji, are dolomites and limestones containing algal reefs, while the overlying Shijia Formation is mainly composed of shales and marls, with glauconitic sandstones, yielding radiometric ages ranging from 738 to 787 Ma. *Multisiphonia hemicirculis* occurs in abundance in the marls. The Wangshan Formation at the top is again a carbonate deposit, containing wurmkalks and calcareous shales, and marked with mudcracks. A clear disconformity separates the Wangshan Formation from the overlying variegated calcareous beds representing the highest Sinian. At the base there is a thin yellow stromatolitic limestone containing profuse *Multisiphonia* and *Gymnosolen*. The Gouhou Formation above, some 110 m thick, is composed of variegated marls and shales, sometimes with rock salt pseudomorphs.

There is little doubt that the Xuzhou–Huainan and the Liaodong regions belonged to the same marine basin, which is limited by the Hebei–Shandong Oldland in the west. Two centres of subsidence may be recognized, one in southern Liaoning and another near Xuzhou. While the base of the Sinian in Liaoning is marked by a sedimentary break and a sudden change in lithology, the sequence is entirely transitional in the Huainan region. There appears to be some discrepancy in correlation and the determination of boundaries. Two interesting fossil horizons other than stromatolites and acritarch remains occur in the Qingbaikouan and Sinian in this region. The lower horizon is in the Liulaobei Formation; it yields *Chuaria* and *Ellipsophysa* (Plate III.1, 3) and can be correlated with a bed in the Qingbaikouan Longshan Formation in western Yanshan. The upper horizon occurs in the Juiliqiao Formation and is probably comparable with the Changlingzi *Sabellidites* beds. But if we regard the Jiuliqiao as Upper Sinian on the basis of this correlation, the thick sequence of 4000 m of carbonates and argillaceous deposits must all be classified as Upper Sinian, and little room is left for the Lower Sinian. On the other hand, the Shijia Formation, which occupies a high position in the sequence, yields a radiometric age of about 750 Ma, which falls within the Early Sinian. Moreover, there is a discontinuity below the Jinzhai Formation, and the advanced algae characteristic of the Late Sinian in the Yangzi Platform have not so far been found. Thus there are two possibilities: either the metazoans are of Late Sinian age and the sedimentary facies and tectonic setting are peculiar; or the metazoans appeared here earlier than elsewhere, and the major part of the sequence belongs to the Early Sinian or even partly to the pre-Sinian. The first interpretation is tentatively adopted here, but further study is necessary.

OTHER PARTS OF CHINA

The regions where Sinian strata are known to occur will be treated in this section. They include North-east China (Fig. 6.1, V), the Kunlun–Qinling region, and western Yunnan (Fig. 6.1, VI).

North-east China

In the Yilehuli region of northern Heilongjiang is a metamorphic sequence called the Jiageda Group, which is unconformable upon the Proterozoic Xinghuadukou Group. It is composed mainly of light grey quartz-schist and quartzite in the lower part and chlorite-schist in the upper part, and its thickness ranges from 1180 to 2500 m. The Group is exposed along the Erguna River and passes upwards into the Lower Cambrian Erguna marbles and dolomites. As the relationship between the Jiageda Group and the overlying Lower Cambrian is transitional, the major part of the Jiageda Group may be assigned to the Sinian. It is probable that it represents the geosynclinal facies in the Palaeo-Hingan sea trough, and that it was folded and metamorphosed in the Xingkaian orogeny.

In eastern Heilongjiang, metazoan fossils, including *Arumberia* and *Glaessnerina* (*Charnia*) and some new forms, have been found near Jixi in a metamorphic carbonate bed assigned formerly to the Mashan Group (see Chapter 5).* This Mashan fauna, as it is called (Liu 1981), is at present the best-known representative of the *Ediacara* fauna in China and is closest to the fauna of the Arumbera sandstone of South Australia. The fossil *Glaessnerina* found here is also akin to the form reported from the Dengying Formation by Ding and Chen in 1981. The fossil bed cannot in any case be older than 700 Ma. It was most probably folded with the older Mashan Group, for deposition had been continuous in the geosynclinal regions of North-east China from late Proterozoic to the Early Cambrian.

The Kunlun–Qinling region

In the axial region and the northern slope of West Kunlun, the metamorphic Saitula Group, some 5000 m thick, which is believed to represent the Sinian, rests unconformably upon the Proterozoic Sangzhutagh Group. In the Tieklik Mountain at the border of the Tarim Basin, Peng Changwen, Gao Zhenjia, and Fang Xilian have discovered a stable type of Sinian sequence no more than 1500 m thick, resting unconformably on Middle and Upper Proterozoic strata, also of limited thickness. The Lower Sinian Qakmaklik Formation contains red sandstones and siltstones with more than one tillite bed and yielding *Taeniatum crassum*. The Upper Sinian includes two formations, the lower Kurkak and the upper Kezirsuhum, both consisting of clastic and argillaceous sediments with dolomite layers. The leading acritarch forms are *Pseudozonosphaera asperella* and *Trachysphaeridium cultum*.

In the Wen Xian–Lueyeng region of southern Kansu, generally known as West Qinling, Precambrian rocks are exposed between Linjiang and Wen Xian. These were formerly called the Bikou Group. Work done by Cao Zhilin *et al.* in 1979 revealed that the Bikou Group at its type locality is Silurian, although

---

*Recently the fossil bed has been separated from the Mashan Group and is called the Zhongsanyang Formation.

the Precambrian is also present. The Precambrian consists of a metamorphosed volcanic sequence in the lower part, the Yangba Formation, which thickens eastwards and probably merges into the Yaolinghe Group of southern Shaanxi. The middle part consists of the Quanjiagou metaconglomerates and tuffaceous slates, which probably correspond to the Nantuo tillites; the upper part, the Linjiang Formation, is chiefly composed of dolomites and intercalated silicolites comparable with the Dengying Formation. This geosynclinal type of the Sinian is also developed further to the south, in the northern Longmen Mountains and south of Pingwu. Here the Sinian rests unconformably on the Proterozoic Tongmuliang Group and includes three formations, the lower purple to red volcaniclastics, the middle variegated slates and sandstones, and the upper dolomites, marls with silicolites, and carbonaceous shales. The thickness is much reduced as compared with the Bikou succession.

Western and Southern Yunnan

In western Yunnan immediately beyond the border of the Yangzi Platform, the Ailaoshan and Cangshan groups are exposed and belong to the pre-Sinian, as pointed out in the previous chapter. Several belts of Precambrian rocks are exposed along the Lancangjiang valley. The Ximeng Group is roughly parallel to the valley and is composed of more than 5000 m of granulitites, marbles, and schists. To the north are distributed the Lancang Group and the Chongshan Group, both underlying the fossiliferous Ordovician. The Lancang Group yields Middle Proterozoic acritarch remains, but recently *Bavlinella* and *Gloeocystoides* have been discovered, indicating a probable Sinian age. Further westwards in the Nujiang valley near Longling, the oldest strata exposed are those of the Lower Gongyanghe Group, which is composed of fine sandstones and black slates with limestone lenticles, amounting to a thickness of nearly 5000 m. This passes upwards into the Upper Gongyanghe Group of flyschoid deposits. Sponge spicules are found in the black banded slates. The Upper Cambrian Haitaoping Formation, which contains trilobites, rests conformably on the Upper Gongyanghe Group, which is thus referred to the Lower to Middle Cambrian; the Lower Gongyanghe Group is consequently referred to the Sinian. It is noticeable that, while near the platform border a remarkable break occurs between the Proterozoic and the overlying Ordovician, marine deposition farther away from the platform was continuous from the Proterozoic to the Cambrian, as is often observed in geosynclinal regions. Similar conditions also occur in the Pingbian area on the eastern side of the Red River in southern Yunnan, where 5000 m of epimetamorphic mudstones and sandstones of variegated colour and rhythmic structure, called the Pingbian Group, are distributed in a NNW–SSE belt. The Pingbian Group is overlain with a slight disconformity by Lower Cambrian shales bearing *Metaredlichia* and *Hsuaspis*; it thus belongs at least partly to the Sinian.

THE MAIN PALAEOGEOGRAPHIC FEATURES OF CHINA IN THE SINIAN PERIOD

The Sinian is the first period from which it is possible to obtain an integrated view of the palaeogeography of the whole of China. We shall now review briefly the main features of the epicontinental seas on the stable platforms and median massifs, and of the continental margins and the intercontinental realms that once separated the old continents.

Considerable changes occurred between the Early and the Late Sinian, but as the demarcation between the two series is not always clear they will here be considered together.

South China

In the Early Sinian, the greater part of the Yangzi Platform was above sea level; only its borders adjacent to the Qinling sea and the south-east China sea were flooded. The platform was then affected by the Kham–Yunnan Uplift in the west, which tilted the whole platform to the east. Thus from the west eastwards, we have successively the Kham–Yunnan oldland of denudation, the central Yangzi lowland and coastal plain, and the Hunan–Guizhou neritic belt. In the piedmont belt of the Kham–Yunnan mountain area the Chengjiang Formation of paramolasse deposits was accumulated. A rift zone was developed in south-western Sichuan, where thick accumulations and volcanic eruptives were formed. Off the northern border of the platform, a thick sequence of volcaniclastic rocks, basic lavas, and spilites representing the northern continental margin was formed. Similar continental margin deposits are also found in western Yunnan. The continental margin on the south-eastern side of the platform was more extensive and complicated. The boundary between the western shelf sea and the eastern marine basin was situated at the Hunan–Guizhou border. Thick clastic deposits, the Jiangkou Group, were accumulated along the rugged coastal zone with islets formed of older rocks. Similar conditions obtained throughout the Jiangnan region, and volcanic activity increased in southern Anhui and western Zhejiang. At the close of the Early Sinian, continental glaciation occurred in the interior of the Yangzi region, and glacio-marine deposits, including erratics, were formed on the outer side of the Jiangnan islands. In the Lower Yangzi, the topography was broadly rugged, volcanism was more active, and the glacial deposits gave way to volcaniclastics. Further to the south-east, subsiding sea troughs separated by submarine uplifts and island groups led to the formation of marginal sea deposits rich in silicolites and tuffaceous layers.

In the Late Sinian, the Yangzi region was dominated by extremely low relief and by an almost universal marine transgression over the whole region. A narrow strip of lowland was left along the western border of the Kham–Yunnan oldland. Strongly subsiding local depressions occurred within the wide epicontinental sea, notably in south-western Sichuan, where thick beds of rock salt were deposited. On the south-eastern side of the Jiangnan islands belt, subsidence evidently exceeded sedimentation, and the Hunan–Guangxi marine basin became an uncompensated, restricted marginal sea basin. In the Lower Yangzi this situation continued from the Early Sinian and silicolite deposition in marginal seas was also prevalent, in Late Sinian times. The extensive region on the outer side of the Jiangnan island groups was in all probability a complicated marine realm with a sea floor mainly of oceanic crust throughout Sinian times.

North China and North-west China

Within the scope of North China and North-west China are included the North China Platform, the Tarim Platform, the Qaidam Massif, and the geosynclinal regions lying between them. They are here treated together because they appear to have formed a united continental domain in Late Proterozoic and Sinian times.

The main parts of North China underwent a general uplifting at the end of the Qingbaikouan Period, and only the borders, especially the eastern and southern borders, were inundated by the sea in the Sinian. Active subsidence and sedimentation were consecutive and pronounced along the belt stretching from Liaoning to Huainan; two sedimentary zones, an inner carbonate and clastic zone of comparatively stable type, and an outer more mobile zone of calcareous and argillaceous flysch deposition may be distinguished. Volcanic activity was practically lacking and the whole region seems to represent a passive continental margin open to the Proto-Pacific oceanic basin in the east.

The southern margin of the North China continent

was on the other hand only partially transgressed by Early Sinian epicontinental seas coming from the Qinling marine realm. Strong uplift probably occurred in the mid-Sinian, when the mountain area of northern Henan and southern Shaanxi was glaciated, and various types of glacio-fluvial to glacio-marine deposits were formed from the piedmont zone in the north to the Qinling seas in the south. The extensive region of southern Alashan to the south of Helan Mountain and Longshou Mountain was also glaciated and later flooded by the Late Sinian sea, coming presumably from the Qilian region. As the Late Sinian glaciation and subsequent marine transgression also occurred in Northern Qaidam, the Qilian marine realm could not have been very wide, and the Qaidam Massif was probably a microcontinent not very far off the southwestern coast of the North China main continent.

Very little is known about the interior of the Tarim Platform in the Sinian. The Kuruktagh was probably a sea trough on the inner side of the Central Tianshan, which was a border uplift of the ancient platform, to the north of which lay the extensive Mongolian open sea. The Kalpintagh and the Huocheng areas may have been connected with the Quruktagh sea trough, but they were probably more related to the Central Asian and Khazakstan region in Sinian times. The discovery of possible Sinian marine and glacial beds in the West Kunlun indicates the southern margin of the ancient Tarim continent. Although the presence of the Sinian System has not been verified in East Kunlun to the south of the Qaidam Basin, it seems reasonable to assume that West Kunlun and East Kunlun marked the southern limit of the northern continental domain, to the south of which was the extensive Proto-Tethys. The presence of median massifs or microcontinents, such as Junggar and Song-Liao in the northern continental margin tract of the North China-Tarim Continent, and such as Gangdise in the Xizang region representing the northern continental margin of the ancient Gondwana continent, has been noted above. For the main pattern of the pre-Sinian tectonic frame see Fig. 14.1 in Chapter 20.

Postscript

Chinese geologists have recommended, in accordance with priority, the use of 'Sinian' for the chronostratigraphic unit of system rank to be established under the Cambrian, covering the time-span of 800–600 Ma. As the Lower Sinian thus defined (800–700 Ma) is very scanty in fossil record and as the lower boundary of the system is probably diachronic, owing to its unconformable relation with underlying strata, Wang once (1973) preferred to use 700 Ma as the lower age limit of the system, in which case the whole Sinian System would be equivalent to the Upper Sinian used in this book. Further study on this problem is certainly needed.

7. THE CAMBRIAN SYSTEM
Yang Zunyi

INTRODUCTION

During the Cambrian period the structural pattern remained more or less the same as that of the Sinian, but Cambrian rocks are spread more extensively than the Sinian with various types of sedimentation and biotas. They are rich in mineral deposits, such as phosphate, iron, mercury, trace elements, gypsum and salts, pyrite, and stony coal. They are especially widespread in the southern part of the North-east, North, Central, and South-west China, with the exception of the Altay, Junggar, North Xizang, and Taiwan.

Based on such factors as sedimentary associations, biotic features, geologic history, and geotectonic settings, the Chinese Cambrian may be divided into three main types and ten stratigraphic regions (Xiang et al. 1981) (see Fig. 7.1). (A) Stable or platform type (North China type), including the Tarim (2), North China (3), Yangzi (8) and Himalayan (7) regions; (B) mobile or basin type (South-east type), including Tianshan–Hingan (1), Qilian (4), South-east (10) regions; and (C) relatively active or transitional type (Jiangnan type), including South of the Yangzi (Jiangnan) (9), Kunlun–Qinling (5), North Xizang–West Yunnan (6) regions.

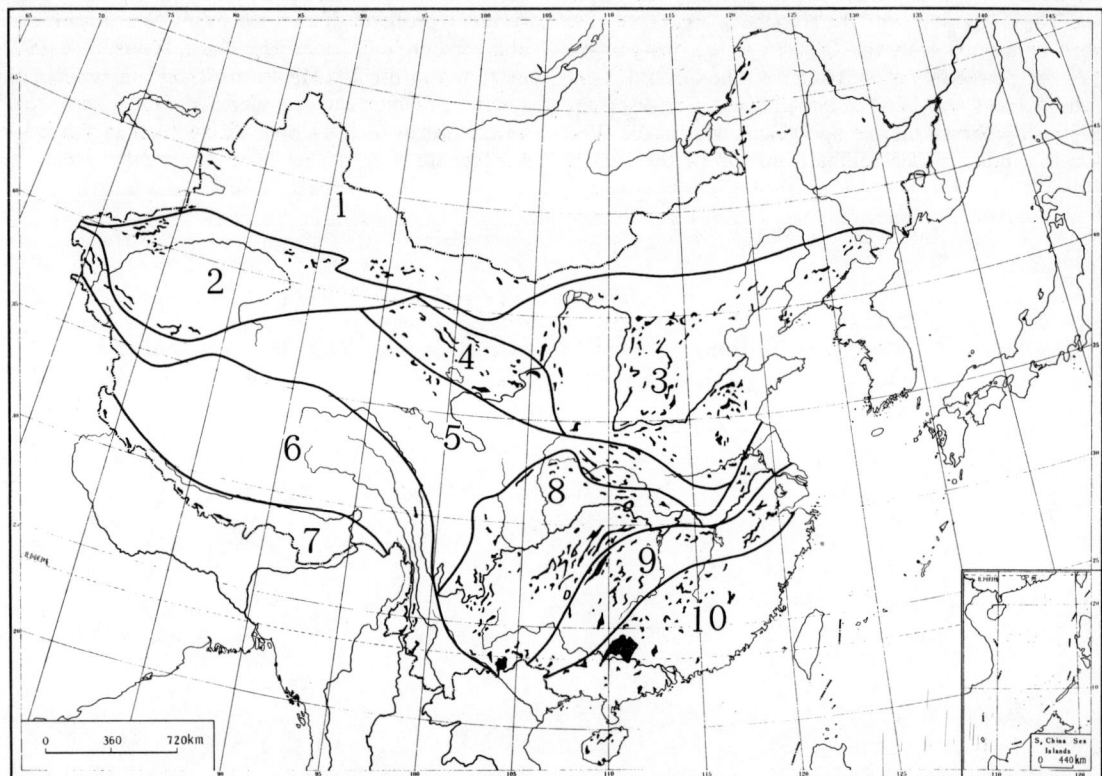

Fig. 7.1. Sketch map showing the distribution and stratigraphic regions of the Cambrian in China (after Xiang et al. (1982)). 1, Tianshan–Hingan region; 2, Tarim region; 3, North China region; 4, Qilian region; 5, Kunlun–Qinling region; 6, North Xizang–West Yunnan region; 7, Himalaya region; 8, Yangzi region; 9, Jiangnan region; 10, South-east region.

STABLE OR PLATFORM TYPE OF SEDIMENTATION (NORTH CHINA TYPE)

The Yangzi region

The Yangzi region occupies the greater part of the Yangzi river basin, covering eastern and southern Yunnan, Guizhou, western Sichuan, southern Shaanxi provinces, the Yangzi Gorges, and the middle and lower reaches of the Yangzi. Over this vast area the Cambrian is chiefly composed of neritic arenaceous shale, limestone and dolomite, with little change in lithology. The strata are of medium thickness, highly fossiliferous, yielding a benthonic fauna, e.g. small shelly fossils, trilobites, archaeocyathids, brachiopods, gastropods, conchostracans, hyolithids, and algae, related to the Pacific realm, all of which reflect deposition on a stable platform. In this region the Cambrian is divisible into three series and ten stages as shown in Table 7.1(3), (4), and (5) (Xiang *et al.* 1981; Lu *et al.* 1974). The Lower Series comprises predominantly clastic rocks, 400–500 m thick; the Middle and Upper Series consist mainly of carbonates intercalated with gypsum and salts, totalling 300–2000 m thick. In the type localities near Kunming and Ichang the basal Meishucunian Stage is marked by numerous small shelly fossils (*Anabarites*, *Circotheca*) and is conformable over the Sinian dolomites. Elsewhere, these fossils may be absent and the Lower Cambrian is disconformably separated from the Sinian.

Eastern Yunnan

In eastern Yunnan (Table 7.1(4)) the Lower Cambrian is divisible into four stages. The base of the Meishucunian Stage is represented by the upper member of the original Dengying Formation, 4–40 m, which consists of phosphatic rock and dolomite or dolomitic collophanite, yielding the *Anabarites–Circotheca* and the *Yunnanotheca–Pupoella* assemblages respectively in the lower and upper parts. The Qiongzhusian Stage of the Lower Cambrian includes the Qiongzhusi Formation which consists of sandstones, sandy shale, and argillaceous siltstone containing the *Eoredlichia* (trilobite) zone, and phosphatic black carbonaceous shale and siltstone; 90–350 m thick. The Canglangpuan Stage consists of the Canglangpu Formation of silty shale intercalated with micaceous sandstone, with the *Palaeolenus* zone at the top and sandstone plus sandy shale and silty sandstone marked by two trilobite zones below (*Drepanuroides* in the upper and *Yiliangella* in the lower parts); 156–412 m thick. The Longwangmiaoian Stage represented by a formation of argillaceous dolomitic limestone and dolomites, occasionally with sandy shale, which yields the *Hoffetella–Redlichia kurakamii* zone; 80–243 m thick.

The Middle and Upper Cambrian are better developed in North Guizhou and the Gorges district. The Maozhuangian or Gaotaian Stage is represented by the Douposi Formation (East Yunnan) or Gaotai Formation (the Gorges district), which consists of patchy dolomitic limestone, sandy dolomite, shale, and sandstone with trilobites (the *Kaotaia–Kunmingaspis* zone); 20–38 m. The Middle Cambrian Xuzhuangian Stage includes the Shilengshui Formation, which consists of dolomites—argillaceous dolomite intercalated with brecciated dolomite yielding *Manchuriella*, *Proasaphiscus*, and *Kaipingella*; 185–315 m. The Middle Cambrian Changxian Stage is represented by the Pingjing Formation made up of limestones and dolomites and yielding *Paranomocare* (trilobite); 225–500 m. Overlying the Pingjing Formation is the Upper Cambrian Houba Formation of crystalline dolomite with *Monkaspis*; 400–415 m thick. This may be correlated with the Gushanian and Changshanian Stages. The last Upper Cambrian Stage is represented by the Maotian Formation which consists of limestones and dolomites, and is marked by the *Calvinella–Metacalvinella* zone; 105–200 m.

North China region

The North China region is even more extensive than the Yangzi region, including as it does the southern part of North-east China, and the southern part of Nei Mongol, Hebei, Shanxi, Shaanxi, Ningxia, Henan, Anhui, and Jiangsu provinces (Fig. 7.1), roughly equivalent in extent to the 'North China Massif', a persistently relatively stable region, where typical platformal Cambrian deposits were laid down. Type sections of the upper part of the Lower Cambrian, Middle and Upper Cambrian are defined there. The Lower Series consists generally of neritic sandy shale and limestone, the Middle Series commonly purplish, greyish-green shale and siltstone, grey or greyish-white oolitic limestone, wurmkalk and dolomitic limestone, with slight lateral change; 400–500 m thick. It is only along the area south of the Huai River, and the western margin of the North China region that the Lower and Middle Cambrian are represented by important phosphate-bearing horizons and the whole Cambrian sequence attains a great thickness, overlying disconformably or unconformably older rocks of different ages.

A complete Cambrian succession for the Yanshan subregion (Table 7.1) in ascending order has been elaborated (Nan Runshang in Xiang *et al.* 1981; Lu *et*

Table 7.1. *Correlation of Cambrian sequences in various stratigraphic regions of China (adapted from Xiang Liwen et al. 1982).*

Series	Stage	Bio-Zones	E. Yunnan–W. Sichuan (4)	Yangzi Gorges (3)	N. China, Yanshan (5)	Tianshan–Beishan (8)
		Overlying	Hualing Fm (D$_3$)	Tongzi Fm	Yeli Fm (O$_1$)	Xingertai Fm (O$_1$)
Upper Cambrian Series	Fengshanian	*Calvinella–Mictosaukia Quadraticephalus–Dictyella Ptychaspis–Tsinania*		Maotian Fm *Calvinella–Metacalvinella* Zone 105–200 m	Fengshan Fm *Calvinella Quadraticephalus Ptychaspis* 50–150 m	Guozigou Fm
	Changshanian	*Kaolishania Changshania Chuangia*		Huaba Fm *Monkaspis*	Changshan Fm *Kaolishania Changshania Chuangia* 8–50 m	*Lotagnostus Xestagnostus Glyptagnostus Kedinaspis Charchaquia Proceratopyge* 14 m
	Gushanian	*Drepanura Blackwelderia*		300–415 m (Cm$_3$; 405–615 m)	Gushan Fm *Drepanura Blackwelderia* 9–85 m (Cm$_3$; 67–285 m)	
Middle Cambrian	Zhangxian	*Damesella Amphoton– Taitzuia Crepicephalina*	Shuanglongtan Fm *Prohedinia* Z.	Pingjing Fm *Paranomocare* Z. 255–500 m	Changxia Fm *Damesella Amphoton Crepicephalina* 100–200 m	Kensair Fm *Diplagnostus Ptychagnostus Damesella Centropleura Xystridura*
	Xuzhuangian	*Bailiella Poriagraulos abrota Sunaspis Kochaspis*	80–200 m	Shilengshui Fm (in N. Guizhou) *Manchuriella, Proasaphiscus, Kaipingella* 190–400 m	Xuzhuang Fm *Bailiella Sunaspis* 22–60 m	
	Maozhuangian	*Shantungaspis*	Douposi Fm *Sinoptychoparia* Z. *Chittidilla– Kumingaspis* Z. (Cm$_2$; 115–370 m)	Gaotai Fm *Kaotaia– Kumingaspis* Z. 20–38 m (Cm$_2$; 465–938 m)	Maozhuang Fm *Shantungaspis Luaspides* 40–60 m (Cm$_2$; 162–320 m)	27–56 m (Cm$_2$; 40–70 m)
Lower Cambrian	Longwang-miaoan	*Hoffetella– Redlichia murakamii*	Longwangmiao Fm *Hoffetella– Redlichia murakami* Zone 93 m	Shilongdong Fm *Redlichia murakamii* Zone 70–150 m	Manto Fm 40–80 m	Linkuanggou Fm *Calodiscus*
	Canglangpuan	*Palaeolenus Drepanuroides*	Canglangpo Fm *Palaeolenus* Z. *Drepanuroides* Zone	Tianheba Fm *Palaeolenus, Retecyathus* 80–110 m	Changping Fm *P. (Mega-palaeolenus)* 30–120 m (Cm1 70–200 m)	
		Yiliangella Yunnanaspis	*Yiliangella* Zone 156–412 m	Shipai Fm *Redlichia Ichangia* 34–570 m		
	Qiongzhusian	*Eoredlichia* (unnamed)	Qiongzhusi Fm *Eoredlichia* Zone 90–350 m	Shuijingto Fm *Hupeidiscus, Tsunyidiscus, Zhenbaspis* 24–140 m		
	Meishucunian	*Allatheca– Yunnanotheca* Ass. *Anabarites– Circotheca* Ass.	Meishucun Fm *Yunnanotheca– Pupoella* Ass. *Anabarites– Circotheca* Ass. 4–40 m (Cm$_1$; 343–895 m)	Meishucun Fm *Allotheca– Lenatheca– Quadrotheca* Ass. *Anabarites– Circotheca* Ass. 1.2–4 m (Cm$_1$; 209.2–973 m)		30–40 m (Cm$_1$; 71–110 m)
		Sinian System (Z)	Dengying Fm (Z)	Dengying Fm (Z)	Jingeryü Fm (Z)	Sinian (Z)

Table 7.1 (cont.)

N. Qilian, Yumen (7)	S.E. Region S. Jiangxi (1)	S. Qinling (6)	Jiangman region, Xiushui, Jiangxi (2)	W. Yunnan, Baoshan (9)
Yingou Gp (O_1)	? Fm	Gaoqiao Fm (O_1)	Yingzhubu Fm (O_1)	Bingdou Fm (O_1)
Xiangmaoshan Fm *Proceratopyge* *Eoorthis* 1210–1830 m	Shuikou Gr *Lingulella liui* *L. manchuriensis* *Homotreta lisani* *Acrothele recta* *Obolus taianensis* >1886 m	Baxianjie Fm 460–900 m	Xiyangshan Fm *Lotagnostus* *Pseudagnostus* *Hedinaspis* *Onchonotina* *Olenus* 89 m	Baoshan Fm *Calvinella* Z. *Quadraticephalus* Z. *Prosaukia* Zone
				Liushui Fm *Kaolishania* Z. *Parachangshania* Z. *Chuangia* Z. 1928 m
			Huayansi Fm *Erixanium* *Glyptagnostus* *Proceratopyge* 134 m (Cm_3; 223 m)	Hetaoping Fm *Bergeronites–* *Cyclolorenzella* Z. 1594 m (Cm_3; 4595 m)
Gelmore Fm *Metagraulos* *Inouyia* *Proasaphiscus* 7227 m Or Heicigou Gr *Hypagnostus* *Huzhuia* >3000 m (Cm_2; 8437–9057 m) (or 4210–4830 m)	Gaotan Group *Homotreta orientalis* *H. lisani* *Paterina lucina* *Obolus luanhsiensis* *Acrothell* sp. *Protospongia* >2047 m	Miaoziba Fm *Linguagnostus* *Peronopsis* *Prohedinia* *Anomocarella* 90–300 m (Cm_2; 550–1200 m)	Yangliugang Fm Assoc. 4 *Guttsiapinga–* *Lejopyge epsa* Assoc. 3 *Lejopyge armata* Assoc. 2 *Phalagnostus ovalis* *Ph.* cf. *majus* Assoc. 1 *Ptychagnostus atavus* 97 m	Gongyanghe Upper Gr *Protospongia*
Lower Cambrian absent	Niujiaohe Gr *Lingulella linei* *Acrothele recta* *Homotreta pileata* *Obolus luanhsiensis* *O. taianensis* >1886 m	Jiangzhuba Fm *Kootenia* 150–338 m Lujiaping Fm *Protospongia* 480–740 m (Cm_1; 630–1078 m)	Dachenling Fm *Arthricocephalus* 30 m Hetang Fm *Hunanocephalus* 97 m (Cm_1; 127 m)	*Protospongia* 2303 m
Pre-Sinian		Yiaolinghe Gr (Precambrian)	Dengying Fm (Z)	Gongyanghe Lower Group (Z)

al. 1974). The Lower Cambrian Changping Formation is made up of limestone and brecciated dolomitic limestone, 30–120 m thick, yielding *Palaeolenus* (*Megapalaeolenus*). Corresponding to the Canglangpuan Stage of E. Yunnan, the Mantou Formation consists of purplish-red shale; 40–80 m. In southern North-east China this formation comprises purple, yellow and greyish-green shale and limestone, marl, and dolomite, 37–1802 m thick, yielding *Redlichia murakamii*, correlatable to the Longwangmiao'ian Stage. The Middle Cambrian is divisible into the Maozhuangian, Xuzhuangian, and Changxian stages. The Maozhuangian Stage is represented by the Maozhuang Formation and consists of purplish-red micaceous shale, and siltstone intercalated with limestone, 40–60 m thick, bearing trilobites (*Shantungaspis* and *Luaspides*). The Xuzhuangian Stage is represented by the Xuzhuang Formation and includes shales and limestones, 22–60 m, with trilobites (*Bailiella* and *Sunaspis* zones). The Changxian Stage or Changxia Formation is made up of oolitic limestone and crystalline limestone, 100–200 m thick, including three trilobite zones (*Crepicephalina–Amphoton-Damesella*). The Upper Cambrian is divisible also into three stages: the Gushanian, Changshanian, and Fengshanian stages. The first stage represented by the Gushan Formation is chiefly shale, wurmkalk and oolitic limestone, 9–85 m, with two trilobite zones (*Blackwelderia* and *Drepanura*). The second stage is represented by the Changshan Formation and consists of shales and wurmkalk, 8–50 m, with three trilobite zones (*Chuangia, Changshania*, and *Kaolishania*). The last stage represented by the Fengshan Formation comprises shale, siltstone, limestone, and wurmkalk, 50–150 m thick, yielding three trilobite zones (*Ptychaspis, Quadraticephalus,* and *Calvinella*).

Tarim region

Within the stable Tarim region, Cambrian rocks are exposed only along the marginal hills, and are especially well-developed in Kalpin, Kuruktag, where the Lower Cambrian disconformably overlies the Sinian and consists of phosphate-bearing siliceous black shale, limestone and dolomite, while the Middle and Upper Cambrian are generally composed of carbonate rocks (limestone and dolomite), totalling 600–800 m and underlying conformably the Ordovician.

Generally, the North China stable or platformal type of deposition reflects a littoral–neritic environment which produced mainly light-coloured sandy shale, limestone, dolomite, oolitic limestone, and brecciated limestone (edgewise limestone), bearing mudcracks and ripple-marks; these are of relatively small thickness.

Within semi-restricted sea basins or lagoons large quantities of dolomite, gypsum and salts were deposited. As the sea was shallow, it was well oxygenated and rich chiefly in varied benthonic life, such as large-sized benthic (both crawling and swimming) trilobites, *Redlichia, Palaeolenus, Shantungaspis, Taitzuia, Drepanura, Kaolishania*, and *Saukia*, bioherms of archaeocyathids, brachiopods, gastropods, bivalves, palaeoconchostracans, hyolithids, and algae. The trilobites are preserved as broken fragments due to wave action. Mineral deposits consist of phosphate, iron, copper, gypsum, and salts.

MOBILE OR BASINAL TYPE OF SEDIMENTATION

Northernmost part of China

This is limited to the northernmost part of China, of Altay, Junggar, Tianshan, Beishan (Ganzu), northern Nei Mongol, the Da (Greater) and Xiao (Lesser) Hingan Mountains, and the northern part of Northeast China. Cambrian outcrops are sporadically distributed and consist mainly of neritic clastic and carbonate rocks. In many districts, especially in eastern Central Tianshan and in the northern part of Northeast China the rocks have been metamorphosed, thereby destroying most fossils. The Cambrian as a whole attains a great thickness, several hundred to several thousand metres. In West Tianshan (typically in Huocheng) and Beishan (Table 7.1) the Cambrian is thin, 110–150 m, but highly fossiliferous. The three Cambrian series form a continuous sequence, but disconformably or locally unconformably overlie the Sinian, showing obvious overlap. In northern North-east China the Cambrian and Sinian form, however, a continuous deposition, but diastrophism is evident within the Cambrian.

Qilian region

The Qilian region covering the whole of the Qilian Mountains is bounded by the Beishan (Gansu Province) in the north and the Qaidam (Tsaidam) in the south. The Cambrian consists entirely of very thick basin deposits of slightly metamorphosed sandstones, phyllites, and slates, intercalated with siliceous rocks and limestone. In many areas volcanic rocks, tuffs, and spilitic keratophyre are exposed though it has not yet been proved that these are definitely Cambrian in age. So far no early Cambrian fossils have been found and the Lower Cambrian is inferred to be absent in this region. Owing to complicated structures, no complete Cambrian sequence is present and both its upper and basal boundaries are not clearly defined (Table 7.1(7)) (Xiang et al. 1981).

South-east region

The South-east region is dominated by the mobile type of sedimentation. It lies in the south-eastern part of China, covering the southern parts of Zhejiang, Jiangsi and Hunnan provinces, a large part of Guangxi, Guangdong, Fujian, Taiwan, and the South China Sea islands. The Cambrian here is characterized by a flyschoid clastic formation, yielding only small inarticulate brachiopods, sponge spicules (*Protospongia*) and a small number of trilobites and molluscs; 8000 m thick. It forms a continuous sequence with the Sinian at the base and the Ordovician at the top. In Hainan Island the Cambrian consists mainly of argillaceous carbonate of small thickness, and is rich in iron and phosphate deposits (Table 7.1(1)) (Zhou Guoqiang in Xiang et al. 1981; Lu et al. 1974).

To sum up, the South-east is an active region frequently marked by flysch and flyschoids and volcaniclastic formations, including rhythmic deposits of sandy shale, marls, limestones, and various extrusive volcanic rocks of great thickness. They are poorly fossiliferous, occasionally containing small chitinous inarticulate brachiopods, and sponge-spicules. Sedimentary deposits consist chiefly of iron, phosphate, rare earths, and stony coal.

The Zhujiang subtype contains little or no volcanic rock or tuff, and is exposed in the South-east region and the greater part of the Kunlun–Qinling region.

RELATIVELY ACTIVE OR TRANSITIONAL TYPE (JIANGNAN TYPE) OF SEDIMENTATION

Qinling Mountains

This type is found between the basin and platform environments in the Kunlun–Qinling region, south of the Yangzi River, and North Xizang–West Yunnan regions. The Kunlun–Qinling region embraces the Kunlun, Karakorum, Bayan Har, and Qinling Mountains and has not yet been closely studied. The best exposed area is the eastern segment of the Qinling Mountains, where the Lower Cambrian is composed generally of carbonaceous shale, sandstones, siliceous rocks, and dolomites, while the Middle and Upper Cambrian consist essentially of carbonate rocks. The thickness varies considerably from 13 800 m in West Kunlun (active portion) to 500–1050 m in Olenbrug, Qinghai (platformal deposits). These strata yield trilobites and *Protospongia* (in West Qinling), and *Circotheca* (in Central Qinling).

The Cambrian in South Qinling is fairly well-developed and consists chiefly of limestones and marls, while the Lower Cambrian consists of thick siliceous rocks and carbonaceous shale, 630–1078 m thick. It overlies the Precambrian unconformably at its base and underlies the Ordovician conformably.

Jiangnan region

Intermediate between the Yangzi platform and the South-east basin regions is the Jiangnan region, where the Lower Cambrian is mainly composed of black (partially yellowish-green) shale, bearing stony coal, phosphorus and vanadium minerals, and is conformable or disconformable over the Sinian. The Upper Cambrian is mainly greyish-black to grey limestone and shale, with a cinnabar-bearing horizon, and conformable with the Ordovician. Fossils include pelagic agnostids and nektonic trilobites, representing the so-called Jiangnan type; 1400–3000 m thick, rarely only 400 m. *Protospongia*, inarticulate brachiopods, hyolithids, palaeoconchostracans, and algae reflect deposition in a marginal sea which deepened from north-west to south-east (Yang Jialu in Xiang et al. 1981).

The Jiangnan type represents an offshore, deeper neritic or gulf environment, which produced generally siliceous and carbonaceous carbonate formations with pyrite-bearing shale and limestone. The reducing or slightly reducing environment (rich in organic matter and H_2S) preserved pelagic agnostids or other trilobites, e.g. *Ptychagnostus, Glyptagnostus, Lotagnostus, Pagetia, Xystridura, Centropleura, Proceratopyge,* and *Olenus* and brachiopods, hyolithids, and sponge spicules. It is rich in P, Mn, rare earths, pyrite, and stone coal. This type has been found in the region south of the Yangzi, North Tianshan (Table 7.1(8)), Beishan, Kuruktag, West Yunnan (Table 7.1(9)).

BOUNDARY PROBLEMS

The Sinian–Cambrian boundary

Formerly the most ancient trilobites were taken to mark the beginning of the Cambrian, but now as very many small shelly fossils have been found beneath them, it is generally accepted that these fossils should indicate the first phase of Cambrian life. Hence, the boundary between the Sinian and the Cambrian is drawn between the Sinian Dengyingian (*sensu stricto*)

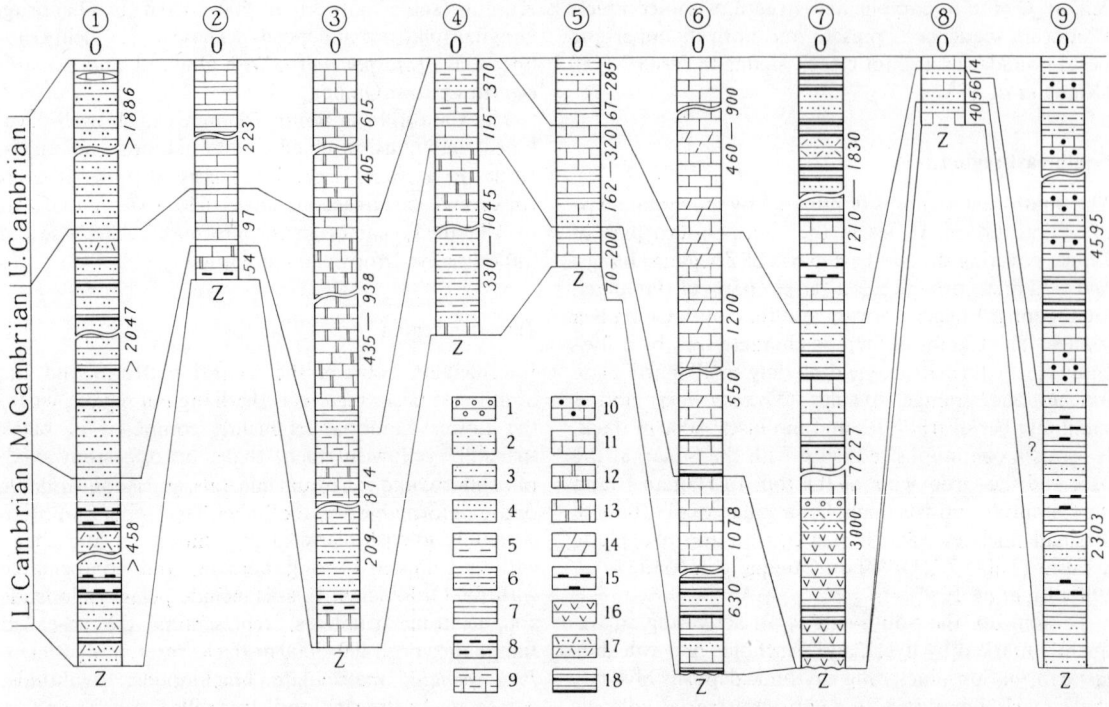

Fig. 7.2. Columnar sections showing development of Cambrian strata in various stratigraphic regions of China. 1 South Jiangxi, South-eastern region; 2 Jiangshan, Jiangnan region; 3 Yangzi Gorges, Yangzi region; 4 East Yunnan, Yangzi region; 5 Yanshan Mts., North China region; 6 South Qinling, Kunlun–Qinling region; 7 Yümeng, Qilian region. 8 North Tianshan, Tianshan 9 Baoshan, Yunnan, North Xizang–West Yunnan region. Legend: 1, Conglomerate; 2, Sandstone; 3, Quartz sandstone; 4, Siltstone; 5, Mudstone; 6, Sandy shale; 7, Shale; 8, Carbonaceous shale; 9, Argillaceous shale; 10, Oolitic limestone; 11, Limestone; 12, Dolomite; 13, Dolomitic limestone; 14, Marl; 15, Silicolite; 16, Volcanic rocks; 17, Tuffaceous sandstone; 18, Slate.

and the Cambrian Meishucunian Stage with its lowest fossil assemblage, i.e. the *Anabarites–Circotheca* assemblage.

The Lower–Middle Cambrian boundary

The crucial problem lies in the placing of the Maozhuanian Stage either within the Lower or the Middle Cambrian. A great majority of Chinese biostratigraphers would regard the Maozhuanian as the first unit of Middle Cambrian, because of distinct biotic change—the Maozhuanian is marked by the appearance of abundant ptychoparids and the practical disappearance of Redlichids. Over the vast South-west area the Lower–Middle Cambrian boundary is therefore drawn between the *Redlichia murakamii–Hoffetella* zone and the *Shantungaspis* zone.

In China the base of the Gushanian Stage, i.e. the base of the *Blackwelderia* zone, is the agreed boundary between the Middle and Upper Cambrian. The upper zone of the Gushanian Stage, the *Drepanura* zone, contains the cosmopolitan Upper Cambrian *Pseudagnostus* and *Homagnostus*; hence it belongs to the Upper Cambrian.

The Middle–Upper Cambrian boundary

The Middle and Upper Cambrian boundary is fixed between the Middle *Damesella* zone and the Upper Cambrian *Blackwelderia* zone. In the Jiangnan region the appearance and disappearance of agnostids serve to separate the Middle Cambrian from the Upper Cambrian. The upper limit of the Upper Cambrian will be treated in the next chapter. A more detailed review of the boundary problems is given by Sheng Xinfu and Xiang Liwen in Xiang *et al.* 1981. The Chinese Cambrian fossil zones can be correlated with those of England, Australia, and the United States of North America as shown in Table 7.2.

Table 7.2. Correlation of Cambrian fossil zones (adapted from Xiang Liwen et al. 1982, by permission).

	Stage	China	England	Australia		USA	
Upper Cambrian	Fengshanian	Calvinella–Mictosaukia Quadraticephalus–Dictyella–Ptychaspis–Tsinania	Peltura	Payntonian	Mictosaukia perplexa	Trempealeauan	Saukia
				Pre-Payntonian	Sinosaukia impages	Franconian	Saratogia
					Pseudagnostus bifax Pseudagnostus clarki		Taenicephalus
	Changshanian	Kaolishania	Leptoplastus	Idamean	Irvingella tropica–Agnostotes inconstans		Elvinia
		Changshania	Parabolina spinulosa				
					Erixanium sentum	Dresbachian	Dunderbergia
	Gushanian	Chuangia	Olenus		Corynexochus plumula		Aphelaspis
					G. recticulatus		
		Drepanura			G. stolidotus		Crepicephalus
			Agnostus pisiformis	Mindyallan	Cyclagnostus quasivespa		Cedaria
					Erediaspis eretes		
Middle Cambrian	Zhangxian	Damesella	Lejopyge laevigata	Lejopyge laevigata		Bolasipidella	
		Amphoton–Taitzuia	Solenopleura brachymetops				
			Goniagnostus nathorsti	Goniagnostus nathorsti			
		Crepicephalus	Ptychagnostus punctuosus	Ptychagnostus puctuosus		Bathyuriscus–Elrathina	
			Hypagnostus parvifrons	Hypagnostus parvifrons			
	Xuzhuangian	Bailiella	Tomagnostus fissus	Ptychagnostus atavus		Glossopleura	
		Poriagraulos abrota	Ptychagnostus gibbus	Ptychagnostus gibbus		Albertella	
		Sunaspis					
	Maozhuangian	Kochaspis	Eccaparadoxides oelandicus	Xystridura		Plagiura–Poliella	
		Shantungaspis					
Lower Cambrian	Longwangmiaoian	Hoffetella–Redlichia murakamii	Protolenid–Strenuellid	Redlichiid		Bonnia–Olenellus	
		Palaeolenus				Nevadella	
	Canglangpuan	Drepanuroides					
		Yiliangella–Yunnanaspis	Olenellid			Fallotaspis	
	Qiongzhusian	Eoredlichia					
		(unnamed)	Non-trilobite				
	Meishucunian	Allatheca–Yunnano theca Ass. Anabarites–Circotheca Ass.					

Table 7.3. *Correlation of Cambrian faunal sequences in China (modified after Xiang Liwen et al. 1982).*

Series	Stage	Condonts	Cephalopods	Others
U. Cambrian	Fengshanian	*Hertzina* *Cordylodus* *Proacodus* *Proconodontus*	*Wanwanoceras* *Protactinoceras* *Plectronoceras*	*Dendrograptus erectus minor* *Dictyonema wutingshanense*
U. Cambrian	Changshanian	*Problematoconites* *Prooneotodus* *Prosagittodontus*		*Callograptus* *Billingsella* *Eoorthis* *Huenella*
U. Cambrian	Gushanian	*Westergaardodina* *Proacodus*		
M. Cambrian	Zhangxian	*Prooneotodus* *Furnishina*		*Tuzoia* *Lingulella* *Acrothela*
M. Cambrian	Xuzhuangian	*Furnishina*		*Homothele* *Nisusia*
M. Cambrian	Maozhuangian			*Girvanella manchurica*
L. Cambrian	Longwangmiaoan			*Kutorgina*
L. Cambrian	Canglangpuan	Hyolithids and Monoplacophora: *Sulcavitus, Linevitus,* *Coleoloides*		*Archaeocyathus, Tuzoia* *Retecyathus, Lingulepsis* *Rotundocyathus, Lingulella* *Coscinocyathus*
L. Cambrian	Qiongzhusian	*Turcutheca* *Scenella*		*Taylorcyathus, Acrothele* *Ajacicyathus, Diandongia* *Plagiogmus* *Hyolithellus*
L. Cambrian	Meishucunian	*Allatheca, Lophotheca,* *Yunnanotheca, Lenatheca,* *Quadratheca, Latouchella,* *Yangtzeconus*		*Sachites* *Chancelloria* *Zhijinites* *Heraultipegma*
L. Cambrian	Meishucunian	*Eosoconus, Protoconus,* *Anabarites, Turcutheca, Circotheca*		*Protohertzina*

8. THE ORDOVICIAN SYSTEM
Yang Zunyi

INTRODUCTION

The distribution of Ordovician sedimentation types and sedimentary regions is similar to that of the Cambrian, thus reflecting the general geotectonic regime prevailing in the Early Palaeozoic. Hence, as in the Cambrian, the Chinese Ordovician may be divided into three main types (stable, mobile, and transitional) and ten stratigraphic regions with some minor changes (see Fig. 8.1):

(1) Stable or platformal and paraplatformal type sedimentation (North China type) is found in the Tarim (2), North China (3), Yangzi (8), and Himalayan (7) regions. The platformal type consists chiefly of carbonate rocks with homogenous lithofacies of moderate thickness and with monotonous faunal make-up, which resembles the North American realm. The paraplatformal deposits consist mainly of carbonate rocks intercalated with thin, well-bedded clastic rocks of irregular thickness, but with marked facies changes.

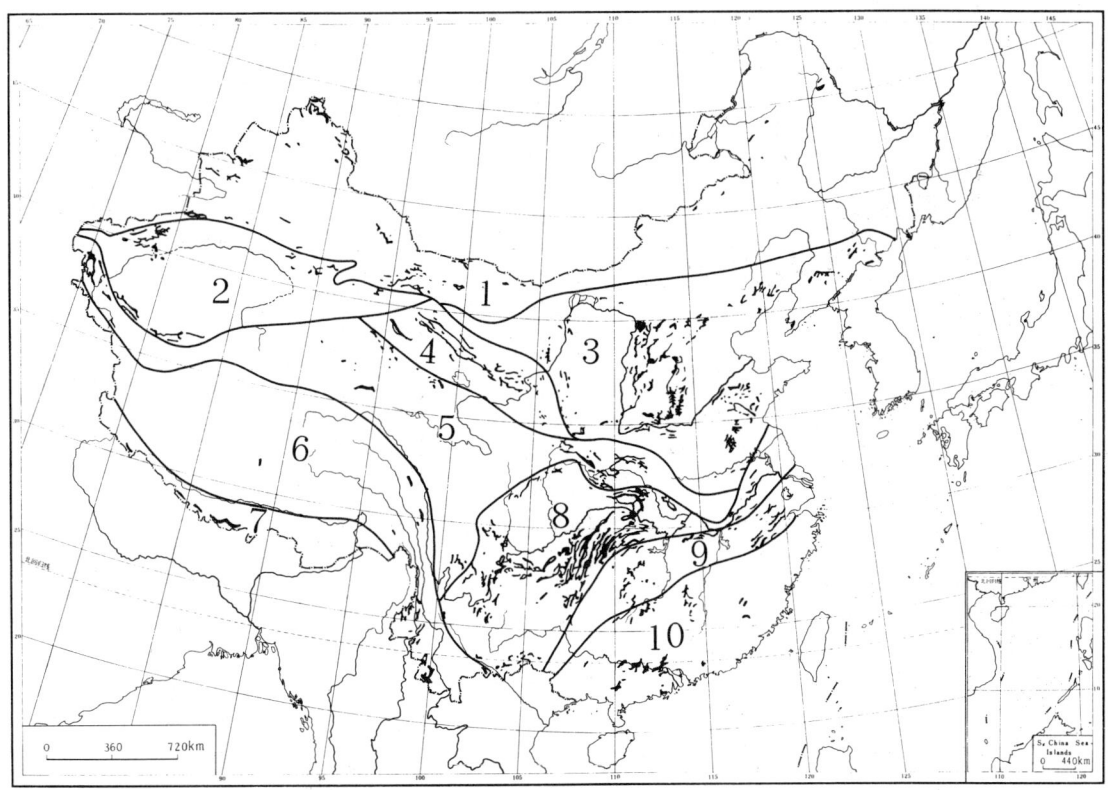

Fig. 8.1. Sketch map showing the distribution and stratigraphic regions of the Ordovician of China. 1, Tianshan–Hingan region; 2, Tarim region; 3, North China region; 4, Qilian region; 5, Kunlun–Qinling region; 6, North Xizang–West Yunnan region; 7, Himalaya region; 8, Yangzi region; 9, Jiangnan region; 10, South-east region. (Outcrops in black.)

74 THE GEOLOGY OF CHINA

(2) Mobile or basin type sedimentation (South-east China type) is found in the Tianshan–Hingan (1), Qilian (4), and South-east (9) regions, the Kunlun–Qinling region (5), which in the Cambrian belongs to the transitional type, and the North Xizang–West Yunnan region. The basin Ordovician deposits are mainly composed of very thick carbonaceous, siliceous detrital rocks (flysch type), of graptolitic facies. No volcanic rocks are present. This typical South-east type is similar to the Australian type. The basin deposits are

Table 8.1. *Correlation of Ordovician sequences in various statigraphic regions of China (chiefly after Wang Xiaofeng 1980).*

Series	Stage	Bio-Zones / Overlying	Yangzi region W. Hubei (3) Longmaxian (S1)	Jiangnan region Qiantangjian–Xiushui(2) Anzi Fm (S$_1$)	South China region (1) Guangdong Guitou Fm (D$_2$)	Kunlun–Qinling Yunxi–Yunxian (5) (S$_1$)
Upper Ordovician	Wufengian	*D. bohemicus/ Hirnantia Kinnella*	Guanyingqiao Fm *Hirnantia–Kinnella* Z. 0.25 m	Yuqian Fm *Diplogr. niushangensis* Z. *Paraorthogr. yuqianensis* Z.		Liangchakou Fm *Agetolites minor A. shanxiensis Favistella intermediata Catenipora robusta* Unfossiliferous in the lower part 179–1300 m
		Paraorthogr. Diceratogr. D. szechuanensis Tanyagraptus typicus	Wufeng Fm *Diplogr. bohemicus* Z. *Paraorthogr.- Diceratogr. mirus* Z. *Tanyagr. typicus* Z. *Dicellogr. szechuanensis– Pleurogr. lui* Z. 5.5–6.6 m	*Dicellogr. szechuanensis* Z. *Pseudoclimacogr. anhuiensis C. leptothecalis– C. venustus* 31–139 m		
	Linxiang-ian	*O. quadrimucronatus*	Linxiang Fm *Nankinolithus* Z. *Hammatocnemis Cyclopyge* 2–3.4 m	Huangnigang Fm *Nankinolithus nankinensis* Z. 45–110 m		
	Baotaan	*D. clingani– D. ramosus/ C. spiniferus, D. baragwanathi var./ C. diplacanthus*	Baota Fm *Richardsonoceras* Z. *Sinoceras chinense* Z. *Hammarodus* cf. *europaeus* 19–25 m	Yanwashan Fm *Xiushuilithus* Z. *Sinoceras chinense* Z. 45–100 m	Longtouzhai Fm Up. part 240 m Mid. part 50–80 m Lower part 688 m	
	Miaopoan	*D. sinensis N. gracilis G. teretiusculus*	Miaopo Fm *Birmanites, Ampyx Nemagr. gracilis* Z. *Lituites miaopopsis Glyptogr. teretiusculus* Z. 2.2–38.55 m (O$_2$ 28.95–38.55)	Hule Fm *Dicranogr. sinensis* Z. *Nemagr. gracilis* Z. *Glyptogr. teretiusculus– Glossogr. hincksii* Z. 6.7–50 m (O$_2$ 127.7–399 m)	Changkengshui Fm *Dicranogr. nicholson diapason* Z. *Nemagr. gracilis– Glyptogr. teretiusculus– Glossogr. hincksii* Z. 53–86 m (O$_2$ 1031–1094 m)	Diaochanggou Fm *Microcoelodus* sp. *Armenoceras jivskense Sactoceras Yokoyamci* 79 m
Lower Ordovician	Guniutanian	*P. elegans/ D. murchisoni A. confertus/ D. artus*	Guniutan Fm *Eoplacognathus suecicus Dideroceras wahlenbergi* Z. *Ambalodus Pseudoplanus* 18 m	Niushang Fm *Pterogr. elegans– Didymogr. murchisoni* Z. *Amplexogr. confertus* Z. *Nicholsonogr. fasciculatus* Subz. 7–170 m	Xiahuangkeng Fm Upper member *Amplexogr. confertus* Z. *Nicholsonogr. fasciculatus* Subz *Paraglosogr. typicalis* Subz 40–110 m Lower member	
	Dawanian	*G. austrodentatus C. amplus/D. nexus Oncograptus D. abnormis/ A. suecicus*	Dawan Fm *Glyptogr. austrodentatus* Z. *Martellia–Lepitorthis Protocycloceras deprati* Z. *Paroistodus originalis Azygogr. suecicus* Z. *Sinorthis typica– Leptella* 54 m	Ningguo Fm *Glyptogr. austrodentatus* Z. *G. sinodentatus Cardiogr. amplus* Z. *Didymogr. abnormis– A. suecicus* Z. *D. vacilans* Z. *D. deflexus*	*Glyptogr. austro- dentatus* Z. *Cardiograptus Didymogy. hirundo* Zone *Isogr. gibberulus Didymogr. abnormis* Basal part unfossilif. 50–70 m	
	Honghua-yuanian	*D. protobifidus/ D. deflexus, T. fruticosus/ D. filiformis, E. approximtus*	Honghuayuan Fm *Coreanoceras– Manchuroceras* Z. *Serratognathus* 17–29 m	*D. filiformis* Z. *Etagr. approximatus* Z. 54–119 m		Shuitianhe Fm *Kaipingoceras styliforme Hopeioceras hupehense* Unfossiliferous in the lower part 495 m
	Yichangian	*Adelogr.–Clonogr./ A. sinensis, Aletogr.–Triogr./ Callogr. Staurogr.–Anisogr./ D. flabelliforme*	Fengxing Fm *Acanthogr.– Tungtzuella* Z. 20–50 m Nanjinguan Fm *Dactylocephalus Finkelnbergia– Asaphellus inflatus* Z *Dictyonema flabelliforme* Z. *Drepanodus simplex* O$_1$ 185–227 m)	Yinzhubu Fm *Clonogr.– Triarthrus* Z. *Dictyonema– Hysterolenus Staurogr.– Anisogr.* Z. *Asaphopsis– Birmanites* Z. 96–700 m (O$_1$ 157–989 m)	Xinchang Fm *Clonogr.–Adelogr.* Zone *C. tenellus A. victoria Aletogr.–Trigonogr.* Z. *Staurogr.– Anisogr.* Z. 55–185 m (O$_1$ 145–360 m)	
	Cambrian System		Cm$_3$ Sanyoutong Gr	Cm$_3$ Xiyangshan Fm	Cm$_3$ Bacun Gr	Cm$_3$

characterized by volcanic rocks intercalated with very thick, regionally metamorphosed detrital and carbonate rocks of unequal thickness, with marked facies changes. All of these features reflect deposition in unstable environments, as found typically in the Hingan subregion, and the Qilian, Kunlun and Qinling regions (Table 8.1) (Lai Caigeng, Wang Xiaofeng et al. 1982a; Lai Caigeng et al. 1982b; Lu, Chu, Qian et al. 1976).

(3) The transitional type (the Jiangnan type) in-

Table 8.1 (cont.)

Himalaya region Qomolangma (8)	Tianshan–Hingan region Hocheng, Guozigou (7)	N. China region N and Southern N.E.China (4)	E. Qilian region Yumen, Gansu (6)	N. Xizang–W. Yunnan region; Baoshan
(S_1) Shiqibo Gr	(S_1) Nilekhe Fm	C_3	(S_1) Shichengzi Fm	(S_1) Xiajenheqiao Fm
Hongshantou Fm 70 m	Hudukdaban Fm Corals: *Rhabdotetradium tianshanense* *R. quadratum* *Agetolites asiaticus* 798 m		Hanshimenzi Gr *Climacogr. putillus* *Climacogr. yumenensis* *Corrugatagnostus* 500–600 m	Wanyaoshu Fm *Dalmanitina mucronata* *Hirnantia* *Climacogr. angustus* *Orthogr. maximus* 12.9 m
Jiacun Gr, Upper Fm *Sinoceras chinense* *Michelinoceras xuanxianense* *Beloitoceras* *Dideroceras nyalamense* 97 m	Aktash Fm *Rhabdotetradium* sp. *Cateniphora* sp. *Zygospira* sp. *Huachengia* sp. *Stereoplasmoceras* sp. 1000 m Koksharexi Fm *Dicellogr. diversicatus* *Climacogr. bicornis* *Nemagraptus* sp. *Glyptogr. teretiusculus* 228 m	Fengfeng Fm *Panderodus gracilis* *Microcoelodus, Badoudus* *Protocycloceras* cf. *eccentrosiphonatum– Sactorthoceras– Fengfengoceras* 145–192 m	Yaomoshan Gr *Dicellogr. sextans exilis* *Nanshanaspis* *Yumenaspis* *Discoceras* 500 m	Up. Pupiao Fm *Nankinolithes nankinensis* 632 m Lower Pupiao Fm *Illaenus* sp. *Basiliella yunnanensis* *Glyptogr. teretiusculus* 217 m
Jiacun Gr, Lower Fm *Paradnatoceras yaliense* *Dideroceras* sp *Ordosoceras* *Wutinoceras* *Eucalymene* *Aporthophyla* *Leptellina* *Manchuroceras* 726 m	Talejihe Fm *Glyptogr. austrodentatus* *Cardiogr. morsus* *Isogr. caduceus* *Oncogr. upsilon* *Didymogr. asperus* 103 m	Majiagou Fm *Multioistodus tangshanensis* *Eoplacognathus* *Armenoceras tateiwai– Discoactinoceras– Selkirkoceras* Ass. 187–295 m		Shidian Fm *Amplexogr. confertus* *Didymogr.* cf. *geminus* *D.* cf. *murchisoni* 464 m
Rouqiecun Gr Upper Fm 40–60 m	Fenggou Fm *Didymogr.* cf. *protobifidus* 172 m	Liangjiashan Fm *Serratognathus bilobata* *Manchuroceras– Coreanoceras– Eothinoceras* Ass. *Archaeoscyphia*	Yingou Gr *Cardiogr. yini* *Paraglossograptus* *Isograptus* *Dictyonema* *Ceratopyge* Fault	Bingdou Fm *Glyptogr. austrodentatus* *Didymogr.* cf. *protobifidus* *D. flabelliforme liaoiungensis*
	Sinertai Fm *Tetragr. fruticosus* *Adelogr. lapworthi* *Etagr. approximatus* 141 m Unfossiliferous in the Lower part 9 m			
Rouqiecun Gr Lower Fm	Cm_3	Cm_3	Cm_3	Baoshan Fm

cludes only the marginal facies of platforms (e.g. North Xizang and West Yunnan (6)) or paraplatforms and of miobasins (e.g. South-east region (9)).

A two-fold division into the Lower and Upper Ordovician is now generally accepted for the Ordovician of China on the basis of two different stages of organic evolution. The graptolites include (1) the Diplograptids–Sinograptids stage and (2) the Dicellograptids stage. Chinese cephalopods may be differentiated: (1) the Ellesmerocerids, Endocerids, and Actinocerids stage; and (2) the Oncocerida, Tarphycerida, and Discocerida stage. Palaeogeographic provincialism was distinct in pre-Llandeilian time, whereas the post-Llanvirnian was characterized by cosmopolitan faunal associations.

The Lower Ordovician includes the Yichangian (Tremadocian), Honghuayuanian, Dawanian (Arenigian), and Guniutanian (Llanvirnian), while the Upper Ordovician includes the Miaopoan (Llandeilian and Caradocian *pars*), Baota'an (Caradocian *pars*), Linxiangian (Upper Caradocian and Lower Ashgillian), and Wufengian (Upper Ashgillian).

STABLE SEDIMENTATION TYPE

Yangzi region

This is best developed in the Yangzi region with the Eastern Yangzi Gorges as its type area and the Huanghuachang section near Yichang, Hubei Province as its stratotype (Table 8.1(3)). It is well-bedded, richly fossiliferous, and divisible into two series, eight stages, and twenty-four zones. The faunas are of mixed character, i.e. in addition to endemic forms, elements of both the North American Pacific Province (North China type) and North Atlantic Province (South-east type) are present (Lai Caigeng, Wang Xiaofeng et al. 1982a, b).

The Lower Ordovician Nanjinguan Formation overlying conformably the Upper Cambrian Sanyoudong Group is composed chiefly of thick-bedded limestone intercalated with bioclastic limestone, and dolomitic limestone yielding trilobites (*Dactylocephalus breviceps, Szechuanella szechuanensis, S. cylindrica, Asaphellus inflatus*), and graptolites (*Dictyonema flabelliforme yichangensis, Callograptus curvithecalis*); 76 m thick.

The Fengxiang Formation is conformable on the Nanjinkuan and comprises dark grey bioclastic limestone or limestone intercalated with greenish shale, 20–50 m thick and rich in graptolites (*Dendrograptus yini, Acanthograptus sinensis, Dictyonema asiaticus*), trilobites (*Tungtzuella*), and brachiopods (*Tritoechia*, *Punctolira orientalis, Imbricatea*). Both the Nanjinguan and Fengxiang Formations are of Tremadocian age.

The Honghuayuan Formation is made up of dark grey to greyish-black, thick-bedded limestone and bioclastic limestone, with occasional chert nodules, containing small amounts of thin-bedded argillaceous striped limestone and yellow shale; 17–29 m. It is rich in Camerocerids, *Archaeoscyphia*, and brachiopods. Two zones have been established: the lower *Serratognathus* and the upper *Coreanoceras–Manchuroceras*, both being of Arenig age.

The Dawan Formation consists lithologically of three parts: the lower part, consisting of greyish-green nodular limestone intercalated with shale, generally with a single glauconite bed at base; the middle part, chiefly purplish-red shale intercalated with greenish nodular limestone, occasionally with thin-bedded shale; and the upper part, dirty greenish shale and nodular limestone or nodular limestone with shale (54 m). There is a rich fauna in the Dawan Formation; among them brachiopods *Yangtzeella, Sinorthis, Lepidorthis,* and *Martellia* are typical of the Yangzi region. This formation can be dated to Arenigian age from the graptolites present.

The Guniutan Formation is composed of thick-bedded nodular limestone and limestone, 18 m thick, yielding nautiloids (*Dideroceras wahlenbergi*) and conodonts (*Ambalodus pseudoplanus* and *Eoplacognathus suecicus* zones). This formation is now given a Llanvirnian age.

The Upper Ordovician Miaopo Formation consists of black shale with limestone lenses (2.2–3.3 m thick); containing the following zones in ascending order: *Glytograptus teretiusculus, Lituites miaopopsia, Nemagraptus gracilis, Birmanites,* and *Ampyx*. It is Llandeilian in age.

The Baota (Pagoda) Formation consists mainly of septarian limestone and nodular limestone, 19–25 m thick. Three zones which have been established are, from the base upwards, *Hamarodus* cf. *europaeus*, *Sinoceras chinense,* and *Richardsonoceras*, all of middle Caradocian age.

The Linxiang Formation contains argillaceous and nodular limestone (2–3.4 m) containing the *Nankinolithus* zone bearing *Hammatocnemis* and *Cyclopyge*. It is upper Caradocian to lower Ashgillian in age.

The Wufeng Formation is made up typically of siliceous rock intercalated with shales (5.5–6.6 m thick). It bears the following zones in ascending order: *Dicellograptus szechuanensis–Pleurograptus lui, Tangyagraptus typicus, Parorthograptus–Diceratograptus mirus,* and *Diplograptus bohemicus*.

The Guanyinqiao Formation is composed of sili-

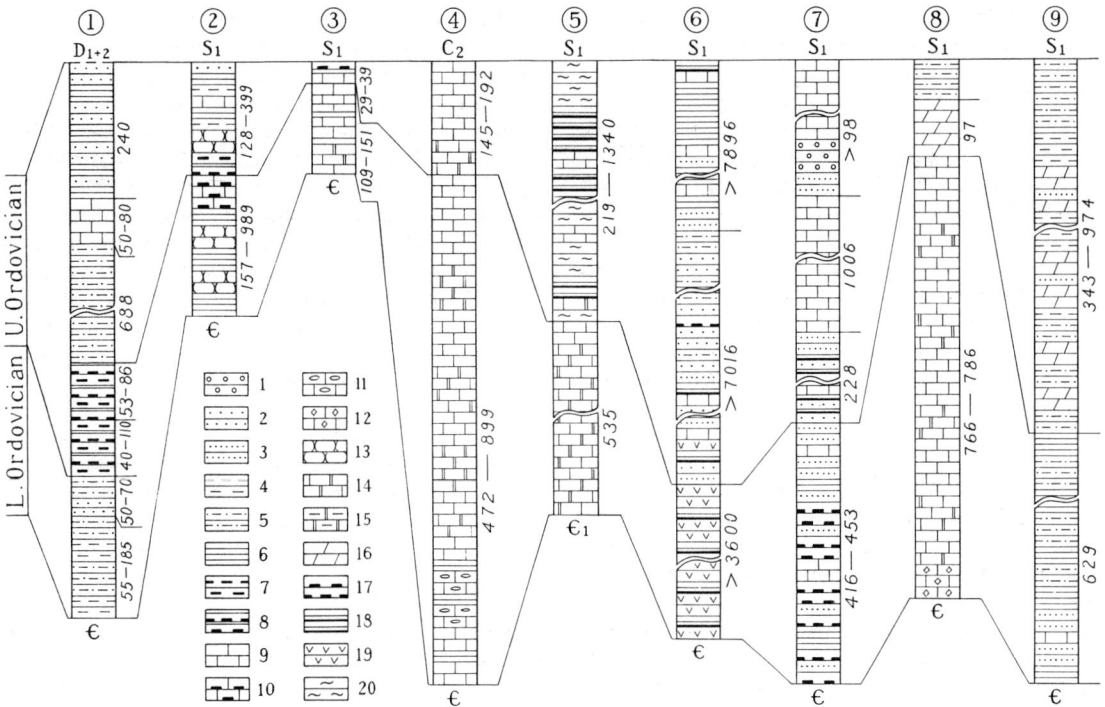

Fig. 8.2. Columnar sections showing Ordovician strata in various stratigraphic regions of China. 1 Central and northern Guangdong, South-east region; 2 Xiushui area, Jiangxi, Jiangnan region; 3 Yangzi Gorges district, Yangzi region; 4 North Hebei, North China region; 5 Kalpin, Xinjiang, Tarim region; 6 North-west Hubei, Kunlun–Qinling region; 7 Yümen, Qilian region; 8 Guozigou, Huocheng, Boroxoro Mt., Tianshan–Hingan region; 9 Nyalam, Himalaya region. Legend: 1, Conglomerate; 2, Sandstone; 3, Siltstone; 4, Mudstone; 5, Sandy mudstone; 6, Shale; 7, Carbonaceous shale; 8, Siliceous slate; 9, Limestone; 10, Siliceous limestone; 11, Wurmkalk; 12, Crystalline limestone; 13, Nodular limestone; 14, Dolomitic limestone; 15, Argillaceous dolomite; 16, Marl; 17, Silicolite; 18, Slate; 19, Andesite; 20, Phyllite. ϵ = Cm.

ceous mudstone, 0.25 m thick, containing the typical *Hirnantia* fauna (*Hirnantia–Kinnella* zone). Both the Wufeng and Guanyinqiao formations are of late Ashgillian age.

North China region

The stable sedimentation type is also well exposed in the North China region (Table 8.1(4)) (Lai Caigeng *et al.* 1982a). It is composed mainly of limestone, dolomite, and brecciated limestone, and is divisible into the following formations in descending order:

Fengfeng Formation equivalent to the Miaopoan.
Majiagou Formation equivalent to the Guniutanian (Llanvirnian).
Beianzhuang Formation equivalent to the Dawanian.
Liangjiashan Formation equivalent to the Honghuayuanian (Arenigian).
Yeli Formation equivalent to the Yichangian (Nanjinguan and Fenxiang) (Tremadocian).

The Yeli Formation is conformable on the Upper Cambrian Fengshan Formation. It consists of limestones and shales, 97 m thick, containing five zones (*Onychopyge–Leiostegium, Dictyonema flabelliforme, Oneotodus reclinatus, Dendrograptus liaotungensis, Dichograptus*).

The Liangjiashan Formation comprises limestones and brecciated limestone intercalated with shale, 157 m thick, yielding *Archaeoscyphia*, the *Manchuroceras––Coreanoceras–Eoactinoceras* assemblage and the *Serratognathus bilobata* zone.

The Beianzhuang Formation contains dolomitic wurmkalk, limestone, and argillaceous dolomite,

187–295 m thick, with nautiloids (*Polydesmia–Wutinoceras–Kogenoceras nanpiaoense* assemblage), and conodonts (*Tangshanodus tangshanensis* zone).

The Majiagou Formation consists of patchy dolomitic limestone, limestone, and argillaceous dolomite, 285–350 m, with nautiloids (*Armenoceras tateiwai–Discactinoceras–Selkirkoceras* assemblage), and conodonts (*Eoplacognathus*, and *Multioistodus tangshanensis* zones).

The Fengfeng Formation is made up of limestones (patchy limestone and dolomitic limestone), 145–192 m thick, also rich in nautiloids (*Protocycloceras* cf. *eccentrosiphonatum–Sactorthoceras–Fengfengoceras* assemblage), and conodonts (*Microcoelodus, Badoudus,* and *Panderodus gracilis*).

Tarim region

In the stable sedimentation type of the Tarim region (Table 8.1(5)) (Lai Caigeng, Wang Xiaofeng *et al.* 1982*a, b*), especially the Kalpin area, are local units equivalent to the Yichangian (Tremadocian), Honghuayuanian and Dawanian (Arenigian), Guniutanian (Llanvirnian), Miaopoan (Llanvirnian), Baota'an (Caradocian), Linxiangian (Upper Caradocian to Lower Ashgillian), and Wufengian (Upper Ashgillian). They are chiefly limestones and shales, totalling 638–1095 m in thickness. The Ordovician section in the Mount Qomolangma area also belongs to the stable type (Table 8.1(9)).

MOBILE OR BASINAL TYPE OF SEDIMENTATION

South-east region

This sedimentation type is exposed in the South-east region, where graptolites and shelly faunas belong respectively to the Pacific and Atlantic provinces. In the Zhujiang (Pearl River) subregion (Shaoguang area), the Lower Ordovician is represented by the Xinchang and Lower Huangkeng Formations, and the Upper Ordovician by the Changkengshui and Longtouzhai Formations.

The Xinchang Formation contains sandy mudstone and mudstone, 55–67 m thick, and yielding three graptolite zones in ascending order (*Staurograptus–Anisograptus, Aletograptus–Triograptus,* and *Clonograptus–Adelograptus*), which are definitely Yichangian (Tremadocian) in age.

The Lower Huangkeng Formation of Arenigian and Llanvirnian age may be differentiated into lower and upper parts: the lower (Arenigian) is composed of 50–70 m of sandy shale and sandstone bearing the graptolite zone (*Didymograptus hirundo,* and *Glyptograptus austrodentatus*). The upper Llanvirnian is made up of 40–110 m of siliceous shale intercalated with carbonaceous shale and felsite–porphyry and feldspar–porphyry. The shales contain the *Amplexograptus confertus* graptolite zone. The formation is 115 m thick.

The Changkengshui Formation consists of 55–80 m of siliceous shale and carbonaceous shale. The graptolite zones (*Glyptograptus teretiusculus–Glossograptus hincksii,* and *Dicranograptus nicholson-diapason*) indicate a Llandeilian age.

The Longtouzhai Formation consists of sandstones, shales and crystalline limestone, totalling 980–1060 m. It is unfossiliferous, but its stratigraphic position gives it a probable Caradocian age.

Tianshan–Hingan region

The Ordovician sequence in Guozigou, Huocheng, north-west of Borohoro Mountain which represents the mobile sedimentation type of the Tianshan–Hingan region, comprises six formations, 2485 m thick, with fossils of Tremadocian, Arenigian, Llandeilian, Caradocian, and Ashgillian ages. Reference may be made to Table 8.1 for other sections in Qilian (Table 8.1(7)), and Kunlun–Qinling (Table 8.1(6)).

TRANSITIONAL SEDIMENTATION TYPE (JIANGNAN TYPE)

The Ordovician sequences in the Xiushui area, Jiangxi Province (Table 8.1(2)) illustrate the development of this sedimentation type. Overlying conformably the Upper Cambrian Xiyangshan Formation the Yinzhubu Formation is composed of 96–700 m of shales intercalated with nodular limestones, yielding five zones from the bottom to top, namely: *Hysterolenus, Staurograptus, Anisograptus, Dictyonema–Hysterolenus, Clonograptus–Triarthrus,* and *Asaphopsis–Birmanites,* which give a Tremadocian age. The Ningguo Formation is composed of 54–119 m of greyish-green shale with six graptolite zones: *Ectagraptus approximatus, Didymograptus filiformis, D. vacilans* (and *D. deflexus*), *D. abnormis–Azygograptus suecicus, Cardiograptus amplus, Glyptograptus austrodentatus* (and *D. sinodentatus*) which indicate an Arenigian age.

The Niushang Formation consists of black siliceous limestone attaining a maximum thickness of 170 m. Llanvirnian graptolite zones include *Amplexograptus confertus* and *Pterograptus elegans–Didymograptus murchisoni*.

The Huluo Formation consists of black siliceous shale intercalated with carbonaceous shale, with a maximum thickness of 50 m, and contains three graptolite zones (*Dicranograptus sinensis*, *Nemagraptus gracilis*, *Glyptograptus teretiusculus–Glossograptus hincksii*).

The Yinwashan Formation is chiefly a nodular limestone unit, 45–100 m thick, bearing the *Sinoceras chinense* and *Xiushuilithus* zones, both of which indicate Caradocian age.

The Huangnigan Formation comprises 45–110 m of calcareous shale and mudstone intercalated with limestones. It contains the *Nankinolithus nankinensis* zone, indicating late Caradocian to early Ashgillian age.

The uppermost Ordovician unit is the Yuqian Formation which is composed of 31–139 m of alternating sandstone and shale, bearing typical Wufengian (Ashgillian) graptolite zones: *Dicellograptus szechuanensis*, *Tangyagraptus typicus*, *Parorthograptus–Diceratograptus mirus*, and *Diplograptus bohemicus*, in ascending order. In Baoshan, West Yunnan, the transitional type

Table 8.2. *Correlation of Ordovician faunal sequences in China (chiefly after Xiang Liwen et al. 1982)*.

	Stage	Brachiopoda	Conodont zones S. Type	Cephalopoda S. Type	Coral faunas	Trilobite faunas or zones
Upper Ordovician	Wufengian	Hirnantia–Kinnella			Sarcinula–Agetolitella–Taeniolites	Dalmanitina mucronatus F.
		Eoconchidium–Rhynchotrema		Tashiceras lamellatum Jiangxiceras yushanense–Yushanoceras serpentinum		
	Linxiangian	Trimerellina–parastrophina		Actinoceras huangnigangense		Nankinolithus nankinensis Z.
	Baotaan	Hallina–Ovalospira Ptychopleurella–Anisopleurella Porambonites–Leptestia	Protopanderodus insculptus Hamarodus europaeus	Sinoceras chinense–Michelinoceras elongatum	Yohophyllum–Ningnanophyllum	Paraceraurus sinicus Z.
	Miaopoan		Protyplacognathus friendsvillensis	Lituites		Tangyaia–Miaopoia–Calymenesun tingi F.
Lower Ordovician	Guniutanian	Horderleyella–Hesperina	Eoplacognathus reclinatus E. foliaceus E. pseudoplanus Amorphognathus antivariabilis	Sinoceras yichangense Meitanoceras–Diderocceras wahlenbergi	Yaoxianopora–Lichenaria–Rhabdotetradium	
	Dawanian	Protoskenidioides Tetraodontella–Lepidorthis–Martellia	Baltoniodus aff. navis		Rhabdotetradium fengfengense	Ningkinolithus–Hanchungolithus F.
			Paroistodus origilis	Cochilioceras yangtzeense Bathmoceras		
		Schedophyla–Leptella	Oistodus multicorrugatus–Peridon flabellum	Anthoceras concavum		
	Honghuayuanian		Oepikodus evaebre Bergstroegnathus extensus Serratognathus Serratognathus diversus	Manchuroceras–Coreanoceras		
	Yichangian	Oligorthis–Syntrophina	Drepanodus deltifer–Scandodus proteus			Tungtzuella Z./Ceratopyge Z.
		Finkelnburgia/Apheorthis	Scolopodus paucicostatus–S. barbatus S. quadraplicatus–S. simplex	Dayongoceras–Retroclintandoceras		Asaphellus trinodosus Z. Leiostegium (Euleiostegium) Aristokinella F.
			Acanthodus costatus Acodus oneotensis Drepanodus simplex	Dakeoceras		Onychopyge–Leiostegium (Alloleiostegium) F. Hysterolenus Z.

Table 8.3. Correlation of Ordovician subdivisions in five countries (chiefly after W. B. Harland et al. 1982).

		China		Britain		Bohemia	Australia	North America	
Upper Ordovician	Wufengian		Diplograptus bohemicus Zone	Ashgrill	Hirnantian	Kosov	Bolindian	Richmond	Cincinnatian
			Paraorthograptus uniformis Zone		Rawtheyan	Králův			
			Diceratograptus mirus Zone		Cautleyan	Dvůr			
			Paraorthograptus typicus Zone						
			Dicellograptus szechuanensis Zone		Pusgillian	Bohdalec		Maysville	
			Amplexograptus disjunctus yangtzensis–Pleurograptus lui Zone						
	Linxiangian		Dicellograptus johnstrupi Zone						
			Orthograptus quadrimucronatus Zone		Onnian		Eastonian	Eden	
	Baotaan		Dicranograptus clingani–Climacograptus spiniferus Zone	Caradoc	Actonian Marshbrookian Longvillian			Sherman	
			Pseudazygograptus chongyiensis–Climacograptus wilsoni Zones		Soudleyan	Zahorány		Kirkfield	
						Vinice			
	Miaopoan		Dicranograptus sinensis–Climacograptus bicornis Zone		Harnagian	Letná	Gisbornian	Rockland	
								Black river	
			Dicranograptus nicholsoni diapason/Nemagraptus gracilis Zone		Costonian	Liben		?	Champlainian
			Glossograptus hincksii/Glyptograptus teretiusculus Zone		Llandeilo	Dobrotivá		Chazy	
Lower Ordovician	Guniutanian		Pterograptus elegans/Didymograptus murchisoni Zone		Llanvirn		Darriwilian	Whiterock	
			Amplexograptus confetus Zone						
	Dawanian		Glyptograptus austrodentatus Zone			Šárka			
			Cardiograptus/Didymograptus nexus Zone						
			Oncograptus magnus Zone				Yapeenian		
			Didymograptus abnormis/Azygograptus suecicus Zone		Arenig	Klabava	Castlemainian	Beekmantown	Canadian
	Hunghuayanian		Didymograptus 'protobifidus'/Didymograptus deflexus Zone				Chewtonian		
			Tetragraptus fruticosus/Didymograptus filiformis Zone				Bendigonian		
			T. (Etagraptus) approximatus Zone				Lancefieldian		
	Yichangian		Adelograptus–Clonograptus Zone		Tremadoc		Warendian	Gasconada	
			Aletograptus–Triograptus Zone						
			Staurograptus–Anisograptus Zone				Datsonian		

consists of two formations (Bingdou and Shidian) in the Lower Ordovician and three formations (Lower Pupiao, Upper Pupiao and Wanyiaoshu) in the Upper Ordovician (Table 8.1(10)).

BOUNDARY PROBLEMS

The lower, Cambro-Ordovician, boundary is marked in most cases by faunal breaks (Lu Yanhao, 1983). In the Yangzi, North China, and South-east regions, the Tremadocian and the Upper Cambrian form mostly continuous sequences. In North China and the southern part of North-east China this boundary proposed by Zhou Zhiyi and Zhang Jinlin (1978) has been followed. It lies between the *Calvinella–Mictasaukia* zone at the top of the Cambrian and the *Onychopyge–Leiostegium* (*Alloleiostegium*) zone at the base of the Ordovician Yeli Formation. In the Yangzi region (the Huanghuachang section) the same boundary is placed below the Lower Ordovician *Asaphellus inflatus* or *Dictyonema flabelliforme yichangensis* zone. In west Zhejiang within the transitional (Jiangnan) type, the boundary lies between the Lower Ordovician *Hysterolenus–Dictyonema* zone and the Upper Cambrian *Hedinaspis* zone, and in South-east region the *Anisograptus–Staurograptus* zone is considered the basal zone of the Lower Ordovician.

In the region where the Ordovician sequence is well-developed, such as in the Yangzi region, the upper limit of the Ordovician is concerned with the age of 'Dalmanitina beds' (the Guanyinqiao Formation). Where these beds contain the *Hirnantia* fauna (Rong Jiayu, 1979), they are treated as latest Ordovician.

According to the proposal of the Second All-China Stratigraphic Conference (1979) a two-fold division of the Ordovician System is now generally accepted, and the boundary between the two divisions is drawn between the Miaoboan and Guniutanian, i.e. between the *Glyptograptus teretiusculus* zone and the *Pterograptus elegans* or *Didymograptus murchisoni* zone. More detailed discussion of boundary problems is to be found in Lai Caigeng, Wang Xiaofeng *et al.* (1982) and Lu Yanhao (1983).

CORRELATION OF ORDOVICIAN SUBDIVISIONS

Referring to the correlation table (Figure 8.1) it is clear that stratigraphic gaps occur in many areas. For example, over North China, southern North-east China, and the Kunlun–Qinling regions, no Caradocian and Ashgillian stages are present and in Kalpin, Tarim region, and the Jiangnan and south-east regions, no Late Ashgillian is represented. These gaps reflect the general crustal uplift and the withdrawal of the late Ordovician (Caradocian–Ashgillian) seas.

9. THE SILURIAN SYSTEM
Yang Zunyi

INTRODUCTION

The uplift that had caused the withdrawal of the Caradocian–Ashgillian seas persisted into the Silurian in the same areas, so a large part of North China, southern North-east China, etc. became land. The Silurian rocks as a result are spread much less extensively than Cambrian and Ordovician rocks, although the Silurian seas still covered quite a large part of China (Fig. 9.1). Again, three sedimentation types (the stable or platformal, mobile or basinal, and transitional) are recognized. Stratigraphic regions are as follows: (A) Stable (Yangzi type) which includes the Tarim (3), Yangzi (8), and Himalaya (7) regions; (B) Mobile or basinal (South-east type) which includes the Xingan (1), Tianshan–Nei–Mongol (2), Qilian (4), Kunlun–Qinling (5), North Xizang–West Yunnan (6), and South-east (10) regions; and (C) the transitional type which includes typically the Jiangnan region (Fig. 9.1).

The biofacies associations include (Lin, Guo, Wang et al. 1982):

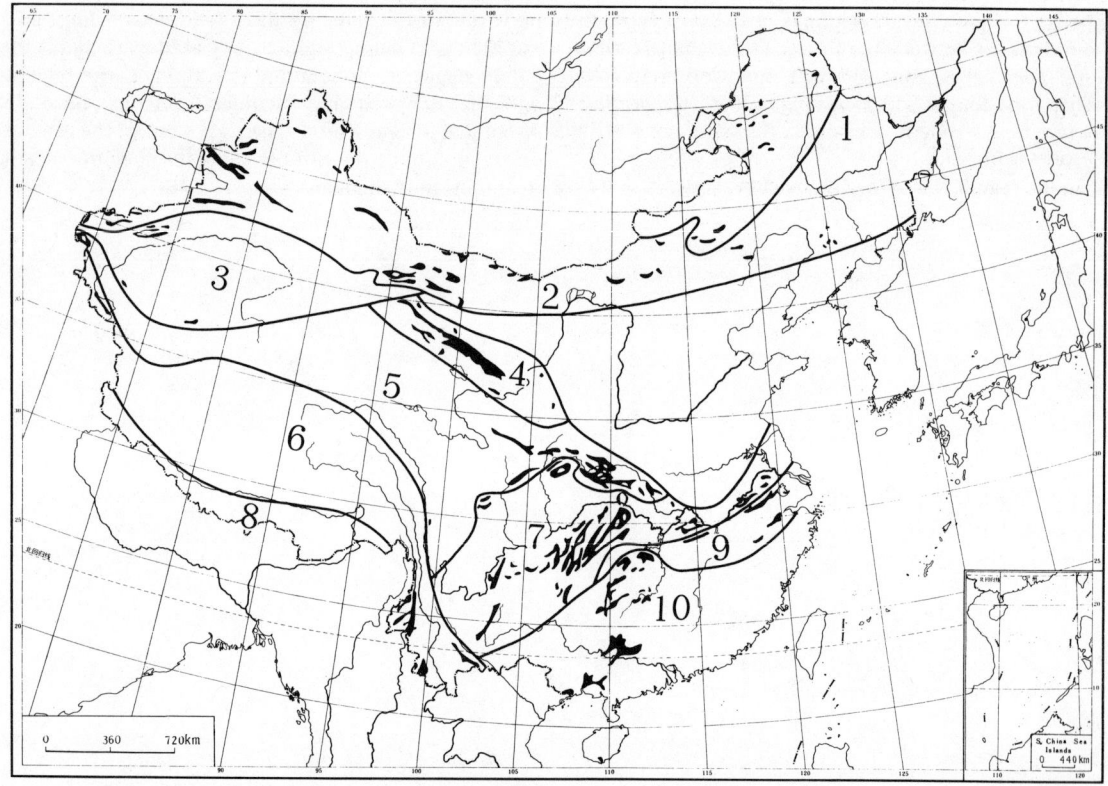

Fig. 9.1. Sketch map showing the distribution and stratigraphic regions of the Silurian of China (chiefly after Xiang et al. (1982)). 1, Hingan region; 2, Tianshan–Nei-Mongol region; 3, Tarim region; 4, Qilian region; 5, Kunlun–Qinling region; 6, Xizang–West Yunnan region; 7, Yangzi region; 8, Himalaya region; 9, Jiangnan region; 10, South China region.

(1) entirely graptolitic facies, mainly graptolite shale in the South-east region;
(2) mixed graptolitic and shelly facies occurring in eight regions: platformal type (Yangzi, Tarim, and Himalaya regions); basin type (Tianshan–Nei Mongol, Qilian, Kunlun–Qinling, North Xizang–West Yunnan and Jiangnan regions);
(3) entirely shelly facies of mainly clastic deposits as found in the Xingan region.

Within the mobile sedimentation type the Lower Silurian generally consists of graptolitic facies with an *Oktavites spiralis* zone at top, whereas the Middle and Upper Series are mostly shelly or mixed facies. Within the stable or platformal sedimentation type, the lower Lower Silurian is graptolitic and the upper Lower, Middle, and Upper Silurian are shelly. In western Yunnan the Silurian sequence is graptolitic except for the upper part of the Middle Series, which is shelly.

STABLE SEDIMENTATION TYPE

Yangzi region

This type is widespread and well-stratified in the Yangzi region, covering eastern and north-eastern Yunnan, central and northern Guizhou, southern Shaanxi, central and eastern Sichuan, central and southern Hubei, north-western Hunan, central Anhui and central Jiangsi. It is characterized by normal neritic to littoral clastic rocks and limestones, reaching a thickness of less than 2000 m, unmetamorphosed, and richly fossiliferous (endemic predominantly). A complete Silurian sequence is divisible into the Lower, Middle, and Upper Series, of which the Lower and Middle are better developed. The Lower Series is graptolitic and the rest shelly, overlying in general the Ordovician conformably, but partly overlapping different Ordovician horizons or even older beds.

Typical sections for the Lower and Middle Series are located in the border areas of Sichuan, Yunnan, Guizhou, and Hubei provinces, and for the Upper Series in Qujing, eastern Yunnan Province. Six stages proposed by the Silurian Research Group (Lin, 1979) are described here (Table 9.1(3), Fig. 9.1): the Longmaxian and Shiniulanian of the Lower Series, the Baisha'an and Xiushanian of the Middle Series, and the Guandian and Miaogao'an of the Upper Series.

The Longmaxian is represented by the Longmaxi Formation in the Yangzi Gorges, Hubei Province, but is best developed in Guanyinchiao, Qijiang, Sichuan Province. It consists of about 161 m of bluish-grey, greyish-green and black sandy mudstone and shale, with numerous graptolites, which are zoned in ascending order as *Glyptograptus persculptus*, *Akidograptus acuminatus*, *Orthograptus vesiculosus*, *Pristiograptus cyphus*, *Pristiograptus leei*, *Demirastrites triangulatus*, and *Monograptus sedgwickii*. The Longmaxi Formation directly overlies the Guanyinchiao Formation which contains the uppermost Ordovician *Dalmanitina–Hirnantia* fauna. It is equivalent to the Liantan Formation of the South-east region and the Xiaoshihugou Formation of the Qilian region — Early Llandoverian.

The Shiniulanian stage has its type section at Shiniulan, Qijiang, southern Sichuan Province, where the Shiniulan Formation is composed of yellowish-grey shale and limestone, 58–510 m thick, containing four faunal assemblages: (1) *Troedssonites–Labyrinthites* (corals), and *Borealis–Virgiana–Camarotoechia fenggangensis* (brachiopods) assemblage; (2) *Baikitolites* (coral), and *Paraconchidium*, *Pleurodium*, *Stricklandia* (brachiopods) assemblage; (3) *Qianbeilites–Meitanopora* (corals), *Pentamerus*, *Clintonella* (brachiopods), and *Yichangoceras* (cephalopod) assemblage; and (4) *Pristiograptus xiushanensis* (graptolite) assemblage. This stage is generally regarded as equivalent to the *Spirograptus turriculatus* to *Streptograptus crispus* zones of the graptolitic facies and approximately corresponding in age (Late Llandoverian) to the lower part of the Angzanggou Formation in the Qilian region and the Wentoushan Formation of the South-east region.

The lower Middle Silurian Baisha'an stage with its type locality at Baisha, Shiqian, east Guizhou Province consists of 180–290 m of purplish-red to greyish-green mudstone, and silty mudstone intercalated with siltstone bearing graptolites (*Hunanodendrum typicum*) which give an Early Wenlockian age.

The upper Middle Silurian Xiushanian stage represented by the Xiushan Formation in Xiushan, Sichuan Province is composed of two members: the lower member is chiefly fine quartz-sandstone occasionally intercalated with sandy limestone and bearing *Ningkiangoceras* sp., *Nalivkinia* cf. *elongata* (Wang), *Loujiashania* sp., while the upper member is made up of calcareous shale and sandstone with calcareous nodules. A supplementary section at Daguan in north-eastern Yunnan consists mainly of limestones, which are richly fossiliferous, yielding mainly trilobites (*Coronocephalus rex*), graptolites (*Monograptus riccartonensis*, *Stomatograptus sinensis*, *M. flemingi*, *Monoclimacis chuhlensis*), brachiopods (*Nalivkinia gruenewaldtiformis*, *Salopina*, *Xinanospirifer*), cephalopods (*Sichuanoceras*, *Yangziceras*, *Trimeroceras*, *Systrophoceras*), and

Table 9.1. Correlation of Silurian sequences in various stratigraphic regions of China (chiefly after Lin et al. 1982).

Series	Stage	Yangzi Region (3)	Qilian Region Yumen (4)	Xizang–W. Yunnan (9)	Jiangnan Region (2)	S.E. China Region (1)	Tianshan–Neimongol Sharburtishan (5)	Xingan Region (6)	Kulun–Qinling Region (7)	Himalaya Region Nyalam, Xizang (8)
Overlying		Yulongsi Fm (D_1)	Shueshan Fm ($D_{1,2}$)	D_1	Wutong Fm (D_1)	Qinzhou Gr (D_1)	Wutubrak Fm (D_1)	Xigulanhe Fm(D_1)	Xiaputonggou Fm (D_1)	Liangquan Fm (D_1)
Upper Silurian	Miaogao'an	Miaogao Fm Schizophoria sp. Aesopomum sp. Spathognathodus crispus Zone 343–492 m	Hanxia Gr	Lichaiba Fm Monograptus scanicus Zone Pristiograptus nilssoni Zone 32–192 m	Tangjiawu Fm Upper part Sinacanthus fancunensis 196–365 m Lower part Sinacanthus fancunensis Orthodonta perlata 580–890 m	Fangchen Gr Pristiograptus transgradiens Zone P. bohemicus P. tumescens P. nilssoni Zone 970–1530 m	Kekexiongkudouk Fm Squameofavosites sp. Encrinurus sp. 1334 m	Gulanhe Fm Lingula sp. 66–500 m	Bailongjiang Gr Squameofavosites sp. Spathognathodus crispus Zone	
Upper Silurian	Guandian	Guandi Fm Heyuncunoceras sp. Morinorhynchus sp. 307 m					Sharburtishan Fm Tuvaella gigantea 150–362 m	Woduhe Fm Tuvaella gigantea 190–370 m		
Middle Silurian	Xiushanian	Xiushan Fm Coronocephalus rex Sichuanoceras sp. Monograptus riccartonensis 170–900 m	Quannaogoushan Gr Nanshanophyllym typicum Mesosolenia biformis Sichuanocerus sp. 2100–2750 m	Upper Renheqiao Fm Upper part Camaracrinus sp. Lower part Cyrtograptus flexilis Zone Cyrtograptus rigidus Zone Monograptus riccartonensis Zone 203–554 m	Daping Gr Modiolopsis sp. Cracianella sp. 805–1070 m	Hepu Gr Monograptus riccartonensis Zone Cyrtograptus murchisoni Zone 154–453 m	Sharbuer Fm Favosites squamatus Subalveolites porosus Favosites gotlandicus 505 m	Bashilixiaohe Fm Tuvaella rackovskii 360 m	Baiai Subfm Subalveolites eichwaldi Antherolites septosus 40–432 m	Shiqipo Gr Upper Fm Upper part Pristiograptus dubius 46 m
Middle Silurian	Baisha'an	Baisha Fm Hunanodendron typicum 8–380 m							Wujiahe Fm Cyrtograptus ramosus Zone C. centrifugus Zone 140–350 m	Lower part Michelinoceras transiens M. capax 30 m
Lower Silurian	Shiniulanian	Shiniulan Fm Pristiograptus xiushanensis Meitanopora sp. Paraconchidium sp. 58–510 m	Angzanggou Fm Oktavites spiralis Zone S. turriculatus Zone 2200 m	Lower Renheqiao Fm Oktavites spiralis Zone Monograptus sedgwickii Zone	Helixi Fm Latiproetus sp. Modiolopsis sp. 490–775 m	Wentoushan Fm Stomatograptus grandis Zone Spirograptus turriculatus Zone 20 m	Bulung Fm Oktavites cf. spiralis Monoclimacis cf. griestoniensis Monograptus sedgwickii	Huanghuagou Fm Chonetoidea laocheensis Hindella sp. 188–550 m	Doushangou Fm S. grandis Zone S. crispus Zone 90–564 m	Lower Fm Oktavites spiralis Zone Streptograptus lobiferus
Lower Silurian	Longmaxian	Longmaxi Fm Monograptus sedgwickii Zone Glyptograptus persculptus Zone 0–504 m	Xiaoshihugou Fm Demirastrites convolutus Zone Glyptograptus persculptus Zone 1400 m	Akidograptus acuminatus Zone 49–280 m	Xiaxiang Fm Pristiograptus leei Zone Akidograptus ascensus Zone Glyptograptus persculptus Zone 778–1564 m	Liantan Fm Streptograptus runcinatus Zone Glyptograptus aff. persculptus Zone 200 m	190–1320 m		Bangjiuguan Fm Spirograptus turriculatus Zone Glyptograptus persculptus Zone 150–300 m	1268–2339 m
Underlying		Guanyingqiao Fm (O_3)	Shichengzi Fm (O_3)	Wanyaoshu Fm (O_3)	Yankou Fm (O_3)	'Sanjian' Gr (O_3)	Bulongguol Fm (O_3)	Chiliathe Fm (O_3)	Quanhekou Fm (O_3)	Hongshantou Fm (O_3)

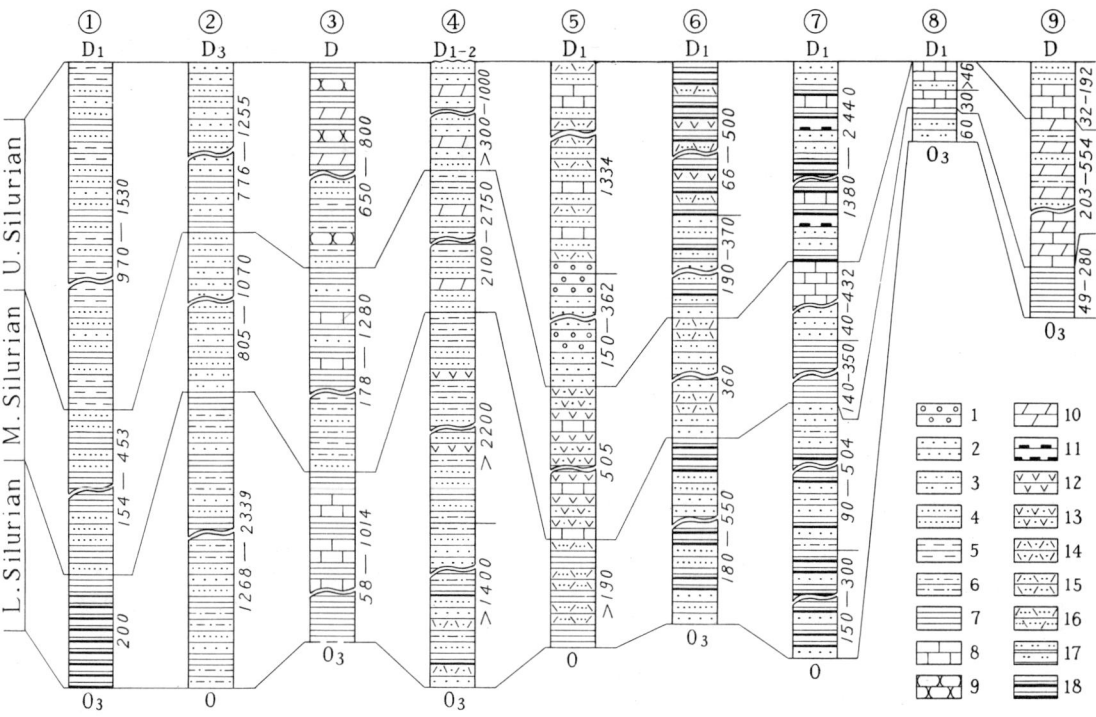

Fig. 9.2. Columnar sections showing the development of Silurian strata in various stratigraphic regions of China. 1 Yunkai area, South-east region; 2 North Jiangxi, Jiangnan region; 3 Yangzi Gorges, West Hubei, South Sichuan, North Guizhou, Yangzi region; 4 Yumen, Qilian region; 5 Saerburte, North Tianshan, Tianshan–Nei-Mongol region; 6 West Heilongjiang, Hingan region; 7 Yunxian-Yunxi; South Shaanxi, Kunlun–Qinling region; 8 Nyalam area, Himalaya region; 9 Baoshan, West Yunnan region. Legend: 1, Conglomerate; 2, Sandstone; 3, Quartz sandstone; 4, Siltstone; 5, Mudstone; 6, Sandy shale; 7, Shale; 8, Limestone; 9, Nodular limestone; 10, Marl; 11, Silicolite; 12, Andesite; 13, Andesitic tuff; 14, Tuff; 15, Tuffaceous sandstone; 16, Tuffaceous siltstone; 17, Quartzite; 18, Slate.

corals (*Nanshanophyllum, Somphopora, Somphoporella* and *Erlangbapora*). The Xiushanian corresponds roughly to the *Monograptus riccartonensis–Monograptus ludensis* zones, belonging to the Late Wenlockian.

The lower Upper Silurian Guandian stage typified by the Guandi Formation near Qujing, eastern Yunnan Province, consists of 307 m of purplish-red siltstone, silty shale intercalated with yellowish-green shale and nodular limestone, yielding cephalopods (*Sichuanoceras, Heyuncunoceras*), brachiopods (*Morinorhynchus, Nikifrovaena, Protathyrisina, Atrypella*), corals (*Squameofavosites*), and conodonts (*Panderodus striatus, Hindeodella* cf. *equidentata*). Trilobites (*Coronocephalus*) have been found from equivalent horizons in north-east Yunnan Province. This stage corresponds to the early and middle Ludlovian, i.e. to the *Pristiograptus nilssoni–P. fragmentalis* zone or lower.

The uppermost Silurian stage, the Miaogao'an, is represented by the Miaogao and Yülongsi Formations also near Qujing. The Miaogao Formation is made up of 343–492 m of dark grey to greyish-black nodular limestone intercalated with yellowish shale and argillaceous siltstone, which contains brachiopods (*Protathyrisina, Howellella, Schizophoria, Aesopomum*), cephalopods (*Yunnanoceras*), and conodonts (*Spathognathodus crispus*). The brachiopod fauna is considered to be of early to middle Pridolian age and the conodont marks the highest conodont zone of the Ludlovian. The Yülongsi Formation consists of alternating thin-bedded limestone and black shale, yielding at its base trilobites (*Warburgella rugulosa sinensis, Encrinuroides*), conodonts (*Trichonedella symmetrica, Panderodus striatus, Pelekyognathus*), and brachiopods (*Protathyrisina plicata, Protathyris* cf. *praecursor* and *Howellella tingi*). These faunas indicate that the Yülongsi Formation should be confined to the Silurian.

Kalpin, Tarim region

In Kalpin, Tarim region, the Silurian is represented by the Kalpin Tag Formation of the Lower Series, which consists of 440–2200 m of greyish-green to purplish-red sandstone, shale, and mudstone. The middle part yields graptolites (*Climacograptus angustus* (Perner), *C. minutus* Carruthers, *Diplograptus modestus parvulus* Lapworth, *Glyptograptus persculptus* var. all of which indicate an early Early Silurian age (Early Llandoverian) (Table 9.1(4), Fig. 9.1).

Himalayan region

In the Himalayan region the Late Ordovician Hongshantou Formation is overlain conformably by the Silurian Shiqipo Group, which is divisible into two formations (Table 9.1(9)). The lower consists of 60 m of brownish-grey, thin-bedded silty shale, and greyish-white quartz sandstone, yielding graptolites (*Oktavites spiralis* zone of late Early Silurian). The upper formation is divisible further into the lower part, which is 30 m of light-grey, medium-bedded limestone, yielding cephalopods (*Michelinoceras transiens* (Barrande), *M. capax* (Barrande); and the upper part, which contains 46 m of greyish-white quartz sandstone and light grey limestone, bearing graptolites (*Pristiograptus dubius* (Suess)) of Middle Silurian age.

MOBILE SEDIMENTATION TYPE

Hingan and Tianshan regions

This type is well-developed in the Hingan and Tianshan–Nei Mongol regions (Lin 1979; Lin, Guo, Wang et al. 1982; Nanjing Inst. Geol. Palaeont. 1982). In western Heilongjiang (Table 9.1(6), Fig. 9.1) overlying the late Ordovician beds is the upper Lower Silurian Huanghuagou Formation which consists of greyish-green slate, siltstone, and graywacke, 180–550 m thick, with brachiopods (*Chonetoidea luoheensis* (Su), *Hindella* sp.). This formation is succeeded by the Bashili Xiaohe Formation of the Middle Silurian, which is composed of greyish-purple, greyish-green tuffaceous sandstone, quartz-sandstone, and graywacke, 360 m thick, yielding brachiopods (*Tuvaella rackovskii* (Tschern.)). The *Tuvaella gigantea*-bearing Woduhe Formation is made up of 190–370 m of clastic rocks (greyish-white, greyish-yellow and greyish-black graywacke, siltstone, slate, and quartzite), and is considered Lower Ludlovian in age. The uppermost unit, the Gulanhe Formation, consists of 66–500 m of greyish-green slate intercalated with tuffaceous siltstone and a small amount of andesite-porphyrite, yielding *Lingula* sp. This is considered the uppermost Silurian unit (Late Ludlovian) in the Hingan region.

Tianshan–Nei Mongol region

In the Tianshan–Nei Mongol region, the Silurian sections at the Saerburte Mountain, Xinjiang are worth mentioning (Table 9.1(5), Fig. 9.1). The Bulong Formation is made up of over 190 m of yellowish-green, tuffaceous siltstone and shale, yielding graptolites (*Oktavites* cf. *spiralis*, *Monoclimacis* cf. *griestonensis* and *Monograptus sedgwickii*) and is Llandoverian in age. The Middle Silurian includes the Shaerbur Formation which consists of greyish-green to dark purple pyroclastic rocks, andesite porphyrite intercalated with limestone, 769 m thick, yielding tabulate corals (*Favosites squamatus* Barr., *Subalveolites porosus* Sharkova, and *Favosites gotlandicus* Lamarck) of Wenlockian age. Overlying conformably is the lower Upper Silurian Saerburte Mountain Formation consisting of greyish-brown medium-grained tuffaceous conglomerate and conglomeratic sandstone, 150–362 m thick, yielding *Tuvaella gigantea*, indicating Ludlovian age. The uppermost Silurian unit is the Kekexiongkuduke Formation, which is composed of 1334 m of purplish-red to greyish-green tuffaceous sandstone, and siltstone intercalated with limestone, yielding *Squameofavosites* sp. and *Encrinurus* sp. most probably of Ludlovian age.

Qilian region

Near Yümen, Gansu Province within the Qilian region (Table 9.1(4), Fig. 2.1) the Lower Silurian is represented by (1) the Xiao Shihugou Formation consisting of over 1400 m of greyish-green, brownish-yellow sandy shale, siltstone, sandstone intercalated in its lower and middle parts with black slate and tuff, yielding graptolites (*Glyptograptus persculptus* to *Demirastrites convolutus* zones of early Llandoverian), overlying conformably the Upper Ordovician Shichengzi Formation; and (2) the Angzanggou Formation composed of over 2200 m of bluish-grey, greyish-green sandy shale, siltstone intercalated, in the lower and middle parts, with black shale grading laterally into basic to intermediate volcanic rocks, yielding graptolites (*Spirograptus turriculatus* to *Oktavites spiralis* zones of Late Llandoverian age).

The Middle Silurian Quannaogou Group is composed of 2750 m of greyish-green to purplish-red sandy shale, siltstone intercalated with mudstone, yielding in its lower part corals (*Paleofavosites,* and *Mesofavosites*), cephalopods (*Sichuanoceras* sp.), and in its upper part corals (*Nanshanophyllum typicum* C. M. Yü and *Mesosolenia biformis* C. M. Yü. The uppermost unit, the Hanxia Group, consists of purplish-red siltstone, finer-grained sandstone and mudstone, more than 300–1000 m thick and is referred to the Upper Silurian from its stratigraphic position, i.e. between the Middle Silurian and the Lower–Middle Devonian.

West Qinling region

In the west Qinling region (Table 9.1(7), Fig. 9.1) the Silurian sequence is fairly well represented, with a mixed fauna of graptolitic and shelly facies. The lower Lower Silurian represented by the Banqiuguan Formation is a sequence of greyish-black to dark grey carbonaceous slate and sandstone, 150–300 m thick, bearing the *Glyptograptus persculptus* to *Spirograptus turriculatus* zones of Early Llandoverian age. The upper Lower Silurian marked by the Doushanguan Formation is composed of grey sandstone intercalated with sandy shale and slate, 564 m thick, containing the *Spirograptus crispus* to *S. grandis* zones of Late Llandoverian age. The lower Middle Silurian Wujiahe Formation consists of 140–350 m of yellowish-green shale, with graptolites (*Cyrtograptus centrifugus* to *C. ramosus* zones), which indicate a Wenlock age for the greater part. The upper Middle Silurian represented by the Baiyaya Formation contains grey to black limestone, and calcareous sandstone, with a maximum thickness of 432 m, yielding corals (*Subalveolites eichwaldi*, and *Antherolites septosus*). The Upper Silurian Bailongjiang Group overlain by the Lower Devonian is composed of dark grey to greyish-black sandstone, slate, limestone, and siliceous rocks, with a maximum thickness of 2440 m, containing corals (*Squameofavosites* sp.), and conodonts (*Spathognathodus crispus*), the latter being the highest zone for the Ludlovian.

South-east region

In the Yunkai area within the South-east region (Table 9.1(1), Fig. 9.1) there is an almost complete Silurian graptolite sequence. The Liantan Formation of the lower Lower Silurian consists of 200 m of black to yellowish-green slates, yielding the *Glyptograptus* aff. *persculptus* to *Streptograptus runcinatus* zones of Early Llandoverian age. The upper Lower Silurian Wentoushan Formation is marked by over 20 m of greyish-black slaty shale, bearing the *Spirograptus turriculatus* to *Stomatograptus grandis* zones of Late Llandoverian age. The Middle Silurian Hepu Group is yellowish-grey to greyish-white argillaceous siltstone and shale, with a maximum thickness of 453 m, and contains the *Cyrtograptus murchisoni* and *Monograptus riccartonensis* zones belonging to the Early Wenlockian. The Upper Silurian Fangcheng Group consists of 970–1530 m of black mudstone and yellowish-grey to greyish-green argillaceous siltstone intercalated with sandstone and shale, bearing the Ludlovian *Pristiograptus nilssoni* and *P. tumescens* to *P. bohemicus* zones, and the Pridolian *Pristiograptus transgrediens* zone.

TRANSITIONAL SEDIMENTATION TYPE

South Anhui

This sedimentation type extends over southern Anhui, western Zhejiang, northern Jiangsi, and northern and south-western Hunan provinces and is represented by clastic rocks which are 3000–4000 m thick, slightly metamorphosed, and rarely fossiliferous. They are therefore not easily differentiated into Lower, Middle, and Upper Series, especially the two latter, although all three are believed to be present (Lin 1979; Lin, Guo, Wang et al. 1982).

The best exposures are found in southern Anhui (Table 9.1(2), Fig. 9.1(2)), where the Xiaxiang Formation, composed of a maximum of 1564 m of greyish-green to yellowish-green shale, silty shale, and siltstone, contains graptolites (*Glyptograptus persculptus, Akidograptus ascensus*), and *Pristiograptus leei* zones of Early Llandoverian age. This is equivalent to the Longmaxian and is underlain by the Upper Ordovician Yankou Formation. The Helixi Formation is an alternation of greyish-yellow to yellowish-green shale and brown to greyish-white sandstone, and siltstone, 490–775 m thick, containing trilobites (*Latiproetus* sp.), and bivalves (*Modiolopsis* sp.), which are inferred to be equivalent to the Shiniulanian (Upper Llandoverian). The Taiping Group consists of 805–1070 m of greyish-green to yellowish-green argillaceous siltstone and quartz sandstone, bearing *Modiolopsis* sp. and *Gracianella* sp. of probable Xiushanian age (Wenlockian). The uppermost unit, the Tangjiawu Group, is separated into the lower part composed of an alternate yellowish-green, purplish-red sandstone and shale, bearing conodonts (*Sinacanthus fancunnensis*), and bivalves (*Orthodonta perlata*), and the upper part of yellowish, greyish-green sandstone and siltstone, 196–365 m thick, bearing conodonts (*Sinacanthus fancunnensis*) of Ludlovian age. This last unit is overlain by the Upper Devonian Wutong Formation.

Baoshan, West Yunnan

In Baoshan, West Yunnan is a Silurian sequence of

graptolite facies. The Lower Silurian Lower Jenheqiao Formation consists of 49–280 m of black shale intercalated with grey, greyish-green or yellow shale, containing three graptolite zones, in ascending order: *Akidograptus acuminatus, Monograptus sedgwickii,* and *Oktavites spiralis*. This is succeeded by the Middle Silurian Upper Jenheqiao Formation composed of two parts. The upper part is red to grey marl intercalated with black sandy shale, also yellow argillaceous siltstone, 203–554 m thick, with *Camarocrinus* sp. The lower part is made up of light grey limestone, reddish marl and grey wurmkalk, bearing another three graptolite zones in ascending order: *Monograptus flexilis, Cyrtograptus rigidus,* and *Monograptus riccartonensis*. The Upper Silurian is represented by the Lichaiba Formation consisting of yellow thin-bedded argillaceous to calcareous siltstone and dark grey sandy limestone, with a maximum thickness of 192 m, and two graptolite zones, the *Pristiograptus nilssoni* zone in the lower, and the *Monograptus scanicus* zone in the upper.

BOUNDARY PROBLEMS

Lower limit

The lower limit of the Silurian, i.e. the boundary between the Ordovician and Silurian, is clearly defined in seven of the ten stratigraphic regions, where the two systems are mostly conformable. In the greater parts of Tarim, Qilian, and a small part of Kunlun–Qinling, however, the boundary is either unconformable or disconformable (Lin, Guo, Wang *et al.* 1982; Mu 1983).

In either the mobile or the stable regions the lower part of the Lower Series is characterized mostly by the graptolitic facies. As the base of the *Glyptograptus persculptus* zone is well-marked at many places, it has been taken as the lower limit of the Silurian. Below it is the *Dicellograptus anceps* zone or *Hirnantia–Dalmanitina mucronata* zone.

Lower–Middle and Middle–Upper Silurian

The boundary between the Lower and Middle Silurian is marked by the top of the *Oktavites spiralis* zone which is present in the mobile regions. In the entirely graptolitic facies, the boundary is defined as being between the *Cyrtograptus centrifugus* or *insectus* or *C. murchisoni* zone, and the *Oktavites spiralis* zone or *Stomatograptus grandis* zone.

In the graptolitic shale facies as in the South-east and Western Yunnan, the base of *Pristiograptus nilssoni* zone is taken as the boundary between the Middle and Upper Silurian. In the Hingan region the appearance of *Tuvaella gigantea* (brachiopod) marks the base of the Upper Series. Over the Yangzi region the appearance of *Squameofavosites* and the top of the *Coronocephalus rex–Sichuanoceras–Stomatograptus sinensis* assemblage in the Upper Xiushan Formation is regarded as the boundary between the Middle and Upper Silurian.

Upper limit

The upper boundary of the Silurian is fairly distinct in five of the ten stratigraphic regions, where the Siluro-Devonian sequence is complete, namely in the Hingan, Tianshan–Nei Mongol, Qinling, North Xizang–West Yunnan and South-east regions, as well as in the western Yangzi region.

CORRELATION OF SILURIAN FOSSIL ZONES

The Silurian subdivisions into stages and their approximate correlation in four countries are shown in Table 9.3.

Table 9.2. Correlation of Silurian faunas of China (modified after Inst. Geol. Palaeont. Nanjing 1982).

Series	Stage	Graptolites	Brachiopods	Corals	Trilobites	Cephalopods	Conodonts and others
Upper Silurian	Miaogao'an	Pristograptus transgrediens, P. tumescens	unnamed Ass.	Mucophyllum	Warburgella rugulosa sinensis	Yunanoceras Ass.	Spathognathodus crispus, Sinacanthus
Upper Silurian	Guand-ian	Monograptus scanicus, P. nilssoni	Protathyrisina plicata–Schizophoria hesta Ass.	Holmophyllum sinensis		Enthyocycloceras Ass.	Sinacanthus
Middle Silurian	Xiushanian	Cyrtograptus ramosus, M. flexilis, C. rigidus, M. riccartonensis	Protathyrisina uniplicata–Atrypoidea qujingensis Ass.	Kyphophyllum primaevum Ketophyllum equitabulatum	Acanthopyge orientalis	Heyuncunoceras Ass.	Dazhucrinus Senticucullus
Middle Silurian	Baish-an	C. murchisoni, C. centrifugus	Salopina–Xinanospirifer Ass.	Nanshanophyllum planocystosum–Shenxiphyllum aggregatum Ass.	Coronocephalus–Chuangianoproetus	Sichuanoceras Ass.	Sinacanthus Pisocrinus pillula
Lower Silurian	Shiniulanian	Stomatograptus grandis, Oktavites spiralis, Monoclimacis griestoniensis, Streptograptus crispus, Spirograptus turriculatus	Nalivkina–Nucleospira Ass.	Somphoporella Carnegiea Ass.	Luojiashania divergensis		Pisocrinus yini
Lower Silurian	Shiniulanian		Pentamerus–Eospirifer Com.	Shanxipora–Neoflectcheriella Ass.		Yichangoceras Ass.	Petalocrinus Spirocrinus
Lower Silurian	Longmaxian	Monograptus sedgwickii, Demirastrites convolutus, D. triangulatus, Pristiograptus cyphus, Orthograptus vesiculosus, Akidograptus acuminatus, Glyptograptus persculptus	Borealis–Kritorhynchia Ass.	Meitanopora–Baikitolites Zone	Ptilillaenus	Songkanoceras Ass.	Petalcrinus
Lower Silurian	Longmaxian			Troedssonites Tetraporella Zone	Raphiophorus guizhouensis		
Lower Silurian	Longmaxian		Protatrypa–Hindella Ass.		Dalmantina nanchengensis		

Table 9.3. *Correlation of Silurian fossil zones in four countries (chiefly after Lin Baoyu 1979).*

		China		M. Europe (Czech.)		W. Europe (England)			North America	
Upper Silurian	Miaogao'an	Monograptus transgrediens / Pristiograptus tumescens		Pridoli	M. angustidens / M. bouceki / P. ultimus	Post L.–Pre G.	Downtonian		Cayugan	
	Guandian				P. tumescens	Ludlow		Whitecliffian		
				Kopanin			Leintwardinian	M. leintwardinensis		
							Bringewoodian	M. tumescens		
		M. scanicus / P. nilssoni			P. nilssoni		Eltonian	M. scanicus / P. nilssoni		
Middle Silurian	Xiushanian			Motoi	P. ludensis	Wenlock	Homerian	P. ludensis / C. lundgreni	Lockportian	Niagaran
		C. ramosus / M. flexilis / C. rigidus / M. riccartonensis			C. rigidus / M. riccartonensis		Sheinwoodian	C. ellesae / C. linnarssoni / C. rigidus / M. riccartonensis	Tonawandan	
	Baisha'an	M. firmus / C. murchisoni / C. centrifugus			C. insectus			C. centrifugus		
Lower Silurian	Shiniulanian	S. grandis / O. spiralis		Zelkovice	S. grandis	Llandovery	Telychian	M. crenulatus		
					M. crenulata			M. crispus	Ontarian	
		M. griestoniensis / S. crispus / S. turriculatus / M. sedgwickii			M. griestoniensis / M. sedgwickii		Fronian	M. turriculatus / M. sedgwickii		
	Longmaxian	D. convolutus / D. triangularis / P. leei			D. convolutus / D. pectinatus		Idwian	D. convolutus / M. gregarious		
		P. cyphus / O. vesiculosus / A. acuminatus / G. persculptus			P. cyphus / A. ascensus		Rhuddanian	P. cyphus / G. persculptus	Alexandrian	

10. THE DEVONIAN SYSTEM
Yang Zunyi

INTRODUCTION

The Caledonian or Qilianian (Guangxian) movement, an important event at the close of the early Palaeozoic, brought about the folding of the Qilian basin into mountain ranges, thus joining the Sino–Korean and the Tarim paraplatforms into an extensive continent, a vast expanse of land under denudation, which accentuated the distinctness of the northern basin from the Tethys to the south. In South China, east of the Kam–Yunnan landmass, the same movement caused the formation of the South-east Caledonian fold-belt and the emergence of the main Yangzi massif from below sea-level. Hence, the Devonian marine domain in South China is limited to Yunnan, Guizhou, Guangxi, and West Guangdong provinces, and both the Lower and Middle Devonian successions are transgressive from south-west to north-east (Fig. 10.1). West of the Kam–Yunnan landmass was the subsiding trough of the eastern Tethys, interrupted by a series of medium-sized massifs.

In South China the Devonian is well exposed, exten-

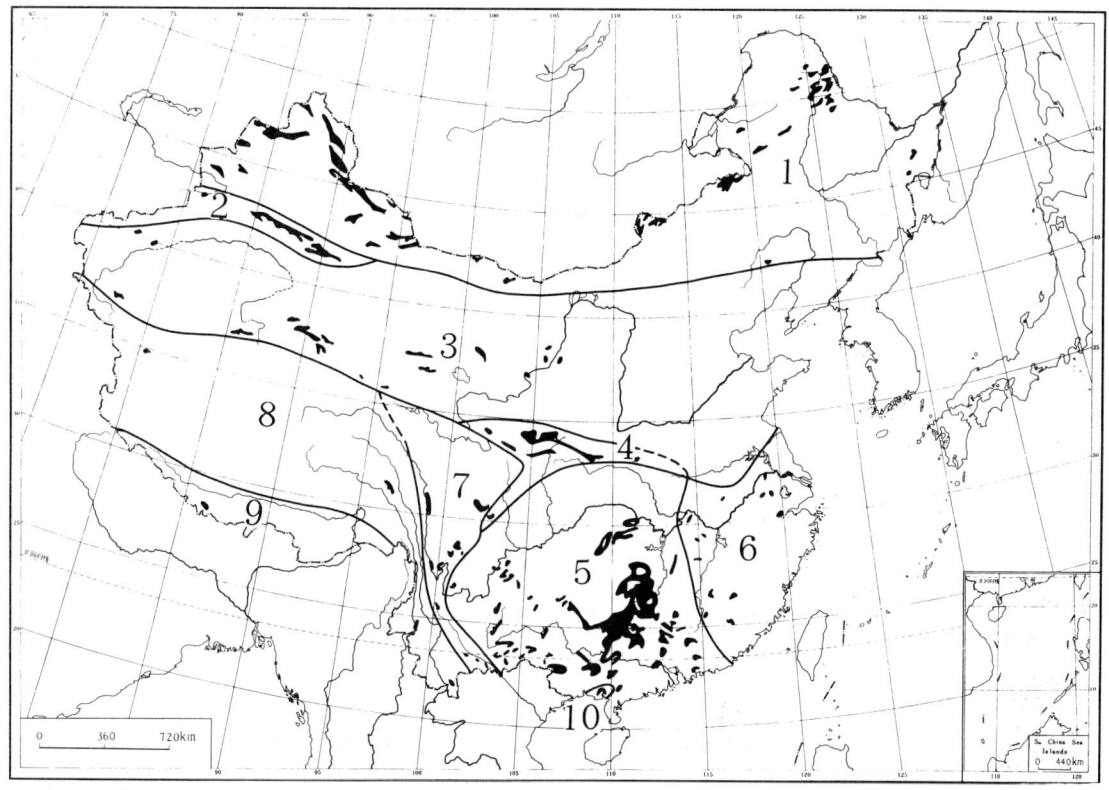

Fig. 10.1 Sketch map showing the distribution and stratigraphic regions of the Devonian of China (adapted chiefly after Xiang *et al.* (1982)). 1, Junggar–Hingan region; 2, South Tianshan region; 3, Qilian region; 4, Qinling–Longmenshan region; 5, South China region; 6, South-east region; 7, Garze–Lijiang region; 8, North Xizang region; 9, Mt. Qomolangma region; 10, Qinzhou region. Outcrops in black.

sively developed and marked by a variety of sedimentary associations and rich biotas. Applying similar criteria to those used in previous periods for differentiating distinctive stratigraphic regions, there are the following two types divided into eight units (Fig. 10.2). (A) Stable or platformal sedimentation type: South China region (5); Himalaya–West Yunnan region (8) (mostly marine facies); South-east region (6); Qilian region (3) (mostly continental facies). (B) Mobile or basinal sedimentation type: Junggar–Hingan region (1), South Tianshan region (2), Longmenshan–Qinling region (4), West Sichuan–North Xizang region (7).

During the Devonian, organic evolution proceeded rapidly along with great physical changes, especially the emergence of land and the laying down of terrestrial deposits. Thus, there appeared quite a number of freshwater and brackish-water fishes and luxuriant land plants (*Psilophytales*). There were also marked changes among the marine invertebrates: namely, the decline of graptolites to only a few monograptid species in the earliest Devonian and the great decline of trilobites; the evolution of rugose corals, brachiopods (Spiriferids, Camaroteochiids, and Terebratulids such as *Stringocephalus*), goniatites, and the profuse increase of microfossils (conodonts and tentaculitids) which are especially useful as zone-markers.

STABLE SEDIMENTATION TYPE

South China region

This is best shown in the South China region, especially in Guangxi and Hunan provinces, where the Devonian regional chronostratigraphic stages proposed by the Devonian research group are as follows (Yang Shipu *et al.* 1979; Hou, Wang, Guo *et al.* 1982*a*, *b*). They propose four stages (Lianhuashanian, Nagaolingian,

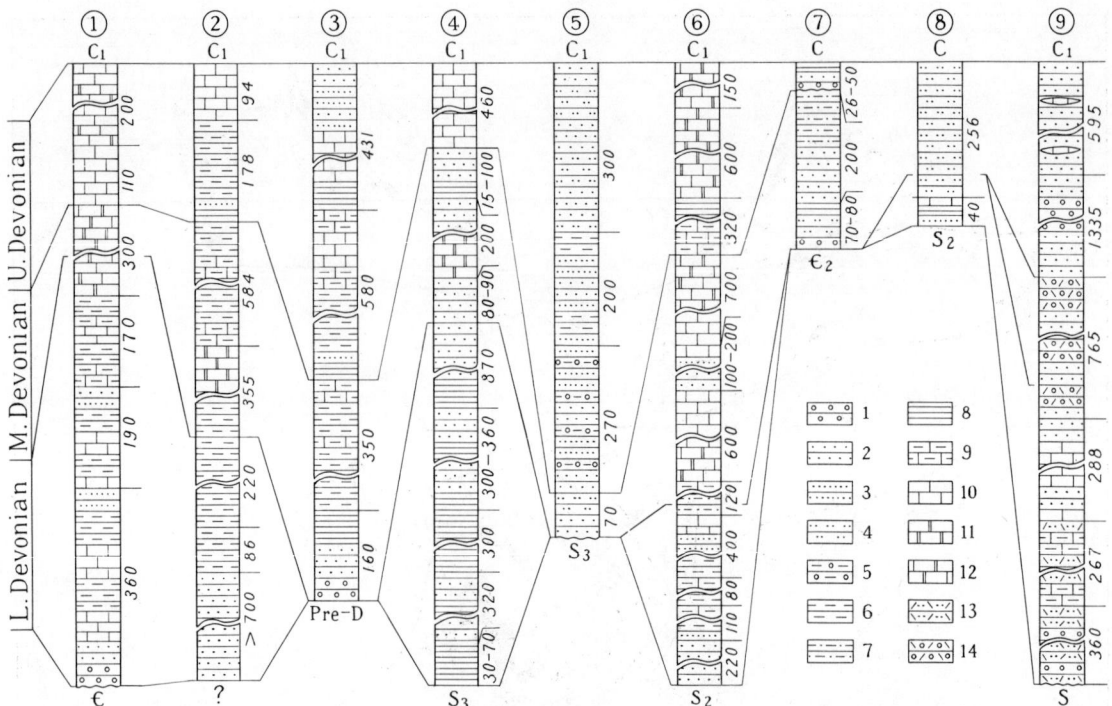

Fig. 10.2. Columnar sections showing the Devonian strata in various stratigraphic regions of China. 1 Liujing, Hengxian, Guangxi, South China region; 2 Nandan, Guangxi, South China region; 3 Central Hunan, South China region; 4 East Yunnan, South China region; 5 Yudu, South-east Jiangxi, South-east region; 6 Jiangyou, Sichuan, Qinling–Longmenshan region; 7 North Qilian, Qilian region; 8 Nyalam, Qomolangma region; 9 North Tianshan, Junggar–Hingan region. Legend: 1, Conglomerate; 2, Sandstone; 3, Siltstone; 4, Quartz sandstone; 5, Conglomeratic mudstone; 6, Mudstone; 7, Sandy shale; 8, Shale; 9, Argillaceous limestone; 10, Limestone; 11, Dolomite; 12, Dolomitic limestone; 13, Tuff; 14, Tuffaceous conglomerate.

Yujiangian, and Tangdingian or Sipainian) for the Lower Devonian Series, two (Nabiaonian or Yintangian and Dongganglingian) for the Middle Devonian Series, and two (Shetianqiaonian and Xikuangshanian) for the Upper Devonian Series in the South China region (Table 10.1(1), (2), (3)).

The Lower Devonian Lianhuashanian is represented by the Lianhuashan Formation with the type section near Liujing, Hengxiang County, Guangxi Region, which consists of 340 m of mainly purplish-red silty mudstone and siltstone, intercalated with argillaceous limestone in the middle part and thick-bedded quartzose sandstone at the base. It bears fishes such as *Asiaspis expansa, Asiacanthus suni, A. kaoi, Galeaspis* sp., *Yunnanolepis* sp., *Lianhuashanolepis liukingensis,* and *Orientolepis neokwangsiensis.* It should be noted that there are as yet no definite time markers for the upper and lower limits of this formation, so the status of this stage is uncertain.

The Lower Devonian Nagaolingian is typified by the Nagaoling Formation, the type section also being near Liujing. It consists of 190 m of greyish-green, calcareous mudstone intercalated with limestone, and fine sandstone at the top, with brachiopods (*Orientospirifer nakaolingensis, O. wangi, Protathyris praecursor, Chonetes ellipticus, Aseptalium guangxiensis, Kwangsirhynchus liujingensis, Chonetes ellipticus,* etc.), corals (*Calicidophyllum nakaolingensis, Eoglossophyllum minor*), conodonts (*Spathognathodus linearis postolinatus, Hindeodella equidentata, Trichonodella excavata*), spores (*Leiotriletes simplex, L. parvus, Calamospora microrugosa, Acantotriletes cuspidatus,* etc.), and Tintinnids (*Ancyrochitina ancyrea, A. spinosa, Sphaerochitina communisa*) which indicate a late Lochkovian age.

Some workers have argued that the Nagaoling Formation is so limited in extent and varied in facies that it cannot be considered to represent a regional stage, and further study is needed.

The Lower Devonian Yüjiangian Stage based on the Yüjiang Formation with its type section at Liujing, Hengxiang, Guangxi is composed of mudstone, calcareous siltstone, and marl intercalated with biogenetic limestone, having a total thickness of 170 m and yielding the following: brachiopods (*Dicoelostrophia–Rostrospirifer tonkinensis* assemblage embracing *Nadiastrophia yukiangensis, Parachonetes nasatus, Parathyrisina tangnae, Howellella papaoensis, Acrospirifer ordinaris,* and *Elymospirifer kwangsiensis* (Wang, Y. 1956)), rugose corals (upper Assemblage Zone of *Heterophrentis angusta–Amplexiphyllum hamiltoniae;* lower Assemblage Zone of *Xystriphylloides nobilis–Heterophalactis semicrassa*), trilobites (*Proetus indosinensis, Dechenella liujingensis,* etc.), conodonts (*Neopriniodus bicarvatus, Panderodus striatus striatus, Spathognathodus exiguus guangxiensis, S. optimus, Ozarkodina denckmanni, Hindeodella priscilla, Polygnathus dehiscens*), which suggest that it is Pragian in age.

The Lower Devonian Tangdingian is based on the stratotype at the village of Tangdin, Nandan County, northern Guangxi region, while the alternative Sipainian is based on the type section at Dale Village, northeast of Xiangzhou County, East Guangxi region. The Tangdingian, represented by the Tangding Formation, includes 376 m of greyish and carbonaceous mudstones, whereas the Sipainian, with the Sipai Formation as type, consists of 453 m of biogenetic limestone, dolomite, and mudstone. The Tangding Formation is characterized by a pelagic fauna: two zones of Tentaculites, the lower zone of *Nowakia praecursor* and the upper zone of *N. barrandei*, and ammonoids (*Anetoceras reticostatum, Erbenoceras solitarius, E. lineare, E. ellipticum, Teicherticeras nantanense,* and *T. ilanense*). The Sipai Formation is characterized by the following three faunal assemblage zones in ascending order, which indicate a Zlichovian age. 1. The *Subcuspidella trigonata–Paramoelleritia xiangzhouensis magna* Assemblage Zone: *Athyrisina plicata, Paramoelleritia miaohuangensis*. 2. The *Otospirifer daleensis–Trapezophyllum cystosum* Assemblage Zone: *Alatiformia* aff. *alatiformis, Cystohexagonaria daleensis, Phacellophyllum daleense, Nathoceras sianghsiensis, Shipaia hexospina, Polygnathus lenzi,* and *P. linguiformis linguiformis.* 3. The *Euryspirifer paradoxus shuijiepingensis–Psydracophyllum cystosum* Assemblage Zone: *Euryspirifer kwangsiensis, Athyrisina yohi, Reticulariopsis indifferens, Lacutkinia lata, Leptoinophyllum subvermiculare, P. irregulare, Zelolasma elegantula, Pseudomicroplasma laticystata,* and *Tryplasma concavotabulata*.

The Middle Devonian Nabiaonian (Yintangian of the Xiangzhou paratype) with type sections respectively at Nabiao, Nandan County and Yintang, Dale, Xiangzhou County, is based on the Nabiao Formation and Yintang Formation. The former includes 355 m of black carbonaceous mudstone, whilst the latter has 218 m of mudstone and limestone. They are characterized by planktonic and benthonic fauna respectively, and the Nabiaonian has the following zones. Ammonoids: 1. *Convoluticeras discordans* zone (*Mimogoniatites fecundus,* and *M. bohemicus*); 2. *Anarcestes (Latanarcestes) noeggerati* zone (*Gyrocerstites gracilis,* and *Subanarcestes macrocephalus*); 3. *Pinacites jugleri* zone (*Foordites platypleura, F. occultus,* and *Agoniatites* sp.). Tentaculites: 1. *Nowakia cancellata* zone; 2. *N. richteri* zone; 3. *N. holynensis* zone; 4. *N. sulcata* zone. Trilo-

Table 10.1. *Correlation of Devonian sequences in various stratigraphic regions of China (adapted chiefly after Hou et al. 1982).*

Series	Stage	South China Region			
		Liujing, Guangxi (1)	Nandan, Guangxi (2)	Central Hunan (3)	Cuifengshan, Chujing, Yunnan (4)
Overlying		R	C_1	C_1	C_1
Upper Devonian	Xikuangshanian	Rongxian Fm *Cyrtospirifer* sp.	Daihua Fm *Wocklumeria sphaeroides* Zone *Clymenia laviegata* Zone *Dzieduszyckia baschkirica* Zone 94 m	Xikuangshan Fm Oujiachong member *Vallatisporites pusillites Bothriolepis cyclostigma* Zone 120 m Magunao limestone *Yunnanella synplicata* 284m Changlongjie shale *Yunnanellina hanburyi* 27 m	Zaige Fm
	Shetianqiaonian	200 m	Xiangshuidong Fm Upper member *Uniconus* sp. *Richterina latior* 98 m Lower member *Beloceras* cf. *acutum* 80 m	Shetianqiao Fm *Tornoceras* *Manticoceras wedekindi* 580 m	460 m
Middle Devonian	Dongganglingian	Donggangling Fm Upper member *Stringocephalus burtini* 80 m Lower member *Nowakia otomari Bornhartina* sp. 30 m	Luofu Fm *Nowakia otomari* Zone 584 m	Qizichiao Fm *Stringocephalus burtini* 350 m Tiaomajian Fm. *Hunanolepis tieni Bothriolepis sinensis* 160 m	Haikou Fm *Bothriolepis sinensis* *Hunanolepis tieni* *Quasipetalichthys haikouensis* 15–100 m Sanshuanghe Fm *B.* sp. 200 m
	Nabiaonian (Yintangian)	Nabiao Fm *Polygnathus* spp. *Ozarkodina denckmanii* *Styliolina* sp. ***Euryspirifer tonkinensis*** *Howellella luomaiensis* 300 m	Nabiao Fm Upper member *N. sulcata–Pinacites jugleri* Zone 120 m Lower member *Anarcestes noeggerati* Zone *N. cancellata–Convoluticeras discordans* Zone 235 m		Chuandong Fm *Xichonolepis qujingensis* *B. tungseni* *Wudinolepis weni* 80–90 m
	Tangdingian (Sipainian)		Tangdin Fm *N. barrandei–Erbenoceras ellipticum* Zone *N. praecursor* Zone *N. zlichovensis* Zone 220 m		Xujiachong member *Galeaspis xiujiachongensis* *Drepanophychus spinaeformis* 870 m
Lower Devonian	Yujiangian (Yujiang Fm)	Liujing Member *Heterophrentis angusta* Ass. Dalian Member 90 m Shizhou Member *Xystriphylloides nobilis–Heterophatactis semicrassa* Zone 80 m	Yilan Fm *Dicoelostrophia annamitica* *Gravicalymene maloungkaensis* 86 m		Guijiatun member *Yunnanolepis* *Chuchinolepis* *Zosterophyllum myretonianum* 300–360 m
	Nagaolingian (Nagaoling Fm)	Xiayiling Member *Howellella* cf. *nucula* 50 m Mahuangling Member 70 m **Nagaoling Member** *Orientospirifer* *Kwangsirhynchus liujingensis* *Protathyris praecusor* 70 m	Danlin Gr 700 m		Xitun member *G. changi* *Nanpanaspis microculus* 300 m Xishancun member *Yunanogalaspis major* *Polybranchiaspis* *Dongfangaspis* spp. 320 m
	Lianhuashanian (Lianhuashan Fm)	Liukankou member *Yunnanolepis* sp. 130 m Hengxian member *Lianhuashanolepis* 110 m *Lingli member* 100 m			Miandiancun Fm *Polybranchiaspis* ***Yunanolepis* sp.** **Macropetalichtyidae** Phlyctaenaspidae 30–70 m
Underlying		Cm		Pre-D	Yulongsu Fm (S_1)

Table 10.1. (*cont.*)

S. China Region Yudu, S. Jiangxi (5)	Longmenshan–Qinling Region Longmenshan, Sichuan (6)	Qilianshan Region Niushoushan, Ningxia	Himalaya–W. Yunnan Region Qomolongma (8)
C_1	C_1	C_2	C_1
Xiashan Fm *Leptophloeum rhombicum* *Sublepidodendron mirable* *Bothriolepis* sp. 300 m	Maoba Fm *Rozmanaria* sp. 150 m	Zhongning Fm Upper part *Remigolepis zhanginensis* Lower part *Leptophloem rhombicum* 26–50 m	Boqu Gr 256 m
Shanmentan Fm *Yunnanella* *Tenticospirifer* 200 m	Shawozi Fm *Hypothyridina lungtungensis* 600 m		
Zhongpeng Fm *Bothriolepis* sp. *L. rhombicum* 270 m	Tuqiaozi Fm *Devonproductus* sp. 320 m		
Yunshan Fm *Bothriolepis lochangensis* *B. kwuangtungensis* *Barrandeina* cf. *duslina* 70 m	Guanwushan Fm Upper member *Stringocephalus obesus* 700 m Lower member 100–200 m	Shixiagou Fm Upper part *Bothriolepis niushoushanensis* 180–200 m Lower part 70–80 m	
	Yangmaba Fm *Zdimir* spp. *Sociophyllum* sp. 600 m Xindianzi Fm *Xenospirifer* cf. *fongi* *Calceola sandalina* *Favosites* spp. 120 m		
	Xiejiawan Fm *Euryspirifer paradoxus* *Teicherticeras* sp. 400 m		
	Ganxi Fm *Dicoelostrophia* *Rostrospirifer tonkinensis* 80 m		Liangquan Fm *Guerichina xizangensis* 11 m Grey siltstone *Nowakia acuaria* Zone *Neomonograptus himalayensis* *Monograptus thomasi* 40 m
	Bailiuping Fm *Orientospirifer* *Howellella* sp. 110 m		
	Pingyipu Fm *Sanqiaspis rostrata* *Lungmenshanaspis kiangyouensis* *Dongfangaspis major* *Xinanpetalichthys shendaowanensis* 220 m		— fault — ?
Cm	S_2	Cm_2	S_2

bites: 1. *Plagiolaria nandanensis* zone; 2. *Phacops guangxiensis* zone; 3. *Cyphaspides orientalis* zone. Brachiopods: *Paranotanoplia faceta–Luofugia delicata* Assemblage Zone.

The following coral and brachiopod zones are represented in the Yintangian of the Xiangzhou area. 1. The *Xenospirifer fongi* Assemblage Zone: *Eospiriferina lachrymosa, Yingtangella sulcatilis, Indospirifer* cf. *padaukpinensis, Productella sinensis, Dalergynchas dingshanlingensis,* and *Squamularina prava.* 2. The *Acrospirifer houershanensis–Utaratuia sinensis* Assemblage Zone: *Desquamatia richthofeni, Uncinulus pentagona, U. goldfussi, U. parallelepipedus, Athyrisina squamosaeformis, Sociophyllum minor, Neospongophyllum tenue, N. glomerulatum, Breviseptophyllum kochanensis, Hexagonaria simplex,* and *Microplasma devoniana.* The Nabiaonian and Yintangian are roughly equivalent to the western European Eifelian in age.

The Middle Devonian Dongganglingian (Tungkanglingian) stage (Givetian) represented by the Daopending section near Xiangzhou County, East Guangxi, comprises 400 m of clastics with limestone in the lower part and limestone and marl in the upper part. It yields brachiopods (*Schizophoria striatula, Stringocephalus burtini, Bornhardtina speciosa, Acrothyris kwangsiensis, Emanuella takwanensis, Rhynchospirifer liujingensis, Undispirifer undiferus,* and *Levibiseptum dushanensis*), and corals (*Temnophyllum waltheri, Dendrostella trigemme, Cyathophyllum expansum, Stringophyllum isactis, Grypophyllum tenue, Cystiphylloides kwangsiensis, Pseudomicroplasma fongi, Macgeea cylindricum,* and *Dialythophyllum crassum*).

The Upper Devonian Shetianiaonian Stage is based on the type section at Shetianiao, Shaodong County, Hunan Province. It contains thin-bedded siliceous rocks or siliceous limestone at its base, mudstone plus argillaceous siltstone in the lower part, limestone and siltstone in the middle, and marl intercalated with thin-bedded limestone in the upper part, and is 580 m thick. It bears corals (*Sinodisphyllum variabile, S. simplex, Hunanophrentis uniforme, Pseudozaphrentis curvatum,* and *Disphyllum* spp.), brachiopods (*Spinatryina douvillei, S. hunanensis, Cyrtospirifer martelli, Tenticospirifer tenticulum, Hypothyridina linglingensis,* and *Ptychomalatechia shetienchiaoensis*), and ammonoids (*Manticoceras wedekindi, M. hunanense, M. zhouguoense, M. tenticulare,* and *Tornoceras* aff. *criibriseptum*). This stage is equivalent to the European Frasnian.

The Upper Devonian Xikuangshanian Stage represented by the Xikuangshan Formation in Xikuangshan, Xinghua County, Hunan, includes 431 m of shale, limestone, haematite, and sandstone, and is marked by three assemblage zones in ascending order: 1. The *Yunnanellina–Sinoproductella–Athyris* zone (*Yunnanellina hanburyi, Y. uniplicata, Cyrtospirifer sinensis, Sinoproductella hemispherica,* and *Athyris gurdini*); 2. The *Yunnanella–Hunanospirifer–Tenticospirifer* zone (*Yunnanella synplicata, Y. hunanensis, Y. supersynplicata, Cyrtospirifer subextensus, Tenticospirifer supervillis, Ptychomalatechia hsikuangshanensis,* and *Hunanospirifer wangi*); 3. The *Bothriolepis–Cyclostigma* zone (*Bothriolepis* sp., *Lepidodentropsis hirmeri, Cyclostigma kiltorkensis, Lepidostrobus grabaui, Hamatophyton verticillum, Sublepidodendron mirabile, S. wushiense, Vallatisporites pusillites, Cymbosporites parvibasilaris,* and *Hymenozontriletes lepidophytus*).

Elsewhere in the Huishui–Changshun area, South Guizhou Province, Famennian ammonoids (*Clymenia* and *Wocklumeria* zones) have been found from beds equivalent to the Xikuangshanian Stage.

A continental–littoral Devonian sequence is well exposed in Chuifengshan, Qüjing County, East Yunnan Province (Table 10.1(4)), where the Lower Devonian overlies conformably the Upper Silurian Yülongsi Formation. There are two Lower Devonian formations, three Middle Devonian, and a single Upper Devonian formation. The Lower Devonian Miandiancun Formation consists of 10 m of black shale containing fishes (*Polybranchiaspis* spp., *Yunnanolepis* sp., *Macropetalichthyidae,* and *Phlyctaenaspidae*). This is succeeded by the Chuifengshan Formation divisible into the Xishancun, Xitun, Guijiatun, and Xujiachong members. The Xishancun Member which is composed of 320 m of yellow sandstone and green shale also bears fishes (*Yunnanogalaspis major, Polybranchiaspis* spp., and *Dongfangaspis* spp.). The Xitun Member is 300 m of grey shale, and contains fishes (*Galeaspis changi,* and *Nanganaspis microculus*); the Guijiatun Member is made up of 300–360 m of red sandstone and shale, with fishes (*Yunnanolepis,* and *Chuchinolepis*), and plants (*Zosterophyllum myretonianum*). The Xujiachong Member composed of red and yellow sandstone and shale is 870 m thick and contains fishes (*Galeaspis xujiachongensis*), and plants (*Drepanophycus spinaeformis*).

The Middle Devonian contains, in ascending order, the Chuandong Formation, the Sanshuanghe Formation, and the Haikou Formation. The first unit is 80–90 m of yellow sandstone, and contains fishes (*Xichonolepis qüjingensis, Bothriolepis tungseni,* and *Wudinolepis weni*); the second formation of yellow sandstone and grey dolomite is 200 m thick with *Bothriolepis* sp.; and the third, the Haikou Formation, composed of yellow sandstone, quartzite and shale, is 15–100 m thick, and yields fishes (*Bothriolepis sinensis,*

Hunanolepis tieni, and *Quasipetalichthys haikouensis*).

The Upper Devonian Zaigen Formation consists of 460 m of grey massive limestone and is so far unfossiliferous.

South-east region

The South-east region also belongs to the stable sedimentation type, though it is characterized by mainly continental deposits, occurring in the Middle and Lower Yangzi and Zhejiang, Fujian, and Eastern Guangdong provinces. In Yudu, South Jiangxi, only Middle and Upper Devonian are present (Table 10.1(5)). The Middle Devonian Yunshan Formation is composed of conglomerate, siltstone, and sandy mudstone and is 70 m thick. It bears fishes (*Bothriolepis lochangensis*, *B. kwangsiensis*, and *Barrandeina* cf. *duslina*) of Dongganglingian age. The Upper Devonian Zhongpeng Formation is made up of 270 m of sandstone and siltstone intercalated with argillaceous conglomerate. It yields fishes (*Bothriolepis* sp.) and plants (*Leptophloem rhombicum*) and is Early Shetianiaoan in age (Frasnian). The Upper Devonian Shanmentan Formation includes 200 m of sandstone, siltstone, and mudstone containing brachiopods (*Yunnanella* and *Tenticospirifer*) and is early Xikuangshanian in age. The uppermost unit, the Xiashan Formation, includes 300 m of sandstone and quartzite bearing plants (*Leptophloem rhombicum*, *Sublepidodendron mirabile*), and fishes (*Bothriolepis* sp.) and is late Xikuangshanian (Famennian) in age.

Qomolangma area

In the Qomolangma area (near Nyalam), South Xizang, the Devonian includes the Liangquan Formation and the Boqu Group (Table 10.1(8)). The former is in faulted contact with the Silurian Shiqibo Group, and is made up of 29 m of grey siltstone in the lower part marked by the *Nowakia acuaria* zone (*N. acuaria*, *Neomonograptus himalayensis*, and *Monograptus thomais*), and in the upper part of 11 m of an alternation of greyish-white shale and black shale, containing *Guerichina xizangensis*. The formation thus belongs to the Lower Devonian Pragian age. The Boqu Group comprises 256 m of quartz sandstone, bearing some plant fragments in its upper part, and is inferred to be of Middle–Upper Devonian age.

Qilian region

In the Qilian Mountain region, which was then incorporated with the Alashan of the North China platform, the Devonian sequence of Niushoushan, Zhongning, Ningxia is represented by only the middle and upper series (Table 10.1(7)). The Middle Devonian Shixiagou Formation has in its lower part yellow sandy shale, quartzite and conglomerate, and is 70–80 m thick and unfossiliferous. In its upper part it consists of 180–200 m of purple sandy shale and sandstone, yielding fishes (*Bothriolepis niushoushanensis* and Actinoleptids). This is unconformably overlain by the Zhongning Formation separable into two parts. The lower one consists of red and purple conglomerate, quartzite and sandy shale, and is 26–50 m thick, with fossil plants (*Leptophloeum rhombicum*, etc.); and the upper part, of purple and red sandstone and shale, contains fishes (*Remigolepis zhanginensis*).

MOBILE SEDIMENTATION TYPE

Saerburte Mountain, West Junggar

This is typically shown in the Junggar–Hingan region, where purely marine, mixed marine and continental, and continental facies are well represented.

In Saerburte Mountain, West Junggar (Table 10.1(9)) the Lower Devonian includes the Utubulake, Mangeer, and Mangkelu formations. The Utubulake Formation is composed of 480 m of tuffaceous sandstone and conglomerate, yielding graptolites (*Monograptus vakatus*), trilobites (*Encrinurus* sp., *Warburgella conica*), and plants (*Sciadophyton pristinum*), which are Pragian in age.

The Mangeer Formation of greyish-green tuffaceous sandstone and marl is 267 m thick and contains trilobites (*Calymene junggarensis*, and *Odontochile sinensis*) which indicate a Zlichkovian age.

The Mangkelu Formation composed of calcareous sandstone and sandy limestone is 288 m thick and yields brachiopods (*Eodevonaria arcuata*, *Leptaenopyxis bouei*, *Gladiostrophia kondoi*, *Paraspirifer gigantea*, also *Crotalocephalus*, etc.) which are also Zlichovian in age.

The Middle Devonian Huerjisite Formation comprising tuffaceous conglomerate and sandstone is 765 m thick, and contains fossil plants, such as *Protolepidodendron scharyanum*, *Lepidodendropsis* sp., and *Barsassia sibirica*, and is possibly Dalejian–Eifelian in age.

The Upper Devonian embraces two formations, the Zhulumite and Hongguleleng. The Zhulumite is a terrestrial deposit (sandstone and conglomerate), 1335 m thick, and it contains fossil plants (*Lepidodendropsis arborescens* and *Lepidosigillaria columnaria*).

The Shetianiaonian or Frasnian Hongguleleng Formation of littoral deposition consists of sandstone and lenses of limestone, which are 595 m thick and yield brachiopods (*Cyrtospirifer sulcifer, Plicatifera alexanderi*), and a fossil plant (*Leptophloem rhombicum*), giving a Xikuangshanian or Famennian age.

Da Hingan Mountains

The Devonian sequence in the Great Hingan Range consists of six formations, two for each series. The Lower Devonian Luotuoshan Formation is composed of 91 m of siltstone and sandstone intercalated with lenses of crystalline limestone, yielding brachiopods (*Cymostrophia alfa, Protathyris praecursor, Howellella* sp., and *Ancilotoechia* sp.), and corals (*Dictyofavosites* sp.) of Zlichovian age. This is followed by the Unuoer Formation of bioclastic limestone, and siltstone, more than 158 m thick, bearing corals (*Lyrielasma, Leptoinophyllum* sp., *Tryplasma hercynica,* and *Amplexiphyllum*), and brachiopods (*Wilsonella grandis,* and *Howellella amurensis*), also of Zlichovian age. The Middle Devonian Beikuang Formation, more than 140 m thick, is composed of calcareous siltstone intercalated with slate and sandstone, and is marked by corals (*Breviphyllum lenense,* and *Cayugaea subcylindrica*) of Yintangian or Dalejian–Eifelian age. This is succeeded by the Hebaoshan Formation of limestone and conglomerate, over 137 m thick, with *Endophyllum abditum,* and *Temnophyllum,* and is Dongkanglingian or Givetian age. The Upper Devonian Xiadamingshan Formation consisting of tuffaceous sandstone and intermediate acid lava, is more than 183 m in thickness, with corals (*Thamnophyllum tomense,* and *Macgeea solitaria*) and brachiopods (*Peneckiella* sp.) of Shetianiaonian or Frasnian age. The uppermost Devonian Shangdamingshan Formation consists of andesitic porphyry in the upper part, and sandstone in the lower. It is over 60 m thick and contains two ammonoid zones, the lower (*Cheiloceras subpartitum*) zone and the upper (*Platyclymenia walcotti*) zone, which indicate the age as being Xikuangshanian or Famennian.

Longmenshan–Qinling region

The Devonian sequence of the Longmenshan–Qinling region is related to the South China region in biotic associations. In Longmenshan the Lower Devonian is composed of the Pingyipu, Bailiuping, Ganxi, and Xiejiawan Formations, in ascending order. The Pingyipu Formation of greyish-white sandstone is 220 m thick and contains fishes (*Sanqiaspis rostrata, Lungmenshanaspis kiangyouensis, Dongfangaspis major,* and *Xinanpetalichthys shendaowanensis*) which are Lianhuashanian to Lower Nagaolingian in age. The Bailiuping Formation is composed of 110 m of yellowish-green mudstone, siltstone, and marl, with brachiopods (*Orientospirifer* sp., and *Howellella* sp.), and is upper Nagaolingian in age. The Ganxi Formation of mudstone and marl is 80 m thick and contains brachiopods (*Dicoelestrophia* sp., *Rostrospirifer tonkinensis*) giving a Yujiangian age. The Lower Devonian Xiejiawan Formation is composed of mudstone, and siltstone with thin-bedded limestone, and is 400 m thick, yielding brachiopods (*Euryspirifer paradoxus*) and nautiloids (*Teicherticeras* sp.) giving a Tangdingian or Sipainian age.

The Middle Devonian consists of three formations. The Xindianzi Formation of sandy mudstone and marl has a thickness of 120 m and the following fossils: *Xenospirifer* cf. *fongi* (brachiopod), *Calceola sandalina* and *favosites* (corals). It is Lower Nabiaonian or Yintangian in age. The Yangmaba Formation composed of massive limestone, and dolomitic limestone is 600 m thick, and contains brachiopods (*Zdimir* sp.) and corals (*Sociophyllum* sp.) of Upper Nabiaonian age. The Guanwushan Formation begins with 100 to 200 m of sandstone intercalated with siltstone and iron beds without fossils, then 700 m of massive limestone and dolomitic limestone, carrying the well-known brachiopod *Stringocephalus obesus,* indicating a Dongganglingian or Givetian age.

The Upper Devonian Tuqiaozi Formation consisting of marl and shale is 320 m thick, and contains brachiopods (*Devonproductus* sp.) of Lower Shetianiaonian age. This is followed by the Sawozi Formation which comprises dolomite and dolomitic limestone, and is 600 m thick and yields brachiopods (*Hypothyridina lungtungensis*) of Upper Shetianiaonian age. The uppermost Devonian unit, the Maoba Formation of dolomitic limestone is 150 m thick, and contains *Rozmanaria* sp., of Xikuangshanian or Famennian age.

Têwo, West Qinling

In Têwo, West Qinling, the Lower Devonian Xia Putonggou Formation composed of slate intercalated with limestone and calcareous sandstone is 300 m thick, and yields conodonts (*Icriodus woschmidti*) and brachiopods (*Protathyris praecursor*) which are important markers for establishing the Silurian–Devonian boundary.

West Sichuan–North Xizang region

The Devonian of the West Sichuan–North Xizang

region may be shown by the sequence in Qamdo of Sichuan Province, where four formations are well exposed. Beginning at the bottom, the Lower to Middle Devonian Haitong Formation, of dolomite, marl, sandy slate, and sandstone, has a thickness of 25 m and carries corals (*Squameofavosites* sp., and *Favosites* sp.) and brachiopods (*Athyrisina squamosa*, and *Acrospirifer* sp.). This is succeeded by the Tingzhonglong Formation of limestone, marl, shale, and dolomite, with brachiopods (*Stringocephalus dorsalis, Schizophoria kutsingensis*) and corals (*Grypophyllum* cf. *tenue, Hexagonaria tungkanlingensis*), which is 55–245 m thick. The Upper Devonian also contains two formations, the Zhuoguodong and the Jiangge. The former composed of limestone, marl, and dolomite is 179 m thick and contains brachiopods (*Spinatrypina douvillei, Hypothyridina linglingensis,* and *Cyrtospirifer sinensis*) and corals (*Sinodisphyllum* sp., and *Temnophyllum heterophylloides*) of Shetianiaonian or Frasnian age. The Jiangge Formation is made up of limestone and marl with a thickness of 321–681 m and also contains brachiopods (*Tenticospirifer tenticulum, Yunnanellina hanburyi,* and *Yunnanella abrupta*) of Xikuangshanian or Famennian age.

BOUNDARY PROBLEMS

Silurian–Devonian boundary of the marine regime

In China the Silurian–Devonian boundary for the marine sequence may be found in Diebu, West Qinling Mountains, where the Early Devonian Xiapudonggou Formation overlying the Silurian Bailongjiang Group is composed of slate intercalated with limestone and calcareous sandstone, and is 300 m thick, and carries conodonts (*Icriodus woschmidti*), and brachiopods (*Protathyris praecursor*), which are crucial for determining the Silurian–Devonian boundary (Wang Yu *et al*. 1974; Hou, Wang, Gao *et al*. 1982*a, b*; Nanjing Inst. Geol. Palaeont. 1983).

In West Junggar the lower limit of the Utubulake Formation with transitional Silurian–Devonian biota (see above) is taken as the Lower Devonian boundary. This boundary can be more satisfactorily fixed in the Nantan facies with planktonic fossils (*Neomonograptus* and *Paranowakia* of Upper Lochkovian age), which mark the onset of a new faunal succession. The boundary of the continental–littoral sequence is temporarily drawn at the base of the Miandiancun Formation on the evidence of fishes (*Polybranchiaspis* spp., *Yunnanolepis* sp., Macropetalichthyidae, and Phlyctaenaspidae), and this formation is conformable on the Upper Silurian Yülongshan Formation.

Lower–Middle Devonian boundary

The boundary between the Lower and Middle Devonian is fixed between the *Nowakia cancellata* zone and *N. barrandi* zone. It is this boundary that marks the extinction of most *Anetoceras* and the appearance of many new taxa of ammonoids, tentaculites, trilobites, and brachiopods. With the Xiangzhou facies, the Lower–Middle Devonian line is provisionally drawn between the Tangdingian or Shipainian and Nabiaonian or Yintangian, i.e., roughly at the top of the *Euryspirifer paradoxus* zone (Hou and Xian 1975; Hou 1978).

The Lower–Middle Devonian boundary of the Junggar–Hingan region is still uncertain, because no characteristic faunas of early Middle Devonian are available. There are also conflicting views in these studies concerning the diagnostic value of brachiopods and corals.

Continental Middle–Upper Devonian boundary

With regard to the continental Devonian, fossil fishes are crucial in determining the Lower–Middle Devonian boundary, i.e. the extinction of Ostracodermi (Galeaspiformes, Polybranchiaspiformes, Yunnanolepidae, and Chuchinolepidae) and the universal appearance of Antiarchi (*Xichonolepis, Wudinolepis,* and *Bothriolepis*) serve as evidence for drawing the boundary between the Lower and Middle Devonian.

The Middle–Upper Devonian boundary is now drawn in most cases above the *Stringocephalus* or *Nowakia otomeri* zone.

The upper limit of the Devonian for the ammonoid-bearing sequence in South Guizhou, South China region is put at the top of the *Wocklumeria* zone, while in North-east China it is drawn below the *Gattendorfia* zone.

INTERNATIONAL CORRELATION

Correlation of the Devonian system, based on biostratigraphical evidence, between China and other countries, is given in Table 10.3.

Table 10.2. *Correlation of Devonian faunas of South China Region (adapted chiefly from Inst. Geol. Palaeont. Nanjing 1982).*

		Conodonts	Tentaculites	Ammonoides	Brachiopods	Corals
Upper Devonian	Xikuangshanian	Bispathodus costatus Zone		Wocklumeria Zone	Yunnanellina hanburyi Cyrtospirifer sinensis Tenticospirifer triplisinosus	
		Polygnathus styriacus Zone				
		Scaphignathus velifera Zone		Clymenia Zone		
		Palmatolepsis guadrantinodosa Zone				
		P. rhomboidea Zone				
		P. crepida Zone				
	Shetianqiaonian	P. triangularis Zone	Striatostyliolina luofuensis–Metastylialina nahaensis Ass.	Manticoceras cordatum Zone	Hypothyridina cuboides Leiorhynchus kwangsiensis Cyrtospirifer sinensis	Peneckiella minima Pseudozaphrentis difficile Sinodiphyllum variabile
		P. gigas Zone				
		P. proversa Zone		Probloceras applanatum Zone		
		Polygnathus asymmetricus Zone				
Middle Devonian	Donggang-lingian	Schimidtognathus hermani Zone	Viriatellina multicostata Zone		Stringocephalus Fauna	Endophyllum yunanensis Ass. Dendrostella trigemme Ass.
		Polygnathus cristatus Zone	V. minuta Zone			
		P. varcus Zone	Nowakia otomari Zone			
	Nabiaonian (Yintangian)	P. pseudofoliatus Zone	N. guangxiensis Zone	Pinacites jugleri Zone	Acrospirifer fongi Eospiriferina lachrymosa Ass.	Breviseptophyllum kochanensis–Ptaratusia sinensis Ass.
		P.c. costatus Zone	N. sulcata Zone			
		P.c. patulus Zone	N. holynensis Zone		Nadiastrophia–Kwangsia–Euryspirifer Indospirifer Ass.	Psydracophyllum cystosum Ass.
		P. serotinus Zone		Anarcestes (Latanarcestes) noeggerati Zone		Trapezophyllum cystosum Ass.
		P. inversus Zone	N. richiteri Zone			
			N. cancellata Zone			
			N. elegans Zone	Convoluticeras discordans Zone		
Lower Devonian	Tangdingian (Sipaian)	P. perbolus Zone	N. barrandei Zone		Howellella–Reticulariopsis Ass.	Lyrielasma guangxiensis Ass.
			N. praecursor Zone	Erbenoceras elegantulum Zone		
		P. dehiscens Zone			Dicoelostrophia Rostrospirifer Fauna	Heterophrentis angusta–Amplexiphyllum hamiltonia Ass.
			N. subtilis Zone			
	Yujiangian	Eognathothus sulcatus Zone	N. acuaria Zone		Orientospirifer Ass.	Xystriphylloides nobilis Ass.
						Chalcidophyllum nakaolingense Eoglossophyllum minor
	Nagao-lingian	Graptolites Neomonograptus hercynicus Zone Monograptus uniformis Zone				
	Lianhua-shanian					

Table 10.3. *Correlation of Devonian fossil zones in four countries (the 3 right columns adapted from Harland et al. 1982).*

	Stage	China	England	E. Australia	North America	
Upper Devonian	Xikuangshanian	*Bispathodus costatus* Zone	Pilton	Hervy	Bradford	Chautauquan
		Polygnastus styriacus Zone	Baggy		Cassadaga	
		Scaphignasthus velifera Zone	Upcott			
		Palmatolepsis guadrantinodosa Zone	Pickwell			
		P. rhomboidea Zone	Down	?		
		P. crepida Zone				
	Shetianqiaonian	*P. triangularis* Zone	Morte		Cohokton	Senecan
		P. gigas Zone				
		P. proversa Zone				
		Polygnasthus asymmetricus Zone			Finger lakes	
Middle Devonian	Dongganglingian	*Schimidtognasthus hermani–Polygnathus cristatus* Zone	Ilfracombe	Condobolin	Taghanic	Erian
					Tioughnioga	
					Cazenovia	
		P. varcus Zone				
	Nabiaonian	*P. pseudofoliatus* Zone	Hangman			
		P.c. costatus Zone	Lynton	Cunningham	Onesquethaw	
		P.c. patulus Zone				
		P. serotimus Zone				
		P. inversus Zone				
Lower Devonian	Tangdingian	*P. perbois* Zone	Breconian			Ulsterian
		P. dehiscens Zone				
	Yujiangian	*Eognathothus sulcatus* Zone	Dittonian	Merions	Deer park	
	Nagaolingian	*Graptolites* Zone			Helderberg	
		Neomonograptus hercynicus Zone *Monograptus uniformis* Zone	Downtonian	Crudine		
	Lianhuashanian					

11. THE CARBONIFEROUS SYSTEM
Yang Zunyi

INTRODUCTION

The Carboniferous System of China is remarkably well-developed and coal-bearing, reflecting the luxuriant growth of vegetation and the consequent formation of the coal measures. It includes a number of sedimentary associations of different facies: normal neritic carbonates, paralic coal-bearing clastics occasionally with lavas and pyroclastics and/or flysch, and continental deposits. It is marked by a new surge of life (Table 11.2), as well as rich mineral deposits, such as iron, manganese, aluminium, phosphate, fire-clay, and gypsum.

The Carboniferous is well-exposed in both the basin and platformal areas. In the Northern basin region lay the Tianshan–Hingan basin or the Northern basin, in which were deposited mainly basinal neritic clastics and various volcanic rocks. To the south was the Northern landmass including the Sino–Korean platform to the east and the Tarim platform to the west, which are marked chiefly by paralic coal-bearing beds and neritic limestones. Further south was the Kunlun–Qinling region, the Transverse Ranges in West Sichuan and West Yunnan, the Yangzi, and the Zhujiang basins. These included, to the east, the South China region associated with normal neritic carbonates, and to the west, the mobile type of complex deposits. In the North China region the Lower Series is entirely absent.

On the basis of the distribution of the various sedimentary associations and fossil content, nine stratigraphic regions may be differentiated in two categories as follows: (A) stable or platformal type—South China region (8), North China region (4), Qinling–Dabie Mountains region (6), Tarim region (2), South Xizang–West Yunnan region (9), West Sichuan–North Xizang region (7), and Qilian–Holan Mountains region (3); (B) mobile or basinal type—Tianshan–Hingan region (1) and Kunlun–Qaidam region (5) (Fig. 11.1).

The Second All-China Stratigraphy Conference in 1979 adopted two divisions of the Carboniferous, instead of the former three, on the grounds that there is a sharp contrast between the Lower Series and the former Middle and Upper Series in terms of organic evolution (see Table 11.2), lithology (Fig. 11.2), sedimentational environment, and palaeogeography. The Lower Series is rich in brachiopods and corals, while the Upper (former Middle and Upper) Series is marked by fusulinids and ammonoids (*Gastrioceratidae*). In South China the lithology of the Lower Series consists of marls, limestones, fine clastic rocks, and coal-bearing beds, reaching a total thickness of 1000 m, whereas the Upper (former Middle and Upper) Series is made up of homogenous carbonate rocks (limestones and dolomites), less than 1000 m in thickness. In North and North-west China the Lower Carboniferous is either absent or consists of neritic limestones, while the Upper (formerly Middle and Upper) Series contains typically swampy and coal-bearing clastic rocks, reflecting two different types of sedimentary environments. In the Northern basinal region the Lower Series is mostly composed of basin clastics and volcanic deposits, whereas the Upper (former Middle and Upper) Series is of terrestrial, or terrestrial to neritic, deposits. During the Early Carboniferous the seas were more restricted than earlier and they no longer covered all the previous basins of deposition. In contrast, during the Late (former Middle and Late) Carboniferous the relief was so much reduced that marine transgression covered all the basins concerned (Hou, Wang, Wu, Yang et al. 1982b).

The Lower and Upper Carboniferous are respectively called Fengningian (equivalent to the Dinantian of Western Europe and the Mississippian of the USA), and the Hutainian (equivalent to the Moscovian and Uralian combined). The Fengningian is derived from the old name of Dushan (Fengning), where the Lower Carboniferous is exceptionally well-exposed. The Hutainian is named after Hutian, Xiangxiang County, Hunan Province.

The subdivisions of the Carboniferous of China are as follows:

Upper Series (Hutianian):   Mapingian Stage
(Uralian of the Russian
subdivision)
Weiningian Stage
(Moscovian of the
Russian subdivision),
further divisible into the

Lower Series (Fengningian): Dala'an and Huashibanian stages.
Dewuian Stage
Datangian Stage (Visean of Western Europe)
Yanguanian (Aikuan) Stage (Tournaisian of Western Europe)

STABLE SEDIMENTATION TYPE
South China

The stable type is widely distributed in South China (Fig. 11.1.8) and is typically exposed in South Guizhou Province, Central Hunan Province, and Central Guangxi Region, especially around Dushan, Guizhou, where the Early Carboniferous stratotype is located (Fig. 11.2.1; Table 11.1.1).

The Lower Carboniferous Yanguanian Stage is based on the type section near Dushan, Guizhou Province, and consists of the Gelaohe Formation (*sensu stricto*) and the Tangbangou Formation.

The Gelaohe Formation, proposed by V. K. Ting in 1931, has been revised recently on fossil as well as lithological evidence to include only its marly upper part (Gelaohe Formation *sensu stricto*) in the Lower Carboniferous, whereas the lower part of the Gelaohe Formation has been renamed the Zewan Formation with *Siphonodella? praesulcata* and is regarded as belonging most probably to the Upper Devonian, according to Yang Jingzhi and Wang Chengyuan (1983). This lower part (Zewan Formation) yielding stromatoporoids (*Platiferostroma kueichowense, P. sinense*, and *Pseudolabechia huanjiangensis*), tabulates, and brachiopods (*Schuchertella galaohoensis*), is now generally correlated with the well-known Shaodong Formation in Shaodong County, Central Hunan Province. The latter is a unit spanning the Upper Devonian Xikuangshanian and the Lower Carboniferous *Cystophrentis* zone, and is composed of dark argillaceous limestone and shale with a maximum thickness of 63 m, bearing brachiopods (*Cyrtospirifer, Mesoplicata, 'Camarotoechia'*, and *Schuchertella gelaohoensis*), rugose corals (*Caninia*, and *Zaphrentites*), and conodonts (*Polygnathus normalis, P. obliquicostatus, Lonchodina* sp., *Apatognathus* sp. and *Ligonodina* sp.), all of which indicate Devonian rather than Carboniferous age.

The Gelaohe Formation (s.s.) is 95 m thick in Dushan and 289 m in Weining and bears the *Cystophrentis–Plicatifera tenuistriata* Assemblage Zone including *Cystophrentis kolaohoensis, Composita ovata, Schuchertella gueizhouensis, Hunanoproductus hunanensis*, and *Paulonia menggongaoensis*.

The Tangbagou Formation is 172–381 m thick, with the *Pseudouralinia–Martiniella* Assemblage Zone containing *Pseudouralinia tangbagouensis, Siphonophyllia* cf. *caninoides, Neozaphrentis sinensis, Martiniella chinglingensis, Eochoristites neipentaliensis*, and *Spirifer geilingensis*.

The Datangian type section is situated near Baijing, Huishui, Guizhou Province and consists of the Jiusi and Shangsi Formations. The Jiusi in the lower part is sandy shale with a thickness of 484 m and is characterized by two assemblage zones: the lower, *Thysanophyllum shaoyangense–Megachonetes zimmermanni* Assemblage Zone and the upper, *Kueichouphyllum sinense–Vitiliproductus gröberi* Assemblage Zone. The Shangsi Formation in the upper part is composed of 473 m of limestones and is marked by the *Kueichouphyllum heishihkuanense–Delepinea comoides* Assemblage Zone, which contains *Gigantoproductus gigantoides*, and (?) *Balakhonia yunnanensis*.

The Dewuian Stage based on the type section at Dewu, Shuicheng, Guizhou Province, contains dolomites, and limestones intercalated with marl, and is 296 m thick. It is fossiliferous, making up the *Palaeosmilia–Aulina–Condolina–Gigantoproductus* Assemblage Zone. It was in this locality that Carboniferous ammonoids were first discovered, such as *Homoceras* cf. *subglobosum, Proshumardites karpinskyi, Cravenoceras dewuense*, and *Homoceratoides* sp. More ammonoids were later found at Nantan, Guangxi region (Yuan Yiping 1978): *Eumorphoceras* sp., *Praedaraelites* sp., *Stenopronorites* sp., *Epicanites, Delepinoceras eothalossoides, Trizonoceras typicale, Kazakhoceras karobinsi*, and *Cluthroceras* sp.

In the Upper Carboniferous (Hutianian) Series, the following stages, Huashibanian, Dala'an, and Mapingian, are briefly described in ascending order.

The Huashibanian is derived from its type section in Huashiban near Panxian, Guizhou Province, where it consists of thick-bedded limestone intercalated with dolomite totalling 545 m. Among its fossil contents there are fusulinids (*Pseudostaffella antiqua posterior* zone), including *Eostaffella kasakhstanica, E. mosquensis, E. prisca ovoidea*, and *Pseudostaffella composita*, brachiopods (*Neospirifera simplex*, and *Kutorginella genicus*), ammonoids of the *Tectiretites* zone in the lower part including *Tectiretites kueichouensis, Gastrioceras* cf. *cumbriense, Billinguites* sp., *Bashkirites* sp., and the *Gastrioceras–Branneroceras* zone in the upper part, having in its lower portion *Branneroceras reticulatum, B. yohi, Syngastrioceras orientale, Gastrioceras* sp., and *Stenopronorites shuichengensis*.

Table 11.1. *Correlation of Carboniferous sequences in various regions of China.*

Series	Stage	South China S.E. Gueichou (1)	Qinling–Dabie Mts. region S. Gansu Têwo (2)	N. China Shanxi (3)	Tianshan Region Boroxoro Mts. (4)
Overlying		Liangshan Fm (P)	Qixia Fm (P₁)	Shanxi Fm (P₁)	?
Hutianian (Upper Carboniferous)	Mapingian	Maping Fm *Pseudoschwagerina–Schwagerina moelleri Rugosofusulina alpina Triticites pusilus* 47–289 m	Gahai Fm *Quasifusulina longissima Pseudoschwagerina moelleri Tritrcites simplex* 275 m	Taiyuan Fm *Pseudoschwagerina texana, Triticites subnathorsti Quasifusulina longissima Choristites pavlovi Neuropteris ovata* 92–118 m	Keguqinshan Fm 1370 m
	Dalaan	Dala Fm *Fusulinella bocki Profusulinella parva* 121–834 m	Minhe Fm *Fusulina shellwieni Profusulinella parva Pseudostaffella sphaeroidea* 363 m	Benxi Fm *Enteletes hemiplicata Marginifera pusilla Fusulina fortissima* 23 m	Dongtujinghe Fm *Profusulinella* cf. *prisca Pseudostaffella* sp. *Choristites crassicostatus Chaetetes lungtanensis Petalaxis stylaxis* 740–1150 m
	Huashibanian	Huashiban Fm *Pseudostaffella sphaeroidea Kionophyllum dibunum* 545 m			
Fengningian (Lower Carboniferous)	Dewuian	Baizuo Fm *Aulina rotiformis Gigantoproductus edelburgensis Gondolina weiningensis* 296 m	Lueyang Fm *Aulina senex Kueichouphyllum sinense Gigantoproductus giganteus Striatifera striata* 1290 m		Aqalehe Fm *Kueichouphyllum heishihkuanense Yuanophyllum kansuensis Palaeosmilia regia Striatifera striata Gigantoproductus sarsimbaii Megachonetes zimmermanni* 1200–1700 m
	Datangian	Shangsi Fm *Kueichouphyllum heishihkuanense Yuanophyllum kansuensis* 473 m			
		Jiusi Fm *Kueichouphyllum sinense Thysanophyllum shaoyangense Vitiliproductus groberi Megachonetes zimmermanni* 484 m			
	Yanguanian	Tangbagou Fm *Pseudouralinia gigantea Martiniella chinglungensis* 172–381 m	Yiwagou Fm *Pseudouralinia gigantea Ptychomarotoechia kinglingensis Mesoplica* sp. 620 m		Meikaluhe Fm *Dictyoclostus robustus Syringothyris altaica Pseudosyrinx mylkensis Welleria subtrigona* 410 m
?	Shaodong Fm	Gelaohe Fm *Cystophrentis kolaohoensis Schuchertella gueichouensis* 95–102 m			
Underlying		Yiaosuo Fm (D₃)	Tieshan Gr (D₃)	Fengfeng Fm (O)	S

Table 11.1. (cont.)

Tianshan–Hingan region Beishan, Gansu (5)	Kunlun–Qaidam region Burhan Buda (6)	S. Qilianshan Qinghai (7)	N. Xizang W. Sichuan Region Qamdo, Xizang (8)	Himalaya, Xizang (9)
Shuanbotang Fm (P_1)	P_1	Bayinghe Fm P_1	P_1	P_1
Ganquan Fm *Choristites pavlovi Eoasianites* sp. *Agathiceras* sp. 6109 m	Sijiaoyanggou Fm *Schwagerina* sp. *Eoparafusulina* sp. *Rugosofusulina* sp. *Pseudofusulina* sp. *Schubertella* sp. *Parafusulina longa Triticites* sp. 840 m	'Maping Fm' *Pseudoschwagerina moelleri Triticites* sp. ?	Licha Fm *Pseudoschwagerina muongthensis Triticites parvulus T. longus Choristites pavlovi Kepingophyllum polythecaloidea* 365 m	Jilong Fm *Stepanoviella gracilis Lissochonetes geinitzianus Attenuatella convexa Trigonotreta* cf. *narsahensis Empodesma* sp. 731 m
Jijitaizi Fm *Fusulinella provecta Pseudostafella* sp. *Pseudowedekindella prolica Choristites* sp. 1038 m Shibanshan Fm *Choristites* sp. *Lithostrotion subtilisum* 1100 m	Tiaosu Fm *Fusulinella* sp. *Neuropteris gigantea Sphenopteris* sp. *Choristites* cf. *niktiniformis* 319 m	Keluke Fm *Pseudostaffella sphaeroidea Profusulinella primitiva Choristites yanghukouensis Rhodea chinghaiensis* 548 m	Aoqu Fm *Fusulina schellwieni Fusulinella obesa Profusulinella parva Pseudostaffella sphaeroidea* 56 m	— — — ? — — —
Baishan Fm *Koueichouphyllum heishihkuanene Lithostrotion irregulare Thysanophyllum grabaui Gangamophyllum hamiense Gigantoproductus edelburgensis* 2300 m	Dagangou Fm *Gigantoproductus edelburgensis G. latissimus Marginifera* sp. *Arachnolasma* cf. *cylindricum Gigantoproductus giganteus Striatifera striata Lonsdaleia crinata Dibunophyllum bristolensis* 132 m	Huaitoutala Fm *Yuanophyllum kansuense Gigantoproductus giganteus Lithostrotion irregulare Thysanophyllum* sp. *Orionastraea* sp. 1100 m	Machala Fm *Eostaffella galinae E. mosquenensis Palaeosmilia regia Aulina carinata Gondolina weiningensis Kueichouphyllum sinense Gigantoproductus moderatus* 1680 m	Naxing Fm *Streblopteria hemisphaerica Wilkingia nyanangensis Fusella yaliensis Marginirugus* cf. *magnus* 1888 m
Lutiaoshan Fm *Syringothyris altaica S.* cf. *texa* 553–1924 m	Wulonggou Fm *Martiniella elongata Alifera expansa Fusella* sp. 38 m	Chengchiangou Fm *Caninia* cf. *cylindrica C.* cf. *juddi* 282 m	Wuqingna Fm *Praewaagenoconcha kiangsuensis Eochoristites* sp. *Cystophrentis kolaohoensis C. flabelliformis Kueichoupora* sp. 570 m	Upper part of Yali Fm *Imitoceras xizangense I. orientalis Gattendorfia yaliana Tylothyris* cf. *pseudopostera Pseudosyrinx keokuk* 60 m
D_2	D_3	D_3	D_3	Lower part of Yali Fm

Table 11.2. Correlation of Carboniferous faunal sequences in China (modified after Yang Jingzhi et al. 1983; Yang Shifu et al. 1980).

Series	Stage	Formation		Fusulinid zones	Coral assemblages	Brachiopod assemblages	Ammonoid zones	Conodont zones	Plant assemblages
Huatianian (C₂)	Mapingian	Upper		Robustoschwagerina Z. Pseudoschwagerina Z. Triticites Z.	Kepingophyllum kueichouwense Ass. Nephelophyllum hexagonum Ass. Pseudotimania delicata Ass.	Dictyoclostus uralicus–Meekella Ass.	Agathiceras Z.	Streptognathodus elongatus Z. Streptognathodus oppletus–Anchignathodus typicalis Z.	Neuropteris ovata–Lepidodendron posthumii Ass.
		Lower							
	Dalaan	U		Fusulina–Fusulinella Z.	Carinthiaphyllum exquisitum–Kionophyllum ovatum Ass.	Choristites mansuyi–Plicatifera chaoi Ass.	Pseudoparalegoceras tzwetaevae Z.	Streptognathodus parvus–S. suberectus Z. Idiognathodus–Neognathodus bassleri Z. Idiognathoides corrugatus–Polygnathodella ouachitensis Z.	Neuropteris gigantea–Linopteris neuropteroides Ass. Neuropteris gigantea–Mariopteris acuta f. obtusa Ass. Cardiopteridium spitsbergense–Triphyllopteris collombiana Ass. Lepidodendron gaolishanense–Eolepidodendron spp. Ass.
		L		Profusulinella Z.					
	Huashibanian	Upper		Pseudostaffella Z.			Branneroceras peronatum–Reticuloceras (Panxianoceras) Z.		
		Lower		Eostaffella Z.	Palaeosmilia regia–Aulina rotiformis Ass.	Gondolina weiningensis–Gigantoproductus edelburgensis Ass.	Homoceras cf. subglobosum Z. Proshumardites Cravenoceras dewuense Z.	Gnathodus bilineatus–G. nodosus Z. (All the above for NW China)	
Fengningian (C₁)	Dewuan	Baizuo			Kueichouphyllum heishikuanense Ass.	Datangia weiningensis–Delepinea comoides Ass.			
	Datangian	Shangsi			Kueichouphyllum sinense–Thysanophyllum shaoyangense Ass.	Vitilproductus groberi–Megachonetes zimmermanni Ass.			
		Jiusi							
	Yanguanian	Tangbagou			Pseudouralinia gigantea Ass.	Fusella shaoyangensis Ass. Martiniella chinglungensis Eochoristites chui Ass.		Polygnatus bischoffi–Pseudopolygnathus Z. Neoprioniodus barbatus–Hindeodella subtilis Z. Siphonodella duplicata–Polygnathus bischoffi Z.	
		Gelaohe			Cystophrentis kolaohoensis Ass.	Schuchertella–Yanguania–Sphenospira–Mesoplica Ass.	Gattendorfia–Eocanites Z.		
D		Shaodong			Caninia dorlodoti	Same as the above (Gelaohe)			

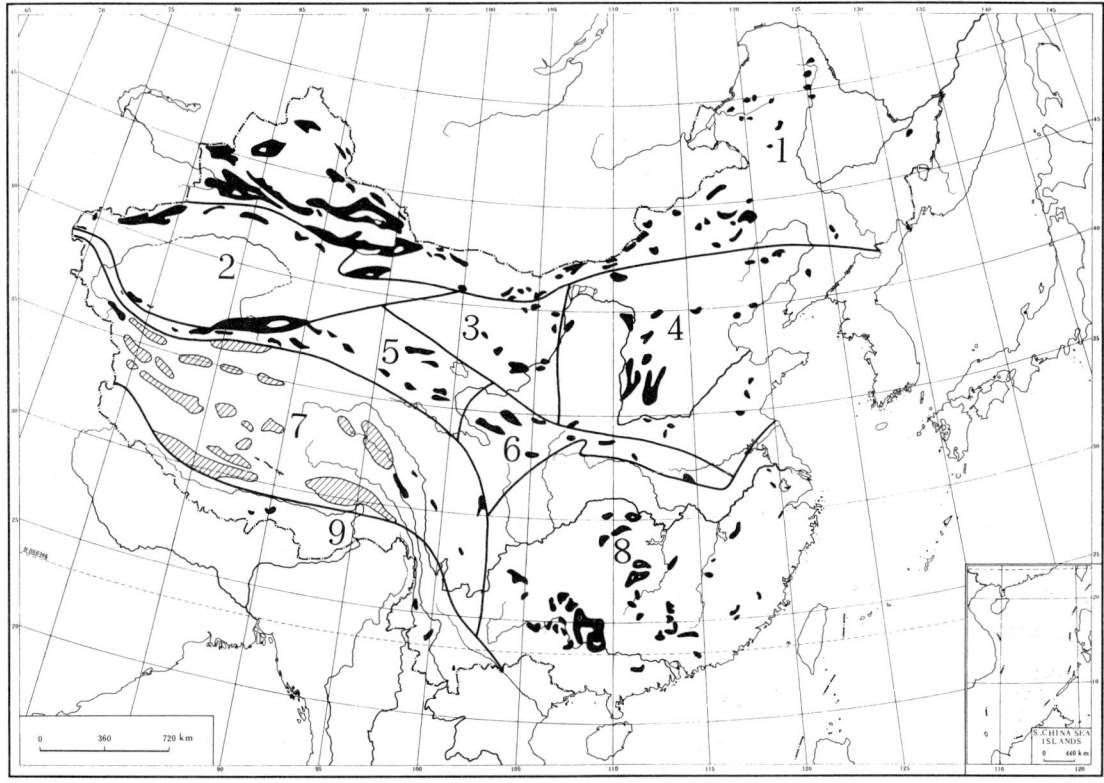

Fig. 11.1 Sketch map showing the distribution and stratigraphic regions of the Carboniferous System in China. 1, Tianshan–Hingan region; 2, Tarim region; 3, Qilian–Helan region; 4, North China region; 5, Kunlun–Qaidam region; 6, Qinling–Dabei region; 7, North Xizang–West Sichuan region; 8, South China region; 9, Qomolangma–West Yunnan region.

The Dala'an Stage is represented by the type section in Dala village, 30 km north-east of Pangxian, Guizhou Province, and made up of crystalline limestone and biogenetic limestone, which is between 121 m and 340 m thick, and yields mainly fusulinids: the lower zone is the *Profusulinella* zone with notable fossils, such as *Taitzehoella* sp., *Profusulinella prisca*, and *Aljutovella succincta*; and the upper zone, the *Fusulina–Fusulinella* zone with *Fusulina cylindrica*, *F. quasicylindrica*, *Fusulinella paracaloninae*, *F. simplicata*, and *F. bocki*. There are brachiopods (*'Muirwoodia' sinensis*, and *Choristites mansuyi*), and ammonoids (*Neodimorphoceras* sp., *Eoparalegoceras* sp., and *Pseudoparalegoceras* sp.).

The Mapingian Stage is named after the Maping Limestone in Liuzhou (formerly Maping) County, but the type section is now at Zhaojiashan, Weining, Guizhou Province. This limestone is light-coloured and 96–126 m thick (reaching 837 m in Shuicheng), containing three fusulinid zones in ascending order: (1) the *Montiparus* zone characterized by a rich occurrence of a single species, *M. weiningensis*; (2) the *Triticites* zone, including *Triticites chui*, *T. chinensis*, *T. simplex*, *Quasifusulina phaselus*, and *Ozawainella praestella*, and corals (*Pseudotimania sinensis*, *Caninia trinkler*, *Antheria abnormis*, and *A. polygonalis*); (3) the *Pseudoschwagerina–Zellia* zone, comprising *Staffella pseudosphaeroidea*, *S. leei*, *Zellia media*, *Z. magnae-sphaerae*, *Pseudoschwagerina moelleri*, and *P. uddeni*, and corals (*Nephelophyllum*, *N. simplex*, *Kepingophyllum weiningensis*, and *K. irregulare*).

Qinling (Têwo, Gansu Province)

The stable Carboniferous sedimentation type is seen also in Têwo, Gansu Province in the Qinling–Dabie Mountains region (Fig. 11.1.6; Table 11.1.2), where the Lower Carboniferous Yanguanian is represented by

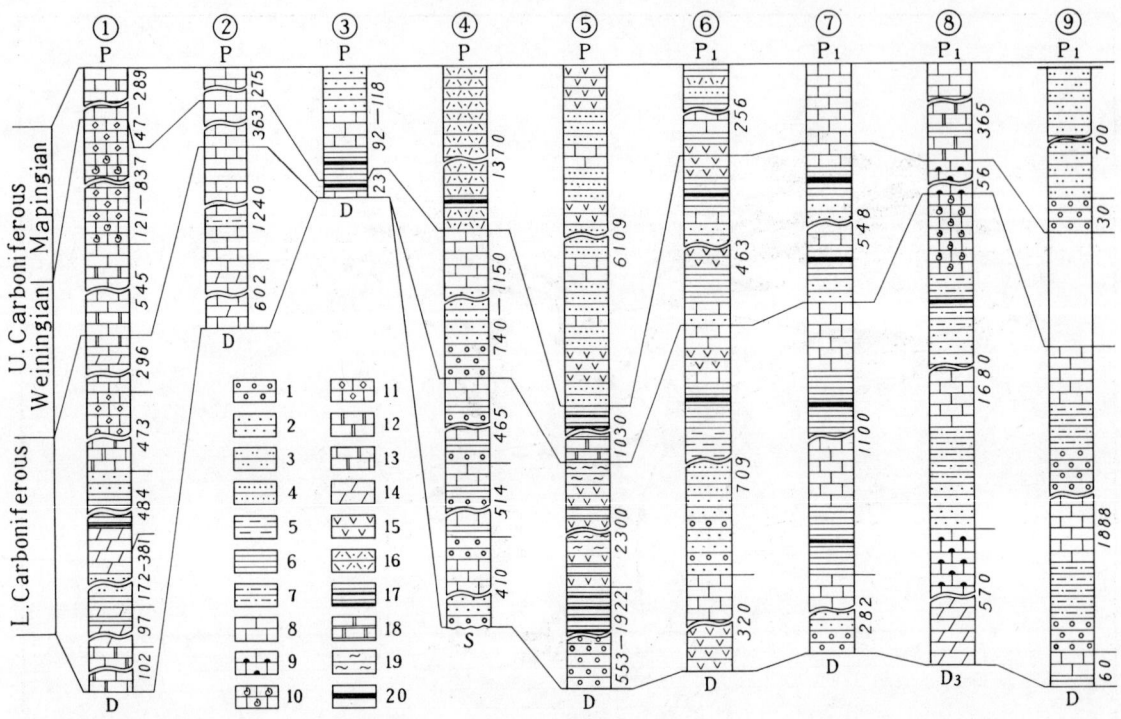

Fig. 11.2. Columnar sections showing development of Carboniferous rocks in various stratigraphic regions of China. 1 Southeast Guizhou, South China region; 2 South Gansu, Qinling–Dabei region; 3 Taiyuan, Shanxi, North China region; 4 Borohoro Mt., Tianshan–Hingan region; 5 Beishan, Helanshan region; 6 Burhan Buda, Kunlun–Qaidam region; 7 South slope of Qilian Mt., Qinghai Province; 8 Qamdo, North Xizang–West Sichuan region; 9 Mt. Qomolangma, Qomolangma–West Yunnan region. Legend: 1, Conglomerate; 2, Sandstone; 3, Quartz sandstone; 4, Siltstone; 5, Mudstone; 6, Shale; 7, Sandy shale; 8, Limestone; 9, Cherty limestone; 10, Biogenetic limestone; 11, Crystalline limestone; 12, Dolomite; 13, Dolomitic limestone; 14, Marl; 15, Volcanic rocks; 16, Tuff; 17, Slate; 18, Marble; 19, Phyllite; 20, Coal.

the Yiwagou Formation. This formation is composed of grey limestone intercalated with argillaceous limestone, which is 602 m thick and bears corals (*Pseudouralinia gigantea*, and *P. tangpakouensis concava*), and brachiopods (*Ptychomarotoechia kinlingensis*, and *Mesoplica* sp.), which are also typical of the Tangbagou Formation in Guizhou.

The Datangian Lueyang Formation consists of grey limestone, and oolitic limestone intercalated with sandy shale, and reaches a thickness of 1290 m. It is rich in corals (*Kueichouphyllum sinense*, and *Aulina* cf. *senex*), and brachiopods (*Gigantoproductus giganteus*, and *Striatifera* cf. *striata*), equivalent to the Juisi Formation of Guizhou Province.

The Upper Carboniferous Weiningian of the Minhe Formation consists of 363 m of grey limestone. It yields fusulinids such as *Pseudostaffella sphaeroidea*, *Fusulina schellwieni*, and *Profusulinella parva*, and corals (*Cani-*

nia sp.), which range from the Huashipanian to the Dala'an stages.

The uppermost unit, the Gahai Formation, is composed of grey limestone which is 275 m thick and carries typical Mapingian fusulinids, such as *Quasifusulina longissima*, *Pseudoschwagerina moelleri*, and *Triticites simplex*.

North China region

In Taiyuan, Shanxi, North China region (Fig. 11.1.4) the Lower Carboniferous is absent and the Upper Carboniferous is represented by the Benxi (Penchi) and Taiyuan formations. The Benxi Formation consists of dark grey shale, sandy shale intercalated with limestone, and coal seams, and reaches a thickness of 23 m. It contains abundant fusulinids (*Fusulina fortissima*), and brachiopods (*Enteletes hemiplicata*, and *Margini-*

fera pusilla). Overlying these deposits conformably is the Taiyuan Formation which consists of black shale, sandy shale, greyish sandstone and limestone, and coal seams. It is 92–118 m thick, and bears fusulinids (*Pseudoschwagerina texana, Triticites subnathorsti*, and *Quasifusulina longissima*), brachiopods (*Choristites pavlovi*), and plants (*Neuropteris ovata*).

Tarim region

In the Tarim region the Carboniferous is composed chiefly of carbonates intercalated with clastic rocks and thin coal seams, locally with volcanic rocks. Lacustrine gypsum and mudstone are found in the western part of North Tarim where the Lower Carboniferous, and the lower part of the Upper Carboniferous are locally lacking. Unlike the Tianshan–Hingan region, neither typical elements of Western Europe nor those of North America are known and no Angara flora has been reported.

South Qilian region

In Yangkang, South Qilian region, the Carboniferous is predominantly of paralic to marine deposits and richly fossiliferous (Table 11.1.7). The Lower Carboniferous Chengqiangou Formation begins with siltstone, sandstone, mudstone, and argillaceous dolomite and passes into siltstone, calcareous sandstone intercalated with arkose, shale and marl, and thin-bedded limestone, to a thickness of 282 m. It yields corals (*Canina* cf. *cylindrica, C.* cf. *juddi, Zaphrentis* sp., and *Lophophyllidium* sp.), and brachiopods (*Ptychomarotoechia kinlingensis*), indicating that this formation can be correlated with the Tangbagou Formation of Guizhou Province.

The Lower Carboniferous Huaitoutala Formation is made up of limestone, shale, and coal seams, and reaches a thickness of 1100 m. It contains many corals (*Yuanophyllum kansuense, Lithostrotion irregulare, Thysanophyllum* sp., and *Orionastraea* sp.), and brachiopods (*Gigantoproductus giganteus*), which indicate the Datangian Stage (Visean).

The Keluke Formation of the Upper Carboniferous is composed of limestone, shale, and sandstone with coal seams, and is 548 m thick. It contains fusulinids (*Pseudostaffella sphaeroidea*, and *Profusulinella primitiva*) belonging to the Weiningian Stage.

The uppermost unit of the Upper Carboniferous is composed of limestone with fusulinids (*Pseudoschwagerina moelleri* and *Triticites* spp.), is of unknown thickness, and belongs certainly to the Mapingian Stage.

Qilian–Helan region

In the Qilian–Helan Mountains region which includes Helan Mountain, Longshou Mountain, Liupan Mountain, and the Central and Southern Qilian Mountains, the Lower Carboniferous is incompletely developed and marked by lagoonal–littoral coal-bearing series.

Qamdo, West Sichuan–North Xizang region

In Qamdo, West Sichuan–North Xizang region (Fig. 11.1.7; Fig. 11.2) the Carboniferous includes four formations, in ascending order, the Wuqingna, Machala, Aoqu, and Licha. The Lower Carboniferous Wuqingna Formation of limestone, and marl with chert-bands, is 570 m thick and yields corals (*Cystophrentis kolaohoensis, C. flabelliformis*, and *Kueichoupora* sp.) and brachiopods (*Eochoristites* sp., and *Praewaagenoconcha kiangsuensis*), all of which indicate that this formation is equivalent to the Gelaohe Formation of the Yanguanian Stage. The Machala Formation, composed of bioclastic limestone and sandy shale, sandstone, and coal seams, reaches a thickness of 1680 m. It yields *Eostaffella galinae* (fusulinid), *Aulina carinata* (coral), and *Gondolina weiningensis* (brachiopod) of the Dequ'an Stage and *Kueichouphyllum sinense* (coral) of the Datangian Stage, both belonging to the Weiningian. The Upper Carboniferous Aoqu Formation is a much thinner unit, 56 m thick, and is mainly limestone with chert bands. It contains the Dala'an fusulinids, such as *Profusulinella parva schellwieni, Fusulina*, and *Fusulinella obesa*. The uppermost Carboniferous unit, the Licha Formation, consists of limestone, and dolomitic limestone intercalated with shale, and is 365 m thick. It is characterized by the two lower fusulinid zones of the Mapingian Stage.

Qomolangma area

In the Qomolangma area the Carboniferous rocks are represented by three formations (Fig. 11.1.9; Table 11.1.9). The upper part of the Yali Formation is limestone and shale with a thickness of 60 m and yields ammonoids (*Imitoceras xizangense, I. orientale*, and *Gattendorfia yaliana*), and brachiopods (*Tylothyris* cf. *pseudopostera*, and *Pseudosyrinx keokuk*), all of which indicate the Lower Yanguanian Stage. The Naxing Formation which is composed of conglomerate, sandy shale, and limestone, totalling 1888 m, is divisible into three on fossil evidence. The lower part contains brachiopods (*Schuchertella* cf. *gueizhouensis, Fusella yaliensis*, and *Composita tibetana*) and bivalves (*Streblopteria hemispherica*, and *Wilkingia nyanangensis*), equivalent to the Tournaisian. The middle part is rich in brachiopods (*Tolmatchaffia* sp., *Marginifera* cf. *magnus*, and *Syringothyris lydekkeri*) and bivalves (*Aviculopecten* sp.), most probably being Lower Visean. The upper part bears numerous bivalves (*Sanguinolites*, *Wilkingia*, and *Clinopiestha*) and some gastropods, neither of which indicate the exact age of the deposits, but an upper Carboniferous age is probable because of their post-Visean and pre-Permian stratigraphic position. Lastly, the Jilong Formation consists of (in ascending order) variegated conglomerate (30 m), siltstone (1 m), and Chaya quartz sandstone (700 m), and yields *Stepanoviella gracilis*, *Lissochonetes* cf. *geinitzianus*, *Attenuatella convexa*, *Trigonotreta* cf. *narsahensis*, and *Empodesma* sp., which show that the Jilong Formation belongs to the Mapingian Stage.

MOBILE SEDIMENTATION TYPE

Kunlun–Qaidam

This type is best shown in the Tianshan–Hingan and the Kunlun–Qaidam regions. In the Borohoro Mountains the Carboniferous reaches a maximum thickness of 4630 m, consisting of the following four formations in ascending order.

The Lower Carboniferous Melukahe Formation is composed of reddish-brown conglomeratic coarse sandstone and limestone, reaching a thickness of 410 m and bearing brachiopods (*Dictyoclostus robustus*, *Syringothyris altaica*, *Pseudosyrinx mylkensis*, and *Welleria subtrigona*) and corals (*Suqiyamaella*), hence belonging to the Yanguanian Stage.

The Akalehe Formation, made up of limestone, shale and sandstone, is 1200–1700 m thick. It is richly fossiliferous, with corals (*Palaeosmilia regia*, *Kueichouphyllum heishihkuanense*, *Yuanophyllum kansuense*, and *Dibunophyllum turbinatum*) and brachiopods (*Fusella ashalaensis*, *Gigantoproductus sarsimbaii*, *Striatifera striata*, and *Megachonetes zimmermanni*), all of which are Datangian to Dewu'an elements.

The Upper Carboniferous Dongtujinghe Formation is composed of limestone, sandstone, conglomerate, and limestone, between 740 and 1150 m thick. It contains numerous fusulinids (*Profusulinella* cf. *prisca*, and *Pseudostaffella* sp.), brachiopods (*Choristites crassicostatus*), and corals (*Chaetetes lungtanensis*, *Petaxis*, and *Stylaxis*), which indicate the Weiningian Stage.

The uppermost Carboniferous Keguqinshan Group is composed of clastic rocks and limestone intercalated with volcanic rocks, and is divisible into two subgroups. The lower subgroup consists of coarse clastic rocks intercalated with volcanic rocks; it is 970 m thick and unfossiliferous. The upper subgroup is composed of clastic rocks and limestone which attain a thickness of 400 m and yields brachiopods (*Choristites* ex gr. *trautscholdi*, *Marginifera* cf. *pusilla*, and *M.* cf. *orientalis*), corals (*Campophyllum schrenki*, and *Caninia* sp.), and fusulinids (*Schwagerina princeps*), which belong to the Mapingian Stage.

Beishan, Gansu

In Beishan Mountain, Gansu (Table 11.1.5, Fig. 11.1, Fig. 11.2.5), the Lower Carboniferous Lutiaoshan Formation consists of a sandstone member in the lower part and a slate member in the upper, totalling 553–1922 m of deposits, and yielding *Syringothyris altaica*, *S.* cf. *texta*, equivalent to the Yanguanian Stage. The Lower Carboniferous Baishan Formation of phyllite and volcanic rocks intercalated with marble, reaches a thickness of more than 2300 m. This contains corals (*Kueichouphyllum heishihkuanense*, *Lithostrotion irregulare*, *Thysanophyllum grabaui*, and *Gangamophyllum hamiensis*) and brachiopods (*Gigantoproductus edelburgensis*), which are identifiers of the Dewu'an Stage.

The Upper Carboniferous Shipanshan Formation consists of clastic rocks and limestone and is over 1100 m thick. It contains brachiopods (*Choristites* sp.), and corals (*Lithostrotion* cf. *subtilisum*), apparently correlatable to the Huashibanian (Lower Weiningian) Stage.

The Upper Carboniferous Jijitaizi Formation composed chiefly of slates and marbles is 1030 m thick and yields fusulinids (*Fusulinella provecta*, *Pseudostaffella* sp., *Pseudowedekindella prolixa*), and brachiopods (*Choristites gobicus*) of the Upper Weiningian Stage.

The Ganquan Formation, the uppermost Carboniferous unit, consists of volcanic rocks, and siltstone intercalated with limestone, reaching a thickness of 6109 m. It is marked by the presence of brachiopods (*Choristites pavlovi*) and ammonoids (*Eoasianites* sp., and *Agathiceras* sp.), which correlate this formation with the Mapingian.

Burhan Buda, Kunlun–Qaidam region

The third mobile-type area of sedimentation is shown by the Burhan Buda section in the Kunlun–Qaidam region (Fig. 11.1.6; Fig. 11.2.6; Table 11.1.6), where the Carboniferous is represented by four formations, in ascending order as follows.

The Lower Carboniferous Wulonggou Formation, composed of quartz sandstone, shale, and limestone, is 38 m thick and yields brachiopods (*Martiniella elongata*, *Overtonia elegans*, *Alifera expansa*, and *Fusella* cf.

metatrigonalis). It can be correlated with the Yanguanian Stage.

The Dagangou Formation, which follows, comprises dark grey limestone, quartz sandstone, crystalline limestone, biogenetic limestone, and argillaceous limestone, forming three cyclothems, and is 132 m thick. It contains brachiopods (*Gigantoproductus latissimus, G.* ex. gr. *edelburgensis, Rhipidomella* cf. *michelini*, and *Marginifera* sp.) and corals (*Lithostrotion* cf. *portlocki, Lonsdaleia crassicona*, and *Arachnolasma* cf. *cylindricum*), all of which are equivalent to the Upper Datangian and Dewu'an stages.

The third is the Tiaosu Formation, which is made up of dark grey massive limestone and grey thick-bedded conglomeratic quartz sandstone, 319 m thick, bearing fusulinids in limestones, such as *Fusulinella* sp., and fossil plants such as *Neuropteris gigantea, Cordaites* cf. *principalis*, and *Sphenopteris* sp.

The Sijiaoyanggou Formation at the top of the sequence is mainly grey thick-bedded limestone, with sandstone and intermediate tuff in subordinate amounts, yielding fusulinids (*Eoparafusulina* sp., *Schwagerina* sp., *Rugosofusulina* sp., *Pseudofusulina* sp., *Schubertella* sp., and *Montiparus* sp.), corals (*Bothrophyllum* sp., and *Arachnastraea manchurica longiseptata*) and brachiopods (*Phricodothyris* sp., *Uncinunellina* sp., *Camarotoechia* sp., and *Neospirifer* sp.). This is equivalent to the Mapingian Stage.

BIOGEOGRAPHIC PROVINCES IN CARBONIFEROUS TIMES

Consideration of the foregoing biostratigraphic data suggests the existence of three biogeographic provinces in Carboniferous times in China, namely, South China, the Northern basin, and West China.

South China province

The South China province is characterized by benthonic life of Indo-Pacific origin with strong regional features. During the Early Fengningian both corals (*Cystophrentis* and *Pseudouralinia*) and brachiopods (*Yanguania, Eochoristites*, and *Martiniella*) are typically Chinese forms, which spread widely over South China—in Yunnan, Guizhou, Guangxi, Hunan, Guangdong, and the border area of Zhejiang and Anhui Provinces. Different individual species have been found in Qinling, West Yunnan, South-east Xinjiang, and as far as the Urals of the USSR, but they are absent from the Northern basin region. In Late Fengningian both corals (*Kueichouphyllum, Heterocaninia, Arachnolasma, Yuanophyllum*) and brachiopods (*Vitiliproductus, Balahonia, Kansuella, Lochengia,* and *Gondolina*) still reflect their strongly regional features and are distributed all over South China. As the late Lower Carboniferous transgression proceeded further, individual species of these genera spread to Japan, Malaysia, Laos, the Pamirs, Iran, Armenia, and Kazakhstan (through Tianshan) and the Kuznetsk Basin, USSR, thus constituting the typical East Tethysan faunas or 'Asiatic faunas'. The South China faunas contain a few elements of West European and North American faunas only at the height of the transgression.

Northern basin province

The Northern basin province has a profusion of ammonoids and brachiopods, but corals are not numerous. Here, the early Lower Carboniferous faunas are closely akin to those of North America and Siberia, while in the western part of the region the faunas are mixed with some Central Asian and West European elements. The main brachiopods are *Tolmatchoffia, Syringothyris,* and *Rotaia,* and the corals are represented by *Enygmophyllum, Zaphriphyllum, Sugiyamaella,* and *Siphonophyllia*. In the Late Carboniferous, as a result of regression occurring in the Northern basin, the marine connection with the North American province was severed; but the Northern seas remained in communication with the west and south, so that in Tianshan there are West European and Fergana elements as well as those of the South China province. With the onset of the Late Carboniferous transgression both the Northern and Southern regions were further submerged, so that the faunal characters are closely related to those of the Urals, the Russian Platform and the Palaeotethys.

West China province

The West China province is characterized by mixed faunas, i.e. both corals and brachiopods found in South China and the Northern basin are known to have flourished side by side here.

In the Qomolangma area, South Xizang, there are colossal Carboniferous detrital deposits which bear faunas with Gondwana affinities. It is interesting to note that here the brachiopods are similar to those of Tianshan and Australia, and that Gondwana-type faunas appeared in the Late Carboniferous.

BOUNDARY PROBLEMS

Devonian–Carboniferous boundary

The boundary between the Devonian and Carboniferous has been under discussion for some years. The traditional boundary is placed below the Shaodong Member or Formation in Hunan and the Gelaohe Member or Formation in Guizhou on the basis of macro-fossils, especially corals and brachiopods. In the late nineteen-seventies evidence from the study of conodonts (Wang Chengyuan, 1978), foraminifers (Wang Keliang), and ammonoids (Yuan Yiping 1978) suggested that the boundary should be moved upwards to the top of the Shaodong Formation and the base of the Upper Gelaohe (Gelache Formation *sensu stricto*). Further study of this problem by Yang Jingzi and Wang Chengyuan (1983) suggests that the Shaodong Member, Gelaohe Formation *sensu stricto* and Wangyu Formation are equivalent to the Etroeungt of Western Europe (see Table 11.4) and the Devonian–Carboniferous boundary is placed above it. This definition of the boundary is used here.

Lower–Upper Carboniferous boundary and Carboniferous–Permian boundary

There is no problem about the boundary between the Lower and Upper Series, and it is placed definitely between the Fengningian and Hutianian. Whether the Carboniferous in China should be divided into two or three is still open to question, but the former is favoured. Another problem is the determination of the boundary between the Carboniferous and the Permian. The conventional practice in China is to place the boundary at the top of the Asselian or Wolfcampian (at approximately 280 Ma), whereas in other countries the same boundary is drawn at the base of the Asselian. It should be noted that Carboniferous boundary problems have been fully treated by Yang Jingshi *et al.* 1982) and Hou Hongfei *et al.* (1982b).

INTERNATIONAL CORRELATION OF CARBONIFEROUS ROCKS

An attempt to correlate the Carboniferous subdivisions of China with those of the British Isles, the Soviet Union, and the United States of America is given in Table 11.3.

Table 11.3. *Correlation of Carboniferous subdivisions of China and other countries.*

China			British Isles	USSR		USA	
Series	Stage	Fossil Ass. or Zone					
Hutianian (Upper Carb.)	Mapingian	*Robusto-schwagerina* Z. *Pseudo-schwagerina* Z. *Triticites* Z.	Cantabrian	Asselian	Upper Carb.	Wolfcampian	P
				Gzelian		Virgilian	
				Kasimovian		Missourian	Pennsylvanian
	Dala'an	*Fusulina–Fusulinella* Z. *Profusulinella* Z.	Westphalian	Moscovian	Middle Carb.	Desmoinesian Atokan	
	Huashi-banian	*Pseudostaffella* Z.	Yeadonian	Bashkirian		Morrowan	
			Marsdenian				
			Kinderscoutian				
Fengningian (Lower Carb.)	Dewuian	*Eostaffella* Z. *Gondolina weiningensis–Gigantoproductus edelburgensis* Ass.	Alportian	Serpukhovian		Chesterian	Mississippian
			Chokierian				
			Arnsbergian				
			Pendleian				
	Datangian	*Kueichouphyllum heishikuanensis–Yuanophyllum* Ass. *Kueichouphyllum sinensis* Ass. *Thysanophyllum shaoyangensis* Ass.	Brigantian	Visean	Lower Carb.	Meramecian	
			Asbian				
			Holkerian				
			Arundian				
			Chadian				
	Yanguanian	*Pseudouralinia gigantea* Ass. *Cystophrentis kolaohoensis* Ass.	Courceyan	Tournaisian		Osagean	
						Kinderhookian	
?	Shaodong Fm	*Caninia dorlodoti Ceriphyllum elegantum* Ass.	Famennian (D₃)			Louisiana (D₃)	

12. THE PERMIAN SYSTEM
Yang Zunyi

INTRODUCTION

In China the Early Permian is similar to the Late Carboniferous in such features as the extent of marine transgression, and the basic tectonic frame and palaeogeographical outline. The North China and Tarim platforms stood as prominent tectonic units that separated the Northern basins and the Northern biogeographic province from the Palaeotethys, with its typical faunas, to the south. South of the Yarlung Zangbo River in the Himalaya region the *Glossopteris* flora and the cold-water fauna (*Stepanoviella*) appeared, which indicate the link with the Gondwanan forms.

By the end of the Early Permian, crustal movements brought about the folding and uplifting of the Northern basins. In Late Permian times the Qinling–Kunlan mountains marked the boundary of the land in the north and the seas in the south, a situation which persisted until the Triassic.

Permian rocks are widely distributed in China (see Fig. 12.1) and are characterized by a considerable range of deposits and a profusion of biotas, both marine and continental (Fig. 12.3). These characteristics arise from the variety in the Permian tectonic frame and palaeo-

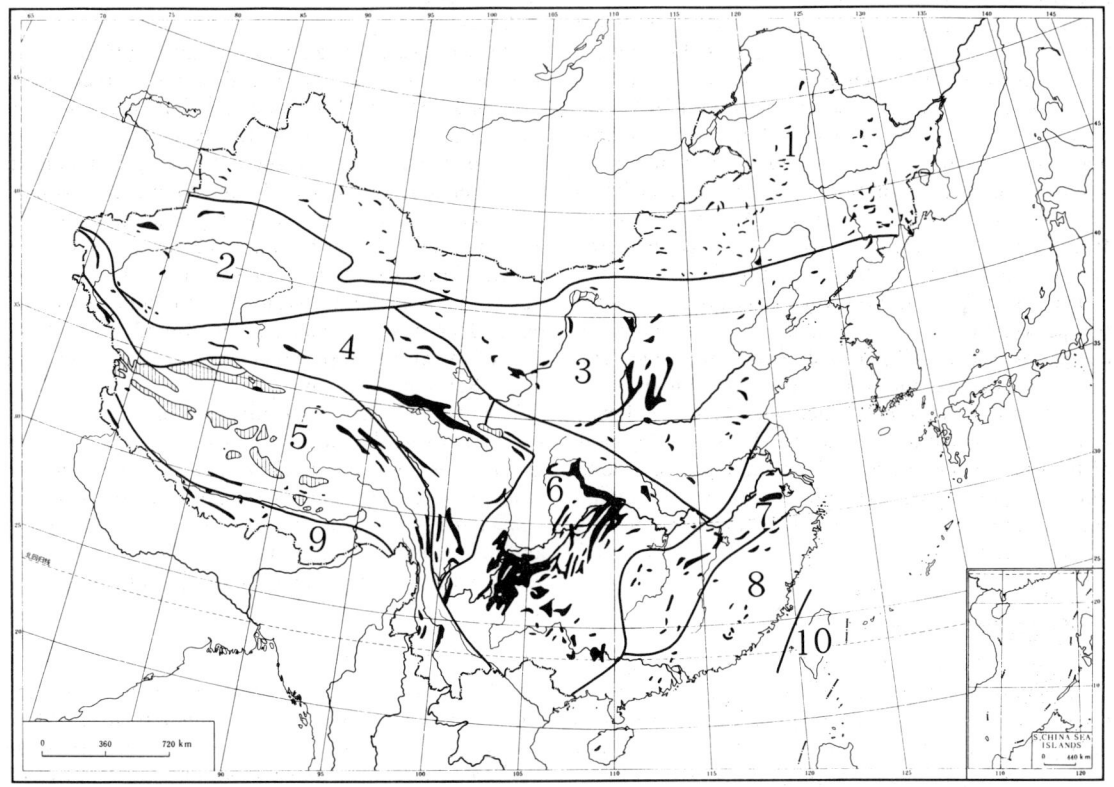

Fig. 12.1 Sketch map showing the distribution and stratigraphic regions of the Permian in China. 1, Tianshan–Hingan region; 2, Tarim region; 3, North China region; 4, Kunlun–Bayan Har region; 5, Xizang–West Yunnan region; 6, Qinling–Yangzi region; 7, Jiangnan region; 8, South-east region; 9, Himalayan region; 10, Taiwan region.

biogeography. On the basis of these factors ten stratigraphic regions as proposed by Zhan and his Permian research group (Zhan et al. 1982) (Fig. 12.1) are followed here, namely (A) the platformal sedimentation type including the Qinling–Yangzi region (6), Jiangnan region (7), South-east region (8), North China region (3), Tarim region (2) and Himalaya region (9); and (B) the basin sedimentation type which includes the Tianshan–Hingan region (1), Kunlun–Bayan Har region (4), Xizang–West Yunnan region (5), and Taiwan region (10).

The Permian rocks of China also contain mineral resources including coal, manganese, phosphates, bauxite, copper, and iron, as well as oil and gas. The Permian coal measures of both North and South China are well-known and deserve our special attention.

PERMIAN CHRONOSTRATIGRAPHIC UNITS OF CHINA

Ever since the study by Huang (1932), and Sheng and Lee (1959) a two-fold division of the marine Permian System of China has been adopted, i.e. the Lower Series (Yangsinian) and the Upper Series (Lopingian). Each of these is now sub-divided by Sheng et al. (1982) into two stages, namely the Qixian (Chihsian) (P_1^1), Maokouan (P_1^2), Wujiapingian (P_2^1), and Changxingian (P_2^2). Zhan et al. (1982), however, proposed to make the Longyin Formation of South-west Guizhou a first Permian stage preceding the Qixian (Chihsian). Reasons for accepting this proposal will be discussed later in connection with boundary problems.

The Lower Permian Longyin Formation, as typified by the section in Longyin, Pu'an, Guizhou Province, comprises 800 m of yellowish silty mudstone, calcareous mudstone, mudstone that is occasionally intercalated with white quartz sandstone, dark-grey limestone, and argillaceous limestone. It is lithologically distinct and easily separable from the Upper Carboniferous Maping Limestone, and merges into the Liangshan member of the Qixian Formation without a distinct break.

Lower Permian Qixian

The Lower Permian Qixian Stage, typified by the section in the Qixia Hills, east of the city of Nanjing, consists of dark-grey, asphaltic limestone, siliceous rocks, and nodular limestone, attaining a thickness of 161 m. It is very rich in fusulinids, corals, brachiopods, bryozoans, and ostracods. Three numbered lithological members with their fossils are placed as in the following list.

Overlying beds: upper siliceous beds of the Gufeng Formation yielding *Parafusulina multiseptata*.
Conformity.
(1) Limestone Member: rich in fusulinids (*Nankinella, Schwagerina*) *S. chihsiaensis* var. *brevis* (Chen), *S. chihsiaensis* var. *regularis* (Chen), *S. pseudochihsiaensis* (Chen), *Yangchienia iniqua* (Lee), *Misellina* cf. *claudiae* (Deprat)); corals (*Hayasakaia elegantula* (Yabe and Hayasaka), *Polythecalis yangtzeensis* (Huang), *P. chinensis* (Girty), *Tetraporinus hanshanensis* (Zhao and Chen), *Protomichellinia* sp.; *Tetraporinus* sp.; brachiopods (*Monticulifera sinensis* (Frech), *Tyloplecta nankingensis* (Frech); and calcareous algae (*Sinoporella* sp.).
(2) Lower siliceous Member: sparsely fossiliferous, yielding only ostracods (*Amphissites* sp., *Kirkbya* sp., *Bairdia* spp.) and brachiopods (*Marginifera obscura* (Chao)).
(3) Swine limestone Member: rich in fusulinids (*Misellina claudiae, Schwagerina chihsiaensis regularis* (Chen)), corals (*Wentzellophyllum volzi*), and brachiopods (*Acrasarina indica* (Waagen), *Orthotichia chekiangensis* (Chao)).
Paraconformity.
Underlying beds: Chuanshan Formation (Mapingian). The Qixian Stage is lithologically simple, either purely carbonate rocks or clastic and carbonate rocks. Carbonate rocks predominate in the type locality and in Wanmo in Southern Guizhou, Desheng in Guangxi, and Zhengan in Shaanxi. In the Permian section at Liangshan Mountain at Hanzhong, Shaanxi, Huanyin Mountain in Sichuan, Landai, and Zunyi in Guizhou, the Qixian consists of the Liangshan coal-bearing clastics at the base and limestone in the middle and upper parts. This type of deposit is distributed over a large area of the South-west and Jiangnan regions, and part of the South-east region.

The fusulinid *Mishellina claudiae* zone is so extensively developed that it serves as the basal unit of the Qixian. The coral *Wentzellophyllum volzi* zone and the brachiopod *Orthotichia chekiangensis–Tyloplecta richthofeni* zone also form the basal unit. The disappearance of *Misellina claudiae* and the appearance of *Cancellina* mark the upper limit of the Qixian.

Lower Permian Maokouan

The Lower Permian Maokouan Stage, based on the type section at Bali, Langdai, Qinglong in Guizhou Province, is composed of greyish and greyish-white thick-bedded and massive limestone, and bioclastic limestone. The section as a whole is characterized by dolomitic patches. It is richly fossiliferous and fusulinids are dominant, with corals and brachiopods being

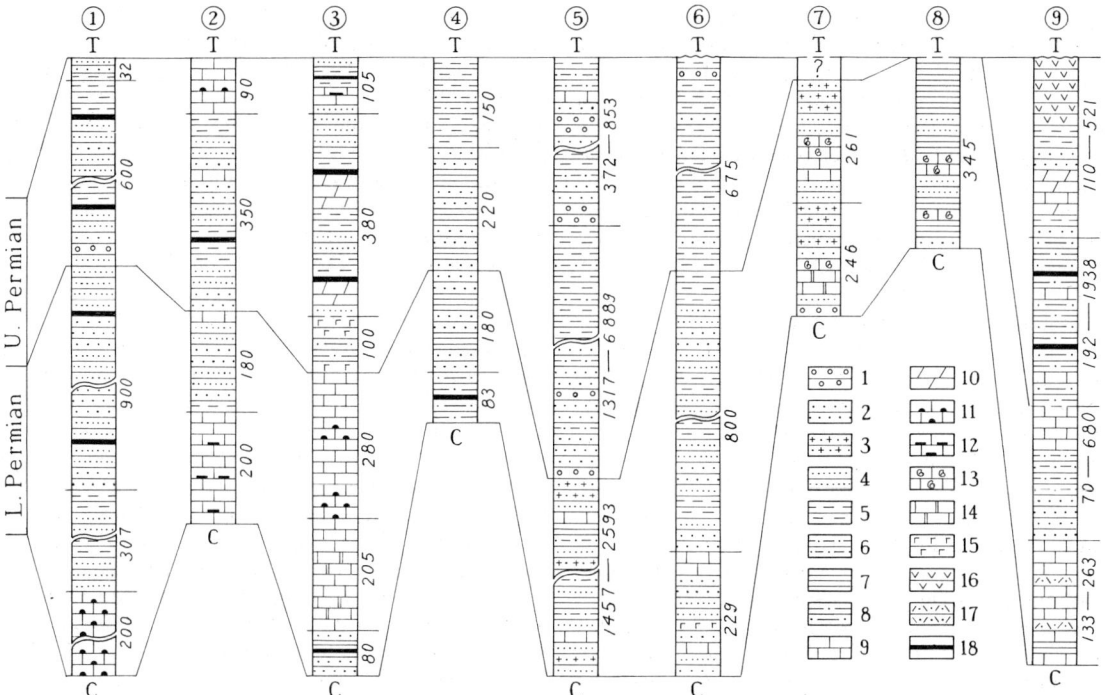

Fig. 12.2. Columnar sections showing development of Permian strata in various stratigraphic regions of China. 1 Yongdin, Fujian, South-east China region; 2 Changxing, Zhejiang, Jiangnan region; 3 Qinglong, South-west Guizhou, Qinling–Yangzi region; 4 Taiyuan, Shanxi, North China region; 5 Junggar–Turpan, Tianshan–Hingan region; 6 Kalpin, Xinjiang, Tarim region; 7 South Qilian, Kunlun–Bayan Har region; 8 Nyalam–Dingri, Xizang, Himalaya region; 9 Qamdo, Xizang, Xizang–West Yunnan region. Legend: 1, Conglomerate; 2, Sandstone; 3, Arkose; 4, Siltstone; 5, Mudstone; 6, Sandy mudstone; 7, Shale; 8, Sandy shale; 9, Limestone; 10, Marl; 11, Concretionary limestone; 12, Siliceous limestone; 13, Biogenetic limestone; 14, Dolomitic limestone; 15, Basalt; 16, Quartzite; 17, Tuff; 18, Coal seam.

less abundant. The fusulinid zones are listed in descending order as follows (Sheng, K. C. 1962):

(3) *Yabeina–Neomisellina* zone
   Neoschwagerina craticulifera (Schwager), *Verbeekina crassispira* Chen, *Neomisellina, Yabeina, Nankinella*;

(2) *Neoschwagerina margaritae* zone
   Schwagerina, Parafusulina, Verbeekina verbeeki (Geinitz), *Chusenella, Pseudodoliolina ozawai* Yabe and Hayasaka, *Afghanella shencki* Thompson, *Neoschwagerina margaritae* (Deprat), *N. douvillei* Ozawa; foraminifers (*Endothyra, Tetrataxis*);

(1) *Cancellina* zone
   Ozawainella sp., *Nankinella* sp., *Parafusulina multiseptata, Cancellina neoschwagerinoides* (Deprat).

The Maokouan is known for its great variety of lithofacies, and at least six of them may be mentioned: (1) marine limestone intercalated with small amounts of sandstone and shale as shown in the section at Xikou, Zhengan, Shaanxi Province, which is rich in fusulinids, corals, and brachiopods; (2) marine carbonates as seen in the Liangshan Mountain of Hanzhong, Shaanxi, Bali, Langdai and Ziyun in Guizhou, Desheng, Yishan in Guangxi, all characterized by fusulinids and corals, but divisible chiefly on the basis of three fusulinid zones: *Cancellina, Neoschwagerina margaritae*, and *Yabeina–Neomisellina*; (3) argillaceous limestone (with augen structure) and limestone as found in Shizipu, Zunyi in Guizhou, Huaying Mountain, and Leshan in Sichuan, Sangzhi in Hunan, and characterized by *Cryptospirifer* and disappearance of *Cancellina*; marked in ascending order by the *Cryptospirifer striatus* zone, *Chusenella douvillei–Ipciphyllum ipeci* zone, and *Yabeina–Neomisellina* zone. In part of south-east Hubei Province the top of this type of deposits changes into *Altudoceras*-bearing siliceous beds; (4) siliceous rocks, shale, and sandstone typified by such sections as the Sanmengjiang, Liuzhou in Guangxi, Kaochuan, Xixuaban in Shaanxi, bearing numerous *Altudoceras, Paragastrioceras, Kufengoceras,* and brachiopods; (5) limestone–siliceous limestone, siliceous rock, shale and locally coal-bearing clastic rocks, as occurring in Xishan, Lake Taihu, Fengcheng in Jiangxi, Jiahe in Hunan Province, where the Maokou Formation is represented by the Gufeng Formation (Dangchong Formation), and the Yangqiao Formation (Guanshan Formation) characterized by ammonoids and brachiopods. Four zones in ascending order have been recognized, namely: (1) *Cancellina* zone (or *Parafusulina* zone), (2) *Gufengoceras* zone, (3) *Waagenoceras* zone, and (4) *Neomisellina–Codonofusiella* Assemblage zone; (6) limestone-siliceous rocks–clastic rocks and coal-bearing clastics, e.g. Tonglu, Western Zhejiang, Qianshan, north-eastern Jiangxi, Jongan, Fujian and Jiahe, Guangzhou. This type includes the topmost member of the original Qixia Formation, the Dingjiashan Formation, the Lengwu Formation, the Lixian Coal Series of Zhejiang, the Wenbishan Formation, the Dongziyan Formation, containing four zones in ascending order: (1) *Cancellina* zone (*Parafusulina multiseptata* zone), (2) *Gufengoceras* zone, (3) *Waagenoceras* zone, (4) *Urushtenia crenulata–Semibrachythyrina* Assemblage zone.

The lower limit of the Maokouan is generally marked by the disappearance of such fusulinids as *Misellina, Nankinella, Pisolina,* and *Sphaerolina* as well as the appearance of *Cancellina* or *Cryptospirifer*. In Jiangnan and South-east regions the upper limit of *Urushtenia crenulata, Shouchangoceras* or the upper limit of the *Neomisellina–Codonofusiella* Assemblage zone is temporarily taken as the top of the Maokouan stage.

Upper Permian Longtanian

The Upper Permian Longtanian Stage is named after its outcrop in Longtan, Nanjing, but its type section is located at Tianbaoshan, a hill also near Nanjing. It is more than 60 m thick and divisible into the following three members based on litho- and biofacies:

An upper member of clastics consisting of shales with marine bivalve fragments.

A middle member of marine deposits composed of shales and sandy shale with impure limestone at the base, rich in brachiopods (*Asioproductus, Leptodus* cf. *nobilis*), and ammonoids (*Pseudogastrioceras* sp., *Pleurodiscoceras* sp., *Prototoceras* sp., and *Anderssonoceras* sp.).

A lower member of coal-bearing series made up of fine sandstone, silt-stone, arkose, and shale with workable coal seams; its basal medium-grained arkose yields numerous *Gigantopteris nicotianaefolia* Schenk, *Pecopteris echinata* Gu and Zhi, *Sphenophyllum,* and *Lobatannularia*.

It should be noted that the Longtanian in its type section is now restricted to the upper three members only, omitting the underlying non-coal-bearing beds, which belong to the Lower Permian Yangqiao Formation.

The Longtanian may be differentiated into four different lithofacies as follows. (1) Basalt–terrestrial coal-bearing clastics, occurring for instance in Leshan, Sichuan and Fuyuan, Yunnan, and equivalent to the lower part of the Xuanwei Formation. It is also found in Zhangping, Fujian (Cuipingshan Formation), Yangchun, Guangdong (Yangchun Formation), and Xinfong, Jiangxi (Dalo Formation). (2) Basalt–paralic coal-bearing clasitcs occurring in Pan Xian, Guizhou and Zhenxiong, Yunnan. (3) Coal-bearing clastic rocks and limestone occurring in Zunyi, Guizhou and Beipei, Sichuan, where coal-beds alternate with limestone and are rich in fusulinids, brachiopods, and plants. Such a facies is a typical Longtan Formation. (4) Limestones, as exemplified by their development in Liangshan, Hanzhong and Xixiang in Shaanxi Province, Guanguyuan, Sichuan Province, Ziyun, Guizhou Province, and Lianxian, Guangdong Province. It is commonly designated the Wujiaping Formation, whose basal part consists of clastics or coal-bearing clastics, yielding numerous fusulinids, corals, and brachiopods, but generally lacking ammonoids. Two zones have been noted in ascending order: (1) the *Tyloplecta yangtzeensis–Edriosteges poyangensis* Assemblage Zone containing *Codonofusiella, Waagenophyllum, Liangshanophyllum, Oldhamina grandis* Huang, *Asioproductus, Permophricodothyris grandis* Chao, and *Araxathyris araxensis* Grant; and (2) the *Codonofusiella kwangsiana* zone containing *Codonofusiella kwangsiana* Sheng, *C. lui* Sheng, *C. schubertelloides* Sheng, *C. asiatica* K. M. Maclay, *Palaeofusulina, Gallowayinella, Reichelina, Chenia, Waagenophyllum lui* Tseng, *Liangshanophyllum wengchengense* Huang, and *Plerophyllum excentricum* Iljina.

The lower limit of the Longtanian has not yet been clearly defined by marine fossils.

Upper Permian Changxingian

The Changxingian, the uppermost Permian unit, is named from the type section at the large coal deposit at Changxing, Zhejiang Province. It consists of grey to dark-grey medium-bedded limestone, asphaltic limestone, and calcareous dolomite, intercalated with thin-bedded cherty limestone, rich in *Palaeofusulina* and associated with important ammonoids, having a thickness of 34–51 m. The faunal sequence suggests subdivision into the lower Baoqing Member and the upper Coal Hill Member. The Baoqing Member yields ammonoids (*Tapanshanites, Pseudostephanites, Mingyuexiaceras, Sinoceltites*) and conodonts (*Neogondolella subcarinata subcarinata* Sweet, *N. subcarinata elongata* Wang, while the Coal Hill Member contains chiefly fusulinids (*Palaeofusulina sinensis* Sheng, *Reichelina changhsingensis* Sheng and Chang, other foraminifers (*Colaniella*), ammonoids (*Rotodiscoceras, Pleuronodoceras, Pseudotirolites, Trigonogastrites, Changhsingoceras*), conodonts (*Neogondolella subcarinata changxingensis* Wang, *N. dicerocarinata*), corals (*Lophophyllum*), brachiopods (*Chonetinella nasuta* (Waagen), *Waagenites*), and fishes (Palaeonisciformis, Platysmoidae, and *Sinohelicoprion changsingensis* (Liu and Chang).

Like the Longtanian, the Changxinian is marked by great changes in lithofacies, five of which may be mentioned. (1) Terrestrial coal-bearing clastic rocks (e.g. Leshan, Sichuan Province, Xuanwei and Fuyuan, Yunnan Province, all bearing *Gigantopteris* flora like the upper part of Xuanwei Formation. (2) The paralic coal-bearing clastic rock facies, e.g. Shuicheng and Pan Xian, Guizhou Province, where beds with *Palaeofusulina* alternate with beds yielding *Pseudotirolites*, and workable coal seams are present. (3) Limestone facies, e.g. Zhengan in Shaanxi Province, Huayingshan in Sichuan Province, Fengcheng in Jiangxi Province, Changxing in Zhejiang Province. This facies is characterized by fusulinids and corals, but with few ammonoids, forming what is commonly called the Changxing Formation. (4) Siliceous and clastic rock facies, described as the Dalong Formation and represented by sections in Li Xian in Shaanxi, Leiyang in Hunan, and Tianbaoshan in Nanjing, all characterized by ammonoids and brachiopods. (5) Limestone–siliceous and clastic rock facies, which are a combination of facies 4 and 5, exemplified by sections in Guangyuan in Sichuan, Luodian in Guizhou, Laibing, and Heshan in Guangxi, all of them characterized by *Palaeofusulina* in the lower and *Pseudotirolites*, and *Pleuronodoceras* in the upper parts.

PLATFORMAL SEDIMENTATION TYPE

Typical of this is the South Qinling–Yangzi region, including the Yangzi (except the Lower Yangzi area), South Qinling, and Guangxi regions, where the Permian is very regularly distributed with stable Lower Permian deposits, which consist mainly of carbonate rocks, except for the Lower Qixian coal-bearing clastic rocks. Basalts (Emeishan basalt) were extruded extensively in Late Maokouan or early Late Permian times. In the Late Permian, differential deposition was so distinct that east of the Kam–Yunnan Massif terrestrial coal-bearing clastic rocks predominate, while to the east there are neritic–littoral swampy coal-bearing clastics and carbonates, and completely carbonate rocks. The eastern Qinling Mountains, the greater part of the Daba Mountains, and the Guangxi region were covered with extensive carbonate deposits and some sandstone and shale. These rocks yield many warm-water fossil groups such as fusulinids, other foraminifers, corals, brachiopods, bivalves, and bryozoans, but ammonoids are relatively rare, though locally fairly abundant. In general, there are 300–1800 m of deposits with a maximum of 3469 m.

Qinglong, South-west Guizhou

The Qinglong section, Guizhou (Table 11.1.3) suggests that the Permian sequence in south-west Guizhou in ascending order is as follows.

The basal Liangshan Formation is composed of 80 m of sandstone, shale, and coal seams. This is succeeded by the Lower Permian Qixia Formation composed of grey limestone and dolomitic limestone, having a thickness of 205 m and bearing fusulinids in the upper horizon (*Cancellina, Prosumatrina, Pseudodoliolina chinghaiensis,* and *Parafusulina* cf. *yabei*) and also fusulinids in the lower horizon (*Misellina ovalis,* and *Parafusulina*).

The Lower Permian Maokou Formation is a 280 m thick unit of grey shale, massive limestone with chert-nodules, yielding in its lower part fusulinids (*Neoschwagerina, Pseudodoliolina,* and *Chusenella sinensis*) and in its upper part (*Neomisellina, Yabeina, Neoschwagerina,* and *Sumatrina* cf. *annae*).

A disconformity separates the above formations from the Upper Permian Emeishan Basalt Formation, which is intercalated with sandy shale and has a thickness of 100 m.

The Upper Permian Longtan Formation is composed of siltstone, mudstone, marl, and coal seams, totalling 380 m, and yielding brachiopods (*Edriosteges poyangensis*) and plants (*Lobatannularia, Gigantopteris*).

The uppermost Permian Liangfengpo Formation consists of siltstone, mudstone intercalated with siliceous limestone, limestone, and coal seams, approximately 100 m thick and containing *Palaeofusulina sinensis, Colaniella, Rotodiscoceras,* and *Pleuronodocer-*

as mapingense in its upper part, and *Palaeofusulina minima, Pseudotirolites, Paryphella sulcatifera*, and *Peltichia zigzag* in its lower part. This is equivalent to the Changxingian.

In this section the Permian is overlain disconformably by the Lower Triassic Feixianguan Formation.

Jiangnan region and South China

The development of another stable type of sedimentation is illustrated in the Jiangnan region and South China (Fig. 12.1.7, Table 12.1.2) covering central Jiangxi, central and southern Hunan, northern Guangdong, the bordering areas between Jiangsu, Zhejiang, and Anhui, and northern Jiangsu. The type section is in Changxing, Zhejiang Province. Here, the Permian sequence starts with the Qixia Formation which is composed of thick-bedded limestone intercalated with siliceous limestone and is 200 m thick. The upper part of this formation typically contains fusulinids (*Schwagerina chihsiaensis*) and corals (*Hayasakaia elegantula, Polythecalis chinensis*), while its lower part yields fusulinids (*Misellina claudiae*).

The Qixia Formation is followed by the Yangqiao Formation, which comprises chiefly siltstone and mudstone with sandy limestone at the top, and siliceous mudstone bearing phosphate nodules in the lower part, and is 180 m thick. It contains ammonoids (*Shouchangoceras, Altudoceras*), and brachiopods (*Neoplicatifera huangi*) in the lower part, and fusulinids (*Neomisellina multivoluta, N. compacta, Kahlerina sinensis*) in the upper part. It is equivalent to the Maokouan Stage.

The Upper Permian Longtan Formation is made up of 350 m of siltstones and mudstones, and sandstone intercalated with thick-bedded limestone and coal seams, and yields fusulinids (*Codonofusiella*), ammonoids (*Konglinites, Araxoceras, Anderssonoceras*), conodonts (*Neogondolella orientalis*), and fossil plants (*Gigantopteris nicotianaefolia*).

The uppermost Changxing Formation, 90 m thick, is the type for the Changxingian Stage as described previously in the discussion of chronostratigraphic units. This is overlain paraconformably by the Lower Triassic Yinken Formation.

South-east region (Yongding, Fujian)

The South-east region covers western Zhejiang, Fujian, southern Jiangxi, and eastern and western Guangdong. In Yongding, Fujian Province (Fig. 12.1.8, Fig. 12.2.1, Table 12.1.1), the Lower Permian Qixia Formation overlying disconformably the Upper Carboniferous Chuanshan Formation consists of dark-grey cherty limestone with siliceous shale at the top. It is 200 m thick, and yields fusulinids (*Cancellina, Parafusulina*, and *Verbeekina*), and brachiopods (*Urushtenia*). It is succeeded by the 307 m thick Wenbishan Formation of siltstone and sandstone, which yields brachiopods (*Pygmochonetes jingxianensis*) and ammonoids (*Paragastrioceras, Altudoceras* and *Paraceltites*). The next unit is the Tongziyan Formation, which comprises dark-grey siltstones, shale, fine-grained sandstone, and coal seams, totalling 900 m of deposits and yielding fusulinids (*Schwagerina*), ammonoids (*Altudoceras, Paraceltites*), and brachiopods (*Urushtenia crenulata, Neoplicatifera huangi*). Both the Wenbishan and Tongziyan Formations are equivalent to the Maokouan Stage.

The Cuipingshan Formation consists of dark-grey siltstone, mudstone intercalated with fine sandstone, and thin coal-seams, with basal chert breccia, 600 m thick, and yields fossil plants such as *Sphenophyllum densinervis, Gigantopteris,* and *Pecopteris*. Hence, this formation can be correlated with the Lopingian Stage.

The uppermost Permian Dalong Formation comprises siltstone, sandstone, and mudstone and is only 32 m thick. It contains ammonoids (*Pseudotirolites, Pleuronodoceras mapingense*), and brachiopods (*Oldhamina squamosa, Orthotetina ruber*), and thus belongs to the Changxinian Stage.

Taiyuan, North China region

The stable type of sedimentation is also well-developed in Taiyuan, Shanxi Province in the North China region (Fig. 12.1.3, Fig. 12.2.4, Table 12.1.4), where the Permian sequence is made up of four terrestrial formations: the Shanxi, Lower Shihezi, Upper Shihezi and Shiqianfeng. Overlying the Taiyuan Formation is the Shanxi Formation which consists of greyish-black sandy shale intercalated with coal seams, having a thickness of 830 m, and yielding fossil plants such as *Emplectopteridium alatum, Emplectopteris triangularis*, and *Taeniopteris multinervis*. The Lower Shihezi Formation, composed of sandstone and shale, is 180 m thick, and bears plants (*Cathaysiopteris whitei, Emplectopteris triangularis*). The Upper Shihezi Formation of sandstone and shale is 200 m thick and contains plants (*Gigantonoclea hallei*, and *Lobatannularia ensifolia*). The uppermost Permian unit is the Shiqianfeng Formation which consists of 150 m of mudstone and sandy mudstone, yielding in the lower part *Rhipidopsis lobata*. It is overlain disconformably by the Lower Triassic Liujiagou Formation.

Kalpin, Tarim region

The Permian in Kalpin, Tarim region also belongs to the stable sedimentation type (Fig. 12.1.2; Fig. 12.2.6; Table 12.1.6). Overlying the Upper Carboniferous Kangkelin Formation is the Lower Permian Balikelike Formation which is composed of a marine sequence of greyish-black limestone or an alternation of limestone and clastic rocks. In the western section of Kalpintag there are basalts, in addition to clastic rocks, which yield brachiopods (*Dictyoclostus, Dielasma, Echinoconchus, Marginifera, Phricodothyris,* and *Notothyris*), bivalves (*Sanguinolites* sp., *Allorisma* sp., and *Aviculopecten* sp.), gastropods (*Bellerophon*), and bryozoans (*Fistulipora* sp., and *Rhombopora*), totalling 229 m. Next is the Kalundal Formation which is an alternation of variegated sandstone and mudstone or siltstone, altogether more than 800 m thick. It yields only fragmentary gastropods and bivalves. The above two formations are assigned to the Lower Permian. The Upper Permian Shajingzi Formation comprises terrestrial variegated sandy mudstone and sandstone topped by conglomerates, together 675 m thick. In some localities freshwater bivalves have been found.

Himalayan region

In the Himalayan region only the Lower Permian of the platformal type is represented in Nyalam and Dingri (Fig. 12.1.9, Fig. 12.2.8, Table 12.1.8). Overlying the Upper Carboniferous Jilong Formation is the Qubu Formation, which consists of 20 m of shale and fine sandstone and yields fossil plants (*Glossopteris communis, Sphenophyllum speciosum, Ranigangia qubuensis*). This formation is overlain conformably by the Quberga Formation, which is composed of 239 m of grey siltstone, shale, and biogeno-clastic limestone in the middle and lower members, containing ammonoids (*Uraloceras xizangensis*), corals (*Lytvolasma asymmetricum*), brachiopods (*Taeniothaerus, Chonetinella nasuta, Costiferina* sp.), and of 86 m of dark-grey shale in the upper member, yielding bivalves (*Phestia darwini* and *Aviculopecten* sp.).

BASIN SEDIMENTATION TYPE

Northern Basin (eastern part)

Typical basin deposits were formed in the eastern part of the Northern Basin, in Nei Mongol, the Songhua River and the Changbei Mountain Trough during the Early Permian when the trough continued to subside and received carbonates and clastics, accompanied by intermediate acidic volcanics and terrestrial clastics, and, locally, volcanic molasse known as Xingan-type deposits. These rocks have been metamorphosed to a certain degree. Biotas consist chiefly of cold-water corals (*Lytvolasma*), brachiopods (*Yakovlevia, Kochiproductus,* and *Spiriferella*), and fusulinids (*Monodiexodina*). There are also warm-water fusulinids (*Parafusulina,* and *Neoschwagerina*), and corals (*Wentzellophyllum,* and *Waagenophyllum*). As to land life, Angaran flora predominated then, but it was mixed with a few Cathaysian elements. So, both marine and land life were close to those of the Boreal Province, although a certain endemism was maintained.

By the end of the Early Permian, the Northern Basins were almost completely folded and uplifted, so in the Late Permian in general deposits were terrestrial or in residual sea waters.

Northern Basin (western part)

In the western section of the Northern Basins, especially north of the Central Tianshan and in the Junggar, intermontane basin deposits predominated. In the early part of the Early Permian, the area eastward from Hami, Xinjiang was covered by seas where deposits of Xingan type as mentioned above were formed. These are colossal deposits of neritic, paralic clastic and carbonate rocks, and intermediate acidic volcanics and pyroclastics, intercalated with carbonates which have all been metamorphosed. The Late Permian is characterized by normal deposits of clastic rocks, limestone, or bioclastic limestone, and volcanic rocks are rarely seen. These beds have a thickness of 2000–4000 m.

In the Junggar–Turpan area especially around Urumchi, Jumusar, and Jiangjunmiao (Fig. 12.1.1, Fig. 12.2.5, Table 12.1.5) the Lower Permian is represented by the Lower Jijicao Group, which overlies the Upper Carboniferous Urumchi Formation and comprises greyish-black sandstone, and siltstone intercalated with mudstone and limestone, varying in thickness from 1457 to 2593 m. It yields ostracods (*Healdia sillens, Moorea oblonga, Cavallina contraria,* and *Jonesina victoria*), foraminifers (*Ammodiscus* sp., and *Glomospira* sp.), and fossil plants (*Calamites* sp., and *Noeggerathiopsis* sp.).

Correlatable to the Longtanian is the Upper Jijicao Group, which is composed of yellowish-grey and greyish-green mudstone, intercalated with conglomeratic sandstone, carbonaceous mudstone and inferior coal, brownish-red in the lower part, being 1317–6889 m thick. It contains bivalves (*Microdontella microdonta, Mrasiella magniforma*), and plant fossils

(*Callipteris* cf. *zeilleri*, and *Cordaites principalis*).

The uppermost unit is the Lower Cangfanggou Group of the Changxinian Stage, which is yellowish-green and dark mudstone, and conglomerate with thin-bedded limestone, totalling 372–853 m, and yields fossil plants (*Calamites* sp., *Pecopteris, Iniopteris sibirica, Fascipteris hallei*) and vertebrates (*Dicynodon tienshanensis, Jimusaria* (*Dicynodon*) *sinkiengensis, Striodon magnus, Kunpania scopulusa, Urumchia lii*).

The Kunlun–Bayan Har region (Fig. 12.1.4) includes the Kunlun Mountains, Qimantag Mountains, South Qilian Mountains, Bulhan Buda Mountains, A'ni Machen Mountains, Yushu–Zhongdian, Songpan–Lixian, etc. It was an active region and is characterized by clastic rocks, volcanic rocks, pyroclastics, and carbonates, exhibiting flysch with strong facies-change, reaching a thickness of from 100 to nearly 6000 m. In South Qilian, Songpan, and Lixian there are normal neritic deposits of clastics and carbonates, the former containing no volcanic deposits, whereas the latter are marked by marine basaltic eruptions. Thus they may be classified as a transitional type of sedimentation. Within this region the Lower Permian is well-developed and characterized by biotas of South China type: fusulinids (*Misellina, Neoschwagerina, Polydiexodina*), corals (*Iranophyllum, Wentzelella, Waagenophyllum*), and brachiopods (*Urushtenia, Callifera, Monticulifera*), locally including ammonoids (*Artinskia, Popanoceras, Perrinites, Waagenoceras*). The Upper Permian is rather limited in its distribution. In some localities of Qinghai, Songpan–Lixian, and Yushu–Zongdian there are Upper Permian outcrops with South China elements: fusulinids (*Colaniella, Palaeofusulina, Gallowaiinella*), and brachiopods (*Permophricodothyris grandi*). Cathaysian floral elements have been found in some plant-bearing intercalations. No coal series is present.

In South Qilian region (Fig. 12.2.7, Table 12.1.7) only the Lower Permian is preserved and is represented by the Bayinhe and Nouyinhe formations.

The distribution of Permian deposits in the Xizang–West Yunnan region is shown on Fig. 12.1.5. Here the Lower Permian consists chiefly of volcanic rocks, pyroclastics and limestone with marked faunal changes, and is over 2000 m thick. The Upper Permian is also very thick, exceeding 2000 m and comprises sandy shale, limestone, siliceous rock, pyroclastic rocks, basalts, and occasionally coal series—with workable coal seams in the Qamdo, and Lanping–Simao areas. The fossil groups are closely related to those of the South China type of the Tethys realm.

In Qamdo, Xizang, the Permian sequence is composed of four formations, two for each series (Fig. 12.2.9, Table 12.1.9). The Lower Permian Mangcuo Formation comprises grey thick-bedded limestone intercalated with tuff, lava, and black shale in the basal part, reaching a thickness of 132–263 m. This is overlain conformably by the Tyokar Formation (Jiaoga Formation) of limestone, argillaceous sandstone, and fine sandstone, having a thickness of 70–680 m. Lying disconformably on the above unit is the Upper Permian Toba Formation, which is made up of sandy shale, coal seams, and small amounts of limestone, and is 192–1938 m thick. It yields fossil plants, such as *Gigantopteris, Gigantonoclea, Lobatannularia*, corals (*Liangshanophyllum*), and brachiopods (*Peltichia, Spinomarginifera*). This formation is thus equivalent to the Longtanian Stage. The uppermost Permian unit is the Zhalagongga Formation, which consists of mudstone, sandstone, and marl, locally intercalated with carbonaceous mudstone and limestone, totalling 110–521 m thick. It is fairly rich in fusulinids (*Palaeofusulina fusiformis, P. subcylindrica*) and is of the Changxinian Stage.

In the Taiwan region, although the Permian is not differentiated from the Dananao schist, its presence is proved by the existence of fusulinids (*Schwagerina? Parafusulina? Neoschwagerina?*) from Nanzi, and corals (*Waagenophyllum*) from Mataianxi. Consequently, the Dananao schist is at least in part of Early Permian age and the Permian in Taiwan Island consists of basin deposits.

PALAEOBIOGEOGRAPHICAL PROVINCES

Southern (warm water) or Tethys province

The above biostratigraphical evidence suggests that there were two chief palaeobiogeographical provinces, the Southern (warm-water) or Tethys province and the Northern (cold-water) or Boreal province, and also a mixed or transitional type. The Southern province has copious deposits of warm-water benthos, such as fusulinids, corals, brachiopods, and bryozoans, which are closely allied to those of the Tethyan province. Among the chief representatives are fusulinids (*Misellina, Cancellina, Neoschwagerina, Neomisellina, Polydiexodina, Palaeofusulina*), non-fusulinid foraminifers (*Colaniella*), compound corals (*Wentzellophyllum, Polythecalis, Ipciphyllum, Iranophyllum, Wentzelella, Waagenophyllum, Liangshanophyllum*), brachiopods (*Monticulifera, Cryptospirifer, Tyloplecta, Permophricodothyris, Oldhamina, Enteletina*), and bryozoans (*Fistulipora, Araxopora*). Among pelagic forms are ammonoids, such as *Shouchangoceras, Pseudotirolites*, and *Pleuronodoceras*. Representative plants include such genera as *Gigantonoclea, Gigantopteris, Lobatannu-*

laria, *Sphenophyllum*, and *Neuropteridium*, all typical of the Cathaysian flora. This is the most extensive province, covering many regions in addition to the Tianshan–Hingan and the Himalayan regions.

Northern (Boreal) province

The Northern (Boreal) province is characterized mostly by cold-water forms with the addition of some warm-water elements. The cold forms are represented by fusulinids (*Monodiexodina*), corals (*Cyathocarinia*, *Lytvolasma*), brachiopods (*Spiriferella*, *Licharewia*, *Yakovlevia*, *Liosotella*, *Horridonia*, *Kochiproductus*, *Jakutoproductus*), and gigantic *Neospirifer*. Among the warm-water forms are corals (*Waagenophyllum*, *Wentzelella*) and brachiopods (*Richthofenia*, *Enteletes*).

Mixed faunal type

The mixed faunal type is confined to Yanbian, Jilin and Qomolangma, Xizang. In the former area a mixture of both warm- and cold-water forms are found and, in addition to the predominating warm-water elements of the Southern Province, such as *Chusenella*, *Neoschwagerina*, *Sumatrina*, *Polydiexodina*, *Yabeina*, *Codonofusiella* (all fusulinids) and *Waagenophyllum* (corals), there occur a number of important Northern representatives such as *Spirifer* and *Kochiproductus* in Central Gilin, Yanbian, and East Heilongjian. Along the southern margin of the Nei Mongol Basin (Kangbao, Hebei) there are also warm-water fusulinids (*Misellina*, *Nankinella*), corals (*Yatsengia*, *Szechuanophyllum*), and brachiopods (*Orthotichia*), which are associated with a few cold-water forms.

Mixed faunas also occur in the Qomolangma area, where the Northern common elements are associated with brachiopods *Costiferina*, *Taeniothaerus*, *Calliomarginatia*, *Fusispirifer*, and *Choristella*, bryozoans (*Streblotrypa*), and ammonoids (*Uraloceras*), which are typical of the Himalayan biota.

BOUNDARY PROBLEMS

In South China the boundary between the Permian and the Carboniferous has been drawn between the *Misellina claudiae* or *Schwagerina tschernyschewi* zone and the *Pseudoschwagerina* zone because of the presence of a widespread disconformity and a sharp lithological difference between the two systems.

The recent discovery in south-western Guizhou of the Longyin Formation is significant in that it spans the gap between the Maping and Qixia formations, and that it comprises both Mapingian *Robustoschwagerina*, *Rugosofusulina*, *Pseudoschwagerina*, *Kepingophyllum*, and earliest Permian elements such as *Pseudofusulina moelleri*, *Propanoceras*, *Wentzellophyllum*, *Protomichelinia*, and *Iranophyllum* according to Yang Jingzi et al. (1979). On the basis of typical Permian elements such as fusulinids (*Pseudofusulina moelleri*) and ammonoids (*Propanoceras*) Zhan et al. (1982) believe that the Longyin should be classified as Permian. Zhan's view reconciles the difference between Chinese and foreign workers about the lower limit of the Permian (see Tables 12.2, 12.3). Thus, for the time being, the Carboniferous–Permian boundary is drawn between the Asselian and Sakmarian ages.

Lower–Upper Permian boundary

The boundary between the Lower and Upper Permian Series is agreed to be between the Maokouan and the Longtanian. However, the lower limit of the Longtanian as yet lacks an accurate bio-horizon, as noted above. The basal part of either the Longtan Formation or Wujiaping Formation consists of clastic rocks or coal-bearing clastic rocks characterized by fossil plants. Neither the marine *Tyloplecta yangzeensis–Edriosteges poyangensis* assemblage, nor the *Anderssonoceras–Prototoceras* zone appearing further up in the sequence, mark the beginning of the Longtanian.

Permian–Triassic boundary

The Permian–Triassic boundary of China is now generally drawn between the Upper Permian Changxinian and the Lower Triassic Feixianguanian or Tayean, but more detailed discussion will be found in Chapter 13. Correlation of Permian sequences in various regions of China as well as that between China and other parts of the world including Germany, North-western Europe, the East Russian Platform, Queensland, Australia, and the Delaware Basin, USA are given on Table 12.1 and Table 12.3 respectively.

THE GEOLOGY OF CHINA

Table 12.1 *Correlation of Permian stratigraphic sequences in various regions of China (compiled from various sources).*

Series	Stage	S.E. China region Yongdin, Fujian (1)	Jiangnan region Changxing Zhejiang (2)	Qinglong–Yangzi Region Qinglong, S.W. Guizhou	N. China region Shanxi, Taiyuan (4)
		Xikou Fm (T_1)	Yingken Fm (T_1)	Feixianguan Fm (T_1)	Liujiagou Fm (T_1)
Upper Permian	Changxinian	Dalong Fm *Pseudotirolites Pleuronodoceras mapingensis Oldhamina squamosa Orthotetina ruber* 32 m	Changxing Fm *Palaeofusulina* cf. *sinensis Rotodiscoceras Tapashanites changxingensis Neogondolella* 90 m	Liangfengpo Fm *Palaeofusulina sinensis Nankinella minor Rotodiscoceras Pseudotirolites Pleurodiscoceras mapingensis* 105 m	Shiqianfeng Fm *Rhipidopsis lobata* 150 m
	Longtanian	Cuipingshan Fm *Sphenophyllum densinerva Gigantopteris Pecopteris* 600 m	Longtan Fm *Conodofusiella Konglingites Araxoceras Anderssonoceras Neogondolella orientalis* 350 m	Longtan Fm *Edriosteges poyangensis Lobatannularia Gigantopteris* 380 m Omeishan Basalt Fm 100 m	Upper Shihhotse Fm *Gigantonoclea halle? Lobatannularia ensifolia* 220 m
Lower Permian	Maokouan	Tongtzeyan Fm *Schwagerina Urushtenia crenulata Neoplicatifera huangi Altudoceras Paraceltites* 900 m Wenbishan Fm *Paragastrioceras Altudoceras Paraceltites* 307 m	Yianqiao Fm *Neomisellina multivoluta N. compecta Kahlerina sinensis Shouchangoceras Altudoceras Neoplicatifera huangi* 180 m	Maokou Fm *Neomisellina Yabeina Neoschwagerina Sumatrina* cf. *annae Pseudoliolina Verbeekina heimi Cryptospirifer striata Ipciphyllum ipci* 280 m	Lower Shihhotse Fm *Cathaysiopteris whitei Emplectopteris triangularis* 180 m
	Chihsian (Qixia)	Chihsia Fm *Cancellina Parafusulian Verbeekina Urushtenia* 200 m	Chihsia Fm *Schwagerina chihsiaensis Hayasakaia elegantula Polythecalis chinmenensis Misellina claudiae* 200 m	Chihsia Fm *Cancellina primigena Pseudodoliolina ozawai Parafusulina yabei Misellina ovalis Nankinella orbicularia* 205 m Liangshan Fm *Monticulifera sinensis Chaoina reticulata* 80 m	Shanxi Fm *Emplectopteridium alatum Emplectopteris triangularis Taeniopteris multinervis* 83 m
		Chuanshan Fm (C_3)	Chuanshan Fm (C_3)	Baomoshan Fm (C_3)	Taiyuan Fm (C_3)

Table 12.1. (cont.)

Tianshan–Hingan region, Jungar–Turpan (5)	Tarim region Kalpin, Xinjiang (6)	Kunlun–Bayanhar region S. Qilian (7)	Himalaya region (8) Nyalam, Xizang	Xizang–W. Yunnan (9) Qambo, Xizang
T_1	R	Junzhihe Gr (T)	T	Jiapila Fm (T_3)
Lower Cangfanggou Gr *Calamites* sp. *Pecopteris* sp. *Iniopteris sibirica* *Fascipteris hallei* *Picynodon tianshanensis* 327–853 m	Shajinzi Fm *Palaeoanodonta parallela* *P. castora* *P. fabaeformis* *Palaeomutela subparallela* 675 m	?		Zhalagongga Fm *Palaeofusulina fusiformis* *P. subcylindrica* 110–521 m
Upper Jijicao Gr *Microdontella microdonta* *Mrasiella magniforma* *Callipteris* cf. *zeilleri* *Cordaites principalis* 1317–6889 m				Toba Fm *Gigantopteris* *Gigantonoclea* *Lobatannularia* *Liangshanophyllum* *Peltichia* *Spinomarginifera* 192–1938 m
Lower Jijicao Gr *Headia sillens* *Moorea oblonga* *Cavallina contraria* *Jonesina victoria* *Ammodiscus* sp. *Glomospira* sp. 1457–2593 m	Kalundal Fm Gastropod Bivalve Fragments 800 m	Nuoyinhe Fm *Buxtonia* *Megaderbyia magna* *Urushtenia crenulata* *Monticulifera* 261 m Bayinhe Fm *Verbeekina crassispira* *Polydiexodina sparsa*	Quburiga Fm *Uraloceras xizangensis* *Lytvolasma asymmetricum* *Chonetinella nasuta* 1145 m	Tyokar Fm *Neoschwagerina* *Verbeekina* *Neomisellina* *Sumatrina* *Wentzelella* *Ipciphyllum* 70–680 m
Also Plants (*Calamites* sp.) *Noeggerathiopsis*	Balikelike Fm *Sanguinolites* sp. *Allorisma* sp. *Aviculopecten* sp. *Echinoconchus* sp. *Marginifera* sp. *Cancrinella* sp. *Notothyris* sp. *Nankinella* cf. *orbicularis* *Krotovia* sp. *Grammysia* sp. *Euphemites* sp. *Lithophaga* sp. 229 m	*Martinia semiconvexa* *Wentzellela* cf. *szechuanensis* *Spinomarginifera sintanensis* *Richthefenia sinensis* *Orthotetina ruber* 246 m	Qubu Fm *Sphenophyllum Speciosum* *Glossopteris communis* 20 m [Selung Gr]	Mangcuo Fm *Cancellina neoschwagerinoides* *Afghanella* *Misellina* *Parafusullina* *Staffella moellerana* 132–263 m
Urumqi Fm (C_3)	Konkeling Fm (C_3)	C	Jilong Fm (C_3)	Lizha Fm (C_3)

Table 12.2. *Correlation of Permian faunal sequences in China.*

Series	Stage		Fusulinids		Corals	Brachiopods	Ammonoid zones	Floras
Upper Permian	Changxingian	*Palaeofusulina* Z.	*P. sinensis* Subzone *Gallowayinella meitienensis* Subzone		*Waagenophyllum–Huayunophyllum* Ass.	*Peltichia zigzag–Spinomarginifera chengyaoyengensis* Ass.	*Rotodiscoceras* Z. *Pseudotirolites–Pleuronodoceras* Z. *Pseudostephanites–Tapashanites* Z. *Parotirolites–Shevyrevites* Z. *Iranites–Phisonites* Z.	*Ullmannia* cf. *bronnii–Gigantonoclea gueizhouensis* Ass.
	Wujiapingian		*Codonofusiella* Zone		*Liangshanophyllum Lophophyllum* Ass.	*Tyloplecta yangtzeensis–Squamularia grandis* Ass. *Edriosteges poyangensis–Alatoproductus trunctus* Ass.	*Sanyangites* Z. *Araxoceras–Konglingites* Z. *Anderssonoceras–Prototoceras* Z.	*Gigantopteris nicotianaefolia–Otofolium* spp.– *Lobatannularia multifolia* Ass.
Lower Permian	Maokouan		*Yabeina–Neomisellina* Z. (*Polydiexodina–Neomisellina* Z.) *Neoschwagerina* Z. (*Chusenella conicocylindrica* Zone)		*Ipciphyllum–Allotropiophyllum* Ass. *Iranophyllum–Ipciphyllum* Ass.	*Neoplicatifera huangi* Ass. *Cryptospirifer striata* Ass.	*Mexicoceras–Waagenoceras* Z. *Kufengoceras* Z.	*Gigantonoclea fukienensis–Tingia carbonica* Ass.
	Qixian		*Parafusulina–Cancellina* Z. *Misellina claudiae* Z. *Schwagerina tschernyschewi* Z.	*Nankinella orbicularia Pisolina excessa* Zone	*Polythecalis yangtzeensis* Ass. *Hayasakaia elegantula* Ass. *Wentzellophyllum volzi* Ass.	*Chaoina reticulata* Ass. *Tyloplecta richthofeni–Orthotichia chekiangensis* Ass.	*Pseudohalorites* Z. *Medlicottia* Z. *Neocrimites* Z.	*Emplectopteris triangularis–Taeniopteris multinervis* Ass.

Table 12.3. *Correlation of Permian subdivisions in six major regions.*

Series	Stage		China	Europe	N.W. Europe (Germany)		E. Russian Platform	Australia (Queensland)	USA (Delaware Basin)		
Upper Permian	Changxingian	*Palaeofusulina sinensis*	Bio-zones *Rotodiscoceras* Z. *Pseudotirolites–Pleuronodoceras* Z. **Pseudostephanites–Tapashanites Z.** *Paratirolites–Shevyrevites* Z. *Iranites–Phisonites* Z.	Tatarian	Buntsandstein		Vyatskiy	?Rewan	Dewey lake	Ochoan	Upper Permian
	Longtanian		*Sanyangites* Z. *Lobatannularia Araxoceras–Konglingites* Z. *Anderssonoceras–Protoceras* Z. *Edriosteges poyangensis–Tyloplecta yangtzeensis* Z. *Gigantopteris nicotianefolia–Lobatannularia multifolia* Ass.				Severodvinskiy	Baralaba			
							Urzhumskiy	Tamaree			
Lower Permian	Maokouan		*Yabeina–Neomisellina* Z. *Urushtenia crenulata–Semibrachythyrina* Ass. *Neoschwagerina margaritae* Z. *Cancellina* Z. *Cryptospirifer–Tyloplecta grandicosta* Ass. *Neoplicatifera huangi* Ass.	Kazanian	Ohre	Zechstein	Upper Kazanskiy	U Curra LST Pelican Creek	Rustler	Guadalupian	
					Aller				Salado		
					Leine			Scottville	Castile		
					Stassfurt Evaporites		Lower Kazanskiy				
					Hauptdolomit-Stinkschiefer				Capitan		
				Ufimian	Werra		Shemshinskiy				
					Zechsteinkalk		Solikamsky	Exmoor	Word		
					Kupferschiefer						
				Kungurian	Weissliegendes		Irenskiy Filippovskiy	Gebbie	Leonardian		
	Qixian		*Misellina claudiae* Z.	Artinskian	Rotliegendes		Ikskiy	Sirius shale			Lower Permian
	Unnamed Stage		*Propopanoceras* Z.	Sakmarian			Sterlitamakskiy	Tiverton	Wolfcampian		
							Tastubskiy	Lizzie creek			

13. THE TRIASSIC SYSTEM
Yang Zunyi

INTRODUCTION

The Triassic, the first period of the Mesozoic, inherited and maintained the Late Palaeozoic palaeogeographic outline, as no strong orogeny took place in the interval. The Triassic rocks here are widespread (Fig. 13.1) with various sedimentation types, complicated lithofacies, and marked by two striking features. First, in terms of space, their outcrops are of tripartite arrangement, i.e. the Qinling–Kunlun mountains separate continental deposits in the north (including some Late Triassic extrusives in north-east Heilongjiang) from marine deposits in the south. The latter is further divided by the Longmen–Kam–Yunan landmass into two types: to the east stable neritic deposits of South China, and to the west mobile basin deposits. Second, in terms of time, a distinct bipartite development may be observed especially in the southern marine regime, where during the Early and Middle Triassic, neritic carbonates predominated, whereas in the Late Triassic paralic clastics prevailed, with the Indosinian Movement intervening between the two. Hence, the Chinese Triassic appears to show a bipartite division, although the world-wide usage of a three-fold division is generally followed here. Furthermore, the Lower Series was formerly divided into two (Induan and Olenekian), but now into three

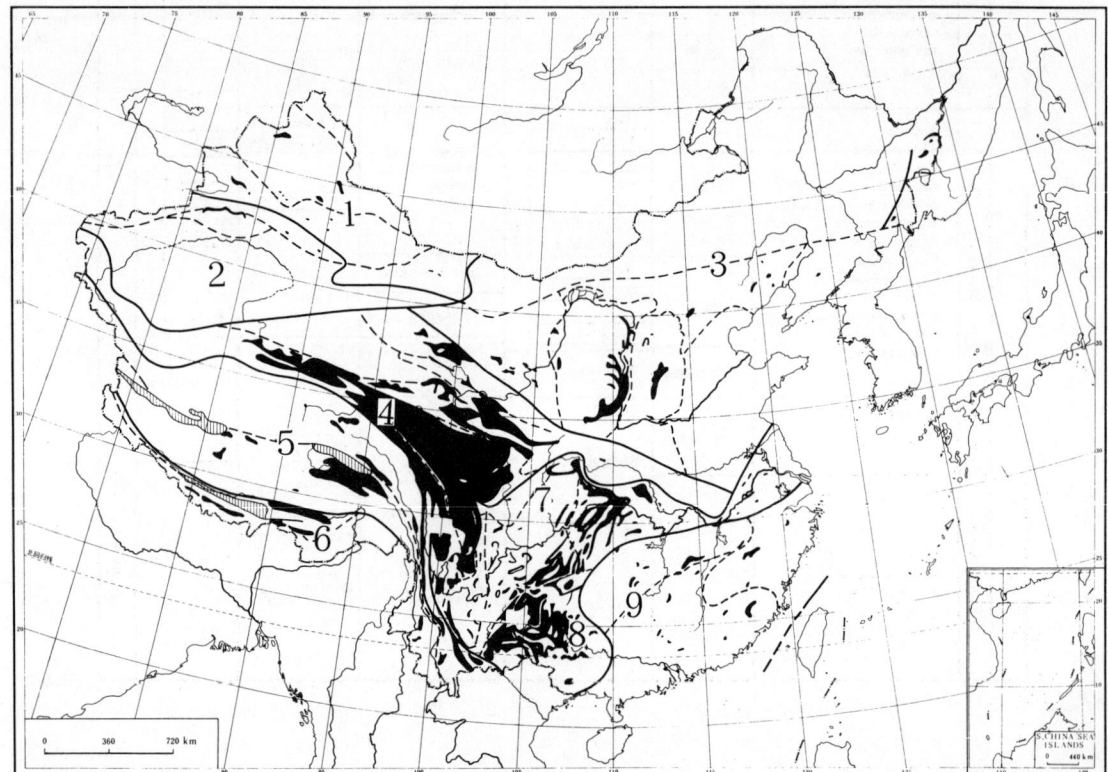

Fig. 13.1. Sketch map showing the distribution and stratigraphic regions of the Triassic System in China. 1, Northern Xinjiang–Beishan region; 2, South Tianshan–Tarim region; 3, North China region; 4, Kunlun–Qinling region; 5, Xizang–West Yunnan region; 6, Himalayan region; 7, Yangzi region; 8, Youjiang region; 9, South China region.

(Griesbachian, Nammalian, and Spathian) stages, the Middle Series into two (Anisian and Ladinian) stages, and the Upper Series again into three (Carnian, Norian, and Rhaetian) stages.

According to Yang and his Triassic research group (Yang et al. 1982a,b), nine stratigraphic regions are recognized. (A) Stable sedimentation type: northern Xinjiang–Beishan region (1), South Tianshan–Tarim region (2), North China region (3), South China region (9), Yangzi region (7), and Southern Himalayan region (6); (B) Mobile sedimentation type: Kunlun–Qinling region (4), Xizang–West Yunnan region (5), and Northern Himalayan region (6); (C) transitional sedimentation type: Youjiang region (8).

STABLE SEDIMENTATION TYPE

This type occurs both in the marine (e.g. Yangzi, South China, and Southern Himalaya) and continental (North Xinjiang–Beishan, South Tianshan–Tarim, and North China) regions.

Marine: Yangzi region

The marine Triassic is well-developed in the Yangzi region (Fig. 13.1.7) where the Lower and Middle series are mainly composed of both normal neritic and lagoonal carbonates including a number of facies: littoral-neritic Feixianguan (chiefly clastics and some limestone), Yelang (mudstone and limestone), and Badong (clastic-carbonate-mudstone). They are richly fossiliferous, with various taxa of the Tethyan realm, dominated by benthonic molluscs. The Lower Triassic consists of three mollusc assemblages (*Claraia wangi–Ophiceras*, *Claraia aurita–Eumorphotis multiformis*, and *Pteria* cf. *murchisoni–Tirolites*), five conodont zones (*Isarcicella isarcicus*, *Neospathodus dieneri*, *Neospathodus pakistanensis*, *Neospathodus waageni*, and *Neospathodus homeri*, and two foraminiferal zones (*Ammodiscus feixianguanensis*, and *Meandrospira insolitus–'Glomospira' sinensis*), all in ascending order. For the early Middle Triassic (Anisian) there are specialized benthic forms, such as the *Eumorphotis* (*Asoella*) *illyrica–Myophoria goldfussi* assemblage which spread over almost all the whole region; the cephalopod *Progonoceratites*, a well-known saline-sea form; two foraminiferal zones (*Lituotuba suevica–Meandrospira waagtsanensis* zone and *Nodosaria leikoupoensis* zone), and conodonts (*Neospathodus germanicus–N. kocheti* zone). The Late Middle Triassic (Ladinian) is represented by *Protrachyceras prinum* and the *P. deprati* zones. In the fluvio-deltaic facies it is characterized by ostracods (*Darwinula*), and estherids (*Euestheria hubeiensis*), and the appearance of freshwater bivalves.

Following the Ladinian marine regression, Late Triassic gulf deposits developed, mixed with paralic to limnic deposits and coal deposits, yielding in the early stage bivalves (*Halobia, Costatoria kuichouensis*) and plants (*Nilssonia*, etc.) and in the middle and late stages Napeng fauna as well as the *Ptilozamites–Lepidopteris* flora. The whole sequence varies from 1000 to 7000 m thick, being thickest along platform margins, lacking locally the Lower and Middle Series, and commonly Ladinian and Carnian. It is usually conformable, and locally paraconformable, with both the Permian and Jurassic (Yang *et al.* 1982; Zhao Jinke, Chen Chuzen *et al.* 1982).

Longmenshan, Sichuan

The Triassic sequence in the Longmenshan, Sichuan (Table 13.1.3; Fig. 13.2.3) is well-developed and conspicuous for the Yangzi region. The Lower Series includes the Feixianguan (Feihsienkuan) and the Jialingjiang (Chialingkiang) Formations. The Feixianguan consists of purplish-red siltstone, and sandy shale intercalated with oolitic limestone and marl, reaching a thickness of 125–475 m. It is rich in bivalves (*Oxytoma scythicum*, *Claraia wangi*, *C. aurita*, *Eumorphotis multiformis*, and *E. hinitidea*), and ammonoids (*Ophiceras* and *Lytophiceras*).

The Jialingjiang Formation comprises limestone, argillaceous dolomite, dolomitic limestone, and brecciated dolomite, totalling 350–650 m and yielding bivalves (*Eumorphotis inaequicostata* and *Pteria* cf. *murchisoni*), ammonoids (*Tirolites, Meekoceras evolutum*, and *Dinarites nudus*), conodonts (*Hindeodella*), and foraminifers (*Glomospira articulosa*).

The Middle Triassic is also made up of two units, the Leikoupo (Anisian) and the Tianjingshan (Ladinian) Formations. The first is composed of dolomitic limestone intercalated with marl and mudstone, 240–395 m thick. It yields bivalves (*Eumorphotis illyrica* and *Myophoria* (*Costatoria*) *goldfussi*), ammonoids (*Progonoceratites* spp. and *Beyrichites* sp.), and foraminifers (*Nodosaria leikoupoensis*). The Tianjingshan Formation comprises thick-bedded limestone which contains chert nodules or stripes intercalated with biogenetic limestone, varying from 0–109 m thick, and yielding bivalves (*Myophoria elegans*, and *Lima chinensis*), brachiopods (*Rhaetina angustaeformis*), and foraminifers (*Nodosaria microniutidana*).

The Upper Triassic Maantang Formation (Carnian) consists of silty mudstone, siltstone intercalated with biogenetic limestone, and oolitic limestone, reaching a

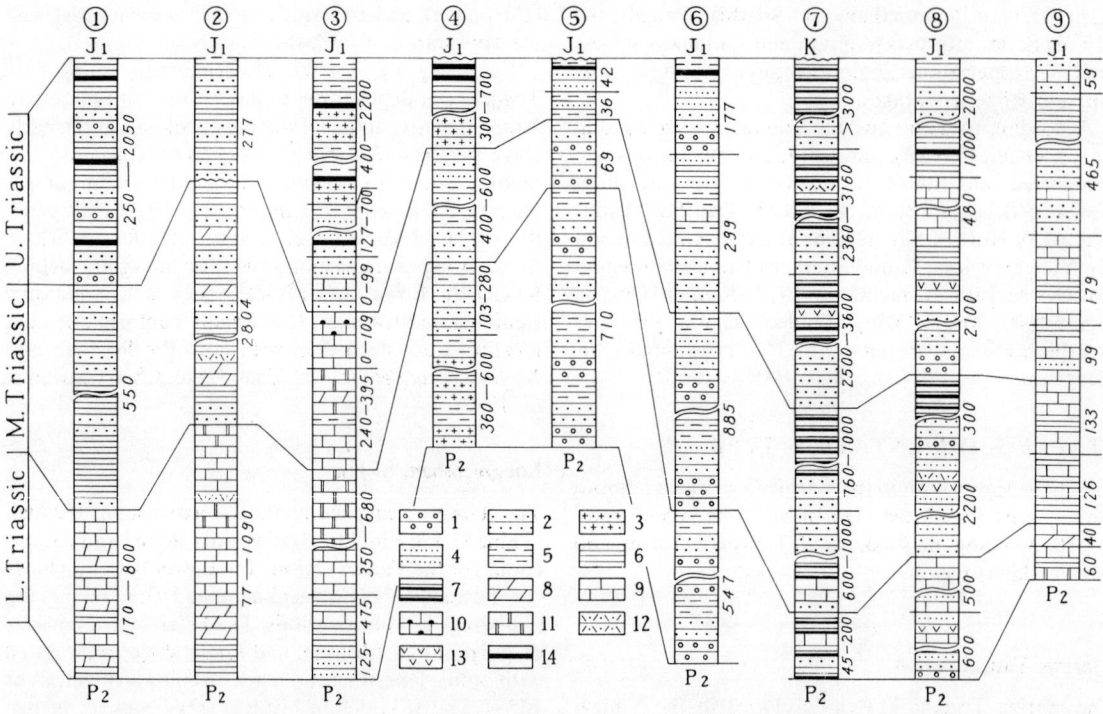

Fig. 13.2. Columnar sections showing development of Triassic strata in various stratigraphic regions of China. 1 North Guangdong, South China region; 2 Youjiang, West Guangxi; 3 Longmenshan, Yangzi region; 4 Shanxi, North China–North-east region; 5 Junggar, Xinjiang–Beishan region; 6 Kuche, South Tianshan–Tarim; 7 Garze–Yajiang, Kunlun–Qinliang region; 8 Qamdo, Xizang–West Yunnan region; 9 Nyalam, Himalaya region. Legend: 1, Conglomerate; 2, Sandstone; 3, Arkose; 4, Siltstone; 5, Mudstone; 6, Shale; 7, Slate; 8, Marl; 9, Limestone; 10, Limestone with chert nodules; 11, Dolomite; 12, Tuff; 13, Volcanic rocks; 14, Coal seam or strings.

thickness of 0–245 m. It carries bivalves (*Burmesia lirata* and *Yunnanophorus boulei*) and ammonoids (*Hoplotropites* and *Trachyceras*) of definitely Carnian age. The Yiaotangxi Formation is composed of fine-grained sandstone, mudstone, quartz sandstone, and coal strings, totalling 700 m thick, and bearing bivalves (*Burmesia lirata, Myophoriopsis quadrata, Yunnanophorus grandi* and *Pteria*) and plants (*Nilssonia*), all of which are of Norian age. The Upper Triassic Xujiahe Formation is made up of arkose, sandy mudstone, and coal seams, varying from 400 to 2200 m. It contains bivalves (*Yunnanophorus boulei*), ostracods (*Darwinula impudica*), fossil plants (*Dictyophyllum nathorsti* and *Daneopsis fecunda*), and spore (*Banksisporites pinguis*), typical of the Norian–Rhaetian age.

South China region

In the South China region (Fig. 13.1.9; Table 13.1) the Triassic rocks outcrop sporadically but are complete with three series. The Lower and Middle Series are of littoral-neritic purplish-red clastics, yielding biotas similar to those of the Yangzi region (also the Youjiang region), except in the eastern part of Hainan Island where terrestrial deposits with *Voltzia–Neuropteridium–Albertia* flora developed. The Upper Series is of paralic to lacustrine, intermontane facies and coal-bearing, even with continental extrusives locally. It is also characterized by circum–Pacific bivalves such as *Bakevelloides, Waagenoperna* and *Palaeopharus,* and the *Ptilozamites–Lepidopteris* flora. Its total thickness varies from 800 to 3500 m, mostly paraconformable with the Permian; but the Middle Series is mostly unconformable with the Upper Series indicating the Indosinian orogeny and the latter series is mostly conformable or paraconformable with the Jurassic.

The Triassic sequence of northern Guangdong may represent the development of the system in the South China region (see Table 13.1.1). The Lower Series is represented by the Daye Formation of limestone and marl intercalated with mudstone, having a thickness of

170–800 m and yielding in its lower part *Claraia wangi* and *C. griesbachi* (bivalves), and *Ophiceras* sp. (ammonoid), and in its upper part *Eumorphotis* cf. *inaequicostata* and *Unionites fassaensis* (bivalves), and *Meekoceras* sp. (ammonoid), equivalent respectively to the Induan and Olenekian stages. These are succeeded by the Huangben Formation of purplish-red sandstone intercalated with mudstone, with a thickness of 550 m, and yielding crinoid (*Isocrinus candelabrum*), bivalves (*Gervillia* sp.), and plants (*Todites* sp., *Pterophyllum* sp. and *Neocalamites* sp.) all of which suggest the Anisian–Ladinian stages.

The Upper Series includes the following formations in ascending order. The Hongweikeng Formation is of greyish-black, sandy shale and coal seams, 45–510 m thick, rich in bivalves (*Guangdongella exquita, G. longimorpha, Bakevellia matsushitai, B. loujiaduensis, B. guangdongensis* and *Myophoriopsis acurus*) of Carnian age. The Xiaoshui Formation is of greyish-white sandstone intercalated with mudstone, totalling 80–120 m thick, and is marked by a prolific number of bivalves (*Palaeopharus lanceolatus, P. oblongatus, Tosapecten, Oxytoma mojsisovicsi, Isocardioides yini, Plagiostoma xiaoshuiensis, Asoella confertoradiata, Chlamys mojsisovicsi,* and *Modiolus*) of Carnian–Norian age. The Ganxi Formation comprises greyish-white to greyish-black sandy shale and coal seams, with a thickness of 100–150 m. It yields such bivalves as *Waagenoperna permoformis, W. lilingensis, W. isognomonnieformis,* and *Jiangxiella* of Norian age.

It is noteworthy that in the Late Triassic the South China region was mainly under the influence of the circum–Pacific ocean currents which spread the large number of bivalve taxa indicated above.

Southern zone of the Himalaya region

The stable marine sedimentation type occurs also in the southern zone of the Himalaya region (Mount Qomolangma) (Fig. 13.1.6; Fig. 13.2.9) where the Triassic is very well represented and characterized by ammonoids. The Lower Triassic Kangshare Formation is divided into lower and upper members. The lower member consists of grey shale and limestone with dolomite at its base, totalling 63 m thick and is rich in ammonoids (*Otoceras latilobatum, Ophiceras, Lytophiceras, Anotoceras, Kymatites, Clypeoceras, Prionolobus lilingensis* and *Gyronites psilogyrus*), all indicating an Induan Stage. The upper member is composed of purplish-red, greyish limestone and argillaceous limestone, 40 m thick; it is also rich in ammonoids (*Owenites, Pseudoceltites, Xenodiscoides, Pseudosageceras, Anasibirites, Gurleyites, Eophyllites, Keyserlingites, Nordophiceras, Albanites,* and *Procarnites*), all of Olenekian Stage.

The Middle Triassic Laibuxi Formation is also divisible into two members. The lower is an alternation of grey, yellowish-green shale and yellow sandy limestone, having a thickness of 126 m, and yielding a rich Anisian ammonoid fauna of 3 horizons: (1) *Leiophyllites, Ussurites, Japonites, Hollandites, Anagymnites, Buddhaites*; (2) *Anacrochordiceras, Beyrichites, Gymnites, Hollandites*; (3) *Ptychites rugifer, Malletoptychites, Beyrichites*). The upper member is chiefly grey limestone intercalated with grey shale in the middle part, totalling 133 m and bearing Ladinian ammonoids (*Protrachyceras, Paratrachyceras, Rimkinites, Israelites, Anolcites, Epigymnites, Joannites, Gymnites, Paralobites, Velebites,* etc.).

The Upper Triassic Carnian is represented by the Zhamure Formation, which comprises chiefly biogenetic limestone, commonly intercalated with 99 m of sandy shale and fine sandstone, and yields 3 assemblages of ammonoids: (1) *Indonesites dieneri, Trachysagenites herbichi*; (2) *Haplotropites, Tropites, Paratropites, Pleuropinacoceras*; (3) *Parahauerites, Carnites, Timortropites, Gymnotropites*.

Next in the succession is the Dashalong Formation, mainly limestone and often intercalated with sandy shale, 179 m thick and yielding ammonoids (*Nodotibetites*). The Qulonggongba Formation is composed chiefly of sandy shale and sandstone, often intercalated with limestone, 465 m thick, and contains ammonoids (*Indojuvavites, Parajuvalites, Distichites, Diffmarites, Parathisbites, Cyrtopleurites, Heliclites,* and *Pinacoceras*). The Derirong Formation is of quartz sandstone with some limestone at the base, totalling 591 m, and yielding bivalves (*Palaeocardita mansuyi, Myophoricardium fulongense,* and *Indopecten*). All these formations are correlated to the Norian (or Norian–Rhaetian).

North China region

Stable continental sedimentation was developed in North China, the South Tianshan–Tarim and Xinjiang–Beishan regions. Continental Triassic rocks are well developed and crop out extensively in North China, where the Lower and Middle Series are composed of purplish-red fine clastics, reflecting their deposition in arid or semiarid conditions, but the Upper Series include greyish-green clastics with coal-bearing deposits reflecting the temperate–humid conditions during their deposition. They reach a total thickness of 1500–4500 m. The Lower Series is characterized by *Pleuromeia*, a plant genus widespread in the southern part of Laurasia, as well as 'glossopterid' type plants, associated with *Fugusuchus* (reptilian), and *Darwinula* (ostracod) fossils. The early Middle Series carries the *Voltzia–Aipteris wuziwanensis* assemblage, typical of North China flora, while the late Middle Series is marked by the newly recognized Tongchuan flora, i.e. *Analepis–Tongchuanphyllum*, associated also with rich *Sinokannemeyeria* fauna. The Upper Series yields the *Thinnfeldia–Danaeopsis fecunda* assemblage which is closely related to the cosmopolitan floras. The whole sequence is commonly conformable with the Permian but disconformable with the Jurassic.

This type of deposit is well-developed in Central Shanxi (Fig. 13.2.3; Table 13.1.4) where the Lower Series is represented by the Liujiagou and Heshanggou Formations, the Middle by the Ermaying Formation, and the Upper by the Yanchang Formation. The Liujiagou Formation, overlying conformably the Upper Permian Shiqianfeng Formation (or Group), consists of purplish-red to brownish-red arkose, and mudstone with well-developed cross-bedding and mudcracks, totalling 360–600 m, and yielding *Pleuromeia* spp. and *Darwinula* spp. The Heshanggou Formation is chiefly brick-red or brownish-red mudstone and sandy mudstone, 103 m thick, and contains plants (*Pleuromeia sternbergi*, *P. rossica*, *Anomopteris mougeoti*, *Equisetites* sp. and *Voltzia* sp.) and vertebrates. The Middle Triassic Ermaying Formation is composed of green and yellow sandstone intercalated with purplish-red, thin, silty mudstone in the lower part, and pinkish-red to greenish medium sandstone intercalated with purplish sandy mudstone in the upper part. It contains numerous plants (*Pleuromeia wuziwanensis*, *Bernoullia zeilleri*, *Todites shensiensis*, *Cladophlebis raciborskii*, *Thinnfeldia* sp., *Protoblechnum wongii*, *Pachypteris* sp., *Aipteris wuziwanensis*, *Nilssonia grandifolia*, *Ctenozamites sarrani*, *Ginkgo* cf. *marginatus*, *Sphenobaiera* cf. *crossinervis*, and *Glossophyllum*? *shensiense*.) and vertebrates (*Sinokannemeyeria*, *Parakannemeyeria* (Fig. 13.3), *Shansiodon*, and *Shansisuchus* (Fig. 13.4).

The Upper Triassic Yanchang Formation is divisible into a lower part of greenish and purplish thick-bedded arkose and green and black shale, occasionally with oil-shale, and an upper part of grey and green arkose and sandy shale with coal seams in the upper part, totalling 300–700 m. It is marked by the *Daneopsis–Bernoullia* assemblage consisting of *Equisetites* spp., *Neocalamites* spp., *Danaeopsis fecunda*, *Asterotheca szeiana*, *Bernoullia zeilleri*, *Todites shensiensis*, *Phlebopteris*? *linearifolia*, *Cladophlebis* spp., *Sphenopteris chowkiawanensis*, *Thinnfeldia* spp., *Protoblechnum hughesi*, *Aipteris* spp., *Lepidopteris ottonis*, *Glossophyllum*? *shensiense*, and *Sagenopteris* spp.

Junggar (Karamay area), North Xinjiang–Beishan region

Continental Triassic rocks are widely distributed also in Junggar in the North Xinjiang–Beishan region (Fig.

Fig. 13.3. *Parakannemeyeria ningwuensis* Sun, Ermaying Formation, Middle Triassic, Central Shanxi ($\times \frac{1}{8}$).

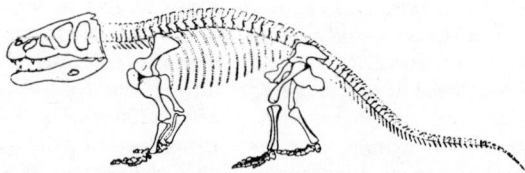

Fig. 13.4. *Shansisuchus shansiensis* Young, Ermaying Formation, Triassic, Shanxi ($\times \frac{1}{50}$).

13.1.1) representing stable to transitional deposits in inland intermontane basins. The Lower Series is made up of red clastics and piedmont molasse, characterized by a *Lystrosaurus* zone (Fig. 13.5) (equivalent to the Middle Beaufort Formation of South Africa), and *Darwinula* spp. comparable to those of the early Lower Triassic of the Russian Platform. The Middle and Upper Series are composed of greyish-green clastics and the Middle Series yields the well-known *Sinokannemeyeria*, *Parakannemeyeria*, *Shansiodon*, and *Shansisuchus*. The Upper Series contains plants typical of Northern Chinese (Yanchang) flora. The whole Triassic sequence here is about 500–2000 m thick, overlying conformably the Permian (or overlapping it at basin margins) and being overlain disconformably by the Jurassic.

In the Karamay area (Fig. 13.2.5) the Upper Cangfanggou Group of the Lower Triassic consists chiefly of fluvial deposits including uneven intercalation of brownish-red to greyish-green conglomerate, sandstone, and mudstone, which are poorly fossiliferous, with only small number of ostracods and spores, reaching a thickness of 710 m. But in Turpan it is 119–561 m thick, divisible into the Jiucaiyuan and Shaofanggou Formations. The former formation yields reptiles (*Lystrosaurus hedini, L. latifrons, L. robustus, L. youngi, Chasmatosaurus yuani*), ostracods (*Darwinula elongata*), and gastropods (*Hydrobia turpanensis*). The Shaofanggou Formation, 83 m thick, is composed of purplish sandy mudstone, and bluish-green, medium to coarse-grained sandstones which are poor in fossils.

The Middle Triassic Karamay Formation in Karamay is 69 m thick and composed of red mudstone in the lower part and greyish-green sandstone in the upper part. It is rich in Yanchang flora. In Turpan it is 265 m thick and composed of clastics (sandstone and mudstone) yielding ostracods (*Darwinula elongata*), reptiles (*Turfanosuchus dabanensis*, kannemeyeriids), amphibians (*Parotosaurus turfanensis*), and plants (*Danaeopsis fecunda*).

The Upper Triassic Huangshanjie Formation in Karamay is 36 m thick, consisting of sandstone, and mudstone with pebbles, yielding in the upper part insects (*Homoptera*), conchostracans (*Jeanrogerum sornayi*), and plant fragments. In Turpan the same formation is 176 m thick, composed of greenish mudstone, sandy mudstone intercalated with carbonaceous mudstone, and thin-bedded sandstone with siderite concretions. Its upper part yields bivalves (*Ferganoconcha sibirica, F. burejensis, F. elongata, F. curta, Sibireconcha anodontoides, S. shensiensis*, and *Utschamiella yenchuanensis*) and plants (*Danaeopsis fecunda*). The uppermost Triassic Hejiagou Formation consists of lacustrine greyish-green to greyish-brown siltstone and mudstone, bearing coal strings with plant fragments.

Continental Triassic rocks crop out rather sporadically in the South Tianshan–Tarim region, but they are similar to those of the North Xinjiang Beishan region both in litho- and biofacies. For example, in Kuche (Fig. 13.2.6, Table 13.1.6) the Lower Triassic Oheblake Group overlies disconformably the Lower Permian and is an alternation of green sandstone–sandy mudstone, and purple conglomerate and conglomeratic sandstone, totalling 547 m and yielding estherids (*Polygrapta* sp., *Liograpta* sp., *Shaerograpta*? sp., *Viliginia* sp., and *Diaplex* sp.) and plants (*Equisetites* sp.).

The Middle Triassic Karamay Formation consists mainly of greyish-green sandy conglomerate, and is 885 m thick. It is rich in plant fossils such as *Danaeopsis, Bernoullia zeilleri, Thinnfeldia* cf. *rhomboidalis, Cladophlebis* cf. *suniana*, and *Glossophyllum*? *shensiensis*, and bivalves (*Utschamiella tungussica*).

The Upper Triassic is represented by the Huangshanjie and Taliqike formations. The former comprises greyish-white to green conglomerate and sandstone in the lower part, greyish-black to greyish-green carbonaceous mudstone intercalated with cone-in-cone limestone in the middle, and greyish-green siltstone intercalated with carbonaceous mudstone in the upper, totalling 299 m and yielding conchostracans (*Almatium gusevi* and *A. elongatum*) and fossil plants (*Cladophlebis* cf. *stenophylla, Marattiopsis mclesteri, Podozamites schenkii, Baiera czekanowskiana, Neocalamites carrerei*, etc.).

The uppermost Triassic Taliqike Formation is an alternation of greenish sandstone, siltstone, sandy mudstone, and carbonaceous shale and is also plant-bearing (*Neocalamites hoeriensis, N. carrerei, Cladophlebis* cf. *tsaidamensis, Asterotheca szeiana* etc.).

MOBILE SEDIMENTATION TYPE

The mobile type of sedimenation is extensively distributed in the Kunlun–Qinling region, where both eubasinal and miobasinal (locally platformal) deposits are

Fig. 13.5. *Lystrosaurus youngi* Sun, Early Triassic, Xinjiang ($\times \frac{1}{12}$).

well represented with mixed biotas of Northern (plus Northern flora) and Tethyan (plus Southern flora) origin. They comprise sandy slate, marble, volcanic and carbonate rocks, with sandy and muddy flysch predominating. During the latest phase of the Triassic, coal-bearing paralic deposits were developed instead.

Garze–Yajiang area, Western Sichuan

In the Garze–Yajiang area, Western Sichuan, the Lower Triassic Bocigou Formation is chiefly black sandy slate, 45–200 m thick. It yields typical Induan bivalves (*Claraia griesbachi, C. clarai, C. wangi*, etc.). The Middle Triassic Zagunao Group is made up of biogenetic limestone bearing crinoids (*Traumatocrinus hsüi*) in its lower part and grey calcareous quartz sandstone intercalated with slate in the upper part, the two being 600 m thick, and also bearing bivalves (*Halobia, Daonella* sp., *Posidonia* cf. *wengensis*). This is of Anisian–Ladinian age.

The Upper Triassic Xiatigu Formation is composed of fine-grained sandstone, siltstone, and grey-black slate, 760–1000 m, yielding bivalves (*Daonella* sp., *Halobia superba*) and corals (*Montlivaltia* sp.) of Carnian age. This is succeeded by the Xinduqiao Formation composed chiefly of dark-grey slate, 760–1000 m thick and yielding bivalves (*Halobia convexa, H.* cf. *yunnanensis*, and *H. talzuana*) and ammonoids (*Trachyceras* sp.). These last two units may belong to the Carnian Stage. They are followed by the Kanzhugou Formation divisible into a lower member of metamorphosed sandstone and slates, 200–2350 m thick, yielding *Trigonodus* and plants, and an upper member of alternating sandstone, siltstone, and slate, containing *Halobia* cf. *ganziensis* and *Pergamidia*. They are believed to belong to the Norian. Finally the uppermost Gedicun Formation is made up of sandy slate, and conglomerate intercalated with coal seams, with a thickness of 300 m, yielding fossil plants (*Cladophlebis* sp., *Clathrophyopsis leeiana, Todites denticulata, Anthrophyopsis leeiana, Pterophyllum prinum*, etc.), all of Norian or Norian–Rhaetian age.

Xizang–West Yunnan region

In the Xizang–West Yunnan region (Fig. 13.1.5; Fig. 13.2.5) the Lower and Middle Series, as in the last region, are composed of basinal facies, including *melange* and ophiolite. They are transgressed by the Upper Triassic composed of stable neritic and coal-bearing paralic deposits, yielding Tethyan biotas, similar to those of the Youjiang region. The early Middle Triassic (Anisian) is marked by ammonoids (*Balatonites, Cuccoceras*, and *Hollandites*), and the late Middle Triassic (Ladinian) by the *Daonella–Protrachyceras* assemblage. The Upper Series yields ammonoids (*Thisbites, Tibetites, Anatibetites*, and *Paratibetites*) and bivalves (*Halobia superba, Burmesia lirata, Yunnanophorus boulei*), totalling 4000–10 000 m in thickness, keeping unconformable contacts with both the Upper Permian and the Jurassic. In Qamdo, Xizang (Table 13.1.8; Fig. 13.2.8) the Triassic is represented by seven formations in ascending order as follows. The Lower Triassic Pushuiqiao Formation comprises chiefly siltstone intercalated with volcanic rocks and limestone, sandstone, and conglomerate, some 600 m thick, and yielding *Eumorphotis* cf. *multiformis, Natiria* aff. *costata* (showing an Induan age). The Lower Triassic Serongsi Formation consists mainly of limestone intercalated with siltstone, is about 500 m thick and is rich in ammonoids (*Owenites, Paranannites, Proptychitoides*) of Olenekian age.

The early Middle Triassic Watasi Formation consists primarily of clastics which are coarser in the lower part, finer with volcanic rocks in the middle, and again coarser with limestone in the upper part, reaching a thickness of 2200 m. It is fossiliferous with ammonoids (*Cuccoceras, Paracrochordiceras, Balatonites*, and *Japonites*) and bivalves (*Posidonia pannenica, Entolioides walashiensis*, etc.), equivalent to the Anisian Stage.

The late Middle Triassic Congla Formation comprising conglomerate, sandstone, and slate is about 300 m thick, yields ammonoids (*Protrachyceras, Gymnites*, and *Balatonites*) and bivalves (*Posidonia*), corresponding to the Latinian Stage.

The early Upper Triassic Jiapeila Formation is a rhythmic sequence of conglomerate sandstone, shale, and volcanics, some 2100 m thick. It is rich in hexacorals (*Distichophyllia yunnanensis, Paradistichophyllum parvus*, and *Margarosmilia zogangensis*) and bivalves (*Cornuicardia timorensis, Neomegalodon* cf. *cassianus*, and *N.* (*Rossiodus*) *columbella*) corresponding to the Carnian.

The Bolila Formation is made up of limestone about 480 m thick and contains ammonoids (*Cyrtopleurites, Ectolicites, Parathisbites*, and *Jellinekites*), bivalves (*Halobia superbescens, Indopecten margariticostatus*, and *Neomegalodon boeckhi*), brachiopods (*Halorella, Koninckina*, and *Amphidrina*), and corals (*Dischophyllia norica, Thamnastropis dronovi*, etc.) of Norian age.

The uppermost Triassic Bagong Formation is composed of shale with thin coal seams (marine) in the lower part, and terrestrial sandstone and shale in the upper part, totalling 1000–2000 m, yielding corals (*Dictyophyllum nathorsti*), and bivalves (*Burmesia* cf. *lirata*) of Norian age.

Northern Himalaya region

The northern zone of the Himalaya region lying along the Yarlungzangbo eubasinal metamorphic belt is marked by flysch slate and radiolarian chert, having a maximum thickness of more than 10 000 m. At least 1500 m of the Upper Triassic is preserved.

TRANSITIONAL SEDIMENTATION TYPE

Youjiang region, West Guangxi

The development of the transitional type of sedimentation may be exemplified in the Youjiang region, West Guanxi (Fig. 13.1.8; Fig. 13.2.8; Table 13.1.2), where the Lower and Middle Triassic are wide-spread, while the Upper Triassic is confined to the Shiwandashan. The Lower Series consists mainly of stable, outer, neritic mixed-rock associations; the Middle Series, of flysch clastics and deep-sea turbidites; the Upper Series contains coal-bearing paralic deposits, as well as piedmont molasse and inland lacustrine red beds, totalling over 6000 m of deposits. Intermediate acidic volcanic rocks are commonly found in various horizons, locally with medium-basic lava, spilite and *melange*. They yield fossils of the Tethyan province but are mixed with some northern elements.

BOUNDARY PROBLEMS

Lower limit of the Triassic

The determination of the lower limit of the Triassic System or the Permo-Triassic boundary is very important, for it is a boundary marking the drastic change from the Palaeozoic to the Mesozoic. It has been and still is a subject of considerable controversial interest (Kummel and Teichert 1970; Newell 1973; Waterhouse 1973; Kozur 1974, 1977). In China, however, the consensus is to fix the limit at the base of the *Otoceras* or *Ophiceras–Claraia wangi* assemblage (Yang Zunyi *et al.* 1981; Liao 1979, Zhao *et al.* 1981). Wherever a Permo-Triassic transitional section occurs, the earliest Triassic biozone (*Otoceras* or *Ophiceras* or *Ophiceras* plus *Claraia wangi* zone) is mixed with some productids (*Waagenites, Fusichonetes*), camarotoechids (*Neowellerella*), and terbratulids (*Araxathyris, Paracrurithyris*). These fossils are taken as Palaeozoic relics, for it should be noted that during the Late Permian Changxingian times there flourished ten groups of biotas (fusulinids, non-fusulinid foraminifers, rugose corals, bryozoans, brachiopods, bivalves, gastropods, cephalopods, trilobites, and crinoids) with over one hundred genera, among which the fusulinids, rugose corals, and trilobites became entirely extinct, whereas the number of taxa of brachiopods, and productids in particular, declined greatly in the Early Triassic. Hence, it seems proper and pertinent to draw the Permo-Triassic boundary at the time of the mass extinction or great decline of Palaeozoic life.

Lower–Middle Triassic boundary

The Lower–Middle Triassic boundary can be observed in the Yangzi region where the Jialingjiang Formation, a key formation for settling this boundary, is well-developed. At the first All-China Stratigraphical Congress (Beijing 1959), this boundary was drawn below the Jialingjiang Formation which was considered Middle Triassic in age, but since the early 1960s when more thorough study of this formation was undertaken by the Geological Institute, Academy of Geological Sciences and No. 201 Team of the Sichuan Bureau of Geology, the Jialingjiang was proved to be of Lower Triassic Olenekian age, based on ammonoids and bivalves: *Tirolites* spp., *T. spinosus* Moj., *Meekoceras* sp., *Xenodiscus* sp., *Xenoceltites* sp., etc., although a number of workers prefer to keep the fourth member of the Jialingjiang Formation in the Anisian. However, in Ebian and Nan Wenquang, Chongqing, the fourth member contains *Dinarites* cf. *posterus* (Diener) of Lower Triassic age. Lately, according to conodont workers (Wang and Cao (unpublished)) the Jialingjiang Formation yields *Gondolella, Parachirognathus, Neogondolella, Hindeolella, Neospathodus,* and *Roundya*, and especially, the fourth member contains *Pachycladina* sp. and *P. tridentata* Wang and Cao, all indicating an early Triassic age. On this evidence it seems appropriate to place the Early–Middle Triassic boundary at the top of the Jialingjiang Formation.

Anisian–Ladinian boundary

The Anisian–Ladinian boundary problem requires discussion of various facies. For example, in the Yangzi region the Middle Triassic is represented by the Leikoupo Formation which was formed in an enclosed marine basin under arid conditions, analogous to the Germanic Middle Triassic. It has been assigned by different workers to the Anisian, the Ladinian, or the Anisian–Ladinian, and the boundary between the two stages has been in dispute for some time. Recently however, support has been given to the view that the Leikoupo Formation belongs to the Anisian on the basis of *Progonoceratites pulcher* (Riedel) and *P. robustus* (Diener) (ammonoids), which are underlain by *Beyrichites* cf. *kesava* (Diener) and *Noetlingites* sp. Yang Zunyi et al. 1981). The first two species should mark the top of the Anisian according to Kozur (1972).

The Alpine-type Triassic is seen in South Guizhou and North Guangxi, where well-developed and continuous Middle Triassic sequences occur with the following ammonoid zones given in descending order:

Falang Formation (T$_2$f)　*Protrachyceras deprati* zone
　　　　　　　　　　　　Protrachyceras prinum zone

Qingyan Formation (T$_2$q)　*Paraceratites trinodosus* zone
　　　　　　　　　　　　　Paraceratites binodosus zone
　　　　　　　　　　　　　Nicomedites yohi zone
　　　　　　　　　　　　　Leiophyllites–Ussurites bed

The Anisian–Ladinian boundary for the time being may be drawn between the *Trinodosus* and *Prinum* zones. The Qingyan and Guanling Formations of Guizhou, the Leijoupo and Badong Formations of Sichuan, etc. are all correlatable and belong to the Anisian, whereas the Falang Formation of Guizhou, the Binyang Formation of Guangxi, and the Tianjingshan Formation of Sichuan may all be correlated to the Ladinian.

The Triassic sequences in West China (South Qilian, West Qinling, and Bulhanbuda, etc.) are marked by Tethyan fauna mixed with northern elements; hence they are very crucial for working out the world-wide correlation of the Middle Triassic. In West China the fossil horizons are as follows.

Upper Anisian–Upper unit: *Gymnotoceras* beds, containing *G.* sp., *Gymnites petilus* Wang and Chen; also brachiopods and bivalves.

Middle Anisian–Middle unit: *Anagymnotoceras* beds and equivalent horizons; *Anagymnotoceras dulanense* Wang and Chen, also *Hollanites hidimba* (Diener), *H.* cf. *visvakarma* (Diener), *Nicomedites osmanni* Toula, *Gymnites toulai* (Arthzber).

Lower Anisian–Lower unit: *Lenotropites* beds, including *L. debilis* Wang and Chen, *L. qinghaiensis* Wang and Chen, *Norites angusellatus* Wang and Chen, *Paracrochordiceras qinghaiensis* Wang and Chen, *Megaphyllites evolutus* Walter, *Japonites* cf. *raphaelis zojae* Tommasi, *Psilosturia mongolica* (Diener), *Ussurites hara* (Diener), *Arctohungarites laevigatus* Popov.

It may be noted that *Gymnotoceras* and *Anagymnotoceras* are characteristic of the North American Upper and Middle Anisian respectively and that *Lenotropites* is a guide for the Lower Anisian of the northern realm, associated with typical Lower Anisian taxa of Tethyan origin. So far no Ladinian fossils have been found in West China.

Upper boundary

As to the upper boundary, three types of contact must be considered. (1) In the case of the Guangdong type (marine sequence) it is placed at the base of the Hettangian *Psiloceras planorbis* zone or *Schlotheimia* zone. (2) For the Hunan–Jiangxi type (littoral–lagoonal deposits) it is put between the Sanqiutian Formation (T$_3$) and Zaoshang Formation (J$_1$). For the central Yunnan type (coal series and red beds) it is drawn between the Late Triassic Yipinglan Group and the Lower Lufeng (or Fengjiahe) Formation (J$_1$).

On the basis of fossil vertebrates, plants, and spores and pollen among the continental Permo-Triassic sequences in North China (Shaan–Gan–Ning Basin), the boundary between the Permian and Triassic is put between the Late Permian Shiqianfeng (Sunjiagou) Formation with *Shihtienfenia* and the Early Triassic Liujiajou Formation with *Pleuromeia*, and the upper limit lies between the Yanchang Formation (T$_3$) and the Fuxian Formation (J$_1$). The Lower–Middle Triassic boundary may be fixed between the Heshanggou with Procolophonidae, Fugusuchus and the Ermaying Formation with *Sinokannemeyeria* and *Shanbeikannemeyeria*. As to the Middle–Upper Triassic boundary the present trend is to draw it between the Tongchuan and Yanchang formations.

In North-west China (Junggar and Turpan basins, Xinjiang Region) the lower limit of the Triassic is at the base of the *Lystrosaurus*-bearing Jiucaiyuan Formation or between the *Lystrosaurus*-bearing Jiucaiyuan and the *Dicynodon*-bearing Wutonggou Formation. On the basis of spore and pollen the Lower–Middle Triassic boundary is drawn between the Shaofangou and Karamay formations. The upper limit is placed between the Upper Triassic Haujiagou Formation and the Lower Jurassic Badaowan Formation.

CORRELATION

For the correlation of Triassic sequences in various regions of China and between China and other parts of the world including Germany, Siberia (USSR), New Zealand and SW Nevada (USA) refer to Tables 13.1 and 13.3 respectively.

Table 13.1. *Correlation of Triassic stratigraphic sequences in various regions of China.*

Series	Stage	South China region N. Guangdong (1)	Youjiang Region W. Guangxi (2)	Yangzi Region Longmenshan, Sichuan (3)	North–N.W. China Region Shanxi (4)
Overlying		J_1	J_1	J_1	Heifeng Fm (J_2)
Upper Triassic	Rhaetian	Ganxi Fm *Waagenoperna permoformis* *W. lilingensis* *W. isognomonieformis* *Jiangxiella* 100–150 m	Fulongao Fm (or Pintong Fm)	Xujiahe Fm *Dictyophyllum nathorsti* *Danaeopsis fecunda*, etc. *Banksisporites pinguis* *Yunnanophorus boulei* *Darwinula impudica* 400–2200 m	Yanchang Gr *Equisetites* spp. *Neoclamites* spp. *Danaeopsis fecunata* *Todites shensiensis* *Cladophlebis* spp. *Sphenopteris choukiawanensis*
	Norian	Xiaoshui Fm *Palaeopharus lanceolatus* *P. oblongatus* *Tosapecten* *Oxytoma mojsisovicsi* *Isocardioides yini* *Plagiostoma xiashuiensis* *Asoella confertoradiata* *Chlamys mojsisovicsi* *Modiolus* 70–120 m	400–2200 m	Xiaotangzi Fm *Burmesia lirata* *Myophoriopsis quadrata* *Yunnanophorus grandi Pteria* *Nilssonia* 122–700 m	*Equisetites* spp. *Neocalamites meriani* *N.* spp. *Danaeopsis fecunda* *Bernoullia zeilleri*
	Carnian	Hongweikeng Fm *Guangdongella exquista* *G. longimorpha* *Bakevellia matsushitai* *B. luojiaduensis* *B. guangdongensis* *Myophoriopis* 45–510 m		Maatang Fm *Trachyceras* *Hoplotropites* *Burmesia lirata* *Yunnanophorus boulei* 0–99 m	*Todites shensiensis* *Annalepis zeilleri* 300–700 m
Middle Triassic	Ladinian	Huangben Gr *Isocrinus candelabrum* *Gervillia* sp. *Todites* sp.	Hekou Fm (or Xilung Fm) *Daonella lommeli* *Halobia* *Protrachyceras archelaus* *Bulogites* *Cuccoceras* *Aploceras* *Pseudoploceras* 2084 m	Tianjinshan Fm *Myophoria elegans* *Lima chinensis* *Rhaetina angustaeformis* *Nodosaria micronintidana* 0–109 m	Ermaying Fm *Todites shensiensis* *Cladophlebis raciborskii* *Aipteris wuziwanensis* *Nilssonia grandifolia* *Ginkgo* cf. *marginatus* *Bernoullia zeilleri* *Pleuromeia wuziwanensis* 400–600 m
	Anisian	*Pterophyllum* sp. *Neocalamites* ? 550 m		Leikoupo Fm *Eumorphotis illyrica* *Myophoria goldfussi* *Progonoceratites* spp. *Beyrichites* *Nodosaria leikoupoensis* 240–395 m	
Lower Triassic	Olenekian	Taye Fm Upper part *Eumorphotis* cf. *inaequicostata* *Unionites fassaensis* *Meekoceras* sp. Lower part *Ophiceras* sp. *Claraia clarai* *C. wangi* *C. griesbachi* 170–800m	Luolou Gr *Claraia wangi* *C. griesbachi* *C. stachei* *Pteria* cf. *murchisonii* *Ophiceras tingi* *Owenites* sp. *Meekoceras* *Dieneroceras* *Glyptophiceras* *Lytophiceras* *Paranorites* 77–1090 m	Jialingjiang Fm *Eumorphotis inaequicostata* *Pteria* cf. *murchisonii* *Meekoceras evolutum* *Dinarites nudus* *Hindeodella* *Glomospira articulosa* 350–680 m	Heshanggou Fm *Pleuromeia sternbergi* *P. rossica* *Anomopteris mougeoti* *Equisetites* sp. *Voltzia* sp. *Fugusuchus Capitosauridea* 103–208 m
	Induan			Feixianguan Fm *Eumorphotis multiformis* *E. hinitidea* *Claraia aurita* *Ophiceras* *Lytophiceras* *Claraia wangi* *Oxytoma scythicum* 125–475 m	Liujiagou Fm *Pleuromeia* spp. *Darwinula* spp. 360–600 m
Underlying		P_2	P_2	P_2	Shiqianfeng Gr (P_2)

THE TRIASSIC SYSTEM

Table 13.1. (cont.)

N. Xinjiang–Beishan Region Karamay (5)	S. Tianshan–Tarim Region Kuche, Xijiang (6)	Kulun–Qinling Region W. Sichuan (7)	Xizang–W. Yunnan Region Qamdo (8)	Himalaya Region Nyalam (9) Xizang
J_1	J_1	K	Chaya Gr (J_1)	Pupuga Fm (J_1)
Hejiagou Fm Plant fragments 42 m	Taliqike Fm *Neocalamites hoeriensis* *N. carrerei* *Cladophlebis* *Asterotheca szeiana* 177 m	Gedicun Fm *Cladophlebis* *Clathropteris* *Anthrophyopsis leeiana* *Todites denticulata* 300 m	Bagong Fm *Dictyophyllum nathorsti* *Burmesia* cf. *lirata* 1000–2000 m	Derirong Fm *Palaeocardita mansuyi* *Myophoricardium tulongensis* *Indopecten* 591 m
Huangshanjie Fm *Todites shensiensis* *Equisetites* sp. *Unio delunshanensis* *Ferganoconcha subcentralis* *Sibiriconcha jenssiensis* *Cuneopsis sichuanensis* *Ketmenia karamaica* *Subioblatta tongchuanensis* 36 m	Huanshanjie Fm *Almatium gusevi* *Cladophlebis stenophylla* *Marattiopsis m'cllesteri* *Podozamites schenkii* *Baiera czekanowskiana* *Ginkgo buttoni* *Neocalamites carrerei* 299 m	Kanzhugou Fm *Halobia ganziensis* *Parahalobia* *Pergamidia attalea* *Neocalamites* sp. 2360–3160 m		Qulonggongba Fm *Distichites* *Dittmarites* *Parathisbites* *Cyrtopleurites* *Helictites* *Pinacoceras* *Indojuvavites* 465 m
		Xinduqiao Fm *Halobia convexa* *H.* cf. *yunnanensis* *H. talauana* *Trachyceras* sp. 2500–3600 m	Bolila Fm *Cyrtopleurites* *Ectolicites* *Parathisbites* *Halobia* 480 m	Dashalong Fm *Nodotibetites* 179 m
		Xiatigu Fm *Halobia superba* *Daonella* *Montlivaltia* sp. 760–1000 m	Jiapeila Fm *Distichophyllia yunnanensis* 2100 m	Zhamure Fm *Parahhauerites* *Carnites* *Timortropites* *Gymnotropites* *Hoplotropites* *Indonesites* 99 m
Karamay Fm *Neocalamites damularioides* *Danaeopsis fecunda* *Thinnfeldia nordenskioldi* *Todites shensiensis* *Sphenobaiera* cf. *crassinervis* *Mesolimnadiopsis karamaica* Palaeoniscidae 69 m	Karamay Fm *Danaeopsis fecunda* *Bernoullia zeilleri* *Rhinnfeldia* cf. *rhomboidalis* *Cladophlebis* cf. *suniana* *Glossophyllum*? *shensiensis* *Utschamiella tungussica* 885 m	Zagunon Fm *Daonella* sp. *Posidonia* sp. *Halobia* sp. *Traumatocrinus hsüi* 600–1000 m	Congla Fm *Protrachyceras* *Gymnites* *Balatonites* *Posidonia* 300 m	Upper member *Paratrachyceras* *Protrachyceras* *Anolcites* *Joannites* *Gymnites* *Paralobites* 133 m
			Walasi Fm *Cuccoceras* *Paracrochor diceras* **Balatonites** **Japonites** **Posidonia pannonica** *Entolioides walashiensis* 2200 m	Lower member *Ptychites rugifer* *Beyrichites* *Japonites* *Holandites* **Anagymnites** **Buddhaiytes** *Ussurites* *Leiophyllites* 126 m
Upper Cangfanggou Gr Some ostracods and spores 15–710 m	Ohebulake Gr *Brachygrapta* sp. *Liograpta* sp. *Sphaerograpta* *Viliginia* sp. *Diaplex* sp. 547 m	Bocigou Fm *Claraia griesbachi* *C. jiajinensis* *C. obliquata* *C. clarai* *C.* cf. *wangi* 45–200 m	Serongsi Fm *Paranannites* *Protychitoides* *Unionites* *Owenites* 500 m	Upper member *Procarnites* *Albanites* *Nordophiceras* *Eophyllites* *Anasibirites* *Owenites* *Pseudoceltites* *Xenodiscoides* *Pseudosageceras* 40 m
			Pushuiqiao Fm *Eumorphotis* cf. *multiformis* *Natiria* aff. *costata* 600 m	Lower member *Cyronites psilogyrus* *Prionolobus liliangensis* *Clypeoceras* *Ophiceras* *Lytophiceras* *Anotoceras* **Otoceras latilobatum** 63 m
P_2	P_2	P_2	P_2	Selong Gr (P_1)

(Laibuxi Fm; Kangshare Fm — right margin labels)

Table 13.2. Correlation of Triassic faunal sequences in China.

Series	Stage	Ammonoid zones	Bivalves Ass.		Conodont zones		Brachiopod Ass.
Upper Triassic	Norian	*Himavatites columbianus Cyrtopleurites socius*	*Yunnanophorus–Permophorus*	*Entomonotis Ochotica* bed			*Himalairhychia media–Eoseptaliphoria tulungensis* (Himalayan region) *Halorella dongqaoensis–Septamphiclina qinghaiensis–Sacothyris sinosa* (Tethyan Northeast marginal region)
		Indojuvavites angulatus Griesbachites–Gonionotites Nodotibetites nodosus	*Burmesia lirata*		*Epigondolella abneptis*		
	Carnian	*Parahauerites acutus Hoplotropites Indonesites dieneri* bed	*Costatoria kweichouensis Heminajas forulata Cassianella beyrichi*	*Tosapecten Palaeopharus Bakevelloides Oxytoma*	*Neogondolella polygnathiformis*		*Neoretzia tibetensis–Oxycolpella oxycolpos–Sanqiaothyris elliptica* (Himalaya & Tethyan Northeast marginal region)
		Sirenites Austrotrachyceras Trachyceras aon T. aonoides					
Middle Triassic	Ladinian	? *Protrachyceras deprati*	*Daonella lommeli–Halobia kui*				*Volirhynchia multicostata–V. himaica* (Himalayan region)
		P. primum	*Asoella illyrica*				
	Anisian	*Paraceratites trinodosus P. binodosus Nicomedites yohi Paracrochordiceras–Japonites–Lenotropites*	*Costatoria goldfussi mansuyi*		*Neospathodus germanicus–N. kockeli Neogondolella constricta N. regale*		*Nudirostralina griesbachi–Tulungospirifer stracheyi* (Himalayan region) *Nudirostralina subtrigonodosi–Diholkorhynchia sinensis* (Gueizhou-Sichuan)
Lower Triassic	Spathian	*Procarnites–Ziyunites–Japonites* ? *Columbites asymmetricus*	*Eumorphotis inaequicostata–Pteria* cf. *murchisoni*		*N. jubata*	*Pachycladina obliqua–Neospathodus homeri*	
						Platyvillosus	
	Nammalian	*Anasibirites kingianus Owenites costatus Koninckites lingyunensis Proptychites kwangsiensis*			*Neospathodus waageni N. pauistanensis N. cristagalli*		
			Eumorphotis multiformis Claraia aurita				
	Griesbachian	*Gyronites psilogyrus Ophiceras (Lytophiceas) sakuntala Otoceras latilobatum*	*Claraia stachei* *C. wangi*		*N. dieneri Anchignathodus isarcicus*		*Fusichonetes pigmaea–Paryphella triquetra? Neowellerella pseudoutahi*

Table 13.3. *Correlation of Triassic subdivisions.*

Series	Stage	China		Germany	USSR Siberia	New Zealand		USA S.W. Nevada
			Ammonoid zone or bed					
Upper Triassic	Rhaetian	Erqiao Fm		Rhatkeuper	Iuosuchanskaya	Otapirian	Balfour	
	Norian	Huobachong Fm	*Himavatites columbianus Cyrtopleurites socius*	Steinmergel	Khedalichenskaya	Warepan		Gabbs
			Indojuvavites angulatus Griesbachites– Gonionotites Nodotibetites nodosus					
	Carnian	Banan Fm	*Parahauerites acutus Hoplotropites Indonesites dieneri*	Rotewand		Otamitan		Luning
				Schilfsandstein				
			Sirenites Austrotrachyceras Trachyceras aonoides, T. aon	Gipskeuper		Oretian		
Middle Triassic	Ladinian	Falang Fm	? *Protrachyceras deprati P. primum*	Lettenkeuper	Tolbonskaya	Kaihikuan	Gore	Grantsville
	Anisian	Guanling Fm	*Paraceratites trinodosus Paraceratites binodosus Nicomedites yohi* **Paracrochordiceras– Japonites– Lenotropites**	Muschelkalk		Etalian		Excelsior
Lower Triassic	Olenekian	Spathian	Yongningzhen Fm	*Procarnites– Japonites– Ziyunites*	Rot	Sygynkanskaya	Olenekian	Malakovian	Candelaria
			? *Columbites asymmetricus*	Solling Folge	Monomskaya			
				Hardegsen				
		Nammalian		*Anasibirites kingianus Owenites costatus Koninckites lingyunensis Proptychites kwangsiensis*	Detfurth	Bunter			
				Volpriehausen				
				Obere Folge				
	Induan	Griesbachian	Feixianguan Fm	*Gyronites psilogyrus Ophiceras (Lytophiceras) sakuntala Otoceras latilobatum*	Untere Folge	Ust'kelterskaya	Indian	?	
				Brockelschiefer				

14. THE JURASSIC SYSTEM
Yang Zunyi

INTRODUCTION

The Indosinian orogeny that took place during the Middle to Late Triassic caused large-scale marine regression, leaving the greater part of China under continental conditions, and only the Island of Taiwan, Xizang, South Qinghai, West Yunnan, and parts of Hunan and Guangdong under marine influence. So, from the beginning of the Jurassic, the general Triassic picture of seas in the South and lands in the North drastically changed and South and North China were united as one continent. Furthermore, from the Jurassic, there appeared an E–W differentiation of sharply contrasting scenes. The mountain ranges such as the Da Hingan, Taihang, Xuefeng may be taken as a dividing line; east of them lay the broad coastal stretch affected by strong diastrophism accompanied by igneous activity, while to the west were large-scale inland basins. Consequently, the Jurassic System of China is predominantly made up of terrestrial deposits.

The marine Jurassic is best developed in Western China, including Xizang, South Qinghai, the Mingtiegai–Chokoli area of Xinjiang, and West Yunnan. In Guangdong and Hunan the Lower Triassic is composed of marine and paralic deposits, whereas in the Nadan Hada (Wanda Hills), east Heilongjiang, the upper Middle and Upper Series are made up of paralic deposits. Following international practice the Jurassic in China is divided into three series: the Lower, Middle and Upper (Tables 14.1, 14.4). Wherever standard biotic zones or assemblages are present, they are correlated accordingly (Table 14.4), and faunal sequences are given (Tables 14.2, 14.3).

Terrestrial Jurassic rocks are distributed in Northwest China (in the Junggar, Tarim, Turpan, Qaidam basins, and the marginal parts of the Shaan–Gan–Ning Basin), South-west China (Sichuan, Central and Eastern Yunnan), in various intermontane basins, and in small faulted basins of the coastal region.

Among the terrestrial deposits, red beds predominate in Yunnan, Guizhou, and Sichuan, whereas coal-bearing deposits of the Lower and Middle Series and red beds or volcano-sedimentaries are known in Northern and Central China, and the Eastern coastal area.

On the basis of their areal distribution, sedimentary characteristics, biotic features, and stratigraphic sequences, the Jurassic rocks may be grouped into six regions as shown on Fig. 14.1: the South-west, Qinghai–Xizang, the North-west, the North-east, the South-east, and the Central South regions.

MARINE JURASSIC

Qinghai–Xizang region

Marine Jurassic rocks find their best expression in the Qinghai–Xizang region with four types of sedimentation in different sub-regions.

(1) The Yarlung Zangbo sub-region (Table 14.1.7) is marked by geosynclinal abysmal-bathyal deposits of fine clastics, flyschoids, and basic-intermediobasic volcanic rocks, characterized chiefly by ammonoids and pelagic radiolarians, showing a Kimmeridgian hiatus. The Lower Jurassic is represented by the Ridang Formation which consists in its lower and middle parts of dark and reddish shale, sandstone, marl, and limestone, intercalated with basalts in its upper part, totalling a thickness of 800 m. It yields *Psiloceras provincialis*, *Arietites* sp., *Prodactyoceras enodum* of Hettangian age. The Middle Jurassic Xiare Formation is made up of three parts: the lower is composed of basalt, and tuff intercalated with siliceous rocks and limestone; the middle, of limestone, marl, and shale; and the upper, of siliceous rocks, siltstone, and sandstone, 6000 m thick. The last two contain Callovian ammonoids (*Macrocephalites, Indocephalites, Reineckeia, Dolikephalites, Guliemiceras*), and Bathonian ammonoids (*Garantiana*), bivalves (*Inoceramus*), and belemnites (*Belemnopsis*). The Upper Jurassic Weimei Formation comprises quartz sandstone, sandy conglomerate in the lower part and shale in the upper, totalling 500 m, and yielding *Virgatosphinctes, Haplophylloceras strigile*, and *Himalayites* of Tithonian age.

(2) In the Mount Qomolangma sub-region (Nyalam for instance) there are platformal neritic-bathyal carbonate rocks and sandy shale without volcanic rocks or stratigraphic gaps. They are characterized by nektonic

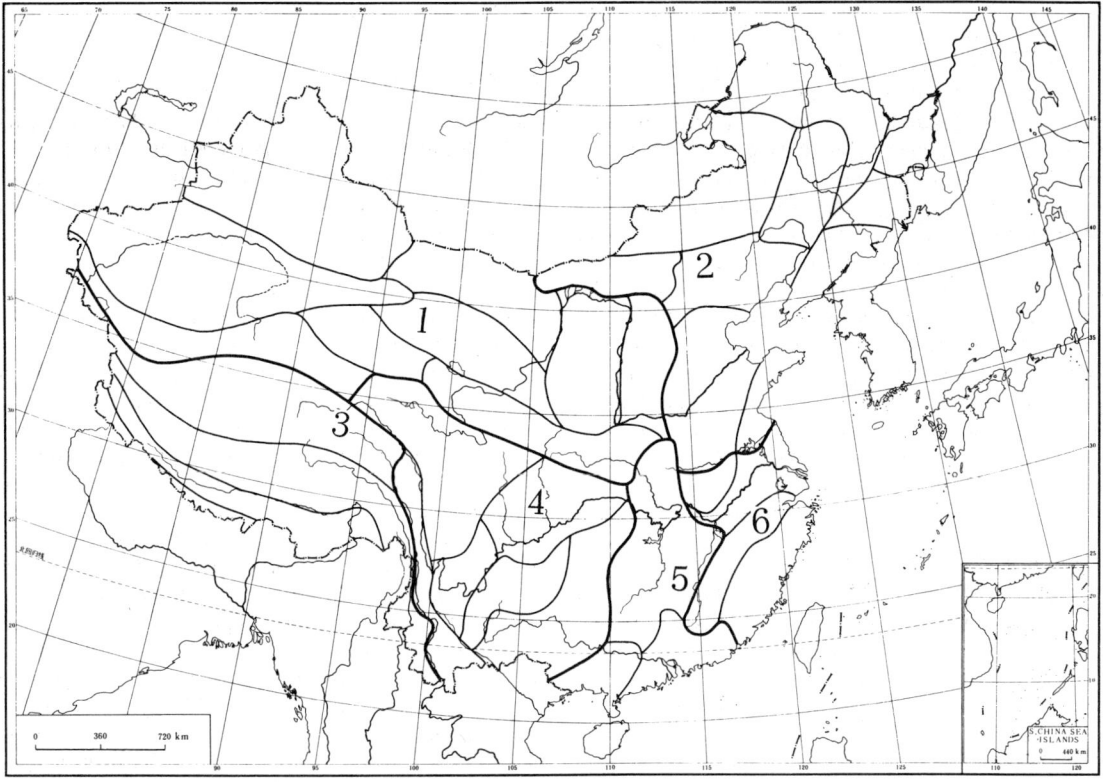

Fig. 14.1. Sketch map showing the stratigraphic regions of the Jurassic System in China. 1, North-west region; 2, North-east region; 3, Qinghai–Xizang region; 4, South-west region; 5, Central–South region; 6, South-east region.

ammonoids and benthic bivalves, brachiopods, corals, and foraminifers. The Lower Jurassic Pupuga Formation comprises yellowish-green shale and limestone, 882 m thick, and yields ammonoids (*Schlotheimia, Sulciferites, Gleviceras,* and *Nyalamoceras*). According to Li Xiao Chi (personal communication), the Middle Jurassic includes the Nieniexiongla and Lanongla formations. The Nieniexiongla is limestone, 775 m thick, with Bajocian ammonoids (*Witchellia, Dorsetensia, Sonninia, Emeleia,* (*Frogdenites*)); and the Lanongla Formation is an alternation of sandstone and shale, 741 m thick, with bivalves (*Grammatodon virgatus* and *Posidonia ornati*), in addition to Callovian ammonoids (*Macrocephalites, Indocephalites, Choffatia, Kamtocephalites, Subkossmatia,* and *Lunnuloceras*). The Upper Jurassic includes two other formations, Menbu and Xiumo. The Menbu Formation, 510 m thick, consists of sandy shale with concretions bearing Tithonian ammonoids (*Virgatosphinctes densiplicatus*) and belemnites (*Belemnopsis geradi*). The uppermost Jurassic Xiumo Formation is composed of sandstone and limestone with a thickness of 1782 m and bears Oxfordian and Tithonian bivalves (*Buchia rugosa, B. piochii, Grammatodon irretans, Entolium demissum, Chlamys, Praeconia,* and *Isocyprina*) and ammonoids (*Aulacosphinctes, Uhligites* sp., *Virgatosphinctes densi-plicatus, Pterolytoceras,* and *Haplophylloceras pingue*).

(3) In the northern Lhasa area the Jurassic sequence is represented by the late Middle (Callovian) and Upper Series, totalling 783 m (Wang Naiwen *et al.* 1983). It begins with an alternation of sandstone and shale, which carries bivalves (*Astartoides dingriensis*) and gastropods (*Pleurotomaria spitiensis*). This is succeeded by medium- to thin-bedded algal limestone bearing foraminifers (*Pseudocyclammina* ex. gr. *lituus* and *Ps. maynci* (?)), and calcareous algae (*Salpingoporella* cf. *johnsoni, S. annulata, Sarfatiella dunbari,* and *Cladocoropsis kotoi* var. *tosaensis*). Above these lie black concretionary shale, siltstone and sandstone which yield marine fossils (*Virgatosphinctes* sp. and *Aulacosphinctes* sp.) in the basal 25 m.

(4) In the Tangula sub-region the Jurassic sequence is

over 6500 m thick. The Lower Jurassic Zamnazu Formation is made up of shale and sandstone and yields ammonoids (*Gleviceras, Angulatinoceras, Baulticeras,* and *Arnioceras*), and bivalves (*Hippopodium ponderosum*). This is followed by the Saiwa Formation of sandy mudstone intercalated with marl carrying Bajocian ammonoids (*Witchellia, Dorsetensia, Oppellia, Zetoceras,* and *Okribites*). It is succeeded unconformably by the Yanshiping Group which is divisible into three formations: (1) the lower Wenquan Formation of clastics, yielding *Burmirhynchia* fauna; (2) the Jiamlechu Formation with purple clastics in the lower and limestone in the upper parts, bearing ammonoids (*Perisphinctes* of Oxfordian age and *Virgatosphinctes* of Tithonian age) and bivalves (*Chlamys superfibrosa*); and (3) the Xueshan Formation of siltstone and limestone yielding bivalves (*Nippononaia* and *Trigonoides*).

Northern Guangdong region

In the Northern Guangdong region (Table 14.1.2), overlying disconformably the Upper Triassic Xiaoping Group, is the Lower Jurassic Jinji Formation or Chujiang Formation, which consists of variegated siltstone and mudstone intercalated with conglomerate, having a thickness of 200–500 m. It yields Hettangian to Pliensbachian ammonoids (*Schlotheimia, Hongkongites, Arnioceras semicostatus, Arietites, Arieticeras,* and *Sulciferites*) and bivalves (*Cardinia toriyamai, Parainoceramus,* and *Meleagrinella*) in the upper part, and *Hiatella arenicola* in the lower part. The Jinji Formation is unconformably overlain by the Middle to Upper Jurassic Baizushan Group, which is purplish-red and greyish-white sandstone and conglomerate, intercalated with rhyolite porphyry of the Upper Jurassic.

Natan Hada (Wanda Hills) region

In the Natan Hada (Wanda Hills) region the Longzhuagou Group, 3700 m thick, is in fault contact with Permo-Carboniferous rocks. It is made up of sandstone and siltstone intercalated with mudstone, with coal series in the middle part; its middle and upper parts are intercalated with pyroclastics. It is marked by two marine horizons: the lower containing ammonoids (*Arctocephalites*) and bivalves (*Yoldia* sp., *Entolium demissum, Nucula* sp., and *Astarte expansa*), correlatable to the Upper Bathonian to Callovian; and the upper with bivalves (*Yoldia* sp., *Corbicella* cf. *eboraciensis, Quenstedtia* cf. *gracilis, Thracia* sp., and *Ostrea* sp.) equivalent to the Upper Jurassic.

NON-MARINE JURASSIC

North-west region

Here the Jurassic rocks belong to inland basin deposits and are typified by their development in the Shaan–Gan–Ning (Ordos) Basin (Table 14.1.5). The Lower Series is represented by the Fuxian Formation which overlies unconformably the Upper Triassic Yanchang Formation and consists chiefly of purplish-red to greyish-black sandy mudstone, about 100 m thick, yielding some fossil plants, spores and pollen, representing middle–upper Early Jurassic deposits. Typical Lower Jurassic also includes the Dashigou Formation of Eastern Gansu, which is composed mainly of grey to greyish-green and yellowish-green sandstone and siltstone, intercalated with mudstone and coal stringers, about 121 m thick. It yields numerous fossil plants, among which are also some Late Triassic elements in addition to abundant Early and Middle Jurassic forms, viz., *Neocalamites carrerei* (Zeiller), *Cladophlebis* cf. *gracilis* Sze, *C.* cf. *kaoiana* Sze, *Todites williamsoni* (Brongn.), *T. denticulata* (Brongn.), *Dictyophyllum* sp., *Clathropteris meniscoides* Brongn., *Swedenborgia cryptorioides* Nathorst, and *Yuccites spathaulatus* Pryn. The fossil plants of the Badauwan and Sangonghe Formations of the Junggar Basin, Xinjiang are similar to those of the Dashigou Formation. The Yongdingzhuang Formation, the Mingxian Group (Lower part) of Qinling, the Xiaomeigou Formation of Qaidam, the Ahe and Yangxia Formations of Kuche, Xinjiang, and the Shalitashi and Konsu Formations of Kashi, Xinjiang are regarded by most workers as being of Early and Middle Jurassic age.

Middle Jurassic rocks of the North-west region include the Yan'an and Zhilo Formations of the Shaan–Gan–Ning Basin, the Yaojie and Xinhe Formations of East Gansu, the Longjiagou Formation and upper Minxian Group of Qinling, the Datong and Yungan Formations of Shanxi, the Dameigou Formation of Qaidam, Xishanyao, the Toudunghe Formations of the Turpan Basin, Kizilnur, the Qiktai Formation of the Kuche Basin, the Yangye and Talga Formations of Kashi, and the middle and upper Yeljiang Group of Kunlun. They are all coal-bearing beds, yielding late *Coniopteris–Phoenicopsis* flora, bivalves (*Ferganoconcha, Tutuella, Sibireconcha,* and *Yananoconcha*), and ostracods (*Darwinula sarytirmensis* Sharapova, *D. impudica* Sharapova, *D. magna* Jiang, and *Timiriasevia* spp.). Among the various formations the Xinhe Formation is by far the richest in its fossil contents, mainly plants (*Coniopteris hymenophylloides* Brongn., *C. tatungensis* Sze, *Phoenicopsis angustifolia* Heer, *Gink-*

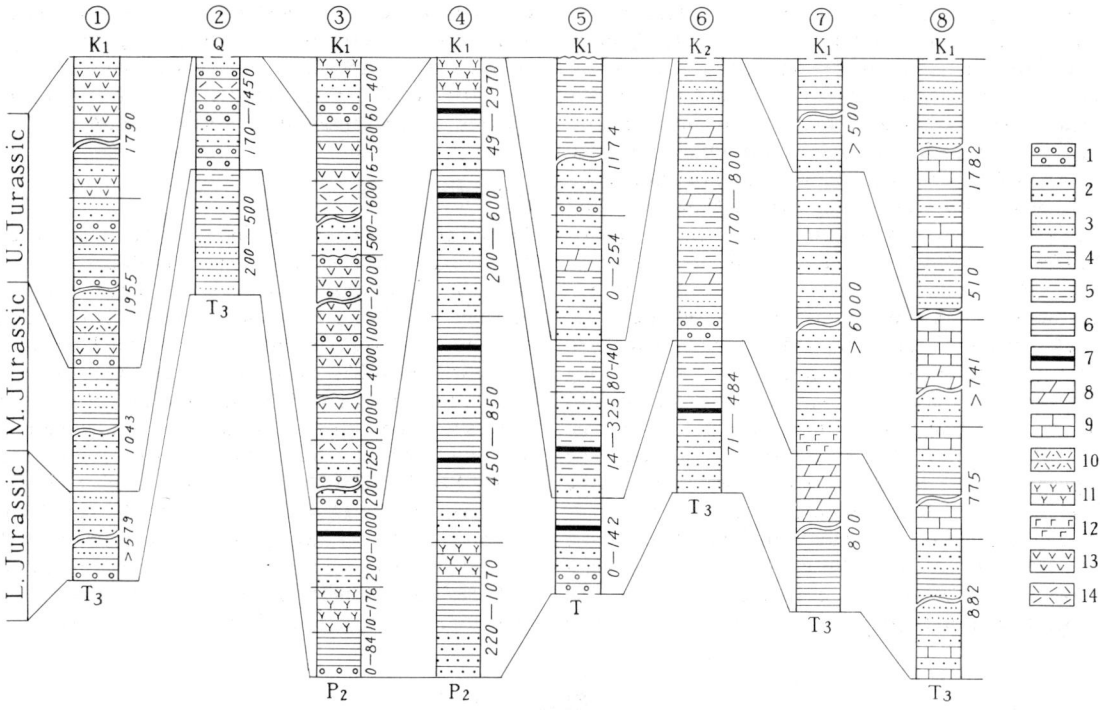

Fig. 14.2. Columnar sections showing Jurassic sequences in various stratigraphic regions of China. 1 Fujian, South-east China region; 2 West and Central Guangdong; 3 Xishan (Western Hills), Beijing, North–North-east China region; 4 W. Heilongjiang, North China and North-east China region; 5 N. Shaanxi, North-west China region; 6 East Yunnan, South-west China region; 7 Yarlung Zangbo R., Qinghai–Xizang region; 8 Nyalam, Xizang. Legend: 1, Conglomerate; 2, Sandstone; 3, Siltstone; 4, Mudstone; 5, Sandy shale; 6, Shale; 7, Coal seams; 8, Marl; 9, Limestone; 10, Tuff; 11, Volcanic rocks; 12, Basalt; 13, Andesite; 14, Rhyolite.

goites cf. *lepidus* (Heer), *Czekanowskia setacea* Heer, *Pityophyllum staratschini* Heer, *P. longifolium* Nath., and *Eboracia lobifolia* (Phillips), bivalves (*Ferganoconcha subcentralis* (Chernyshev), *Lamprotula* (*Eolamprotula*) *gansuensis* Ma, *Margaritifera isfarensis* Chernyshev, *Psilunio trigonus* Ma, *P. ovalis* Ma, *Cuneopsis linjiagouensis* Ma, *Pseudocardinia carinata* Martinson), ostracods (*Darwinula sarytirmensis* Sharapova, *D. impudica* Sharapova, and *D. magna* Jiang), and estherids (*Euestheria shandanensis* Chen, *E. ziliujingensis* Chen, and *E. haifanggouensis* Chen). The Xinhe Formation is thus of late Middle Jurassic age.

The Upper Jurassic of the North-west can be divided into two parts: the lower comprises mostly red sandstone, mudstone, and conglomerate, while the upper part consists of red or variegated sandstone and mudstone. To the lower part belong the Anding and Fengfanghe Formations of the Shaan–Gan–Ning Basin, the Hengtang, Tieyegou, and Kushuixia Formations of Gansu, the Tianhechi Formation of Shanxi, the Chigu and Karaza Formations of Xinjiang, and the Chaishiling Formation of the Qaidam. The Anding and Fengfanghe Formations are especially representative.

There is some uncertainty about the age of such terrestrial units as the Zhidang, Liupanshan, Hekou, Donghe, Turpan, and Kapsalian Groups, and the Hongshuigou Formation. A number of workers consider them Early Cretaceous (Hao *et al.* 1982), but according to the Jurassic Research Team (Wang Si'en *et al.* 1982), and Gu Zhiwei (1982*a,b,c,d*), they should be regarded as uppermost Jurassic. The former proposal is followed here.

North-east region

The Jurassic of this region is characterized by volcanic rocks including pyroclastics and sedimentaries, which are more completely developed in the Yanliao (Yanshan and Liaodong) area and are richly fossiliferous. For instance, in the Western Hills of Beijing (Table 14.1.3; Fig. 14.2.3) the Lower Jurassic is represented by three formations in ascending order as follows. The

Xingshikou Formation is made up of greyish-yellow sandstone, and shale with basal conglomerate. It varies from 0 to 84 m thick and contains fossil plants, such as *Todites denticulata, Ctenis chinensis,* and *Czekanowskia rigida.* The Nandaling Formation is of intermediate–basic volcanic rocks and agglomerates, and varies from 10 m to over 760 m. The Mengtougou Formation consists of coal-bearing clastics, 200–1000 m thick, and yields fossil plants (*Neocalamites carrerei, Coniopteris burejensis, Raphaelia* sp., and *Phoenicopsis angustifolia*). The Middle Jurassic includes the Jiulongshan Formation which is of purplish-grey to red, green and black tuffaceous sandstone, and siltstone intercalated with conglomerates, 200–1250 m thick, and bearing fossil plants (*Podozamites lanceolatus*). It is succeeded by the Tiaojishan Formation of andesite and its clastics, varying from 2000 to 4000 m in thickness. The Upper Jurassic is divisible into three formations: (1) the Houchang Formation composed of purplish-red sandstone and conglomerate intercalated with trachyandesite, 1000–2000 m thick, and yielding plants (*Neocalamites* sp. and *Coniopteris hymenophylloides*); (2) the Donglingtai Formation of quartz porphyrite, rhyolite, quartz andesite, and sandstone, totalling 500–1600 m; and (3) the Dahuichang Formation composed of pyroclastics and black shale, intercalated with andesite, 16–560 m, yielding *Neocalamites* sp., *Eosestheria* spp., *Ephemeropsis trisetalis*, and *Lycoptera davidi*.

South-west region

The Jurassic rocks of this region consist almost entirely of red beds, except in the north and north-east margins of the Sichuan Basin where the Lower Jurassic is coal-bearing. The Lower Jurassic includes the Baitianba Formation of North Sichuan, the Xiangxi Formation (*sensu stricto*) of East Sichuan, the Ziliujing Formation of Central Sichuan and North Guizhou, the Yimen Formation of South-west Sichuan, the Fongjiahe Formation of Central Yunnan, the Lower Lufeng Formation of East Yunnan, the Yangjiang Formation of West Yunnan, and the Daye Formation of Qamdo. These are classifiable into two types, coal-bearing and red-bed-bearing, represented by the Baitianba Formation and the Lower Lufeng Formation respectively. The first type yields fossil plants (*Todites denticulata*

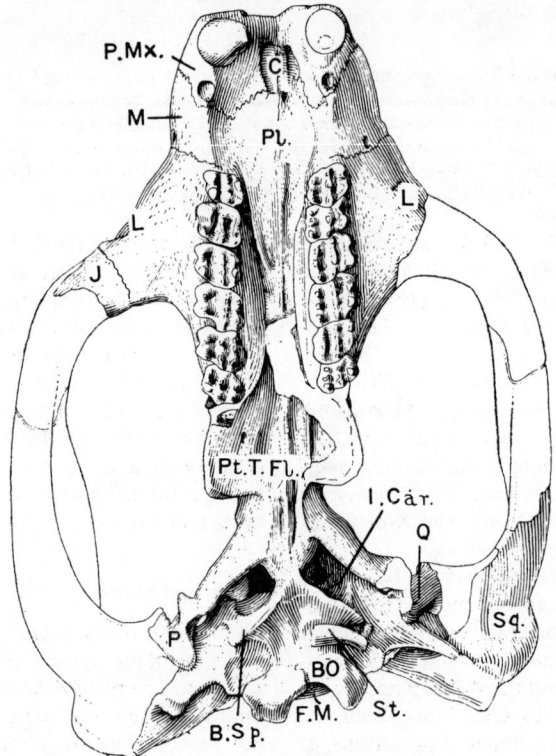

Fig. 14.3. *Bienotherium yunnanense* Young. Skull in palatinal aspect × 0.77 diam. Lower Jurassic, Yunnan.

(Brongn.), *Ptilophyllum pecten* (Phillips), *Anomozamites incinatans* (Braun), *Coniopteris hymenophylloides* Brongn., *Baiera muensteriana* (Presl), and *Czekanowskia rigida* Heer), estherids (*Palaeolimnadia baitianbaensis* Chen, and *P. chuanbeiensis* Shen), and bivalves (*Apseudocardinia* spp.). The red-bed type is composed mainly of purplish-red sandstone and mudstone, and contains fossil vertebrates, such as *Bienotherium yunnanensis* Young (Fig. 14.3), *Lufengosaurus changduensis* Chao (Fig. 14.4), *Plesiosaurus changduensis* Chao, *Megalosaurus tibetensis* Chao, and *Ichthyosaurus changduensis* Chao. The presence of *Ichthyosaurus* in the Daye Formation indicates a possible temporary marine connection.

To the Middle Jurassic belong the Qianfuyan, and Upper and Lower Shaximiao Formations of the Sichuan Basin, the Xincun Formation of South-west Sichuan, the Changhe Formation of Central Yunnan, the Upper Lufeng Formation of East Yunnan, the Huakaizuo and Hepingxiang Formations of the Lanping–Simao area in West Yunnan, and the Dabuka Formation of Qamdo, all of which are characterized by variegated clastic rocks—sandstone, siltstone, mudstone and marl, totalling 2599 m, and richly fossiliferous: e.g. the Changhe Formation of Central Yunnan yields estherids (*Euestheria xiazhungensis* Chen and *E. exilis* Chen); ostracods (*Darwinula* sp., *Metacypris monosulcata* Ye and *M. xianyunensis* Ye); insects (*Yunnanocaradus litus* Lin), and bivalves (*Pseudocardinia minuta* (Tscherny.), *P. elliptica* Kolesnikov, and *P. khadjalanensis* (Cherny)), all testifying to a Middle Jurassic age.

The Upper Jurassic is represented by the Suining, Fenglaizhen (Lianhuakou) formations of the Sichuan Basin, the Niugundang and Guangou Formations of South-west Sichuan, the Shedian and Tuodian Formations of Central Yunnan, the Bazulu Formation of West Yunnan, and the Kenzoga Formation of Qamdo, all of which consist mostly of brick-red to purplish-red mudstone, sandstone, and coarse sandstone, yielding ostracods (*Darwinula oblonga–Djungarica postiacuminata* assemblage), estherids (*Eosestheria* sp.), and charas (*Aclistochara* and *Protochara*), all pointing to an Upper Jurassic age.

Central–South region

Here the Lower Jurassic is comparatively well-developed, the Middle Jurassic less so, and the Upper Jurassic occurs only in a few basins. As mentioned above the Early Jurassic seas invaded parts of Guangdong and Hunan, forming marine paralic deposits. The Lantang Formation (*sensu stricto*) in East Guangdong consists typically of paralic deposits reaching a thickness of 3000–4000 m. In addition to ammonoids and marine bivalves in marine intercalations, the terrestrial beds contain fossil plants (*Cladophlebis meniscoides* Brongn., *Equisetites* cf. *sarrani* (Zeiller), *Todites* cf. *denticulata* (Brongn.), and *Coniopteris hymenophylloides* Brongn.). Lower Jurassic terrestrial deposits occur also in South and Central Hunan, North-east Guangxi, West Hubei (Danyang Basin), and South-east Hubei; they are called respectively the Xiangxi and Wuchang (s.s.) Formations in the last two areas, and are coal-bearing.

The Middle Jurassic, though not well-developed, is distributed in East Guangdong, South-east and North-east Guanxi, South Hunan and South-east Hubei. The Zhangping Group of East Guangdong consists mainly of purplish-red to greyish-green tuffaceous sandstone, siltstone, shale, and conglomerate, over 1000 m thick, and yielding bivalves (*Pseudocardinia* sp., *Tutuella rotunda* Ragozin, and *Ferganoconcha* sp.) and plants.

Upper Jurassic rocks are seen only in Shiwandashan, and Central and Eastern Guangdong. There the Gaozhiping Group comprises chiefly intermediate and intermedio-acidic volcanics, intercalated with sedimentaries, reaching a thickness of 1000–5000 m, and bear-

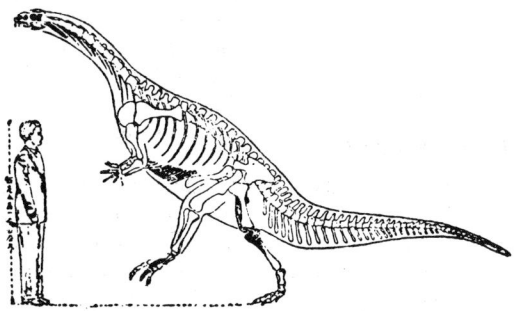

Fig. 14.4. *Lufengosaurus huenei* Young (Restoration). Lower Jurassic, Yunnan.

ing plants (*Cupressinocladus gracilis* (Sze), *Sphenopteris* sp., and cf. *Klukia browniana* (Dunker)).

South-east region

The Lower Jurassic of this region is generally coal-bearing, as in the Fanghu Formation of north Anhui, the Lower Xiangshan Group of Jiangsu, the Linshan Formation of Jiangxi, the Majian and Huaqiao Formations of Zhejiang, and the Lishan Formation of Fujian. For example, the Majian Formation of Zhejiang is chiefly grey, yellow and greenish sandstone, siltstone, conglomerate, and black shale with coal-seams, reaching a maximum thickness of 1135 m and yielding plants (*Coniopteris hymenophylloides* Brongn., *Cladophlebis raciborskii* Zeiller, *Todites denticulata* (Brongn.), *Clathropteris meniscoides* Brongn., *Swedenborgia* cf. *cryptomerioides* Nathorst, *Nilssonia* sp., *Baiera* sp., and *Otozamites* sp.).

The Middle Jurassic outcrops in north Anhui, Jiangsu, Jiangxi, Zhejiang, and Fujian may be exemplified by the Yushanjian Formation. It consists of yellowish-green to purplish-red conglomerate, sandy conglomerate, siltstone, fine-grained sandstone, and sandy mudstone, grading upward from a yellowish-green to purplish-red colour, and is more than 3000 m thick; it bears plants (*Coniopteris hymenophylloides* Brongn., *Cladophlebis raciborskii* Zeiller, *Todites verticulata* (Brongn.), *Ptilophyllum* sp., *Baiera gracilis* (Bean), and *Pagiophyllum* sp.) and bivalves (*Tutuella* sp., *Pseudocardinia* sp.).

The Jurassic of this region is similar to that of the North-east region in that it is an alternation of volcanics and sedimentaries. For example, the Laocun Formation of West Zhejiang (in Jiande) is chiefly of red clastics, and mudstone, intercalated with pyroclastics, over 2000 m thick, and yielding ostracods (*Rhynocypris* sp.), estherids (*Yanjiestheria sinensis* (Chi) and *Y. chekiangensis* (Novojilov)), and insects (*Ephemeropsis* cf. *trisetalis* Eichwald and *Chironomapteris meramira*).

BOUNDARY PROBLEMS

Triassic–Jurassic boundary in the marine regime

In the marine regime the Triassic–Jurassic boundary is fairly well-defined in South Xizang (Loza to Lancazhi for example), where the Noric–Rhaetic ammonoids (*Griesbachites* cf. *kastneri, Buchites* sp., *Clionites* sp., *Arcestes* sp., and *Cladiscites* sp.) together with sea-urchins from the Upper Triassic Kabo Group are found to underlie the Lower Jurassic Dalong Group with *Arietites*, though the Hettangian *Psiloceras* has not yet been discovered. According to Wang Yigang (1976) the Ridang Formation with *Psiloceras* was found in Longzi, and five km away to the east, grey shale and marl with Triassic ammonoids (*Placites, Cladiscites* sp.) were discovered. Hettangian *Psiloceras* and *Schlotheimia* also occur in Central Guangdong, where they are, however, underlain by continental deposits. On the basis of this evidence the lower limit of the marine Jurassic is fixed at the base of the Hettangian *Psiloceras* zone.

Upper limit of the Jurassic: the Jurassic–Cretaceous boundary

The upper limit of the Jurassic has been, and still is, in dispute. First, one group of geologists (Gu Zhiwei 1982*a,b,c,d*; Wang Si'en *et al.* 1982) include the Berrisian in the Upper Jurassic on the ground that Berrisian ammonoids (*Berrisia*, etc.) have closer affinities with the Jurassic ones. Another group of geologists (Hao *et al.* 1982), following European usage, would treat the Berrisian as the first Cretaceous unit, based on the evolutionary stages of many biotic groups including invertebrates, vertebrates, and plants (see Chapter 15). Furthermore, there is an even wider divergence of opinion concerning the Jurassic–Cretaceous boundary for the continental regime, and the unresolved criterion is the exact age of the so-called Rehe (Jehol) fauna (Grabau 1928) which characterizes the Rehe (Jehol) Formation (Group) distributed chiefly in North and North-east China.

In West Liaoning the 'Rehe fauna' originally referred to as the *Bairdestheria middendorfii* (estherid)–*Ephemeropsis* (insect)–*Lycoptera* (fish) assemblage is now extended, according to Gu Zhiwei (1983), to include molluscs, reptiles, and mammals, giving a Late Jurassic age. According to Hao and her Cretaceous research team (Hao *et al.* 1982), the Rehe Group is of Early Cretaceous age because of the presence of reptiles, insects, ostracods, plants, spores and pollen, and chara, though they acknowledge that a part of the evidence provided by a number of researchers on molluscs, fishes, and estherids suggests otherwise, i.e. making the Rehe Group Late Jurassic. Gu Zhiwei (1983, pp. 65–82) argues at length that the Jurassic paralic Longzhuagou Group might be correlated with the Jixi Group and the Rehe Group with the Rehe fauna, so they should belong to late Middle to Late Jurassic. Li Zishun *et al.* (1983) also correlated the greater part of the Longzhuagou Group with the Rehe and Fuxin groups, giving a Jurassic instead of a Cretaceous age. However, the actual stratigraphical relation between the Longzhuagou Group and the Jixi Group needs further study, but for the time being Hao's analysis of the age of the greater part of the Rehe Group is accepted, and hence the Jurassic–Cretaceous boundary for the non-marine regime in North and North-east China is drawn through, but not above, the Rehe Group.

CORRELATION OF JURASSIC SEQUENCES

The correlations of Jurassic sequences in various regions of China and between those of China and other countries (Dorset, England; West Siberia, USSR; New Zealand, and California, USA) are given in Tables 14.1 and 14.4 respectively.

Table 14.1. *Correlation of Jurassic stratigraphic sequences in various regions of China.*

Series	Stage	S.E. China Region Fujian (1)	S. China Region W. & Central Guangdong (2)	North China–North-east China Region	
				Xishan, Beijing (3)	W. Heilongjiang (4)
		Shaxian Fm (K₁)	Q	Changxindian Fm (E)	Kz
Upper Jurassic	Tithon.	Nanyuan Fm *Onychiopsis* sp. *Pagiophyllum* sp. *Yanjiestheria* cf. *sinensis* 1790 m	Baizushan Gr *Onychiopsis elongata Coniopteris* sp. *Phoenicopsis* sp. *Ferganoconcha* sp. *Eosestheria* sp. 170–1450 m	Dahuichang Fm *Neocalamites* sp. *Lycoptera davidi* 16–560 m	Xinganling Gr *Coniopteris* sp. *Phoenicopsis manchuricus Pseudocardinia ovalis Ferganoconcha* spp. *Lycoptera davidi* 40–2970 m
	Kimm.	Changling Fm *Phoenicopsis* sp. *Cupressinocladus elegans Ferganoconcha* spp. *Cypride* sp. *Mesoclupea* sp. 1955 m		? Donglingtai Fm 500–1600 m	
	Oxford			Houcheng Fm *Neocalamites* sp. *Coniopteris* 1000–2000 m	
Middle Jurassic	Callov.	Zhangping Fm *Neocalamites* sp. *Coniopteris hymenophylloides Ferganoconcha* aff. *jorekensis Pseudocardinia busimensis Tutuella* spp. 1043 m		Tiaojishan Fm 2000–4000 m	Nanping Fm *Coniopteris hymenophylloides Coniferus Pagiophyllum pollenites* 200–600 m
	Bath.			Jiulongshan Fm 200–1250 m	
	Baj. Aalen.				Taipingchuan Fm *Coniopteris* cf. *burejensis Phoenicopsis angustifolia* 450–850 m
Lower Jurassic	Toar.	Lishan Fm *Coniopteris hymenophylloides Clathropteris* sp. *Cladophlebis shansiensis* 579 m	Jinji Fm *Cardinia toriyamai Parainoceramus matsumotoi Retroceramus heyuanensis Meleagrinella japonica Oxytoma kobayashii Arnioceras semicostatus Hongkongites hongkongensis Hiatella arenicola* 200–500 m	Mentougou Fm *Neocalamites carrerei Coniopteris burejensis Phoenicopsis angustifolia* 200–1000 m	Chayihe Fm *Pityophyllum longifolium Neocalamites carrerei* 220–1070 m
	Plien.				
	Sinem.			Nandaling Fm 10–760 m	
	Hett.			Xingshikou Fm *Todites denticulata Ctenis chinensis Czekanowckia rigida* 0–84 mn	
		Wenbinshan Fm (T₃)	Xiaoping Gr (T₃)	Shuangquan Fm (P)	Laolungtou Fm (P)

THE JURASSIC SYSTEM

Table 14.1. (cont.)

N.W. China Region	S.W. China Region	Qinghai–Xizang Region	
N. Shaanxi (5)	E. Yunnan (6)	Nyalam, Xizang (8)	Yarlunzangbo R. (7)
Zhidan Gr (K₁)	K₂	Gucuocun Fm (K)	K
Fenfanghe Fm 1174 m		Xiumo Fm *Buchia rugosa* *B. piochii* *Aulacosphinctes* *Uhligites* *Virgatosphinctes* *Haploceras* *Haplophylloceras pinque* *Grammatodon (Indogrammatodon) irretans* *Entolium demissum* *Chlamys* *Praeconia* *Isocyprina* 1782 m	Weimei Fm *Himalayites?* *Virgatosphinotes* *Haplophylloceras strigile* 550 m
		Menbu Fm *Aulacosphinctoides* *Ptychophylloceras lalonglaensis* *Katroliceras* *Mayaites* *Reineckeia* *Reineckeites* *Macrocephalites* *Buchia concentrica* *B. spitiensis* 510 m	
Anding Fm *Darwinula* 6–254 m			
Zhiluo Fm *Ferganoconcha* 180–140 m	Upper Lufeng Fm *Darwinula sarytirmenensis* *Theiosynoecum?* sp. (or *Rhinocypris*) *Bairdestheria* sp. 170–800 m	Lanongla Fm *Grammatodon virgatus* *Posidonia ornati* 741 m	Xiare Fm *Macrocephalites* *Indocephalites* *Garantiana* *Inoceramus* *Belemnopsis* 600 m
Yanan Fm *Tutuella crassa* *Ferganoconcha* 14–325 m		Nieniexiongla Fm *Witchellia tibetica* *Dorsetensia haydeni* *Sonninia* *Emileia (Frogdenites)* 775 m	
Fuxian Fm *Equisetites sarrani* *Coniopteris hymenophylloides* *Phoenicopsis angustifolia* 0–142 m	Lower Lufeng Fm *Kunmingosaurus wutingi* *Lufengosaurus huenei* *L. magnus* *Sinosaurus shawanensis* *Yunnanosaurus huangi* *Bienotherium elegans* *Podozamites chinensis* 71–484 m	Pupuga Fm *Weyla* sp. *Astarte* cf. *volzii* *Inoceramus* sp. *Protocardia* sp. *Schlotheimia* sp. *Sulciferites* sp. *Nyalamoceras nyalamensis* *Orbitopsella praecursor* 882 m	Ridang Fm 800 m *Prodactyoceras enodum* *Arietites* sp. *Psiloceras provincialis*
Yanchang Gr (T)	T₃	T₃	T₃

Table 14.2. *Correlation of Jurassic marine faunal sequences in China.*

Series	Stage	Ammonoids	Foraminifera	Bivalves	Brachiopod
Upper Jurassic	Tith.	*Virgatosphinctes–Haplophylloceras pingue* Ass.		*Buchia rugosa–B. piochii* Ass.	*Rhynchonella* aff. *paucicosta* *Rutorhynchia–Monticlarella* Ass.
Upper Jurassic	Kimm.	*Aulacosphinctoides–Ptychophylloceras lalonglaenli* Ass.		*Grammatodon (Indogrammatodon) irritans* *Entolium demissum–Chlamys–Praeconia–Isocyprina* Ass.	*Thurmanella rotunda–Kutchithyris dengqenensis* Ass.
Upper Jurassic	Oxf.	Mayaitids bed		*Buchia concentrica–B. spitiensis* Ass.	
Middle Jurassic	Call.	*Macrocephalites–Indocephalites* Ass.		*Grammatodon (Indogrammatodon) virgatus* *Posidonia ornati–Pinna–Palaeoneilo–Mytilus* Ass. *Camptonectes lens–Liostrea birmanica* Ass.	*Burmirhynchia–Holcothyris* Ass. *Nyalamurhynchia mirifica–Rhactorhynchia lanta* Ass.
Middle Jurassic	Bath.				
Middle Jurassic	Baj.	*Sonninia, Dorsetenia Witchella, Frogdenites*			
Middle Jurassic	Aalen.				
Lower Jurassic	Toar.	*Nyalamoceras nyalamense*	*Orbitopsella* Fauna	*Weyla ambongoensis–Entolium nieniexionglaensis* Ass.	*Cirpa himalaica–Homoeorhynchia bolinensis* Ass.
Lower Jurassic	Pliens.				
Lower Jurassic	Sinem.	*Sulciferites*			
Lower Jurassic	Hett.	*Psiloceras planorbis Schlotheimia Hongkongites Arnioceras semicostatus*			

Table 14.3. Correlation of Jurassic non-marine biotic sequences in China.

Stage		Conchostracans	Ostracodes	Bivalves	Charophytes	Spores and pollen	Plants NW, N, NE China	Plants S. China
Upper Jurassic	Tith.		Wolburgia-Darwinula-Damonella Ass.					
	Kimm.	Pseudograpta F.	Darwinula-Cetacella Ass.	Ferganoconcha-Mengjinaria Ass.	Porochara cf. hildesuensis Euestheriacharu cf. nuquihanensis	Upper Denosisporites Classopollis Ass. Lower Classopollis Concavissimisporites Ass.	Annulariopsis simpsoni Hausmannia leeiana Coniopteris hymenophylloides C. burejensis C. tatungensis Clenis huneharai Cl. chinensis Psendoctenis lanei Ps. lucasta	Ginkgoites lepidus Sphenobaiera longifolia Baiera ahnerti B. gracilis Phoenicopsis speciosa Ph. angustifolia Czekanowskia rigida Cz. setacea Sagenopteris phillipsi	Coniopteris spp. Czekanowskia setacea Cupressinocladus spp. Pagiophyllum spp. Brachyphyllum spp.
	Oxf.							
Middle Jurassic	Callo.			Eolamprotula-Psilunio Ass.			Equisetites lateralis Neocalamites nathorsti Coniopteris hymenophylloides C. spectabilis Eboracia lobifolia Ginkgoites sibiricus Sphenobaiera pulchella Sph. furcata	Phoenicopsis speciosa Ph. angustifolia Czekanowskia rigida Cz. setacea Nilssonia compta Elatocladus manchurica Podocarpites menuukuoensis	Neocalamites nathorsti Todites princeps Coniopteris hymenophylloides C. cf. murrayana Phlebopteris polypodioides Tyrmia nathorsti Ptilophyllum contiguuum Ptilo. hsingchuense Otozamites hsingchuensis O. mixomorphus
	Bath.	Euestheria ziliujingensis F.	Darwinula F.	Yunanoconcha-Ferganconcha-Sinomurganifera Ass.	Euestheriachara Flora				
	Baj.								Marattiopsis asiatica Todites goeppertianus Clathropteris obovata Coniopteris hymenophylloides Nilssonia inouyei Ginkgoites tawakuoensis Phoenicopsis spp. Swedenburgia cryptomerioides Ferganiella sp.
Lower Jurassic	Aalen						Neocalamites carrerei Marattiopsis asiatica Thaumatopteris hissarica Dictyophyllum sp. Hausmannia sewariensis Coniopteris hymenophylloides	Cladophlebis sulakiensis Cl. magnifica Sagenopteris sp. Ginkgoites obrastichewi Sphenobaiera spectabilis Czekanowskia rigida Phoenicopsis spp. Ptyrophyllum spp. Cycadocarpidium sp.	
	Toar.	Eosestheriopsis F.	Gomphocythere Darwinula F.	Qiyanga-Apseudocardinia Ass.	Stellatochara xiangjiensis Stenochara tuanmenensis			Marattiopsis asiati Dictyophyllum muensteri Clathropteris meniscioides Gonopteris microsipnoides Cladophlebis vulgarius Stachypteris alata Ctenozamites lanshaensis Clenis stewartiensis Otozamites spp. Sagenopteris cf. hallei Sphenobaiera spectabilis Ferganiella sp.	
	Pliens.								
	Sinem.	Palaeolimnadia baitianhaensis F.							
	Hett.								

Table 14.4. Correlation of Jurassic subdivisions between China and other countries.

Series	Stage	China Marine — Qomolangma Region	China Non-marine S.W. China	S.E. China	N.E. China	N.W. China	England Dorset	USSR (W. Siberia)	New Zealand	USA California
Upper Jurassic	Tithonian	Xiumo Fm	Penglaizheng Fm	Huangjian Fm	Yixian Fm	Fenfanghe Fm	Purbeck / Portland	Bagenov	Puaroan (Oteke)	Knoxville Franciscan
Upper Jurassic	Kimmeridgian			Laocung Fm	Dabeigou Fm	Fenfanghe Fm	Kimmeridge Clay	Georgiev	Ohauan (Oteke)	
Upper Jurassic	Oxfordian	Menbu Fm	Suining Fm		Zhangjiakou Fm		Sandsfoot / Trigoma Clavellata / Osmington Oolite / Nothe Clay etc	Barabin	Heterian	Mariposa Amador
Middle Jurassic	Callovian		Upper Shaximiao Fm		Baiqi Fm	Anding Fm	Oxford Clay / Kellaways / U Cornbrash	Tatar	? (Kawhia)	
Middle Jurassic	Bathonian	Lanongla Fm	Lower Shaximiao Fm	Yushanjian Fm	Tuchengzi Fm		Forest Marble / Boveti Bed		Temaikan (Kawhia)	
Middle Jurassic	Bajocian	Nieniexiongla Fm	Xintiangou Fm		Lanqi Fm	Zhiluo fm	Fullers Earth Clay / Zigzag Bed	Tumen — Upper		
Middle Jurassic	Aalenian						Upper Inferior Oolite / Middle Inferior Oolite		Ururoan (Herang)	
Lower Jurassic	Toarcian	Pupuga Fm	Ziliujing Fm	Majian Fm	Haifanggou Fm	Yanan Fm	Lower Inferior Oolite / Bridport Down Cliff	Tumen — Middle		
Lower Jurassic	Pliensbachian				Baipiao Fm	Fuxian Fm	Junction bed / Marlstone Rock / Green Ammonite / Belemnite Maris etc	Tumen — Lower		
Lower Jurassic	Sinemurian				Xinlunggou Fm		Black Ven Marle / Shales with Beef		Aratauran (Herang)	
Lower Jurassic	Hettangian				Kungtoubolo Fm		Blue Lias			

15. THE CRETACEOUS SYSTEM
Yang Zunyi

INTRODUCTION

Cretaceous rocks were formed under almost the same palaeogeographic conditions as those of the Jurassic, i.e. they are distinguished by three main types of deposits: (1) the inland basin deposits occurring chiefly in large or medium basins of West and North-east China, characterized by dark, red, and variegated deposits, often with gypsum and salts, containing plentiful non-marine fossils; (2) volcanics and pyroclastics distributed mainly in the medium and small basins of East and North-east China, with intermedio-basic or intermediate volcanics intercalated with many layers of pyroclastics, intermedio-basic or intermedio-acidic volcanic rocks, and many layers of clastics, locally coal-bearing and gypsum-bearing, which are also marked by a great variety of non-marine life; (3) marine or paralic deposits, chiefly cropping out in West Kunlum, the Himalayas, North Xizang, and Taiwan composed mainly of sandstone, shale, and limestone with various taxa of marine life, as well as terrestrial plants.

A three-fold subdivision of the Cretaceous proposed by Hao and her Cretaceous research team (Hao *et al.* 1982) (Table 15.2, and 15.3) on the basis of three conspicuous evolutional stages of terrestrial life has been accepted by a great number of geologists including the present authors, although the conventional two-fold subdivision is still being used by some other geologists (Gu 1983; Chen *et al.* 1982).

The marine Cretaceous in China is also subdivisible into three series (Tables 15.1, 15.2, 15.4).

For the non-marine sequences the Lower Series is represented by the Zhidan Group (equivalent to the Rehe Group of North China), and the Jiande Group of the South-east region, roughly correlatable to the Neocomian. The Middle Series is represented by the Jongkang Group (equivalent to the Lower Songhuajiang Group or Huashan Group of North-east China), correlatable to the Aptian–Turonian. The Upper Series represented by the Wangshi Group (equivalent to the Upper Songhuajiang Group, the Chüjiang Group of the South-east region and the Nanxiong Group of Guangdong), correlatable roughly to the Coniacian–Maastrichtian.

In North China non-marine Cretaceous fossil groups indicate their origin in a humid temperate zone, whereas the southern fossils reflect their arid or semi-arid tropical or subtropical origin. Starting from the coastal region inland, i.e. from east to west, the palaeoclimate gradually changed from a humid marine to an increasingly dry continental regime. Furthermore, the northern biotas are in general homogeneous, flourishing in fresh-water lakes. The southern biotas, on the other hand, are distinctly heterogeneous, showing various ecological patterns in the fresh and brackish environments of the Tethys tropical marine province (for example, South Xinjiang and Xizang). No certain evidence of Boreal life has been registered.

In the light of sedimentation types and biotic compositions, eleven stratigraphic regions have been differentiated as follows: North-east region (1), North China region (2), South-east region (3), Gobi region (4), North-west region (5), South Xinjiang–Qing-Zang Plateau region (6), Sichuan–Yunnan region (7), Yangzi region (8), Youjiang-Zhujiang region (9), Xinjiang–Xizang region (10), and Taiwan and South Seas region (11).

NON-MARINE LIFE

Early Cretaceous

According to Hao and her research team (Hao *et al.* 1982), during the Early Cretaceous there were large basins in North-east, North, North-west, South-west, and South-east China in which terrestrial life flourished. The bivalves include primitive trigonioidids, unionoids and corbiculids, such as the widespread *Nakamuranaia, Nippononaia, Ferganoconcha, Corbicula (Mesocorbicula),* and *Sphaerium* found in the basins of North-east, North, North-west and South-east China. In Sichuan and Yunnan there developed only primitive trigonioids (*Koreanaia, Peregrinoconcha,* and *Nakamuranaia*), but no *Ferganoconcha, Corbicula (Mesocorbicula),* and *Sphaerium*.

Chief among the estherids are *Eosestheria, Yanjiestheria, Orthosestheria,* and *Orthestheropsis.* In the Yan-

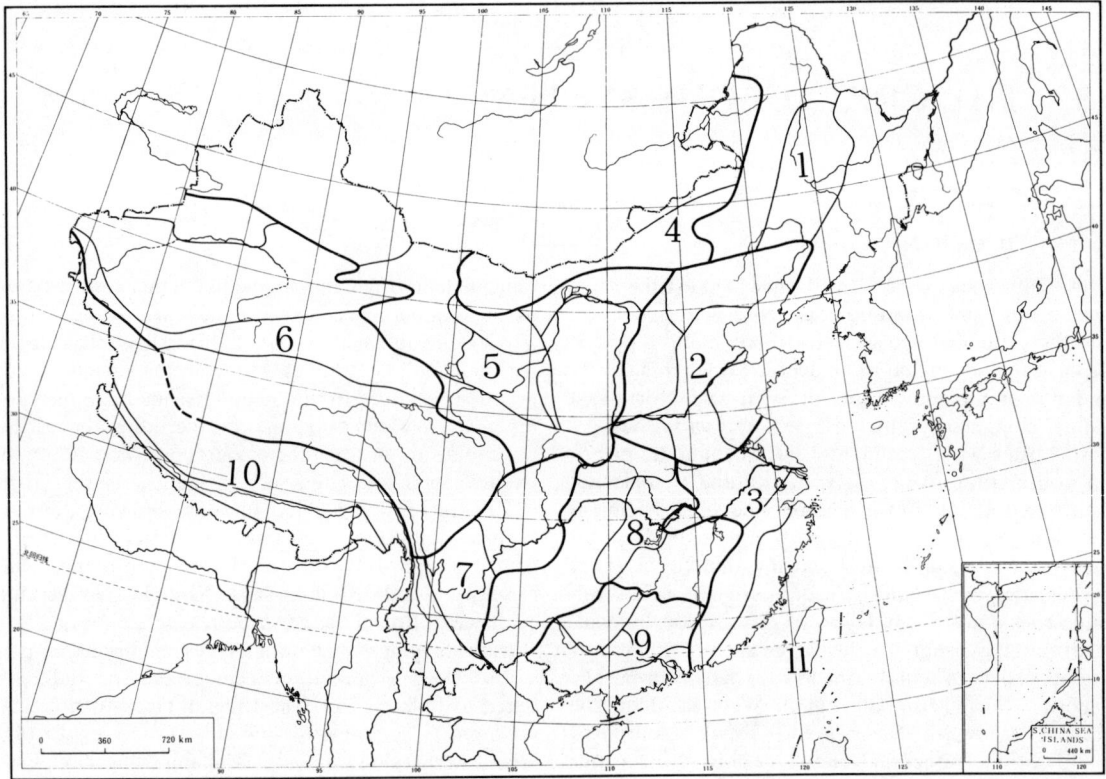

Fig. 15.1. Sketch map showing the stratigraphic regions of the Cretaceous in China. 1, North-east region; 2, North China region; 3, South-east region; 4, Gobi region; 5, North-west region; 6, South Xinjing–Qinghai–Xizang Plateau region; 7, Sichuan–Yunnan region; 8, Yangzi region; 9, Youjiang–Zhüjiang region; 10, Xinjiang Xizang region; 11, Taiwan and South China Sea Isles.

shan–Liaodong area *Eoestheria* is associated with the local genus *Liaoningestheria*, and occasionally *Orthestheria*. In the South-east region (Zhejiang and Fujian) and North-west region (Zhaoshui–Yablai and Jiuquan basins) *Yanjiestheria* is the principal genus, accompanied by a small number of *Eoestheria*, i.e. a mixed community of the two genera is characteristic in these regions. In the South-east, associated with these two genera are *Migransia* and *Congestheria*, Gondwana elements which have not been reported in the Northern Continent. Elsewhere in the Shaan–Gan–Ning Basin, the Liupanshan area, and the Junggar, Xinjiang area, *Yanjiestheria* alone predominates (no *Eoestheria*). In Central Yunnan, South-west region, the estherids are represented by *Orthestheria* and *Orthestheropsis*.

Ostracods are extensively developed in the South-east, North, North-east and North-west including the Junggar and Turpan basins of North Xinjiang and Tarim Basin of South Xinjiang. They are represented by *Cypridea koskulensis*, *C. unicostata*, *Rhinocypris cirrita*, *R. echinata*, and *Darwinula contracta*; they are also associated with *Cypridea vitimensis*, *C. faveolata*, *C. sulcata*, *C. pauglovensis*, *Lycopterocypris infantiles*, and *Clinocypris scolia*. In the South-east (Zhejiang and Anhui) the above faunule is associated with *Damonella*. In the Junggar Basin, Guyang Basin, and Shaan–Gan–Ning Basin the same faunule is associated with *Djungarica*.

Insects include *Ephemeropsis trisetalis*, *Coptoclava longipodo*, *Chironomaptera gregaria*, and *Mesolygaesus rotundocephalus*, the first three of which are widespread in West Liaoning, North Hebei, and the Jiuquan Basin. Most widely distributed is *Ephemeropsis trisetalis*, which occurs also in West Zhejiang, the Laiyang Basin or Shandong, and the Guyang Basin of Nei Mongol.

Among Early Cretaceous freshwater fishes may be mentioned *Lycoptera*, *Sinamia*, *Sunolepis*, *Kuyangichthys*, and *Kuntulunia* which form the typical fish

THE CRETACEOUS SYSTEM

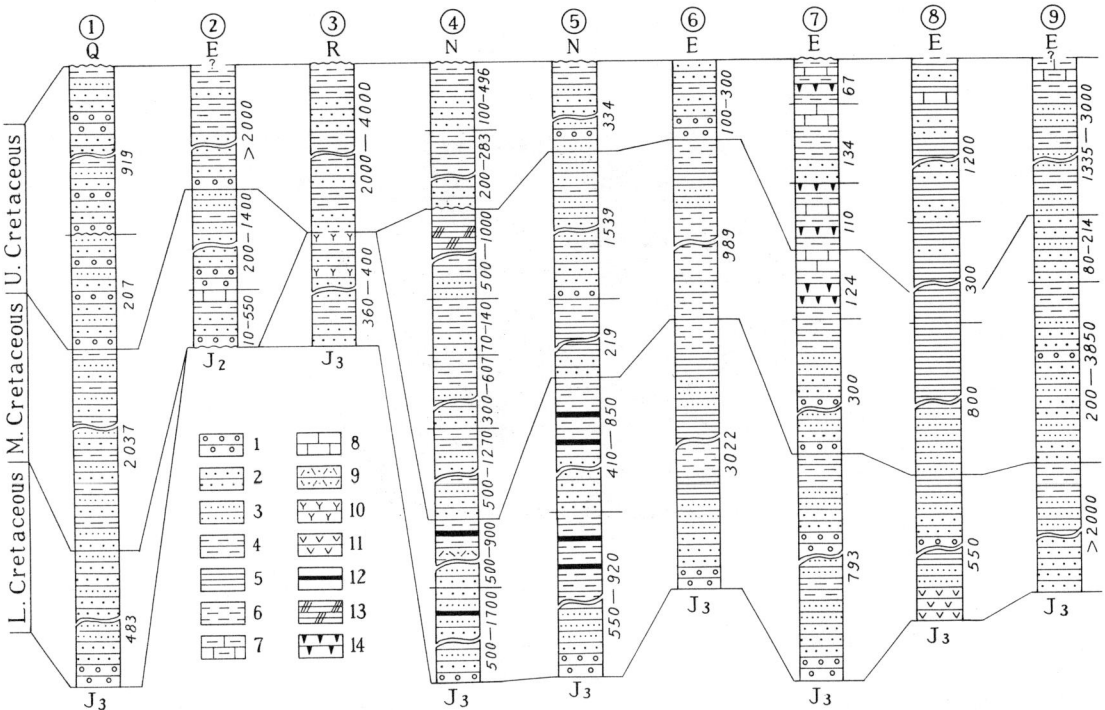

Fig. 15.2. Columnar sections showing Cretaceous sequences in various stratigraphic regions of China. 1 Su-Wuan Area, Southeast region; 2 Hengyang, Yangzi region; 3 Shandong, North China region; 4 Songliao, North-east region; 5 Hailar, Gobi region; 6 East Gansu, North-west region; 7 South Xinjiang–Qinghai–Xizang, Tarim; 8 Yangzhouyong Lake, Xizang; 9 West Yunnan, Sichuan–Yunnan region. Legend: 1, Conglomerate; 2, Sandstone; 3, Siltstone; 5, Shale; 6, Clay; 7, Argillaceous limestone; 8, Limestone; 9, Tuff; 10, Volcanic rocks; 11, Andesite; 12, Coal seams; 13, Oil shale; 14, Gypsum.

fauna for Northern China. *Lycoptera* has been reported from Laiyang, Mengyin, Yanliao, Guyang, Shaan–Gan–Ning, and the Liupanshan and Jiuquan basins, and it is always associated with *Sinamia*. In South China the corresponding fish fauna is composed of *Mesoclupea, Fuchunkiangia, Paraclupea, Huashia,* and *Sinamia,* the last being present also in North China and Anhui.

Important Early Cretaceous reptiles include dinosaurs (*Psittacosaurus*, Fig. 15.3, *Euhelopus, Wuerhosaurus,* and *Tugulusaurus*), pterosaurs (*Dsungaripterus*), Crocodilia (*Edentasuchus*), and Chelonia (*Peishanemys* and *Sinemys*). The genus *Psittacosaurus* is widely distributed, from Shandong in the east to Junggar in the west, and is often associated with *Lycoptera* fauna east of the Jiuquan Basin, Gansu; hence it is important for correlation purposes.

Early Cretaceous land plants include the *Acanthopteris–Ruffordia* flora in the Northern province (the Siberian phytogeographic province), which is richest in Pteridophytes, less so in Ginkgoales, and least developed in Cycadales and Coniferales, reflecting a humid temperate climate.

The Southern flora is distributed chiefly in the Southeast (Zhejiang, Fujian, Jiangxi) and has numerous Pterdophytes, conifers, and cycads, but lacks Equisetales and Ginkgoales. It has no *Acanthopteris* (typical of the Northern flora), but is associated with *Weichselia,* an important genus of the Wealden flora.

Among the Charophytes, *Mesochara stipitata* is widely distributed (Yan-Liao, Junggar, West Zhejiang, West Jiangsu, and South Anhui). The florule of *Flabellochara xiangyunensis* and *Nodosochara puckangheensis* is limited to Sichuan and Yunnan.

In the case of Lower Cretaceous spores and pollen, the Northern province is marked by the *Cicatricosisporites–Rouseisporites–Piceaepollenites* assemblage, whereas the southern province is distinguished by the *Cicatricosisporites–Classopollis* assemblage.

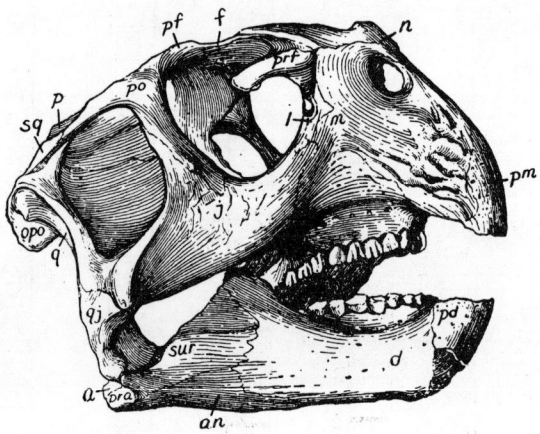

Fig. 15.3. *Psittacosaurus Youngi* Chao, Lower Cretaceous, Shandong (× ½).

Middle Cretaceous

During the Middle Cretaceous all the basins in Xinjiang, Sichuan–Yunnan, the North-east, and the South-east continued to receive sediments rich in various taxa. Among bivalves the *Trigonioides–Plicatounio–Nippononaia* (TPN) assemblage plays a conspicuous part in the biotic make-up. Also found are *Nakamuranaia* and *Martinsonella*, and trigonioidids and *Sphaerium* (corbiculids) are extensively developed in the North-east (Songliao and Yanji basins), the South-east (Zhejiang, Anhui, and Fujian), Nanling Ranges, the Jianghan Basin of the Middle Yangzi reaches, the Hengyang Basin and South-west, West and Central Yunnan.

Estherids evolved rather rapidly. In the early Middle Cretaceous they are represented by the *Orthestheria* and *Orthestheriopsis* assemblage associated in the South-east with less important taxa such as *Yanjiestheria, Ellipsograpta, Cratostracus*, and *Aglestheria*; in the North-east *Yanjiestheria* is more important, but the other genera are absent. Next flourished the *Nemestheria, Jilinestheria,* and *Plectestheria* assemblage in the North-east (Songhuajiang) and South-west (Central Yunnan). Finally, the *Halysestheria, Estherites, Calestherites,* and *Mesolimnadiopsis* assemblage has been found only in the Songliao Basin.

Middle Cretaceous ostracods reached their climax in the *Cypridea* faunule associated with such important taxa as *Ziziphocypris* and *Monosulcocypris*.

Among fishes endemism is very marked, as no widespread forms like *Lycoptera* and *Sinamia* are known. In North China three fish faunas, *Manchurichthys* (K_1), *Plesilycoptera* (K_2), and *Sungarichthys* (K_3), appeared in succession in the Songliao and Janji basins (NE). In South China (Central Zhejiang) there are *Paralycoptera, Pingolepis,* and *Huashia*; and in Xizang a marine fish (*Ceratodus tibetensis*) has been reported.

Reptiles are represented by dinosaurs (*Probactropsaurus, Asiatosaurus, Monkonosaurus, Microvenator, Chilantaisaurus*) and Crocodilia (*Eotomistoma*). No widespread genus like *Psittacosaurus* is known.

Plants experienced rapid evolution in the Middle, especially late Middle Cretaceous when seed plants predominated. In the early part of the Middle Cretaceous there survived such Early Cretaceous elements as *Onychiopsis elongata, Ruffordia goepperti* and *Asplenium, Gleicherites*, with the addition of a new taxon like *Manica*. In South China *Weichselia* remained, but no angiosperms appeared. In North China there appeared Angiospermae, such as *Sapindopsis* cf. *variabilis, S.* cf. *obtusifolia,* and *Rogersia angustifolia*. In late Middle Cretaceous times the Northern flora is dominated by angiosperms (*Platanus, Trochodendroides, Trapa?,* and *Aralia*) which are chiefly broad-leaved deciduous plants, reflecting a more arid tropical climate (in Heilungjiang, Jilin). Correspondingly, the Southern flora flourished in the Shiwandashan, Guangxi, predominated by angiosperms (*Cinnamum* and *Nectandra*) reflecting a more arid tropical climate. South of the Yarlung Zangbo River there are *Populus*, *Salix*, *Juglandites*, *Ficus*, *Viburnum*, *Aralia*, *Laurophyllum*, and *Dicotylophyllum*, indicating a humid subtropical climate (K_2–K_3).

Middle Cretaceous charophytes include mainly *Mesochara symmetrica*, *Euaclistochara mundula*, and *Atopochara trivolvis*, including such important elements as *Obtusochara cylindrica*, and *Flabellochara jurongica*, which occur chiefly in the Jianghan Basin (Hubei), the Jiangsu–Zhejiang–Anhui area and Hainan Island, Guangdong.

Two palynological assemblages are the *Nevesisporites radiatus–Classopollis–Tricolporopollenites* assemblage of the Northern flora and the *Schizaeoisporites–Classopollis–Tricolporopollenites* assemblage of the Southern flora.

Late Cretaceous

Late Cretaceous rocks occur in medium to small basins, where they yield many fossil groups. Among them, bivalves are dominated by *Pseudohydria*, specialized trigonioidids, associated with some species of *Sphaerium* and *Margaritifera*. Ostracods include *Cristocypridea* (= *Talocypridea*) derived from *Cypridea* which is by then secondary in importance. These are associated with *Cantona*, *Candoniella*, and *Limnocythere* which became important only in the Tertiary.

Late Cretaceous reptiles are represented by Hadrosaurs, mostly specialized types, such as *Bactrosaurus*, *Mandchurosaurus*, *Tanius*, *Tsintaosaurus*, *Shantungosaurus*, and *Microhadrosaurus*; also *Ornithomimus*, *Protoderatopus*, *Tyranosaurus*, *Tarbosaurus*, *Nanshiungosaurus*, *Wannanosaurus*, and *Megacervixosaurus*, discovered most abundantly from the Wangshi Group of Shandong and the Dabsu Formation of Erlian, Nei Mongol respectively. Dinosaurs are varied and widespread—so far reported from Xinjiang, Ningxia, Nei Mongol, Shandong, Guangdong, and Jiangxi. They, together with *Pseudohydria* (bivalve) and *Cristocyprina–Cypridea* (ostracods), are regarded as the important guides to the Upper Cretaceous.

Late Cretaceous Charophytes include two floras: (1) the early flora consisting of Mesozoic types (*Porochara anluensis*, *P. gonganzhaiensis*, *P. oblonga*, *P. stipitata*, and *Latochara curtula*), and Cainozoic types (*Gyrogona*, *Peckichara*, *Charites*) all found in the Jianghan Basin, Hengyang Basin, Jiangsu, Central and Western Yunnan; (2) the late flora represented mainly by *Latochara curtula* and *Charites tenuis*, discovered so far in the Jianghan, Hengyang, Nanxiong, Sanshui, and Minghe (in Qinghai) basins, as well as in West Yunnan. The *Aquilapollenites–Schizaeoisporites* assemblage is associated with some gymnospermous pollen such as *Tsugaepollenites*, *Keteleeria*, and the Southern *Ptersisporites–Rugubivesiculites–Morinoipollenites* assemblage.

MARINE CRETACEOUS

Yarlung Zangbo area

The marine Cretaceous rocks of China crop out chiefly in South and North Xizang, East Karakorum, and Taiwan. Near the Yangzhouyong Lake of the Yarlung Zangbo area the Lower Cretaceous Yulangbaijia Group consists of three formations listed upwards (Table 15.1.8). (1) The Kadong Formation comprises grey quartz sandstone and sandy conglomerate in the lower part (200 m thick), which is unfossiliferous, and grey siltstone, shale, sandstone, and limestone lenses in the upper part (150 m thick) yielding ammonoids (*Haplophylloceras strigile*, *Spiticeras*). (2) The Sangxiu Formation consists in the upper part of black carbonaceous siltstone, shale and limestone with oriented belemnites, yielding also ammonoids (*Calliptychoceras*, *Neocomites*, *Haplophylloceras*), and is 150 m thick. The lower part is green tuffaceous sandstone and tuffaceous sandy conglomerate, reaching a thickness of 100 m. (3) The Rimova Formation is made up of two members: the lower one is of grey and greyish-yellow sandstone and limestone lenses, yielding a great quantity of large-sized ammonoids (*Olcostephanus*), brachiopods, and belemnites, having a thickness of 50 m; the upper member is composed of grey to greyish-yellow silty shale, calcareous shale, and marl, 100 m thick. In the calcareous shale and marl are ammonoids (*Eulytoceras*) and bivalves (*Parainoceramus*), mostly imprints.

The Shadui Group contains two formations, the Zhawangzi and the Duojiu. The Zhawangzi Formation, 200 m thick, is of dark grey siliceous shale and siliceous siltstone intercalated with chert nodules and limestone concretions. The calcareous shale always contains small ammonoids and small belemnites, mostly impressions. The succeeding Duojiu Formation is made up of two members. The lower one consists of dark grey calcareous shale and siliceous shale, intercalated with large amounts of limestone, marl lenses or concretions, yielding ammonoids (*Oxytropidoceras*, ?*Bhimaites*), and small bivalves, and is 300 m thick. The upper one is composed of dark grey calcareous shale and marl, and limestone, occasionally with brownish thin-bedded siltstone and biogenetic limestone. It is 200 m thick and yields ammonoids (*Turrilites* and other small forms).

Kashi, Tarim Basin

In Kashi in the Tarim Basin, Xinjiang (Table 15.1.7) the Cretaceous Kezilesu Group is divisible into a lower part of 793 m of clastics (siltstone, quartz sandstone, and conglomerate), and an upper part of 300 m, also of clastics (conglomerate, quartz sandstone, sandy shale, and mudstone), in some horizons yielding ostracods (*Rhinocypris*), and assigned to the Lower and lower Middle Cretaceous. The Kezilesu Group is followed by the Yinjisa Group of four formations in ascending order. (1) The Kukebai Formation is made up of normal marine green mudstone and shelly limestone, totalling 124 m; it yields bivalves (*Ostrea delettrei, O. vatonnia,* and *Trigonia ferganensis*), gastropods (*Aporrhais (Helicaulax) tarimensis*), ostracods (*Brachycythere turonica*), and cephalopods (*Placenticeras placenta,* and *Thomasites koulabicus*), all of which point to a Turonian age (upper Middle Cretaceous). (2) The Wuyitake Formation begins with variegated mudstone with gypsum, and grades upwards into greyish-green mudstone and limestone yielding marine bivalves, and is topped by red mudstone and gypsum. It has a thickness of 110 m and yields ostracods (*Centrocythere circinocostata*). This is assignable to the Coniacian–Santonian age. (3) The Yigeziya Formation is of massive red-grey limestone, yielding foraminifers (*Globigerina, Cristellaria, Triloculina, Nodosaria,* and *Tylostoma*), Rudistae, and other fossils forming bioherms. It is 130 m thick. (4) The Tuyilok Formation, the uppermost unit, is composed of brownish, gypsiferous mudstone, occasionally with thin-bedded limestone. It is so far unfossiliferous and is 67 m thick.

North Xizang–East Karakorum

In North Xizang–East Karakorum the Cretaceous is of neritic facies. In the lake district of that area the Homoshan Formation with Berrisian ammonoids (*Spiticeras, Neocosmoceras*) is unconformably overlain by the Lower Cretaceous Doba Group which consists in the lower part of the mottled Doba Formation yielding *Orbitolina lenticularis* (Lam.) and in the upper part of the Langshan Formation.

NON-MARINE CRETACEOUS

South-east region

In the Leping area the Lower Cretaceous is represented by the Lower Kueiling Formation of terrestrial red clastics (conglomerate and sandstone), 483 m thick (Table 15.1.1; Fig. 15.2.1). It is rich in estherids (*Sinoestheria shehsianensis*), gastropods (*Viviparus* sp., *Bithynia* sp., and *Valvata* sp.) and chara. The Middle Cretaceous Upper Kueiling Formation consists mostly of purplish-red clastics (coarse- and fine-grained sandstone, siltstone, and mudstone), 2037 m thick. It yields bivalves (*Plicatounio multiplicatus, Nippononaia* sp., *Trigonioides kodairai*), and estherids (*Nemestheria yunnanensis, Orthestheria* cf. *hungshuikouensis*). The Upper Cretaceous includes two formations. (1) The Qiyunshan Formation consists of cyclothems of red conglomerate, sandstone, and siltstone, with a maximum thickness of 207 m. It bears plant fossils (*Cladophlebis* cf. *exiliformis*) and spores (*Lygodiumsporites* sp. and *Hymenophyllumsporites* sp.). (2) The Upper Cretaceous Xiaoyan is of brick-red, cross-bedded clastics, without fossils.

Hengyang, Yangzi region

In Hengyang, Central Hunan in the Yangzi region (Table 15.1.2; Fig. 15.2.2) only the Middle and Upper Cretaceous are represented. The Middle Series includes the Dongjing and Shenwangshan Formations. The Dongjing comprises purplish-red sandstone, mudstone, and limestone, with conglomerate in the lower part, containing 24 per cent Pteridophytes and 76 per cent Gymnospermae. There are also bivalves ('*Unio*' *purengensis, Nakamuranaia yongkangensis, Trigonioides kodoirai,* and *Nippononaia* sp.). The Shenwangshan Formation is composed of brownish-red sandstone, silty fine-grained sandstone, and green mudstone with basal conglomerate, 200–1400 m thick; yielding 42.4 per cent Pteridophytes, 56.5 per cent Gymnospermae, 11 per cent Angiospermae, and ostracods (*Cypridopsis torsuosus,* '*Lycopterocypris*' cf. *multifera*).

The Upper Cretaceous also includes two other formations. (1) The Daijiaping Formation comprises purplish-red, greyish-green mudstone, intercalated with sandstone and purplish-red sandy conglomerate in the lower part, together more than 2000 m thick; containing charophytes (*Porochara anluensis, Latochara curtula, Gyrogona hubeiensis, Peckichara paomagangensis, Charites tenuis, C. guanpingensis,* and *Maedlersphara minuscula*), and ostracods (*Cypridea cavernosa, Cristocypridea amoena,* and *C.* spp.). (2) The Dongtang Formation is composed of brownish-red to greyish-white argillaceous sandstone and siltstone, intercalated with marl, having a thickness of 220–400 m. It yields fossil plants: (Pteridophytes 39 per cent, Gymnospermae 36 per cent, Angiospermae 25 per cent), bivalves (*Pseudohydria* spp. and *Plicatounio* (P) *hunanensis*), and ostracods (*Cyclocypris* sp.). This formation is either conformably or disconformably overlain by the Palaeocene Chejiang Formation.

East Shandong, North China

Non-marine Cretaceous rocks are well-known in East Shandong (Table 15.2.3; Fig. 15.2.3). Here, overlying the Jurassic Laiyang Group unconformably or disconformablyis the Cretaceous Qingshan Formation made up of intermedio-basic volcanic rocks and their pyro- and sedimentary clastics, totalling 360–400 m. It yields fossil plants (*Brachyphyllum obesum*); spores (*Cicatricosisporites*); bivalves (*Nakamuranaia chingshanensis* and *Nippononaia tetoriensis*); estherids (*Yanjiestheria sinensis*); ostracods (*Darwinula leguminella*) Chelonia (*Peishanemys latipons*); and dinosaurs (*Psittacosaurus sinensis* and *P. youngi*). This is followed disconformably by the Wangshi Group which consists of red clastics, 2000–4000 m thick, and yields plants (*Cupressinocladus gracilis* and *Sagenopteris mentelli*); gastropods (*Campeloma liui*); bivalves (*Pseudohyria cardiiformis*, *P.* spp., and *Sphaerium shantungense*); ostracods (*Cypridea gigantica*, *Cristocypridea amoena*, *Candona habros*, and *Candoniella candida*); and dinosaurs (*Tsintaosaurus spinorhinus*, *Chingkankousaurus fragilis*, *Tanius* spp., *Shantungosaurus gigantus*, and *Oolites* spp.).

Songliao, North-east region

Another non-marine Cretaceous sequence is well developed in the Songliao area, North-east region (Table 15.1.4; Fig. 15.2.4). It starts with the upper part of the Huoshiling Formation which consists of coal-bearing rocks with tuff, and intermedio-basic volcanic rocks containing plants (*Nilssonia sinensis* and *Elatocladus manchuricus*). This is succeeded disconformably by the coal-bearing Shahezi Formation of greyish-white to black sandstone, mudstone, and siltstone, with tuffaceous mudstone and acidic tuff, 216–690 m thick. It bears plant fossils (*Ruffordia goepperti*, *Acanthopteris gothani*, *Nilssonia sinensis*, *Sphenolepis kurriana*, and *Elatocladus manchuricus*), bivalves (*Ferganoconcha* spp.), and fish (*Lycoptera* sp.). Above this is the Yingcheng Formation of intermedio-acidic volcanic rocks with sediments and unstable coal-seams, 250–970 m thick, yielding plant fossils (*Acanthopteris gothani* and *Neozamites*) and Conchostracans (*Cratostracus*). Next is the Denglouku Formation which is made up of variegated sandstone, conglomerate, dark sandy mudstone, and greyish-white sandstone intercalated with some thin-bedded tuff, reaching a thickness of over 1500 m; it yields plants (*Asplenium dicksonianum* and *Sphenolepsis sternbergianum*): Pteridophyta 80.00 per cent, Gymnospermae 16.19 per cent, Angiospermae 0.52 per cent, and Charophytes (*Aclistochara yanjiensis*). All these formations are assigned to the Lower Cretaceous.

The Middle Cretaceous of the same sequence includes four formations. (1) The Quantou Formation comprises dark brownish-red to variegated sandy mudstone, locally with tuff, and is 700–1900 m thick. It is rich in plant fossils (*Onychiopsis* sp. and *Platanus* spp.), Charophytes (*Atopochara trivolvis* and *Euaclistochara mundula*), bivalves (*Nippononaia? jilinensis* and *Plicatounio* cf. *multiplicatus*), and ostracods (*Cypridea subtuberculisperga* and '*Lycopterocypris*' *torsuosus*). (2) The Qingshankou Formation is dark green sandy mudstone, 300–500 m thick, containing bivalves (*Plicatounio latiplicatus* and *Martinsonella paucisulcata*); estherids (*Nemestheria qingshankouensis* and *Jilingestheria nonganensis*); ostracods (*Cypridea gibbosa*, *C. adumbrata* and *C. dekhainensis*); fishes (*Manchurichthys* sp.); dinosaurs (*Chilingasaurus chingshankouensis*); plants (*Onychiopsis* sp. and *Diospyros rotundifolia*); and charophytes (*Aclistochara songliaoensis*). (3) The Yaojia Formation consists of greyish-green to black mudstone intercalated with siltstone and brownish-red mudstone, having a thickness of 70–200 m, and yielding bivalves (*Plicatounio (P.) latiplicatus* and *Martinsonella paucisulcata*); estherids (*Dictyestheria elongata*); ostracods (*Cypridea songhuajiangensis*, *C. tuanshanensis*, and *Ziziphocypris concta*); fishes (*Plesiolycoptera daqingensis*); spores (*Balmeisporites*); and algae (*Maedlerisphaera minuscula*). (4) The Nenjiang Formation is dark mottled argillaceous sandstone and oil shale, totalling 500–1000 m, and contains plants (*Trapa? microphylla* and *Lioplacodes sungariana*); bivalves (*Pseudohyria* aff. *gobiensis*, *Musculus manchuricus*, and *Fulpioides orientalis*); estherids (*Estherites mitsuishii*); ostracods (*Cypridea gunsulinensis*, *C. (Pseudocypridina) tera*, *Cristocypridea amoena*, and *Ilyocyprimorpha netchaevae*) and fishes (*Jilinichthys rapax*, *Hama macrostoma*, and *Sungarichthys longicephalus*).

The Upper Cretaceous includes two formations. (1) The Sifangtai Formation is made up of variegated mudstone, sandstone, siltstone, and reddish sandy conglomerate, varying from 200–282 m thick; and it yields charophytes (*Obtusochara* sp. and *Latochara yuananensis*), ostracods (*Cypridea cavernosa*, *Cristocypridea amoena*, *Candoniella candida*, and '*Lycopterocypris*' *cuneata*), and bivalves (*Pseudohyria cardiiformis* and *Sphaerium rectiglobosum*). (2) The Mingshui Formation comprises mottled sandy mudstone and siltstone, varying from 100–496 m thick, containing charophytes (*Latochara guangdongensis* and *Atopochara* cf. *trivolvis*); bivalves (*Pseudohyria uralica*, *P. robusta*); ostracods (*Cypridea cavernosa*, *Cristocypridea amoena*, *Candoniella candida*, and '*Lycopterocypris*' *cuneata*);

gastropods (*Physa kuhuensis*); and estherids (*Daxingestheria distincta*).

Hailar, Gobi region

In the Hailar area, Gobi region, the Lower Cretaceous is represented by the Zalainor Group (Table 15.1.5; Fig. 15.2.5) including the Dameguaihe and Yimin Formations. The Dameguaihe Formation is composed of greyish-black coal-bearing sandstone and mudstone, with basal conglomerate, reaching a thickness of 550–920 m; it yields plant fossils (*Coniopteris nympharum, Onychiopsis elongata, Acanthopteris gothani*, and *Phoenicopsis* sp.) and bivalves (*Ferganoconcha sibirica*). The Yimin Formation consists also of greyish-black coal-bearing mudstone, siltstone, and sandstone, attaining a thickness of 400–850 m; it bears plant fossils (*Phoenicopsis* sp., *Pityophyllum lindstroemi, Coniopteris burejensis*, and *Ginkgo digitata*).

The Middle Cretaceous includes the Xianggang and Badatu Formations, respectively 219 m and 1539 m thick, and both are poorly fossiliferous.

The Upper Cretaceous Qinyuangang Formation is an alternation of pink and grey mudstone and siltstone. It yields ostracods (*Candona prona, Cristocypridea* cf. *amoena*, and *Lycopterocypris cuneata*).

East Gansu, North-west region

The non-marine Cretaceous of the North-west region may be illustrated by the sequence in East Gansu (Table 15.1.6; Fig. 15.2.6). The Hekou Group of Lower and Middle Cretaceous age is composed of brownish sandstone, conglomerate, and mudstone, having a thickness of 130–2000 m, and yielding plant fossils (*Brachyphyllum spinosum, Classopollis* (max. 70 per cent)—*Schizaeoisporites*, and Angiospermae (max. 19 per cent)); charophytes (*Aclistochara huihuibaoensis, Mesochara stipitata*, and *Spherochara verticillata*); gastropods (*Valvata transbaicalensis* and *Bithynia leachioides*); bivalves ('*Unio*' *grabaui* and *Nippononaia tetoriensis*); estherids (*Yanjiestheria kyongsangensis* and *Y. huanjenensis*); ostracods (*Cypridea yumenensis, C. vilimensis, C. kosculensis, Rhinocypris tugriquensis, Jingguella* spp., and *Djungarica* spp.), and fishes (*Sinamia* sp.).

The Upper Cretaceous Minhe Formation is of brownish-red to orange-red sandstone, conglomerate, and mudstone, 100–300 m, yielding charophytes (*Latochara cylindrica, L. curtula, Gyrogona hubeiensis*, and *G. xindianensis*).

Lanping–Simao area, West Yunnan

The mainly non-marine Cretaceous is also developed in the Lanping–Simao area, West Yunnan (Table 15.1.9; Fig. 15.2.9). The Lower Cretaceous is represented by the Jingxing Formation which is composed of light coloured, thick-bedded quartz sandstone and an alternation of mottled mudstone and fine-grained sandstone in the upper part, and light thick-bedded sandstone in the lower, totalling 300–1000 m. It bears bivalves (*Nippononaia* (*Eonippononaia*) *diana, Peregrionoconcha yunnanensis, P.* spp., *Quenstedtia* cf. *laevigata*, and *Falcimytilus* aff. *diettrichi*), ostracods (*Rhinocypris jurassica, Damonella ovata, Jingguella extensa, J. ovata*, and *Darwinula oblonga*), and reptiles (*Lanpingosaurus magnus*).

The Middle Cretaceous includes the Mangang and Shuicheng Formations. The Mangang Formation comprises light-grey to purple siltstone, sandstone intercalated with brownish-purple mudstone, and sandstone, with conglomerate in the lower part, varying in thickness from 200–3850 m, containing plants (*Thallites yunnanensis*); bivalves (*Yunnanoconcha pupengensis, Nippononaia carinata*, and *Trigonioides* (T) *sinensis*); estherids (*Orthestheria simplica* and *Orthestheriopsis scutulata*), and ostracods (*Rhinocypris jurassica, Monsulcocypris subovata, M. gigantea, Cypridea* (C) *angustocaudata*, and *C.* (*Pseudocypridina*) *pulvinata*). The Shuicheng Formation comprises yellow to purple quartz sandstone and arkose, 80–214 m, but is unfossiliferous.

The Upper Cretaceous Mankuanhe Formation is a series of argillaceous siltstone, mudstone, sandy mudstone, and fine sandstone of alluvio-lacustrine origin, 1335–3000 m thick, yielding charophytes (*Porochara anluensis, Peckichara dangyangensis*, and *Charites tenuis*) and ostracods (*Eucypris* cf. *anluensis, Hemicyprinotus* sp., *Sinocypris jinghongensis, S.* cf. *favosa, Cristocypridea* cf. *amoena*, and *Cypridea* cf. *cavernosa*).

BOUNDARY PROBLEMS

Marine sequences

For the marine sequences the Jurassic Cretaceous boundary is well-defined in the Northern Himalayan Geosyncline (including the Yangzhoyong Lake, Gyangze, south of Lhasa), where the upper part of the Karton Formation carries the Tithonian (*Himalayites*) and the Berrisian *Spiticeras–Blandordiceras* Zones, and therefore the boundary should be drawn between them.

The Cretaceous–Tertiary boundary is not so clearly fixed. In the Gambo–Dingri area the Zhongshan Formation yields foraminifers, sea-urchins, and algae, showing that its upper part belongs to the Maastrichtian. This follows upwards into the Jidula Formation which contains only algae (*Cymopolia tibetica*) and ostracods (*Uroleberis* sp.), making the age most probably Maastrichtian. This is followed by the member of the Zhongpur Group with gastropods (*Bernaya expansa, Campanile ganesha, Diconomorpha elegans,* and *Confusiscala indica*), all of which serve to correlate this member to the *Cardita beaumonti* Bed in Sind, Pakistan, and it is thus assignable to the Danian. The Cretaceous–Tertiary boundary here can therefore be temporarily put between the top of Jidula and the base of the first member of the Zhongpur Group.

Non-marine sequences

For the non-marine sequences different opinions have been expressed regarding the basal boundary of the Cretaceous in North, North-west and North-east China, and the crux lies in treating the Rehe fauna as either Late Jurassic or Cretaceous. Recently, Gu Zhiwei (1983) has reviewed this problem fully and he strongly advocates the assigning of the Rehe Group and its equivalent units to the Upper Jurassic. Also on biotic grounds Hao and her research team put the large part of the Rehe Group (at least the Jiufutang and Fuxin Formations) in the Cretaceous, though other geologists would include even the upper part of the Jingangshan, Jianchang, Jiufutang, and Fuxin Formations in the Cretaceous. Several suggestions have been made including putting the boundary below either the Fuxin, or Jiufutang, or Jianchang, or Jingangshan, each position being based primarily on the presence of one or several fossil groups.

The Zhidan Group of the North-west region and the Jingxing Formation of West Yunnan are also assigned to the Early Cretaceous, following Hao and her team. Nevertheless, some doubt remains about the position of the boundary between the Cretaceous and Jurassic.

CORRELATION OF CRETACEOUS SUBDIVISIONS

Correlation of Cretaceous subdivisions in various stratigraphic regions of China and between those of China and other parts of the world—England, USSR (Far East), Japan, New Zealand, and USA (Gulf Coast) are given in Tables 15.1 to 15.4.

Table 15.1. *Correlation of Cretaceous stratigraphic sequences in various regions of China.*

Series	S.E. China Region Leping, Jiangxi (1)	Yangzi Region Hengyang, Hunan (2)	N. China Region E. Shandong (3)	N.E. China Region Songliao (4)
	Q	Xialiushi Fm (E)	R	N
Upper Cretaceous	Xiaoyan Fm 919 m	Dongtang Fm *Pteridophyta* 39% *Gymnospermae* 36% *Angiospermae* 25% *Pseudohydria* *Plicatounio (P)* *hunanensis* *Cyclocypris* 220–400 m	Wangshi Gr *Cupressinocladus gracilis* *Sagenopteris mentelli* *Campeloma liui* *Cypridea gigantica* *Cristocypridea amoena* *Candona habras* *Candoniella candina* *Tanius* spp. 2000–4000 m	Mingshui Fm *Latochara guandongensis* *Atopochara* *Pseudohyria uralica* *Cypridea cavernosa* *Candoniella candida* 100–496 m
Upper Cretaceous	Qiyunshan Fm *Cladophlebis* cf. *exiliformis* *Cicatricosisporites* sp. 207 m	Daijiaping Fm *Porochara anluensis* *Latochara turtula* *Gyrogona hubeiensis* *Peckichara paomagangensis* *Charites tenuis* *Cypridea cavernosa* *Cristocypridea amoena* 2000 m		Sifangtai Fm *Obtusochara* sp. *Cristocypridea amoena* *Lycopterocypris cuneata* 200–283 m
Middle Cretaceous	Gueiling Fm Upper member *Plicatounio multiplicatus* *Nippononaia* *Trigonioides kodairai* *Nemestheria yunnanensis* 2037 m			Nenjiang Fm *Fulpioides orientalis* *Ilyocyprimorpha netchaeva Mesolanistes globiformis* *Sphaerium* sp. 500–1000 m
Middle Cretaceous				Yaojia Fm *Plicatounio latiplicatus* *Martinsonella paucisulcata* *Ziziphocypris concta* 70–200 m
Middle Cretaceous		Shenwangshan Fm *Pteridophyta* *Cypridopsis torsuosus* *Lycopterocypris* cf. *multifera* 200–1400 m		Qinshankou Fm *Plicatounio* spp. *Nemestheria qinshankouensis* *Cypridea gibbosa* *Chilingosaurus chishankouensis* 300–607 m
Middle Cretaceous		Dongjing Fm '*Unio*' *purengensis* *Nakamuranaia yongkangensis* *Trigonioides kodairai* *Nippononaia* sp. 10–550 m		Quantou Fm *Onychiopsis* sp. *Plicatounio multiplicatus* *Lycopterocypris* 500–1270 m
Lower Cretaceous	Gueiling Fm Lower member *Sinoestheria shehsianensis* *Viviparus* sp. *Valvata* sp. 483 m		Qingshan Fm *Brachyphyllum obesum* *Nakamuranaia* 360–400 m	
	J₃	J₂	J₃	J₃

Table 15.1. (cont.)

Gobi Region Hailar (5)	N.W. China Region E. Gansu (6)	S. Xinjiang–Qinghai–Xizang Region Tarim (7) Kashi	Xinjiang–Xizang Region Yangzhuoyong Lake, Xizang (8)	Sichuan–Yunnan Region W. Yunnan (9)
E	E	E	E	E
(unconformity) Qinyuangang Fm *Candona* sp. *Cristocypridea* cf. *amoena* *Lycopterocypris* 334 m	Minhe Fm *Latochara cylindricum* *L. curtula* *Gyrogona hubeiensis* *G. xindianensis* 100–300 m	Tuyilok Fm 67 m	Yangzhuoyonghu Gr *Globotruncana linnieana* *G. carinata* *Pseudotextularia* *Heterohelix* 1200 m (Yinjisa Gr)	Mankuanhe Fm *Porochara anluensis* *Peckichara dangyangensis* *Charates tenuis* *Eucypris* cf. *anluensis* *Hemicypprinotus* sp. *Sinocypris jinghongensis* *Cristocyprida* cf. *amoena* *Cypridea cavernosa* 1335–3000 m
		Yigeziya Fm *Globigerina* *Gristellaria* *Triloculina* *Nodosaria* *Tylostoma* 134 m		
		Wuyitake Fm *Centrocythere circinocostata* 110 m		Shuicheng Fm 80–214 m
		Kukebai Fm *Ostrea delettrei* *Brachycythere turonica* *Trigonia ferganensis* *Placenticeras placenta* *Thomasites koulabicus* 124 m	Duojiu Fm *Turrilites* *Protexanites* *Venezoliceras* *Leymeriella* *Oxytropidoceras* 300 m	
Badatu Fm *Aralia* sp. 1539 m	Shangyan Fm *Selaginella* sp. *Schizaea* sp. *Bennetites* sp. *Cycas* sp. *Ginkgo* sp. *Podozamites* sp. *Podocarpus* sp. *Pinus* sp. *Picea* sp. 989 m	Upper part 300 m	(Keziesu Gr)	Mangang Fm *Thallites yunnanensis* *Nodococlayator puchangheensis* *Nippononaia carinata* *Trigonoides sinensis* *Monsulcocypris subovata* *Cypridea angusticaudata* *C. pulvinata* 200–3850 m (Shadui Gr)
Xuexianggang Fm 219 m			Zhawangzi Fm *Pseudohaploceras* *Acanthohoplites* *Desmoceras* 800 m	
Yimin Fm *Ruffordia goepperti* *Coniopteris burejensis* 410–850 m	Xiayan Fm *Mohria* sp. *Cycas* sp. *Ginkgo* sp. *Picea* sp. *Margaritifera lacustris* *Unio grabaui* *Brachygrapta* 3022 m	Lower part 793 m *Rhinocypris*	Rimowa Fm *Olcostephanus* *Pleurotomaria spitiensis* 200 m	Jingxing Fm *Nippononaia diana* *Plicatounio* sp. *Perigrionoconcha yunnanensis* *Darwinula obolonga* *Damonella ovata* *Lanpingosaurus magnus* 2000 m (Yulangbaijia Gr)
			Sangxiu Fm *Neocomites* *Kilianella* 200 m	
Dameguehe Fm *Coniopteris onychioides* *Ferganoconcha sibirica* 550–920 m			Kadong Fm *Thurmanniceras* *Kilianella* 350 m	
J₃	J₃	J	J₃	J₃

Table 15.2. *Correlation of Cretaceous marine faunal sequences in China (chiefly after Hao et al.).*

	Stage	Ammonoids	Foraminifers	Brachiopods	Bivalves
Upper Cretaceous	Maest.		Orbitoides–Omphalocyclus Ass.	Xenothyris tuilaensis Ass. Rectothyris sinkiangensis–Ornatothyris–Carneithyris Ass.	Bournonia Fauna
Upper Cretaceous	Camp.		Globotruncana elevata–G. stuartiformis–Orbitoides tissoti Ass.	Xenothyris tuilaensis Ass. Rectothyris sinkiangensis–Ornatothyris–Carneithyris Ass.	Bournonia Fauna
Upper Cretaceous	Sant.		Globotruncana schneegansi–G. renzi Ass.	Xenothyris tuilaensis Ass. Rectothyris sinkiangensis–Ornatothyris–Carneithyris Ass.	Bournonia Fauna
Upper Cretaceous	Con.		Globotruncana schneegansi–G. renzi Ass.	Xenothyris tuilaensis Ass. Rectothyris sinkiangensis–Ornatothyris–Carneithyris Ass.	Bournonia Fauna
Middle Cretaceous	Tur.	Mammites, Placenticeras Thomasites Acanthoceras Calycoceras Turrilites	Praeglobotruncana helvetica–Hedbergella murphi Ass.	Orbirhynchia henpolaica–Alithyris shiquanheensis Ass.	Bournonia Fauna
Middle Cretaceous	Cenom.	Mammites, Placenticeras Thomasites Acanthoceras Calycoceras Turrilites	Rotalipora–Praeglobotruncana Ass.	Orbirhynchia henpolaica–Alithyris shiquanheensis Ass.	Bournonia Fauna
Middle Cretaceous	Alb.	Diploceras Oxytropidoceras	Ticinella roberti–Hedbergella trocoidea Ass.		
Middle Cretaceous	Apt.				
Lower Cretaceous	Barr.			Sellithyris mayuensis–Platythyris xanzaensis Ass. Peregrinella Ass.	
Lower Cretaceous	Haut.			Sellithyris mayuensis–Platythyris xanzaensis Ass. Peregrinella Ass.	
Lower Cretaceous	Valag.	Neocomites 'Neohoploceras' Berriasella oppeli			Buchia shuomoensis–B. mankamanensis Ass.
Lower Cretaceous	Berr.	Neocomites 'Neohoploceras' Berriasella oppeli			Buchia shuomoensis–B. mankamanensis Ass.

Table 15.3. *Correlation of Cretaceous non-marine biotic sequences in China (after Hao et al.).*

Series	Stage	Conchostracans				Ostracods	Charophytes	Bivalves	Spores and pollen
		S.W. China	S.E. China	N.E. China					
Upper Cretaceous	Maest.	*Aglestheria* F.	*Tenuestheria* F.	*Euestherites* F.		Talicypridea–Cypridea–Candona Ass.	Lotochara cylindrica–Charites tenuis Ass.	*Pseudohyria–Sphaerium shandongensis* Ass.	N. China Schizaeoisporites–Aquilapollenites–Wodehouseia Ass.
	Con.	?	?	D Zone			Porochara anluensis Ass. Atopochara restricta–A. trivolvis qidongensis Ass.	Acclino plicatounio Plicato trigonioides Ass.	S. China Pterisisporites–Rugubivesiculites–Classopollis–Morinoipollenites Ass.
				M Zone		Eucypris–Quadracypris–Cypridea Ass.			
Middle Cretaceous	Tur.				Cratostracus F.	S. China Cypridea (Morinia)–C. (Bisulcocypridea)–Darwinula Ass. N. China: Cypridea 'Lycopterocypris'–Candona Ass.	Atopochara trivolvis trivolvis–Flabellochara hangzhouensis Ass. A. trivolvis triquetra–F. hebeiensis Ass.	T.P.N. Fauna Trigonioides Plicatounio Nippononaia	N. China Upper Balmeisporites–Lythraites–Aquilapollenites Ass. Lower Cicatricosisporites–Schizaeoisporites Beaupreaidites Ass. S. China Schizaeoisporites–Welwitschiapites–Cranwellia Ass.
	Apt.				*Orthestheria* F.				
Lower Cretaceous	Barr.				*Yanjiestheria* F.	S. China: Cypridea–Mongoianella–Darwinula Ass. Jingguella–Pimnocypridea–Darwinula Ass. N. China: Cypridea–Yumenia–Mongolianella Ass.	Perimneste ancora–Clypeator jinquanensis Ass. F. xiangyuensis–C. zongjiangensis Ass.	Peregrinoconcha–Koreanaia–Damlengicocha Ass.	N. China Cicatricosisporites Pilosisporites Piceaepollenites Ass. S. China Cicatricosisporites Cicatricosisporites Ass.
	Berr.								

Table 15.4. *Correlation of Cretaceous subdivisions between China and other countries (non-Chinese data after Harland et al.).*

Series	Stage			China		
				Non-marine		Qomolangma region (marine)
				N.E. China	S. China	
Upper Cretaceous	Maest.	Wangshi Gr	Songhuajiang Group	Mingshui Fm	Laijia Fm	Jidula Fm
	Camp.					Zungshan Fm
	Sant.			Sifangtai Fm	Tangshang Fm	Jiubao Fm
	Con.					
Middle Cretaceous	Tur.	Yongkang Gr		Nengjiang Fm	Fangyan Fm	Xiawuchubo Fm
	Cenom.			Yaojia Fm	Chaochuan Fm	Lengqinre Fm
	Alb.			Qinshankou Fm		Chaqiela Fm
	Apt.			Quantou Fm	Guantou Fm	
Lower Cretaceous	Barr.	Zhidan Gr	Rehe Group	Fuxin Fm	Moshishan Fm	Dongshan Fm
	Haut.			Jiufutang		
	Valag.			Jianchang Fm Jingganshan Fm (Upper part)		
	Berr.				Gucuocun Fm	
J	Tith.			Jinggangshann Fm (Lower part)	Xiumo Fm (J₁)	Shimudi Ls

Table 15.4. (cont.)

England	USSR Far East	Japan	New Zealand		USA Gulf Coast	
Upper Chalk (vertical hatching above)	Orochenian	Hetonian	Haumurian	Mata	Navarro	Gulf
		Urakawan	Piripauan		Taylor	
Upper Chalk			Teratan	Paukumara	Austin	
Middle Chalk	Gilyakian	Gyliakian	Mangaotanian		Eagleford	
Melbourne rock			Arowhanan			
Plenus marls	Ainusian		Ngaterian		Woodbine	
Grey Chalk						
Chalk marls		Miyakoan			Washita	Comanche
Up. Gr. Sand			Motuan	Taitai Clarence		
Gault			Urutawan			
Folkstone Beds			Korangan			
Sandgate Beds					Trinity	
Hythe Beds						
Atherfield Clay	Suchanian				Nuevoleon	
Weald Clay (Wealden Beds)		Aritan			Durango	Coahuila
Hastings Beds	(Beds with Aucella)					
		Kochian				
Durlston Beds (Purbeck)						
Lulworth Beds						

16. THE CAINOZOIC
Yang Zunyi

THE TERTIARY SYSTEM

Introduction

During the Tertiary, terrestrial sediments continued to accumulate in basins, large and small, over the vast territory of China, with the exception of a few areas—the Himalayas in Southern Xizang, the western part of the Tarim Basin, and Taiwan and its nearby isles, where the sea lingered, leaving marine deposits (Fig. 16.1). Geotectonically, the eastern part of China behaved differently from the western part. At the end of the Early Tertiary, owing to the Himalayan Movement, the continental part of China was further uplifted, so that seas drained from the Himalayas and the sea in South-western Tarim became shallower, forming lagoonal and terrestrial deposits. But marine conditions remained in Taiwan and short ingressions occurred in the Leizhou Peninsula of southern Guangdong and the Lüda area beside the Bohai Gulf.

During the Early Tertiary the Himalayas and the Qinghai–Xizang Plateau stood much lower than they are at present, so that warm and humid air currents penetrated easily into South-west China, bringing about the distribution of diverse vegetation zones in China. From the south to north the following zones were established: the tropical evergreen broad-leaved forest, tropical savanna and desert, and subtropical mixed forest zones. Under such climatic conditions terrestrial deposits dominated by red beds developed.

During the Late Tertiary, as a result of further uplift of the Himalayas and the North-east mountainous region, atmospheric circulation was affected to such an extent that climatic differentiation brought about different types of deposits in East and West China. The eastern part was under the influence of the monsoon, and various vegetation belts appeared. From north to south these ranged from the temperate forest and savanna of the North-east, to warm temperate forest and forest grassland in North China, to subtropical evergreen broad-leaved forest in the South-east, to tropical monsoon forest in South China. Such an arrangement is similar to the present distribution, except that the climate was much warmer. In Western China deposits were formed in dry grassland and desert grassland, and in the Qinghai–Xizang Plateau dry grassland to high mountain desert developed, which reflected the gradual uplift of the plateau that prevented the warm Indian air currents from penetrating northwards.

Based on tectonic features, palaeogeographic position, sedimentation type, and biotic make-up, the Tertiary of China may be grouped as shown in Fig. 16.1. In general terms, Tertiary marine deposits in China include (1) geosynclinal bathyal to neritic–littoral swampy deposits, as found in Taiwan; (2) platformal marine neritic deposits, such as in the Himalayas; and (3) platformal and enclosed bay-lagoonal deposits, such as in Tarim. Tertiary continental deposits are preserved in: (1) great depressions, for example the Songliao, North China, northern Jiangsu and Jianghan basins of Eastern China, with deposits of great thickness that have been subject to repeated marine ingressions; (2) medium to small intermontane basins or fault basins in a generally rising area, for example those basins in the Da Xingan Mountains and Nei Mongol in the north-east, up to the Yunnan–Guizhou Plateau, and in South-east China; (3) large and medium inland intermontane basins, for example the Junggar, Hami––Turpan, and Qaidam basins of the North-west. In Late Tertiary times the well-known 'Hipparion red clay' deposits were formed in Central China, and distributed from Eastern Qinghai, Gansu to Shanxi, Shaanxi, Henan, the southern part of Nei Mongol, and over a great part of North China. Under the control of palaeoclimatic and palaeogeographic factors, different types of deposits were formed in various intermontane basins. For instance, in two water-rich zones (in the north-east and south-west) where precipitation exceeded evaporation, vegetation flourished repeatedly, so as to form dark organic rocks containing lignite and oil shale. On the other hand, the intermediate area between these two zones came under the influence of temperate and tropical monsoons, and alternately dry and humid tropical climates prevailed so that fluvio-lacustrine deposits were produced.

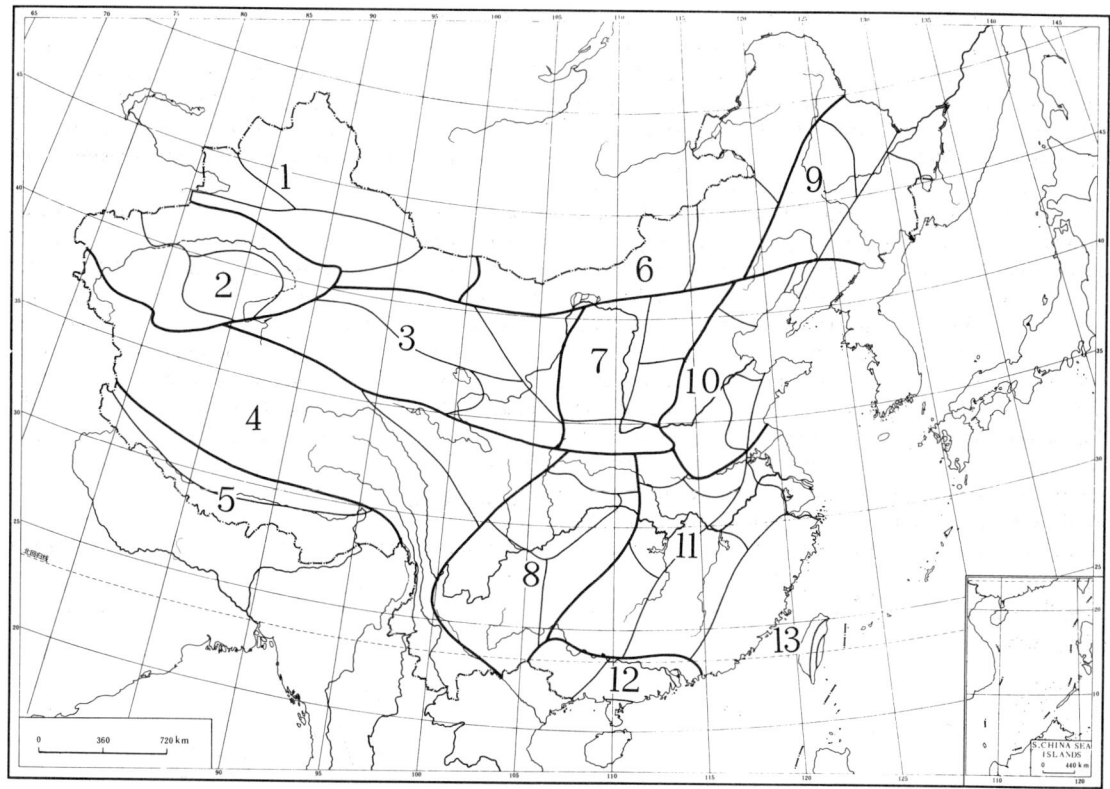

Fig. 16.1. Sketch map showing the stratigraphic regions of the Tertiary System in China. 1, North Xingjiang region; 2, South Xingjiang region; 3, Qilian–Helan region; 4, Qing–Zang–West Yunnan region; 5, South Xizang region; 6, Nei Mongol–Da Hingan region; 7, Qinling–Taihang region; 8, Yunnan–Guizhou region; 9, North-east China region; 10, North China region; 11, South China region; 12, South of Nanling Mountains; 13, Taiwan and South China Sea Islands.

Marine Deposits

Taiwan. The marine Tertiary is developed in Taiwan and its neighbouring isles (Table 16.1.10; Fig. 16.2.10) (Ho Chungsun 1976). The presence of the Palaeocene is not reliably proved here, but the Lower Tertiary is mainly distributed in the northern part of the Central Range. The Eocene Xicun Formation is composed of dark grey slate and phyllite. The Eocene–Oligocene Siling Formation contains light quartz sandstone, quartzite intercalated with phyllite, slate, and poor coal, yielding foraminifers (*Discocyclina* sp. and *Nummulites* sp. *Pitar* (*Costollipitar*) sp.). The Upper Tertiary is better developed, and widely distributed, especially in the western part. Here, the Miocene consists of coal-bearing marine sandy, and muddy, deposits of great thickness: the Yeliu Group in the lower part consists of sandstone, shale, and mudstone with coal seams; the Ruifang Group in the middle is of calcareous sandstone, mudstone, and siltstone with coals, and the Sanxia Group in the upper part is lithologically similar to the lower unit, but much richer in foraminifers. In the Central Range Late Oligocene to Early Miocene deposits include the Gangou, Datong, Aodi, and Lushan Formations, but lack a unit equivalent to the Yeliu Formation.

The Pliocene and Pleistocene form a continuous sequence, chiefly in the western strip, which consists mainly of marine clastics (sandstone, shale, and mudstone) locally intercalated with andesite and basalt, yielding very numerous foraminifers and bivalves. In the eastern part of Taiwan there are many exotic blocks of *melange*.

Xizang. Prior to the Eocene the region south of the Gandise and the western Nyanqen–Tanglha Mountains in Xizang was part of the Tethys, in which were deposited the Zongpu Formation (limestone, intercalated with shale and mudstone) and the Late Eocene Zhepure Formation (black calcareous shale, marl, and limestone intercalated with light quartz sandstone) bearing foraminifers (*Nummulites, Orbitulites,* and *Miscellanea*) (Table 16.1.1; Fig. 16.1.1; Fig. 16.2.1) (Li Yuntong *et al.* 1982).

During the Late Tertiary (Neogene) the sea retreated from Xizang and only fluvio-lacustrine deposits were formed: for example, the Bulong Formation of North Xizang which yields *Hipparion xizangensis* of latest Miocene age; in the Gyirong Basin of South Xizang there are also fluvio-lacustrine deposits (sandstone and mudstone) which bear however *Hipparion gyirongensis* and *Chilotherium xizangensis* of Late Miocene to Early Pliocene age. This reflects the gradual uplift of the Himalayas, for by the Late Miocene the Bulong Formation and its fauna show an environment of sedimentation approaching that of the Siwalik Formation of South Xizang and South Asia, a plateau area with an altitude of around 500–1000 m; by the Early Pliocene the Gyirong fauna is quite different from the Siwalik fauna of South Asia owing to the gradual uplift of the Himalayas that prevented free migration of the mammals. The Bulong Basin is now 4560 m above sea level, and the Gyirong Basin 4100–4300 m; that is to say, they have already risen about 4000 and 3000 m respectively.

Kashi, South Xinjiang. The Lower Tertiary Kashi Group of South Xinjiang (Fig. 16.1.2; Fig. 16.2.2; Table 16.1.2) is composed of marine and lagoonal deposits, chiefly sandstone, siltstone, mudstone, limestone, and dolomite, intercalated with gypsum and salts. It is richly fossiliferous, including foraminifers, oysters, bivalves, and gastropods, reflecting the lingering sea in S. Xinjiang. The sea withdrew later, and the Upper Tertiary is made up of non-marine, brownish-red clastics.

Non-marine deposits

Large and medium intermontane basins. Basins of this

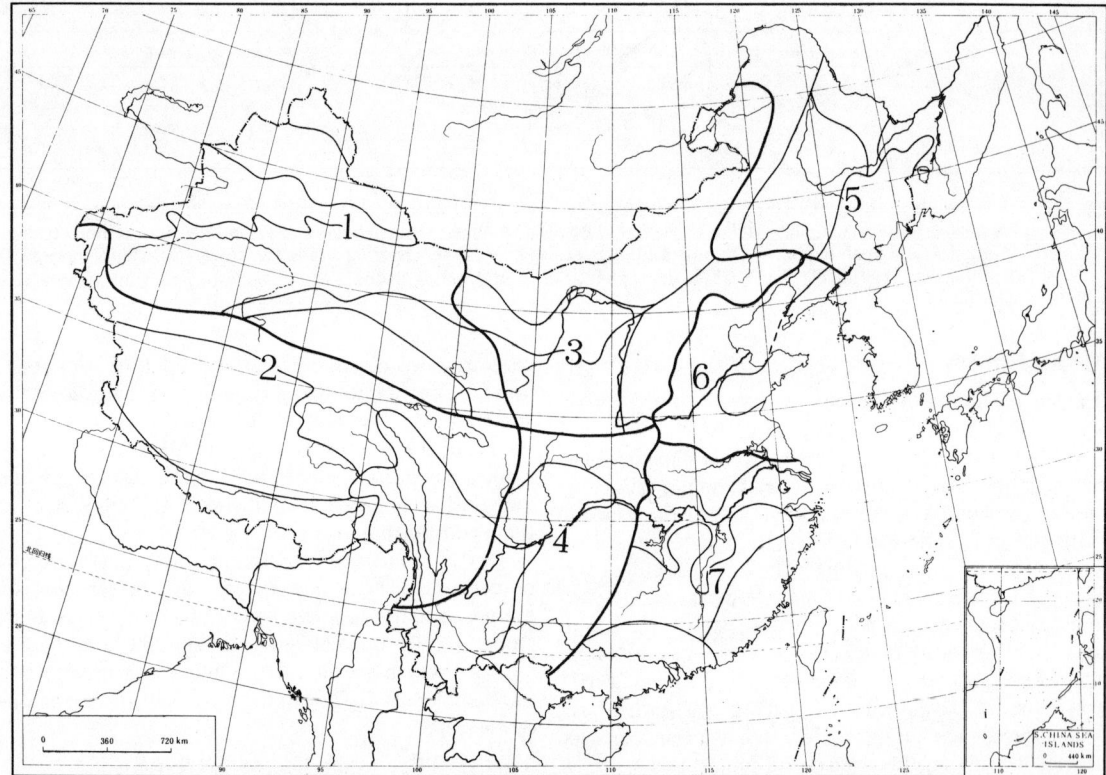

Fig. 16.2. Sketch map showing the stratigraphic regions of the Quaternary System in China. 1, North-west region; 2, Qing–Zang region; 3, North China region; 4, South-west region; 5, North-east China region; 6, North China–Jiangsu–North Huai Plain region; 7, South-east region.

type in the North-west region, such as the Junggar, Hami-Turpan, and Qaidam were surrounded by mountains, and in their central portion there accumulated under arid conditions a thick sequence of red clastics intercalated with beds or lenses of gypsum and salts. The Lower Tertiary section in the Hami-Turpan Basin has been studied in great detail. Here, overlying disconformably the Subashi Formation of middle–late Late Cretaceous age, the Palaeocene Taizicun Formation bears reliable fossil mammals. It is here that such mammals were first discovered. The Eocene and Oligocene do not form a continuous sequence, because they were formed in different areas of the basins. From the basal and upper parts of the Taoshuyuanzi Formation were discovered *Cadurcodon ardynensis* and *Dzungariotherium*, both being rhinoceroids, which roughly indicate the Early and Late Oligocene respectively. The formations mentioned above (Shubashi and Taizicun) and the Dabu, Shisanjianfang, and Liankang Formations have been differentiated from the former Early Tertiary Shanshan Group. The Upper Tertiary fluvio-lacustrine deposits of the Junggar Basin deserve mention (Fig.16.1, Fig. 16.2.3; Table 16.1.3). The Shawan Formation with mammals (*Dzungariotherium orgosensis* and *Lophiameryx* sp.) may be assigned to the Oligocene, but ostracods (*Paracandona euplectella* and *Limnocythere cinctura*) suggest a Miocene age; hence this formation is considered Oligocene–Miocene. The Tashihe Formation belongs to the Miocene, and the Dushanzi Formation bearing *Hipparion* is taken to be Miocene–Pliocene. The latter is conformably overlain by the Xiyu gravel of Pleistocene age.

Eastern part of China. In the eastern part of China, the North China Plain is of interest. It is here that great depressions were formed with thick deposits which yield numerous fossil invertebrates, spores, and pollen, with marine intercalations reflecting repeated marine ingressions. Further south in the Jianghan and North Jiangsu basins there are Palaeocene deposits with microfossils, which are absent in North China. As a result of tectonic activity within this basin there are uplifts

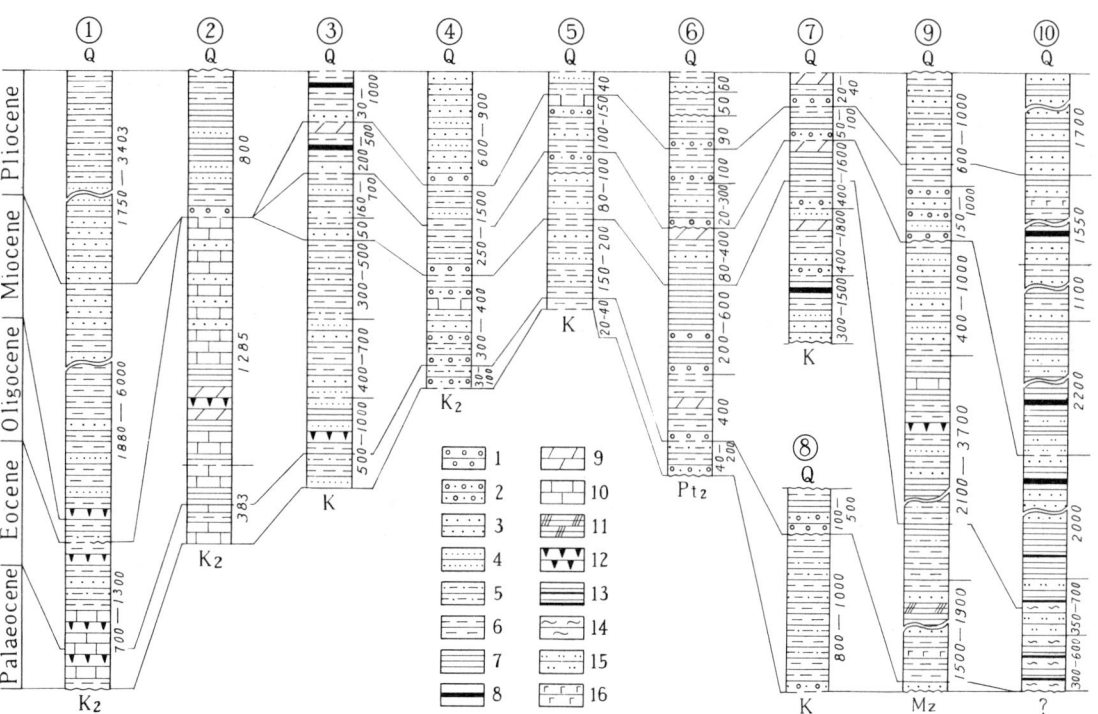

Fig. 16.3. Columnar sections showing Tertiary sequences in various stratigraphic regions of China. 1 Tarim; 2 Himalayas; 3 East Yunnan; 4 Turfan; 5 Nei Mongol; 6 Shanxi-Shaanxi; 7 West Shandong; 8 Nanxiong, North Guangdong; 9 North China Plain; 10 Taiwan. Legend: 1, Conglomerate; 2, Sandy conglomerate; 3, Sandstone; 4, Siltstone; 5, Sandy mudstone; 6, Mudstone; 7, Shale; 8, Slate; 9, Marl; 10, Limestone; 11, Oil shale; 13, Gypsum; 13, Slate; 14, Phyllite; 15, Quartzite; 16, Basalt.

and downwarps of second order, which are favourable for the accumulation of oil and gas.

Tertiary deposits in medium- and small-sized intermontane and fault basins. These are distributed in a belt stretching from North-east China south-westwards to the Yunnan–Guizhou Plateau and the West Yunnan valleys, and also in the south-east coastal region of Fujian and Zhejiang, south of the Nanling Ranges (Guangdong and Guangxi). Under tectonic control, deposition and erosion proceeded alternately in these basins, so complete stratigraphic sequences are lacking in any single locality. Palaeogeographic and palaeoclimatic influences over a range of 40° of latitude have produced considerable variation in the composition and biotic make-up of deposits. However, according to recent studies on the Tertiary 'red beds', correlation of Tertiary formations in different parts of the country is possible on the basis of vertebrate fossils, and microfossils including spores and pollen.

Palaeocene. The most detailed study has been made on the Palaeocene of the Nanxiong Basin (Fig. 16.1.9; Fig. 16.9; Table 16.1.9), where the Shanghu and Nongshan Formations are composed of red mudstone and argillaceous sandstone, yielding fossil vertebrates, invertebrates, and spores and pollen. It is generally considered that the Shanghu Formation should be assigned to the Lower and Middle Palaeocene, roughly equivalent to the North American Puercan and Torrejonian, and the Nongshan Formation to the Upper Palaeocene, being equivalent in its fauna to the Gashanto Formation of the People's Republic of Mongolia. Besides the Taizicun Formation of North Xinjiang (Table 16.1.3; 16.2.3), Palaeocene deposits with reliable vertebrate remains occur in the Shizi and Chikou Formations of Jiangxi. The Naomugen Formation of Siziwangchi, Nei Mongol (Table 16.1.5; Fig. 16.2.5) yields Late Palaeocene fauna different from that of the Gashanto. In addition, Palaeocene deposits with fossil mammals have been found in the Zaoshi Formation of East Hunan, the Wanghudun and Doumu Formations in Qianshan, Huaining in W. Anhui, and lately in the Shimen Formation in Luonan, Shaanxi (Table 16.1.4; Fig. 16.2.4). Palaeocene vertebrates number nearly 20 orders, more than 20 families and over 100 species, among which are Notoungulates (*Palaeostylopus*) discovered in the Gashanto Formation, *Sinostylopus* found in Anhui, and especially noteworthy is a complete skeleton of the Edentates (*Ernanodon antelios*) discovered in the upper part of the Nongshan Formation. Plesiadapiformes (fossil Primates) have also been found in Anhui. The basal and upper parts of the Xuantasi Formation are probably of Late Palaeocene and Eocene age respectively. The Laohutai and Lizigou Formations in Wushun in Liaoning, the Chijiachuan Formation of Qinghai, the Yunlong Formation of West Yunnan, the Dalangshan and Buxin Formations of Guangdong, and the Mingshan Group of western Sichuan Basin all yield Palaeocene microfossils, spores, and pollen.

Eocene deposits have much wider distribution, including the Eocene–Oligocene sequence of Siziwangchi, Nei Mongol (Table 16.1.5; Fig. 16.2.5). The Late Eocene Nadu Formation and the Oligocene Kongkang Formation in Baise, Guangxi are characterized by *Anthracoheryx* (Table 16.1.9; Fig. 16.2.9), *Anthracothema*, *Propterodon*, and *Huananodon*, which resemble the Bangtang fauna of Burma. The early Late Eocene fossils of the Dongjun Formation (Table 16.1.9) in Guangxi are quite similar to fossil mammals of Nei Mongol, and the similarities may be related to further changes in palaeoclimatic and palaeogeographic conditions. The oil-shale bearing Youganwo Formation of Mouming also indicates effects of palaeoclimatic changes. In South-east, Central, and South China there are basins lacking Oligocene deposits. Oligocene mammals have been found however in the South-west (Lunan and Shizong Formations of East Yunnan) (Table 16.1.8; Fig. 16.2.8), in addition to their well-established occurrence in Nei Mongol, Ningxia, Gansu, and Xinjiang in North and North-west China. Recently it has been discovered that the Early Oligocene Chaganbulage Formation of Ningxia yields *Embolotherium grangeri* and *Harpagolestes alsaensis*; the Middle Oligocene Wulanbulage Formation of Nei Mongol bears *Desmatolagas goviensis* and *Tataromys signodon*; the Late Oligocene Yikebulge Formation contains *Amphechinus kansuensis* and *Sinolagomys major*. From the extensive depressions of eastern China only Oligocene ostracods, charas, spores, and pollen have been reported.

In East China Oligocene rocks are relatively scarce because of lack of deposition resulting from the Himalayan uplift. Miocene deposits are represented by the Shanwang Formation of Shandong, the Dongxuanguan Formation of Nanjing, the Xiacaowan Formation of Sihong, the Zhangwu Formation of Ci-Xian, the Luoyang Formation of Henan, and the Lengshuigou Formation of Lintong, Shaanxi all belonging to the Middle Miocene; the Kujiafeng Formation of Lantian, Shaanxi, the Hongliugou Formation of Ningxia, the Xianshuihe Formation of Gansu, the Shaping Formation of Fangxian, Hubei, and the Xiaolongtan Formation of Yunnan are assigned to the Late Miocene. Early Miocene deposits are scarce, and have been found only from the Minhe Basin of East Qinghai where the upper part of the Xiejia Formation yields Rodentia.

As a result of recent research by a number of Chinese vertebrate palaeontologists and Cainozoic geologists, a new subdivision of the Miocene–Pliocene stratigraphy has been proposed. Thus, the Shanwang Formation of Shandong is considered to date from the early part of the Middle Miocene, and to be equivalent to the MN4–MN6 Unit of the European mammalian fauna; the Tungur Formation is considered to date from the upper part of Middle Miocene, and to be equivalent to MN7–MN8; and both are correlatable to the middle and upper parts of the Aragonian. The Bahe Formation of the Lantian area, Shaanxi, is correlated with the European Vallesian of Late Miocene age and is no longer regarded as Early Pliocene. The well-known Baode Formation (Red Clay) is now assigned to the latest Miocene and compared with the European Turolian, but it is still correlated with the Middle–Upper Pontian of the Tethys region. The Bahe Formation is equivalent to the whole Pannonian and the basal Pontian of the Tethys. The Shihuiba Formation of Lufeng, Yunnan, which is rich in *Sivapithecus*, *Ramapithecus*, and other mammals can be correlated with the Upper Bahe–Baode Formation, while the Shahegou Formation (=basal Yuanmou Formation), Yunnan is equivalent to the Baode deposits (=Lower Bahe Formation).

'Hipparion red clay' and Pliocene. As the widespread 'Hipparion red clay' (=Baode Formation) is now assigned to the Late Miocene, the area of Pliocene deposits is correspondingly smaller. The Jingle and Yushe Formations of Shanxi, and the Leijiahe Formation of Lingtai, Ganxu are of Early Pliocene age. The Upper Pliocene is represented by the Yuhe Formation of Weinan, Shaanxi, the Sanmen Formation of Henan, and the Lower Nihewan Formation of Hebei. Also the lignite-bearing Middle Yuanmou Formation (=beds above the Shagou Formation) and the Sanying and Ciying Formations of West Yunnan probably include Pliocene deposits. In short, more work is needed to clarify the age of the Pliocene deposits in different regions.

Tertiary boundary problems

The Cainozoic–Mesozoic boundary is clearly shown in the non-marine sequences of South China where the Cretaceous and Palaeocene form a continuous sequence, or are marked by a short gap. In the marine regime, the Cainozoic–Mesozoic boundary lies between the Maastrichtian and the Danian.

The boundary between the Lower and Upper Tertiary is usually put between the Aquitanian and the Chattian, but in most parts of China there is a gap between the Oligocene and Miocene. In the large and medium basins of North-west China these two units may form a continuous sequence, but the exact boundary remains obscure. In the great depressions of East China an unconformity exists between the Oligocene and Miocene, each of which yields distinctive microfossils; hence the distinction between them is clearly marked.

The Miocene–Pliocene boundary proposed above is very different from the conventional one, but the stratigraphic succession in each locality remains the same. This sharp change is made on the basis of new detailed studies of the previously known Pliocene Pontian Hipparion fauna and the so-called Pre-Nihewan earliest Pleistocene fauna, which are assigned to the Late Miocene and Pliocene in age respectively.

THE QUATERNARY SYSTEM

Introduction

General remarks. The Quaternary, the most recent chronostratigraphic unit, is also extensively developed in China (Fig. 16.2). Its distribution is controlled, as in the case of the Tertiary, by the NNE tectonic trend, with an alternating arrangement of upwarps and downwarps. The latter, represented by the Songliao Basin of the North-east, the North China Basin, and the Jianghan Basin, continued sinking with more, and more widespread, sedimentation of fluvio-lacustrine deposits. To the east and west of these basins are uplifted regions. Further uplift of the Palaeozoic folded belt in North-west China brought about intermontane basins where, especially in the piedmonts, piedmont gravel accumulated and at the centre of the basins fluvio-lacustrine and earthy deposits were formed. Fossil evidence shows that marine ingressions happened more than once in Beijing, Tianjin, Weixian in Hebei, Yuncheng in Shanxi, North Jiangsu, and in the Yangzi Delta. Marine Quaternary deposits also occur in Taiwan and the coastal area of Guandong.

Cave deposits. In South China rock solution (karst) produced many caves in which cave deposits accumulated. Similar deposits also developed in North China and are important localities for Quaternary fossils including fossil man.

The Qinghai–Xizang Plateau was uplifted to such an extent that it prevented the humid and warm Indian air current from reaching the North-west, thus causing increasingly serious aridity in this region.

Glaciation. The world-wide Quaternary glacial and interglacial climates also exerted their influence in China, bringing alpine glaciers to the west. Mount Yushan in Taiwan, Mount Taibai in Qinling, and Mount Lushan in Jiangxi were affected by glaciation, whereas ice-margin landscape and permanently frozen earth have been discovered in Zalainor and Harbin in Heilongjiang, the northern part of Da Xingan Mountains, Daihai in Nei Mongol, Wutai in Shanxi, and in the Qilian Mountains in Qinghai.

Climatic zones. In addition to the control by geological and geographical factors mentioned above, Quaternary deposits are also influenced by climatic zones arranged latitudinally. Thus, the northernmost cold temperate zone is marked by a frozen earth belt, the temperate zone by a chernozem belt, the warm temperate zone (for example North China) by loess and red clay, and South China by subtropical and tropical red earth.

Quaternary stratigraphy and palaeontology

Quaternary studies have been made in a number of areas. In some, long-term intensive studies have been made but in other areas only preliminary research has been undertaken. Recent Quaternary research has revealed many new problems requiring further study. A correlation table of Quaternary subdivisions for different important regions of China is given for general reference. (Table 16.2.)

Lower limit of the Quaternary

Much attention has been focused on the lower limit of the Quaternary in China. For a number of years Chinese geologists have adhered to the resolution of the Eighteenth International Geological Congress (1948) in assigning the Nihewan Formation to the Early Pleistocene, and have put *Homo erectus pekinensis* (*Sinanthropus pekinensis*) (Loc. 1 of Zhoukoudian) into the Middle Pleistocene. Since 1970, with the rapid advance of Chinese science and culture, there has been much discussion on the lower limit of the Quaternary.

According to some geologists, the lower limit of the Quaternary is determined by the beginning of Quaternary glaciation, that is, it should be drawn below the first glacial stage, the Longjiang (Longchuanjiang) glacial stage of the Yuanmou Basin of Yunnan and the Hongya glacial stage of Yangyuan, Hebei. Other geologists prefer to take the Gauss normal as the first unit of the Quaternary, roughly 3 Ma old. Vertebrate palaeontologists, however, prefer to take the appearance of *Equus, Elephas,* and *Bos* (or *Leptobos*), or any one of them, as marking the beginning of the Quaternary.

The present authors are inclined to take the beginning of the Matuyama palaeomagnetic reversal (1.8 Ma) as the lower limit of the Quaternary, because it marks both the beginning of the Quaternary glaciation and sharp changes in both land mammals and pelagic foraminifers, and therefore indicates a significant geological event. This is equivalent stratigraphically to the boundary separating the Upper Nihewan Formation from the Lower–Middle Nihewan Formation.

The problem of Quaternary glaciation in China

The Quaternary glaciation of China has been and is a considerable problem. A group of geologists led by Professor Li Siguan (Lee, J. S. 1934a, b) believe that glacial and outwash deposits as well as glacial geomorphology exist widely in East China. Others emphasize that mountains in East China could not have provided suitable conditions for glaciation (Shi Yafeng, Deng Yangxin 1982), bearing in mind their relief (altitude) and the climatic conditions prevailing then. Despite lengthy discussions, both verbal and written (in symposia and publications), no conclusions have been reached. Although intensive study on glacial distribution in West China has been made, no thorough work has been done in East China, except in the Lushan area, and until further studies have been made the problem will not be solved.

Correlation

A correlation scheme of the Quaternary of China is given primarily on a biostratigraphic basis, but Quaternary glaciation has also been considered. Under the world-wide influence of the alternation of glacial and interglacial stages, air temperature in East China probably suffered drastic changes, and biotic make-up altered in consequence. The origin of alpine glaciers is closely related to atmospheric humidity, and the amount of snowfall in winter and thawing in summer, which are ultimately related to the activity of atmospheric circulation. Isolated alpine glaciers might have appeared on comparatively high mountains such as the Huangshan and the Taibai in East China, where glaciated features as well as moraines have been reported.

Since the beginning of the Quaternary, atmospheric circulation has experienced sharp changes as the uplift of the Qinghai–Xizang Plateau has continued. During the Quaternary ice age the sea-level of the East China Sea and Yellow Sea was much lowered, so that both Lushan and Huangshan were quite distant from the sea. The East Asian monsoon is characterized by a south-east wind in summer, bringing with it warm and rainy weather, and by a north-west wind in winter which is dry and produces less precipitation.

The authors do not favour the hypothesis of widespread glaciation in East China; nevertheless, the possibility of scattered, intermittent alpine glaciation in the higher mountains of East China (e.g. Mount Huangshan in Anhui and Mount Taibai in Qinling), cannot be dismissed. It is important that serious research should be undertaken so as to try to resolve this vexing glaciation problem.

Table 16.1. *Correlation of Tertiary sequences in various regions of China (after Li Yuntong et al.; Hou Youtang et al.; Li Chuankui et al.).*

	European standard	Xizang	South Xinjiang	North Xinjiang	Shaanxi Shanxi Honan and M. reaches of Yellow R.
	L. Villafranchian	Q₁	Q₁	Q₁	Nihewan Fm
Late Tertiary (Neogene) — Pliocene	E.M. Villafranchian	Woma Fm *Hipparion gyirongensis Gazella gaudryi Metacervulus capreolinus*	Atushi Fm	Dushanzi Fm *Hipprion sp.*	Youhe Fm *Hipparion houfenense Archidiskodon youheenis* 60 m
	Ruscinian	450 m	1750 3403 m	1500 2000 m	Jingle Fm *Hipparion houfenense Gazella blacki Antilospira licenti* 20 60 m
Late Tertiary (Neogene) — Miocene	Turolian	Bulong Fm *Hipparion xizangense Chilotherium tanggulaense*	Pakabulake Fm *Ammonia* 1100 3000 m	Changjiha Group	Lantian and Baode Fm *Hipparion placodus Chilotherium hahereri Gazella gaudryi* 50 m
	Vallesian	>41 m	Anjuan Fm *Hemicyprinotus valvaetumidus* 500 2000 m		Bahe Fm *Hipparion weihoense Chilotherium gracile* 90 m
	Aragonian		Wuqia Group	Tasihe Fm *Gomphotherium sp. Hemicyprinotus valvaetumidus*	Koujiacun Fm *Platybelodon grangeri Listriodon gigas* 100 m
				100 320 m	Lengshuigou Fm *Listriodon lishanensis Oioceras lishanensis* 20 300 m
	Aquitanian		Kaziluoyi Fm *Paijenborchella tricostata*	Shawan Fm *Lophiameryx sp. Dzungariotherium orgosensis Paracondona euplectella* 44 800 m	Xiejia Fm 52 m *Plesiosminthus xiningensis Eucricetodon youngi*
	Chattian		280 1000 m		Qingsuiyeng Fm *Indricotherium grangeri Tsaganomys altaicus*
Oligocene	Stampian		Suweiyi Fm	Lower Taoshuyuanzi Fm *Cadurcodon ardynense Sinolagomys kansuensis* 150 1800 m	40-420 m
	Sannaisian		200 400 m		Baishuicun Fm *Brachyodon hui* 80 400 m
Early Tertiary (Palaeogene) — Eocene	Ludian		Bashibulake Fm *Platygena asiatica Anomalinoides vialovi* 350 m	Liankan Fm *Lophialetes expeditus Amynodon mongoliensis* 80 m	Heti Fm *Deperetella depereti Anthracokeryx sinensis* 600 m
	Bartonian	Zhepure Fm *Nummulites gallensis* 1285 m	Wulagen Fm *Liostrea kokanensis Reassela secans* 20 130 m	?	Lushi Fm *Breviodon minutus Gobiohyus orientalis* 400 m
	Luteitian	Zongpu Fm V *Fasciolites ellipsoidalis*	Kalataer Fm *Ostrea turkestanensis Cibicides celebrus* 40 120 m		Dacangfang Fm *Breviodon sp. Euryodon minimus* 200 1000 m
	Cuisian		Qimugen Fm *Globigerina varianta Ostrea bellovacina*	Shisanjianfang Fm *Hyopsodus sp.* 200- Dabu Fm 300 m *Coryphodon dabuensis* 20 m	Yuhuangding Fm *Asiocoryphodon conicus Rhombomylus sp.* 400-600 m
	Sparnacian	Zongpu Fm IV *Orbitolites complanatus*	Kashi Group		
Palaeocene	Thanetian	Zongpu Fm II III *Miscellanea miscella Velates tibeticus*	150 200 m	Taizicum Fm *Prodinoceras turfanensis* 65 m	Shimen Fm *Bemalambda sp. Hukotherium shimenensis*
	Montian		Altashi Fm *Modiolus jeremejewi Corbula biangulata* 150 300 m		
	Danian	Zongpu Fm I *Keramosphaera tergestina Confusiscala indica* I V 383 m			150 m
		K. Jidula Fm	K₂ Yingjisha Gr	K Subashi Fm	P₁–M₄

Table 16.1. (cont.)

Nei Mongol Plateau	North China Plain	West Shandong Mountains	East Yunnan	Guangdong Guangxi	Taiwan
Q	Q	Q₁	Upper Yuanmou Fm Q₁ (S.S.)	Q	Dananwan Fm Q₁
Baogedawula Fm *Chilotherium* sp. *Hipparion* sp.	Minhuazhen Fm *Candoniella suzini Cyprinotus formalis* 600–1000 m	?	Yuanmou Fm (Lower) *Stegodon* sp. *Muntiacus nanus* 80–100 m	Wangyougang Fm *Pulenniatina* sp. *Sphaeroidinella* sp. 360 m	Toukeshan Fm 1200 m
			Shagou Fm *Enhydriodon* cf. *falconeri Stegotetrabelodon primitum Chilotherium yunnanensis* 40–60 m		Zhuolan Fm *Asterorotalia* sp. *Textularia* sp. *Fusinus* sp. 200 m
					Jinshui Fm 300 m *Burasa* sp. *Turritella* sp. *Pseudorotalia* sp.
		Balouhe Fm *Sinohippus zitteli Hipparion ptychodus Gervavitus demissus* 20–40 m	Shihueiba Fm *Sivapithecus Hipparion* cf. *nagriensis Ramapithecus Ictitherium gaudryi* 10–20 m	Felou Fm *Turborotalia* sp. *Globigerina* sp.	Sanxia Fm *Ammonia* sp. *Textularia* sp. 1550 m
			Xiaolungtan Fm *Dryopithecus* sp. *Gomphotherium* sp. *Listriodon* sp.	130–300 m	Rueifang Fm *Operculia* sp. *Globorotalia* sp. 110 m
8–40 m				Jiaowei Fm *Cassigerinella* sp. *Globorotalia* sp.	
Tungur Fm *Listriodon mongoliensis Stephanoceras thomsoni Hemicyon teilhardi* 20–60 m					Yeliu Fm *Globigerina* sp. *Globigerinoides*
		Shanwang Fm *Plesiaceratherium gracile Lagomeryx colberti* 50–100 m	200–600 m		
Zhaowuda Fm	Guantao Fm			40–450 m	2166 m
Damiao Fm *Anchitherium* sp. 40 m	150–1000 m			Xiayang Fm 40–500 m	Lushan Fm 1000m ?
Upper Naogangdai Fm Yikebulage Fm 60 m *Indricotherium grangeri Sinolagomys major*	Dongying Fm *Chinocythere unicuspidata Dongyingia inflexicostata Maedlerisphaera ulmensis*		Shizong Fm *Indricotherium intermedium*	Weizhou Fm	Aodei Fm 700 m *Gaudrina* sp. *Cyclammina* sp.
Wulanbulage 200 m	400–1000 m		160–700 m	150–500 m	Datongshan Fm *Crassatellites* sp. *Pholadomya* sp. 1500 m
					Gangou Fm 600–1200 m *Chilostomelloides Ceratobulimina* sp.
Lower Naogangdai Fm *Embolotherium andrewsi* 34 m *Gigantamynodon* sp.	Shahejie Fm *Huaheinia Phacocypris Cyprinotus Grovesichara*	Dawenkou Fm	Xiaotun Fm *Gigantamynodon gigantus Brachyodus hui* 50 m	Gongkang Fm *Anthracokeryx* sp. *Huananodon* sp. *Heothema* 1300–1450 m	Silen Fm *Discocyclina* sp. *Nummulites* sp.
Sharamurun Fm *Pterodon hyaenoides Gobiolagus andrewsi Rhinotitan mongoliensis* 8 m			Lumeiyi Fm *Lophialetes expeditus Gobiohyus* sp. *Helaletis mongoliensis*	Nadu Fm *Anthracothema* sp. *Propterodon* sp. 150–650 m	300–700 m
Irdin Manba Fm *Miacis invitus Andrewsarchus mongoliensis* 40 m				Dongjun Fm *Eudinoceras* sp. *Andrewsarchus crassum* 20–50 m	Xicun Fm *Pitar* sp. *Taiwancorbicula* sp.
Arshanto Fm *Schlosseria magister Metacoryphodon luminis* 38 m	Kungdian Fm *Eucypris wutuensis Neochara sinuolata*	Guanzhuang Fm *Manteodon flerowi* 400–1800 m	500 m	Liuniu Fm *Eolacerta* sp.	600 m
Bayanwulan Fm *Mimotona borealis Mongolotherium efremovi Heptodon* sp. 65 m	1500 m	Wutu Fm 300–1500 m *Homogalax wutuensis Heptodon niushanensis*		350 m	
Naomugen Fm *Palaeostylops iturus Pseudictops lophiodon Lambdopsalis bulla* >20 m			Xiangpuoshan Fm 550–1400 m	Nongshan Fm *Petrolemur brevirostre Haltictops mirabilis Planocrania datangensis* 200 m	
				Shanghu Fm *Bemalambda nanhsiungensis Hukoutherium ambigum Mongolemys australis* 600 m	
K₂	M₁	O₂		K₂ Nanxiong Fm	

Table 16.2. *Simplified correlation table of Quaternary stratigraphy, faunas and human fossils (compiled by Li Fenglin).*

		Xinjiang	Hexi	North-east China	North China		Northern fauna
					Fluvio-lacust dep.	Earthy deposits	
Holocene	Post Glacial	Aeolian lacustrine alluvium diluvium	Aeolian lacustrine alluvium diluvium	Alluvium & lacustrine deposits	Recent alluv. diluv.	Secondary loess	
Quaternary (Q_3)	Dali Glacial Stage Lushan–Dali Intergl. Stage Lushan Glacial Stage	Xinjiang Group	Malan loess	Guxiangdun Fm	Shandingdong deposits Shara usu Fm Qian'an Fm Dingcun Fm	Malan loess	Shandingdong fauna Sharausu fauna Dingcun f.
Quaternary (Q_2)	Dagu–Lushan Intergl. Stage Dagu Glacial Stage	Usu Group	Jiuquan Gravel	Huangshan Fm	Zhoukoutian dep.	Lishi loess	Zhoukoutian f. Chenjiawo f. Gongwangling f.
Quaternary (Q_1)	Boyang–Dagu Interglac. Stage Boyang Glacial Stage	Xiwu Gravel	Wumeng Gravel	Baitushan Fm	Nihewan Fm Sanmen Fm	Wucheng loess	Yangguo f. Loc. 18 Nihewan f.
Underlying		(N) Kuche Group	(E) Shulehe Fm		L. Nihewan L. Sanmen	Youhe Fm	

Table 16.2. (cont.)

Southern fauna	South China		Taiwan	Human fossils	Evolution stages	Stone age
	Fluvio-lacust. dep.	Earthy deposits				
	Reticulated Redclay	Alluvium	Recent deposits	Banpo Man Zalainor Man		
Ziyang fauna Liujiang fauna Maba fauna	Ziyang Fm Liujiang Man Cave deposits	Xiashu Fm	Michang Fm Tainan Fm	Shandingdong Man Hetou Man Xujiayao Man Dingcun Man	*Homo sapiens sapiens* *Homo sapiens neanderthalensis*	Neolithic
Yanjinggou f. Bama f. Guanyindong Up. Nabeng	Yanjingou cave deposits Upper Nabeng (Yuanmou 4)		Dianzihu Fm	Dali Man He Xian Man Beijing Man Lantian Man Yuanmou Man	*Homo erectus*	Palaeolithic
Gaoping f. Luicheng f. Yuanmou f. (upper)	Liucheng (Yuanmou 3)	Yühuatai Fm	Danangou Fm			
Yuanmou f. (middle)		Pujiang Fm	Toukeshan Fm			

Table 16.3. *Correlation of Quaternary faunas of South and North China (compiled by Li Fenglin).*

	South China		North China	
Q_3	Ziyang fauna (Sichuan)		Shandingdong fauna (Beijing) (Upper Cave)	
	Liujiang fauna (Guangxi)		Sarawusu fauna (Nei Mongol) (Shara Osu)	
	Maba fauna (Guangdong)		Dingcun fauna (Shanxi)	
Q_2	Yanjinggou fauna (Sichuan)		Zhoukoudina fauna (Beijing)	
	Bama fauna Guangxi (*Gigantopithecus* fauna)	Guanyindong fauna (Guizhou)	Chenjiao fauna (Shaanxi) Gongwangling fauna (Shaanxi)	
Q_1	Gaopin fauna (Hubei) (*Gigantopithecus* fauna)		Yangguo fauna (Shaanxi)	Loc. 18 fauna (Beijing)
	Liucheng fauna (Guangxi) (*Gigantopithecus* fauna)		Nihewan fauna (Hebei)	
	Yuanmou fauna (Yunnan)			

THE CAINOZOIC

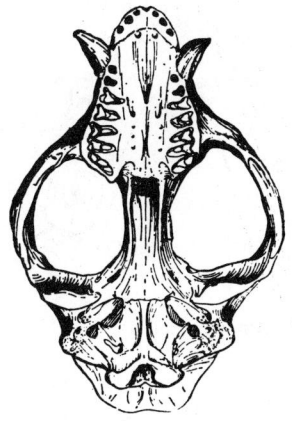

Fig. 16.4. *Bemalambda nanhsiungensis* Chow *et al.* × ¼, Shanghu Fm, Palaeocene, Guangdong.

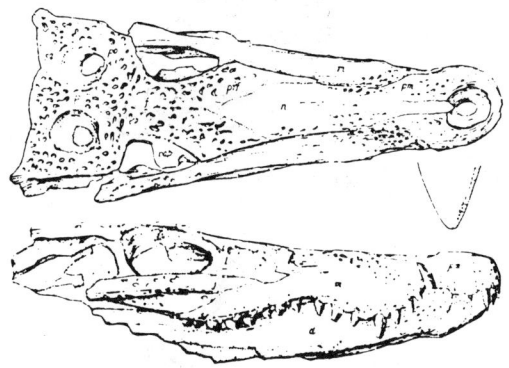

Fig. 16.5. *Planocrania datangensis* Li, × ¼, Palaeocene, Guangdong.

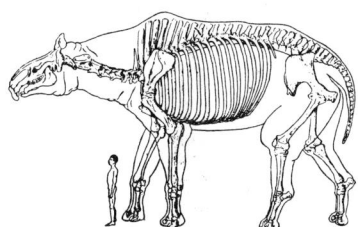

Fig. 16.6. *Indricotherium grangeri* Osborn, Oligocene, Nei Mongol.

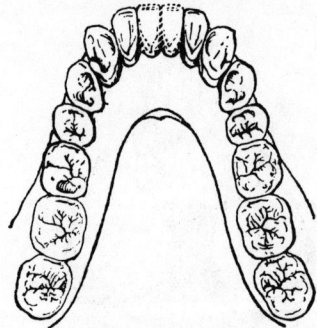

Fig. 16.7. *Ramapithecus lufengensis* Xu et al. ×0.7, Miocene, Yunnan.

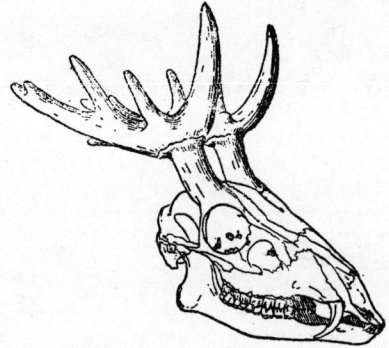

Fig. 16.8. *Stephanocemas thomsoni* Colbert, ×¼, Tungur Fm, Miocene Shandong, Nei Mongol.

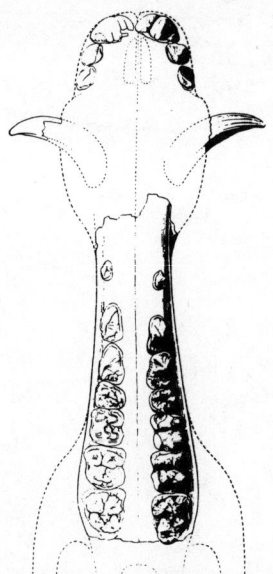

Fig. 16.9. *Listriodon lantienensis* Lin and Lee, ×1/6.6, Konjiacun Fm, Miocene, Shaanxi.

THE CAINOZOIC

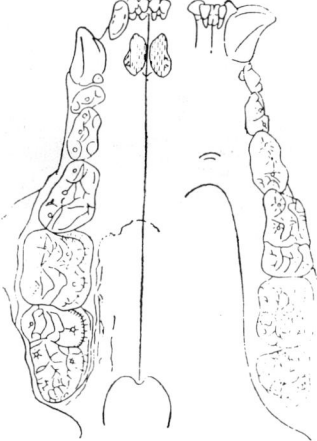

Fig. 16.10 *Ailuropoda* sp. ×⅓, Pleistocene, Zhoukoudian, Shanxi, South China.

Fig. 16.11 *Probosaidipparion sinense* Sefre, ×⅗, Lower Pleistocene, North China.

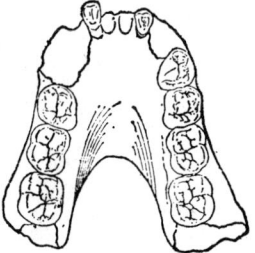

Fig. 16.12. *Gigantopithecus blacki* Koenigswald, ×0.45, Pleistocene, Guangxi.

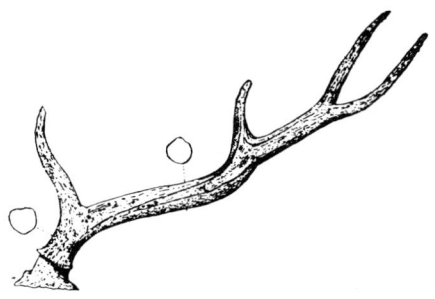

Fig. 16.13. *Cervus (Pseudaxis) grayi* Zdansky, ×$\frac{1}{12}$, Pleistocene, Heilongjiang, Shaanxi, Zhoukoudian.

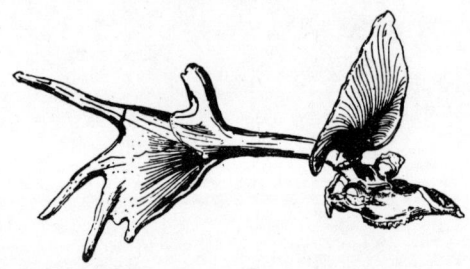

Fig. 16.14. *Megaloceros (Sinomegaceros) pachyosteus* Young, $\times \frac{1}{25}$, Pleistocene, Shaanxi, Zhoukoudian.

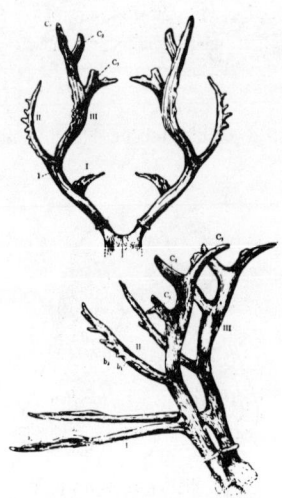

Fig. 16.15. *Elaphurus davidianus* Milne-Edwards, $\times \frac{1}{30}$, Pleistocene–Holocene, Hebei, Henan, Jiangsu, Zhejiang.

Fig. 16.16. Reconstruction of a female *Homo erectus pekinensis* skull (from Weidenreich 1943), Pleistocene, Zhoukoudian, Beijing.

Fig. 16.17. Reconstruction of the physiognomy of a female *Homo erectus pekinensis* (from Woo *et al.* 1959), Pleistocene, Zhoukoudian, Beijing.

PART III
MAGMATIC AND METAMORPHIC ROCKS OF CHINA

PART II

17. MAGMATIC ROCKS AND MAGMATISM IN CHINA

Cheng Yuqi

INTRODUCTION

Igneous rocks occur in great variety in China and are widely distributed outside the Mesozoic and Cainozoic basins and the regions of dominant late Proterozoic and Palaeozoic sedimentation. They were formed, either at or below the Earth's surface, during most or all of the geological past by various and complex geological processes known collectively as igneous activity or magmatism. They were the outcome of the evolution of this particular portion of the crust and its underlying mantle, especially during the culminating stages of geological events termed 'crustal movements', which are to be regarded as an important aspect of the evolution of the earth rather than being merely local phenomena.

Our knowledge of igneous rocks and igneous activity in China was rather meagre before the late 1940s. Abundant data obtained through systematic, comprehensive geological surveying at 1:200 000 and other scales, extensive prospecting and exploration for all kinds of mineral resources, and other subsequent geological and related work have provided an overall picture of the distribution of the major types of igneous rocks of China both in time and space, and have given an insight into the geological setting of the related igneous activity leading up to their formation (Song, in press). This will constitute the main content of this chapter, with a few regions described, or related problems analysed in a more detailed way, to serve as examples. It will be a general geological treatment rather than a specifically petrogenetic, petrological, or petrographic study. However some tectono-magmatic and plate tectonic topics will be considered when necessary.

CHRONOLOGICAL SEQUENCE OF IGNEOUS ACTIVITY

While there has been a great deal of igneous activity in China throughout geological history, events are concentrated within certain periods.

Basic and ultrabasic intrusives

Basic and ultrabasic rocks in China were formed at frequent intervals in time but are of quite limited distribution. They occur especially in Palaeozoic geosynclinal fold belts and fracture zones of many geological ages. Since isotopic dates for such rocks are scanty, the age of many intrusions can only be inferred geologically. This is especially true for those found in Precambrian metamorphic terrains.

There are numerous rather small plagioclase-amphibolite and pyroxene-granulite bodies in Archaean and early Proterozoic formations, which are metamorphosed basic intrusions forming integral parts of the protolithic basic volcano-sedimentaries. In addition to these there are discrete ultrabasic bodies (chiefly peridotites and serpentinites), and also basic ones (chiefly plagioclase-amphibolite) which can reasonably be regarded as products of a separate intrusive phase, possibly of a later period, but still of Archaean or early Precambrian age. These occur in almost all the Archaean and Early Proterozoic regions mentioned in Chapters 4, 5, and 18 (Cheng *et al.* 1973; Cheng *et al.* 1982*a*). Apatites from serpentine–magnetite ore and apatite-bearing peridotite in the Zhaoanzhuang Formation of the Taihua Group, Henan on the northern slope of eastern Qinling Range region (Fig. 4.1, VII), give U–Pb ages of 2580 and 2435 Ma respectively (Yichang Inst. 1980). There are a few exposures of Middle to Late Proterozoic or Late Proterozoic serpentinite intrusions, giving a K–Ar isotopic date of 600^+ Ma in the Yanshan Archaean region (Fig. 4.1, V). There are also ultrabasic rocks in the Hannan Complex, Shaanxi and pyroxenite and olivine–pyroxenite associated with the Fanjingshan Complex, Guizhou, of Late Middle Proterozoic age, and gabbroic and peridotitic rocks earlier than 1100 Ma in the Yuanbaoshan region, Guangxi. It is evident not only that the Early Precambrian ultrabasic and basic intrusions were the products of at least two periods of igneous activity, but also that those of the Late Precambrian were also the outcome of more than one phase or period of magmatism.

Basic and ultrabasic intrusive activity in China was probably best displayed in the Palaeozoic. In the geosynclinal belt, it occurred probably a little later than the marine volcanism, and was also multi-period or multi-stage. They can be roughly grouped into Early and Late Palaeozoic stages which were related to the Caledonian and Hercynian crustal movements respectively. The first one is further subdivided into an earlier Cambro-Middle Ordovician stage, and a later Middle Ordovico-Silurian stage. In the Qilian Mountains, where such intrusions are well developed, the latter was dominant. Basic and ultrabasic intrusives found in the south Qinlin Mountains and Pengxian, Sichuan, and fairly large basic ones, mainly diabase-gabbroic, in the north Tapa Mountain near the Sichuan–Shaanxi border were formed at about the same time.

Basic and ultrabasic intrusions of Late Palaeozoic age, similar to those of the early Palaeozoic, came into existence slightly later than the closely associated marine extrusives. In general, they are mostly either of Devono-Carboniferous or Permo-Carboniferous age, the later ones have a wider distribution, for example those exposed in the northern Tianshan belt. The ultrabasics of the Jishi Mountain of south-eastern Qinghai are of Permian age. Also belonging to this category, and probably of Permo-Carboniferous age, are those which outcrop in the Da Hingan and Xiao Hingan Mountains, and to the west of the Junggar Basin, Xinjiang, and along the N–S fracture zone in the Xichang–Huili region, south-western Sichuan.

Mesozoic basic and ultrabasic intrusions are of more limited distribution than those of the Palaeozoic. The early phase that probably related to the Indosinian Movement is represented by the small ultrabasic bodies of the dissected plateau in western Sichuan, near Kanzi and further south, which intruded into Middle to Upper Triassic formations. Middle-phase ultrabasics such as those occurring in the Dongqiao and Nachu area to the south of the Tanggula Mountains, northern Xizang, are probably of post-Early Jurassic age; and the late-phase bodies found in West Yunnan and in the vicinity of Dingquing and further south-east along the Nujiang River, eastern Xizang, were emplaced after the Late Jurassic, or in the Early Cretaceous. Both are related to the Yanshanian. The ultrabasic belt composed of a series of intrusives along the Yarlung Zangbo River, South Xizang was perhaps formed still later, possibly in the Late Cretaceous to Early Tertiary, and was related to both the Yanshanian and Himalayan. Tertiary ultrabasic intrusions are found in eastern Taiwan.

Granitic rocks and granitoids

Granitic rocks and granitoids are widely distributed in China, with exposures covering an area close to 9 per cent of the whole country. They have come into being during crustal movements at almost every stage of geological history. Permo-Carboniferous rocks of the Hercynian cycle and those of Jurassic–Cretaceous age related to the Yanshanian, are by far the most developed and widespread, and are frequently connected with the migration and enrichment of various mineral substances and the formation of many related mineral deposits.

Different chronological schemes for the ages of these rocks formed in Precambrian times have been suggested by a number of geologists and the authors recommend the following:

(2) Late Precambrian–Middle and Late Proterozoic (Fig. 17.1, γ_1).
  2c, Chengjiangian, c. 700 Ma (Nanjiang, northern Sichuan; Paoxing, western Sichuan; Yuanmo, Yunnan; etc.).
  2b, Jinningian, c. 850 Ma (Huangling, western Hubei; Kangding and Xichang, western Sichuan; Jiuling, Jiangxi; southern Shaanxi; etc.).
  2a, Sibaoan, c. 1050 Ma (Locheng, Guangxi); c. 1400 Ma (Miyun, Beijing).
(1) Early Precambrian–Archaean and Early Proterozoic (Fig. 17.1, γ_1).
  1d, Luliangian, 1700–1800 Ma; 1800–2000 Ma (for localities, see Chapters 5 and 18).
  1c, Wutaiian, 2200–2300 Ma; 2500 Ma (for localities, see Chapters 5 and 18).
  1b, Fupingian (Tiepuan)—Archaean B, 2500–2600 Ma (for localities, see Chapters 4 and 18).
  1a, Archaean A, c. 3000 Ma (for localities, see Chapters 4 and 18).

The formation of the Early Precambrian, especially of the Archaean rocks, is generally associated with regional migmatization and occasionally also with marginal migmatization, which will be discussed briefly in Chapter 18. Some of the Late Precambrian types are also accompanied by migmatization, mostly of a marginal character.

Early Palaeozoic or Caledonian acid to intermediate intrusive activity is well displayed in the marine eugeosynclinal belts, of which the Qilian region serves as an excellent example. So far three sub-stages have been recognized: an earlier one (post-Middle Ordovician) of 520–530 Ma, a middle one (probably Early Silurian) dated to 430–460 Ma, and a late sub-stage (Early to Middle Silurian) of 380–410 Ma and geologically of pre-Devonian age, all being slightly later than the corresponding volcanic epochs. Granitic rocks partly accompanied by migmatization of similar age are also found in the Guangxi–Guangdong borderland, Jiangxi, Hunan, the Qinling Mountains, Nei Mongol, the Da Hingan Mountains, etc.

Rocks of Late Palaeozoic or Hercynian age are well developed in the Tianshan Fold Belts where three sub-stages have been distinguished: those intruded into Middle Devonian or older rocks giving an isotopic age of $350 \pm$ Ma; those invading both Devonian and Carboniferous formations but overlain by Permian rocks; and lastly, the post-Permian granitic rocks unconformably overlain by the Triassic and characterized by 240 Ma, 260 Ma, and similar dates. Rocks of similar ages are also distributed extensively in other regions in the north-west as well as in the north, north-east and south of China. For instance, granites of the first sub-stage are found in the Qinling and Kunlun Mountains, of the second sub-stage in the Altay, Peishan, and Da Hingan Mountains, and granite and alaskitic granite of the third sub-stage, in northern North-east China, Nei Mongol, western Yunnan, and the Qilian Mountains.

While the influence of the Mesozoic acid to intermediate intrusive activity can be detected in almost every province or other major administrative unit in China, exposures of Mesozoic intrusions cover a smaller area than do those of the Late Palaeozoic. Chronologically such activity falls into two stages, an earlier Indosinian stage of Triassic age and a later Yanshanian stage of Jurassic–Cretaceous age. The rocks of the former are primarily granite and diorite, intruded into the Triassic—Middle or Lower—or older strata, and overlain by the Upper Triassic or Lower Jurassic, with isotopic ages of 180–220 Ma and also of 230 Ma. They are mainly found in southern Qilian and the Qinling Mountains, the eastern part of the Qinghai–Xizang Plateau, western Sichuan, south-western Yunnan and part of the Nanling region, including Guangxi and Hunan. Their occurrence in the north of North-east and North China has also been reported.

Rocks of the Yanshanian stage are of much wider distribution than those of the Indosinian, and are even more widespread than those of the Hercynian, especially in the circum-Pacific regions of the eastern part of China. They are closely related genetically to the formation of many mineral deposits, such as tungsten, tin, copper, iron, and rare elements. The intrusives may be generally subdivided into an early and a late phase, related to two phases of the Yanshanian Crustal Movement, roughly in the Jurassic and Cretaceous respectively. The rocks of the earlier phase, an isotopic age of 140–190 Ma, are exposed chiefly in South-east China and in many places in the Qinghai–Xizang Plateau, and also in North and North-east China, Qinling ranges, the lower Yangzi valley, and various other districts. The chief rock types include biotite-granite, granodiorite, and diorite, with the first predominating and often appearing in large intrusions. There are also igneous bodies consisting of a granite or granodiorite core which grades outward into diorite. Small alkaline intrusions are exposed in some of the south-east coastal regions. The late phase intrusives, with an isotopic age of 80–130 Ma, occur mainly in East China, as well as in Xizang, West Yunnan, the Taihang Mountains in North China, etc. In south-eastern China the rocks are more varied, including biotite-granite, granodiorite, monzonitic granite, hypabyssal rocks and alkaline rocks.

Tertiary granitoids and related rocks are primarily distributed in southern Xizang, the Sichuan–Yunnan region, and Taiwan (see below).

Volcanic rocks

Extrusive rocks are extensively distributed both in time and space throughout the country. They may be of submarine or continental origin. Rocks of the first category constitute the dominant group, and was mostly formed in geosynclinal, especially eugeosynclinal, belts. The exact geological setting of the formation of the rocks of early Archaean age is still not entirely clear, and is a matter for further investigation. A great part of the rocks have been considerably metamorphosed. The continental type becomes dominant only from Jurassic times onwards, and is exposed mainly in the eastern part of the country, that is in the circum-Pacific region.

Since volcanic (and volcano-sedimentary) rocks of the Early Precambrian have without exception been metamorphosed and form integral parts of the stratigraphic units in which they occur; their chronological sequence is treated in two other relevant chapters (4 and 5). It should be noted that there are as many volcanic periods as there are epochs for the formation of relevant major stratigraphic units. One can provisionally adopt a four-fold chronological scheme for Early Precambrian volcanic activity similar to that for granitic rocks, namely Early Archaean (before $c.$ 3000 Ma), Late Archaean (3000 to 2500–2600 Ma), Wutaiian (2500–2600 Ma to 2200–2300 Ma), and Hutuoan (2200–2300 Ma to $c.$ 1850 Ma).

The late Precambrian volcanism is of much more limited extent both in time and space (Chen et al. 1981). In the Yanshan region east of Beijing, there occur several beds of neritic K-rich trachyte in the Dahongyu

Formation of the Lower–Middle Proterozoic Changcheng System, giving a K–Ar date of 1621–1643 Ma on glauconite. This is probably the earliest volcanic marker for this major chronological division. Approximately of similar age, but not quite of the same petrochemical character, are the volcanic rocks of the Xiong'er Group in West Henan, and the basic rocks of the Zhulungguan Group of Qilian Mountains, and probably also the partly soda-rich volcanics of the Huili Group of south-western Sichuan. Those found in the formations correlated with the Upper–Middle Proterozoic Jixian System (1050–1400 Ma), and the Lower– Upper Proterozoic Qingbaikou System (850–1050 Ma), are the slightly metamorphosed, partly spilitic–keratophyric volcanic rocks of the Sibao Group, and of the unconformably overlying Banxi Group, and of corresponding formations of Hunan and Guangxi (also partially in Guizhou, Jiangxi, and south-western Fujian); various types of volcanic and tuffaceous rocks of the calc-alkaline series of certain formations of the Kunyang Group in Yunnan; spilitic and basaltic rocks of the Shennongjia Group of western Hubei, and basic to intermediate volcanic rocks in some formations of the Qilian Mountains and a few other districts. Some of the volcanic rocks of these regions are slightly metamorphosed. There are also occurrences of different types of volcanics, usually in minor amount, in the lower part of the uppermost Proterozoic member, the Sinian System (615–800 Ma), in certain districts. These include Quruktagh in Xinjiang (andesite and rhyolite); localities in the Kunlun Mountains (spilitic and keratophyric rocks to the south of Nuoqiang); the Qilian–Qinling Mountain belt and the Qaidam region (except for the south-western Sichuan district near the Omeishan Mountains, where volcanics and volcano-clastics of dominantly acid–intermediate composition up to several thousand metres thick are exposed). To sum up, there have been epochs of volcanism generally of local or very local extent for each of the four periods of the Middle and Late Proterozoic that give rise to the relevant stratigraphical units mentioned, and some of them are of spilitic–keratophyric character.

As in the case of granitic activity, Palaeozoic volcanism is chronologically divided into the earlier Caledonian phase and the later Hercynian. Activity occurred mainly in geosynclinal belts, especially during the early stages of the eugeosynclines. The first one was well developed only in the north Qilian belt and the Lagi Mountains region in the eastern part of south Qilian Mountain. There the extrusions were chiefly concentrated in the Middle Cambrian and Early Ordovician, reaching a climax in the latter. There was declining activity in Middle and Late Ordovician, and even very locally influence in the Early Silurian. All of this was probably a continuation of the Late Precambrian volcanism of the same eugeosynclinal belt. The volcanic rocks are largely basalt and andesite of the calc-alkaline series, and spilite together with quartz-keratophyre, occasionally also keratophyre, of the peralkaline type.

The Late Palaeozoic volcanics, which are almost exclusively of submarine origin, are well developed in the long belt extending from the Tianshan Mountains roughly eastward through part of Nei Mongol, and then east-north-eastward to the Da Hing'an Mountains in north-eastern China. Most of these volcanics date to Late Devonian, Upper Carboniferous, and Early Permian times, fewer of them to the Early Devonian and Early to Middle Carboniferous, and they extend only locally to Late Permian times. The rocks are mostly andesite and andesitic basalt with minor spilite–keratophyre, accompanied by subvolcanic diabase. In the north Kunlun–Qimantak–North Qinling belt, lying to the south of Qilian Mountain and its continuation, there are also exposures of Late Palaeozoic extrusives. This volcanism covers the time-span from the Late Devonian to the Early Permian in Kunlun in the west, and exists only in the Late Devonian in Qimantak and in the Early Carboniferous in north Qinling, thus showing an eastward decline in its magnitude. Areas of prominent extrusive activity of the same general period are also present in the south-west of the country, including Xizang, Yunnan, Sichuan, Guangxi, and Guizhou, with the formation of quite widespread Permian Omeishan Basalt (β_3 in subregions IV, and V_1, Fig. 17.1) and other volcanic series.

As products of the continuation of the Late Palaeozoic submarine volcanism, there are exposures of Triassic, and local Cretaceous, extrusive rocks in some marine sedimentary belts in places in the south-west and north-west of China, such as Qinghai, Xizang, Yunnan, and Sichuan. This indicates that Mesozoic submarine volcanic activity was also prevalent in this part of the country, and it can be divided into Indosinian and Yanshanian epochs, the details of which have not yet been investigated.

Probably the most widely distributed and most intense continental Mesozoic volcanism prevailed from Jurassic, especially Late Jurassic, to Cretaceous times in the eastern part of the country in a NNE–SSW belt extending from Heilongjiang to Guangdong provinces for more than 3000 km. The rocks are mostly rhyolites, including pearlite (Plate XIV.6) and alkaline rhyolite (Plate XIV.5), and andesites (Plate XIV.2), and subordinately andesitic–trachytic, latitic, and trachytic and other types. A consideration of their salient geological characters will be deferred to a later section of this chapter.

Tertiary volcanism was prominent in the Oligocene and Miocene in Taiwan Province. In other provinces and regions it was rather limited and was represented chiefly by basalt eruption.

Quaternary basaltic eruption was of even more local a character.

MAJOR REGIONS OF IGNEOUS ROCKS

Since igneous rocks and activity are the outcome of the evolution of the crust as well as of the upper mantle in general, and of the regions concerned in particular, the geological character and background of magmatic activity in China throughout the ages can appropriately be discussed in terms of a systematic description of the occurrence and other features of the three major rock types of different ages mentioned above in five major regions, each of which is marked by certain salient geological and evolutional characters. These regions have been divided mainly on a geotectonic basis (Chapter 19), with special consideration of their tectono-magmatic development over the various main periods of magmatism. The boundaries between them, and between those of their sub-regions, are thus mostly geotectonically significant; many of them are in fact fractures of various kinds, some coinciding with ancient suture zones between plates or micro-plates.

Tianshan–Yinshan–Da Hing'an Mountains region (Fig. 17.1, I_1, I_2 and I_3)

This region occupies most of China lying to the north of 41°N and is characterized by a preponderance of Late Palaeozoic Hercynian volcanism and intrusive activity. It is dotted with scanty relics of Early Palaeozoic magmatism, further supplemented in the eastern section (Fig. 17.1, I_3) by rather extensive Mesozoic Yanshanian volcanics and volcano-sedimentaries. There are many exposures of Yanshanian granitic rocks and flat-lying Tertiary and Quaternary basalts, and several occurrences of old granites, most probably of Early Proterozoic age. It may be divided into three sub-regions.

The western or Tianshan sub-region (Fig. 17.1, I_1), with its approximate eastern boundary at 101°E, includes the Altay Fold Zone, the Junggar Massif, the Transjunggar Fold Zone, the Tianshan (Hercynian) Fold Belt, and the Quruktagh Aulacogen (Fig. 19.3, ID_1, IA, IE_1, IIC, and IIA_2). The igneous rocks are dominated by Hercynian intermediate, basic, and acid volcanics, mostly metamorphosed to some extent, and granitic and migmatitic rocks (Altay) and some ultrabasics, including harzburgite, olivinite and even dunite, with very local Caledonian and Yanshanian granitoids and granitics of probable Early Proterozoic age ($\gamma_{(1)}$). The details of the Hercynian igneous activity may be well exemplified by that which occurred in the eastern section of the Tianshan geosynclinal belt (Song, in press) at the border of Xinjiang and Gansu (Fig. 17.1, A). This belt is located to the north of the northwestern Qilian Early Palaeozoic eugeosynclinal system and separated from it by the intervening Tarim Platform and Alxa Massif (Fig. 19.3, IIA and IIB_3). It is characterized by the presence of a very narrow discontinuous central uplifting zone (Fig. 19.3, IIC_2, the Median Tianshan Uplift), assigned to a Middle to Upper Proterozoic age, which contains scanty intermediate volcanics. The Palaeozoic succession is marked by quasi-miogeosynclinal clastics and carbonates up to Lower Ordovician horizons. Some volcanics start in the Middle to Upper Ordovician and Silurian sediments, and there follow Carboniferous or Permian eugeosynclinal volcanics and volcano-sedimentaries. For instance, north of the uplift zone, the mostly Lower Carboniferous volcanic sequence is generally composed of basalt or other basic rocks in the lower part, and rhyolite and andesite in the upper. In addition to the calc–alkaline rocks, there are also keratophyric porphyries richer in either potash or soda, and quartz–porphyry of peralkaline type, especially in the regions with iron mineralization. But to the south of the uplift zone, as in the district extending from Hongliuyuan, Gansu, westward to easternmost Xinjiang, Lower, and even Upper, Permian rocks including andesite, dacite, basalt, and subvolcanic diabase are well developed. The eastern Tianshan region is further characterized by the rather frequent, yet scattered occurrence of basic and ultrabasic intrusives, including gabbro, harzburgite, and dunite, the ultrabasics being usually smaller and of slightly later age. The intermediate to acid intrusions are rather widely distributed, occurring as batholiths or stocks in the uplift belt, and as volcanics and basic–ultrabasic intrusives on both its flanks. Both the Proterozoic and Palaeozoic volcanic rocks are slightly metamorphosed or belong to the greenschist facies (Fig. 18.1, III_2').

The middle or Yinshan sub-region (Fig. 17.1, I_2), extending roughly to the western limit of the extensive area of Mesozoic continental volcanics of the third sub-region, is characterized mainly by Hercynian granites and volcano-sedimentaries, including an ophiolite sequence (Bei Wenji, personal communication) composed of harzburgite, olivinite, dunite, gabbro, diabase, and siliceous rocks. Subordinately there are Yanshanian granites, Early Palaeozoic volcanics, and Tertiary basalts.

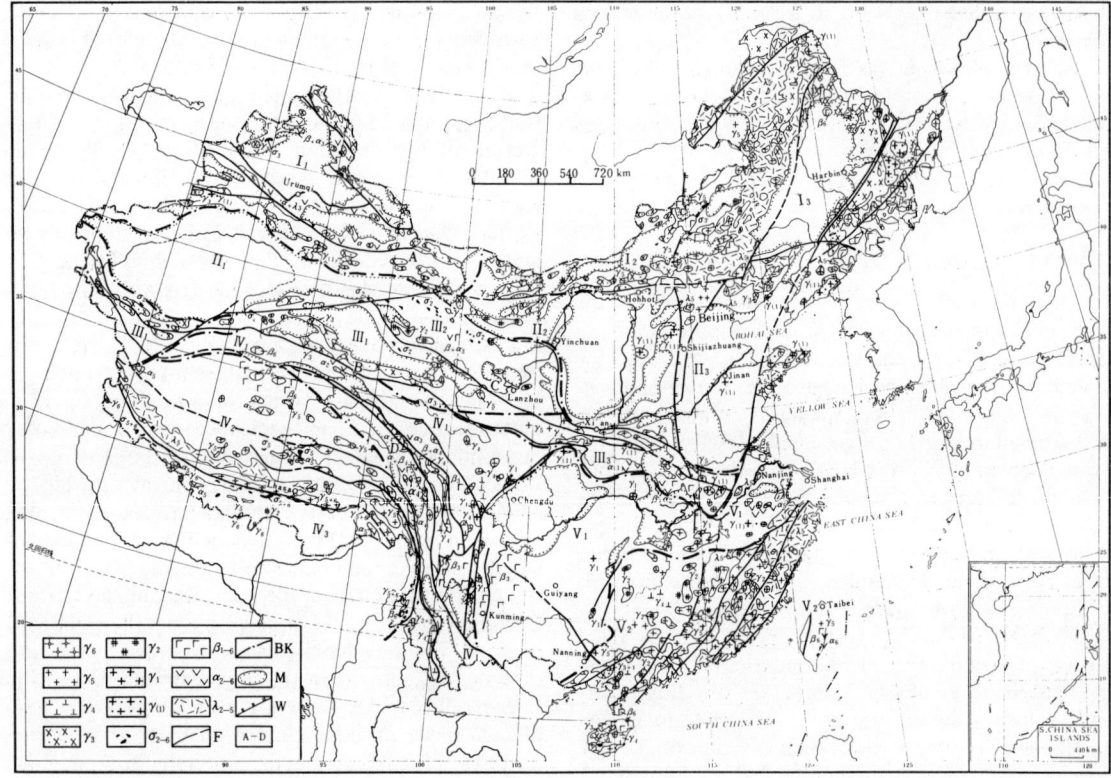

Fig. 17.1. Sketch map showing the distribution of major types of igneous rocks of China (compiled by Cheng Yuqi and Xu Huifen (1982)).

Granitoids (γ), with occasional migmatitic rocks: γ_6, Himalayan; γ_5, Yanshanian; γ_4, Indosinian; γ_3, Hercynian (Tianshanian); γ_2, Caledonian (Guangxian, Qilianian); γ_1, Late Precambrian (Middle and Late Proterozoic); $\gamma_{(1)}$, Early Precambrian (Archaean and Early Proterozoic).

Ultrabasic intrusions (σ): Arabic numbers for geological ages same as for granitoids.

Volcanic rocks, with associated rocks: β, basic composition (Arabic numbers for geological ages same as for granitoids); α, intermediate composition (Arabic numbers for geological ages same as for granitoids); λ, acid composition (Arabic numbers for geological ages same as for granitoids).

F, Fractures of various nature; BK, Approximate boundary of major regions of igneous rocks; M, Mesozoic and Cainozoic sedimentary basins; W, Approximate western boundary of the outcrops of Yanshanian continental volcanic rocks; A–D, Localities.

Key to major regions and sub-regions of igneous rocks. I, Tianshan–Yinshan–northern North-east China region: I_1, Tianshan sub-region; I_2, Yinshan sub-region; I_3, Northern and central North-east China sub-region. II, Tarim–Alxa–North China region: II_1, Tarim sub-region; II_2, Alxa sub-region; II_3, North China sub-region. III, Kunlun–Qilian–Qinling region: III_1, Kunlun sub-region; III_2, Qilian sub-region; III_3, Qinling sub-region. IV, Qinhai–Xizang Plateau and Transverse Ranges region: IV_1, Southern Qinghai and Transverse Ranges sub-region; IV_2, Central and northern Xizang sub-region; IV_3, Southern Xizang sub-region. V, South China region (South Central and South-east China region): V_1, Yangzi Platform sub-region; V_2, South-eastern China sub-region.

Key to localities: A, Tianshan geosynclinal belt at the border of Xinjiang and Gansu; B, Kunlun Pass; C, Median Qilian Uplift; D, Yushu district of the Qamdo (Chamdo)–Yushu–Baiyu region.

The eastern sub-region, comprising northern and central North-east China (Fig. 17.1, I_3), which includes the Xingkaian Fold Zones, the Nengjiang Fold Zone, the Zhangguangcailing Fold Zone, the Nadanhada (Yanshanian) Fold Zone (Fig. 19.3, IC, ID_2, IE_3 and VD), the eastern part of the Xing'an–Nei Mongol Fold Zone, and the South Nei Mongol (Palaeozoic) Fold Zone (Fig. 19.3, IE_2 and IID), is marked chiefly by widespread Hercynian granitoids and volcanics, Yanshanian extrusives, and less extensive Tertiary basalt. It also contains, subordinately, dispersed Yanshanian granitoids in many regions, old granitics probably of Early Proterozoic age in its eastern part, and also sporadic relics of metamorphosed Early Palaeozoic volcanics. Indosinian granitoids have been reported in the Da Hing'an Mountains and in the south-western part of this sub-region.

Tarim–Alxa–North China (Platform) region

The region as a whole is marked by a relative scarcity of igneous rocks as compared with the other major regions and is divided into three sub-regions. In the western Tarim sub-region (Fig. 17.1, II_1) only Hercynian granites, with or without late Palaeozoic volcanics, and some ultrabasic intrusions are found at some places along the periphery. In the middle Alxa sub-region (Fig. 17.1, II_2; Fig. 19.3, IIB_3) there are both Early and Late Palaeozoic granites, partly accompanied by volcanics of corresponding ages. There are also localities of Proterozoic metamorphic rocks of volcanic or volcano-sedimentary parentage. In the eastern North China sub-region (Fig. 17.1, II_3, North China Platform), there are various types of early Precambrian metamorphic rocks of protolithic volcanic character, and granitic and migmatitic, or even basic and ultrabasic, rocks (referred to elsewhere in this book). There are also a few localities of Middle Proterozoic volcanic horizons and Rapakivi granite in the vicinity of Beijing. Exposures of Yanshanian volcanics (Plate XIV.5) and granitoids, Hercynian and Indosinian granites and pegmatites, and Tertiary and Quaternary basalts (Plate XIV.1) are found in many places in the north-eastern and eastern part of the region. In fact, type localities of the Yanshanian Orogeny, characterized by a relatively wide distribution of intermediate to acid extrusives and abundant granitoid intrusions, are situated around Beijing and in the regions further to the north-east.

Kunlun–Qilian–Qinling region

This region is in general characterized by the dominant effect of Palaeozoic igneous activity in many districts, with evident influence of Yanshanian magmatism at many places in the east, and the presence of Precambrian igneous activity in the metamorphic terrains of the same age. It may also be differentiated into three sub-regions. As the area is extensively covered by marine Triassic and later sedimentaries, the boundaries between the three sub-regions south of Lanzhou (as shown on Fig. 17.1) are rather arbitrarily drawn.

The Kunlun (Mountains) sub-region (Fig. 17.1, III_1) in the west, roughly including the Tieklik–Altyn Uplift, the West Kunlun Fold Zone, the East Kunlun Fold Zone, the Qaidam Massif (Fig. 19.3, IIA_4, IIH_1, IIH_2 and IIE), and the western part of the West Qinling (Indosinian) Fold Zone (Fig. 19.3, IIJ_1), is dominated by Hercynian magmatism, particularly shown by the granitoids. This is supplemented by Indosinian and Yanshanian granitoidal activity in its eastern part and Late Palaeozoic volcanic activity in some districts. The eastern portion in the Qinghai province, viz. the Burhanbuda Mountain and the area further south, provides a well investigated example (Song, in press). In the Burhanbuda Mountain and its western extension, the slightly metamorphosed geosynclinal sediments are intercalated with volcanics accompanied by basic and ultrabasic intrusive complexes, similar to the rocks produced by typical eugeosynclinal magmatic activity. These may also be of Early Palaeozoic age, but this age is not yet confirmed. In the Kunlun pass–Majiu area further south (Fig. 17.1,B) there is a Permo-Carboniferous geosynclinal belt with Lower Permian volcanics, mostly of andesitic, dacitic, and even more acid, types, which are intruded by somewhat sporadic ultrabasic bodies such as harzburgite, lherzolite, and augite-peridotite, and less basic intrusions. These volcanics and the sedimentaries are further invaded by granitic to dioritic bodies. It is thus evident that this region also exhibits intensive Late Palaeozoic intermediate to acid intrusive activity, but weaker volcanism and basic–ultrabasic intrusive activity when compared with the eastern Tianshan (Hercynian) Fold Belt (Fig. 19.3, IIC).

Lying along the border of Gansu and Qinghai provinces, the WNW-ESE striking Qilian (Mountains) sub-region (Fig. 17.1, III_2) corresponds to the Qilian (Caledonian) Fold Belt (Fig. 19.3, IIG) which occupies approximately what was the extent of the Early Palaeozoic eugeosynclines and part of the Lanzhou–Xining Massif (Fig. 19.3, IIF). It now consists of three parallel zones (Song, in press). The central one (Fig. 19.3, IIG_2; Fig. 17.1, C; Fig. 18.1, II) is an uplift composed of Late Precambrian, slightly metamorphosed formations that may be correlated with the Proterozoic Changchengian, Jixianian and Qinbaikoun Systems, often with intercalations of marine volcanic rocks towards the upper part of the sequence. In its eastern section, the Proterozoic neritic volcano-sedimentaries in the Xinglong Mountain district, Gansu, consist of a lower formation of dacitic tuff and dacite with intercalations of basalt and phyllite, and an upper one of basalt, andesitic basalt, and also dacitic and andesitic tuffs, interbedded with phyllite. They are, however, not widely distributed and were apparently extruded from some local centres of activity. In its western section in the districts of Tolainanshan, Shulenanshan, etc. the rocks are basalt and intermediate types, and even felsite. Most of the volcanics have already been metamorphosed.

Both the northern and southern zones are characterized by Lower Palaeozoic volcanic rocks, and by eugeosynclinal magmatism which was much more intense than that of the Late Precambrian. Since no unconformity has ever been found between the Early Palaeozoic and Late Proterozoic formations, it is inferred that the Early Palaeozoic volcanics were more or less continuous with those of the Late Precambrian with a certain interval of quiescence.

The characteristics of the igneous activity of the northern zone (Fig. 19.3, IIG_1) can be seen in the geology of the district south-west of Jiuquan, in the west. Here, the eugeosynclinal stratigraphical sequence is composed of Cambrian, Ordovician, and Silurian clastics, siliceous and carbonate rocks over 10 000 m thick, in addition to the volcanics. The Middle and Upper Cambrian consist of a lower member of quartzite and limestone with intermediate to acid tuff; a middle one of siliceous rocks intercalated with andesite, basalt, and related tuffs; and an upper one of slate, tuff, and basalt, with a gradual increase in volcanic components upward in the sequence. Ordovician rocks are not only better developed than the Cambrian ones, but also contain more volcanics which are most abundant in the lower part. This indicates that igneous activity reached its climax in the Ordovician period. The extrusive rocks are mostly of intermediate and intermediate–basic composition, but are very occasionally acidic, and include lava as well as tuff. The sedimentaries are slate and limestone. Silurian formations occur chiefly along the northern flank of this northern zone and are composed mainly of sandstone and slate, associated only locally with volcanic rocks. In general, the Early Palaeozoic submarine volcanism was mostly concentrated near the central sub-zone and its southern side, close to the central uplifting zone of the Qilian Fold Belt; it decreased in intensity northward up to the Alxa region, where Early Palaeozoic lava is practically absent. Groups of small basic and ultrabasic bodies, having elongated or lenticular shapes at the surface, are closely associated with the volcanic rocks, but some are also found to have been intruded into the Late Precambrian fault blocks in the Lower Palaeozoic. They are of more frequent occurrence in the central section than in the eastern and western portions of the zone, often showing a northern sub-zone of dominantly ultrabasic rocks and a southern one of preponderantly basic types. The ultrabasics are chiefly dunite and harzburgite with varying amounts of pyroxenes. Small intermediate to acid granitoid bodies are mostly found as intrusions in these fault blocks, and chiefly occur along the margins of the clustered basic–ultrabasic bodies. The larger granitoids are, however, exposed in both the Lower Palaeozoic and Precambrian formations to the north and south of the zones of volcanic rocks. Such a mode of distribution is typical of the entire Qilian eugeosynclinal belt.

Though the conditions of magmatism of the eastern section of the northern Qilian zone were in general similar to those of the western one, they exhibited certain differences. For instance, in the region north of Xinglong Mountain, the Cambro-Ordovician volcano-intrusive sub-zone extends discontinuously for hundreds of kilometres just north of the central Precambrian zone of the Qilian Mountains. The Middle Cambrian volcanic rocks are chiefly spilite, keratophyre, quartz-keratophyre, and related tuffs. The Ordovician, especially Early and Middle Ordovician, extrusives, which are more developed than the Cambrian, consist mainly of andesitic basalt, spilite, andesite, dacite, and keratophyre. These rocks are usually found as intercalations in a quite thick sequence of phyllite, slate, tuff, and limestone. There are only scanty volcanic rocks in the Upper Ordovician, and no indication of volcanism is present in the Silurian formations, which are composed mostly of flysch, flysch-like rocks, and phyllite. It is noteworthy that the volcanics in this section are more acidic than in the western one, and are further characterized by the appearance of soda-rich peralkaline types, especially at volcanic centres. The intrusions in this section are also less basic; they are devoid of dunites and peridotites bearing orthorhombic pyroxene. There are also fewer and smaller intermediate to acid intrusions, which are mostly small stocks in Cambro-Ordovician formations.

The Early Palaeozoic igneous activity of the southern zone of the Qilian sub-region (Fig. 19.3, IIG_3) was also of eugeosynclinal character, but of less intensity and somewhat different from the northern one described above. For instance, in the Laji Mountain in the east, the Middle and Upper Cambrian volcano-sedimentary formations are about 2000 m thick, consisting in ascending order of: (1) basalt, tuff, and phyllite; (2) basalt, tuff, siliceous rock, and carbonates; (3) basalt, andesitic basalt, andesite, etc., and (4) basalt and tuff, with carbonate rock intercalations, these being locally accompanied by dacite, and more acidic volcanics. The Ordovician is represented mainly by terrigeneous sediments, interbedded with intermediate to basic extrusives, even occasionally by spilite, and the Silurian is absent. The ultrabasic intrusions are small, mainly of dunite and harzburgite, and the basic ones are also rather scanty. There are more intermediate to acid intrusives, of which the larger granodiorite bodies are mostly distributed along the periphery of the ultrabasic zone. These large bodies are usually intruded into the Precambrian formations, but a few small granodiorite ones are intruded into the ultrabasic bodies and have an isotopic age of 443 Ma, approximately Late Ordovician. The western section of the southern zone is more complicated and as yet much less investigated. In the slightly metamorphosed Lower Palaeozoic pelitic and carbonate rocks there are many submarine volcanic beds that are invaded by basic and ultrabasic rocks. The ultrabasics, though mostly small bodies, occur more frequently, and contain mainly harzburgite and subordinate pyroxenite. It is interesting to note that, in contrast to the eastern section, the intermediate to acid stocks, and especially the batholithic intrusions, are chiefly distributed to the north of the discontinuous zone of basics and ultrabasics. Most of the Lower Palaeozoic rocks have been metamorphosed in one way or another (Chapter 18).

The Qinling sub-region (Fig. 17.1, III_3) extends from the extreme eastern part of Qinghai province through southern Gansu, Shaanxi, and Henan to the Tancheng–Lujiang deep fracture in Anhui province, including the entire Qinling Mountains and their extensions, and neighbouring regions. Geotectonically it corresponds roughly to the West Qinling (Indosinian) Fold Zone, the East Qinling (Palaeozoic) Fold Zone, the Ankang–Tongbo (Caledonian) Fold Zone, and the Dabie Uplift (Fig. 19.3, IIJ_1, IIJ_2, IIID, and $IIIA_2$). In the western portion, lying approximately to the west of the Gansu–Shaanxi provincial boundary, tuffaceous rocks occur locally in the very weakly metamorphosed Carboniferous and Devonian clastic and carbonate rocks. They are accompanied by sporadic small basic and ultrabasic bodies such as dunite and pyroxenite, still smaller ones of peridotite, olivine-gabbro, etc., and more frequently by small stocks of granodiorite, quartz-diorite, and granite. All these were products of the Hercynian igneous activity which was much weaker than that of the Tianshan region (Song, in press). Some Yanshanian, and probably also Indosinian granitic intrusions, have also been found. The history of igneous activity in the middle portion, which is situated in Shaanxi and part of Henan, is much more complicated. In addition to the protolithic volcanic rocks in both the Proterozoic and Lower Palaeozoic metamorphic formations, Devonian extrusives also occur in a few localities. The granitoid intrusions of Yanshanian age are the most frequent and are represented by batholiths such as the Huashan and Mangling bodies, numerous small stocks including potash-rich granite and granite porphyry, and also quartz-diorite and syenite. However there also occur Proterozoic, Caledonian, Hercynian, and Indosinian granitoid bodies and pegmatite dykes. The basic and ultrabasic intrusives are mostly of Caledonian and Hercynian ages. It is noteworthy that the effect of magmatism, especially that of volcanism, is greatly decreased in the Hercynian. The eastern portion shows even more influence of the Yanshanian granitoidal activity in many districts. It is locally associated with slightly earlier volcanism, with a strong imprint of Early Precambrian magmatism. This includes Archaean basic volcanism and Early Proterozoic spilitic to keratophyric activity, and related events such as regional migmatization* in the (Archaean to Early Proterozoic) metamorphic terrain of the Dabie Uplift and the neighbouring region. Although the area has evidently been affected geotectonically by Palaeozoic crustal movements, no positive evidence for corresponding igneous activity is found, except for some K–Ar dates of granitics possibly representing modified values of older rocks, and for some possible Caledonian acid intrusive bodies and pegmatites in its western part.

Qinghai–Xizang Plateau and Transverse Ranges (Fig. 17.1, IV_{1-3})

Situated in south-west China, this area includes the greater part of the Qinhai–Xizang Plateau lying to the south of the Kunlun Mountains and the western part of the Qinling Range and the longitudinal Transverse Ranges. Geotectonically it corresponds roughly to the Songpan–Ganzi Fold Belt, the Qiangtang Massif, the Qamdo (Chamdo)–Simao Fold Belt, the Karakhorum (Yanshanian) Fold Zone, the Himalaya Massif, the Gangdise Massif, the Lhasa–Nagqu (Yanshanian) Fold Zone, the Ngali (Yanshanian) Fold Zone, the Yarlung Zangbo (Himalayan) Fold Zone, and the Kham–Yunnan Uplift (Fig. 19.3, $III_{B, C, E, F}$, IV_{A-E} and III_{A3}). Except for the northern and eastern parts, and certain localities in Yunnan, where Late Palaeozoic or even older igneous rocks are still present, this region is characterized by the strengthening of Mesozoic magmatic activity at the beginning of this era. There is a broadly zonal distribution of igneous rocks of different ages, and the younger zone is generally on the southern or south-western side until the youngest Himalaya zone is reached. The area may be divided into three sub-regions.

*See Chapter 18.

The first sub-region (Fig. 17.1, IV_1), approximately coinciding with the goetectonic regions III_{A3}, III_C, and II_E of Fig. 19.3, appears in a somewhat distorted, inverted triangle or funnel-shaped area, extending from the region west of the Kekexili Mountain and the Bayanhar Mountains to the Songpan–Ganzi district in western Sichuan, and then southward to the Transverse Ranges up to the national border. Scattered localities of weakly or moderately metamorphosed Middle Proterozoic volcanic rocks of varying composition are found near its eastern boundary. Some of these are soda-rich, and there are also Late Proterozoic (Sinian) acid-intermediate volcanics, and more extensive areas of Late Proterozoic granitic intrusives, including those associated with migmatitic rocks or ultrabasic intrusions. North of 28 °N lies a large area where the influence of Indosinian magmatism is most marked; there are granitoids, sometimes accompanied by basic and ultrabasic intrusions, and volcanic and volcano-sedimentary rocks. This magmatism was preceded in some places by Permian volcanism, was followed by Yanshanian magmatism, mainly in the form of granitic and ultrabasic intrusions. For instance, in the Qamdo (Chamdo, Changdu)–Yushu–Baiyu region (Fig. 17.1, D) in the borderland of Xizang, Qinghai, and Sichuan, the Lower Permian carbonate rocks are intercalated by basic volcanic rocks, and the Upper Permian contains andesitic lava and agglomerate. The Permian is unconformably overlain by the Upper Triassic, highly fossiliferous, Batang Group, nearly 10 000 m thick, which, according to a paper by Zhao Rongli (Zhao 1982), is of Carnic–Noric age and contains up to 50 per cent of volcanic rocks belonging to five eruption cycles. The volcanics are considered to be products of linear extrusion under the sea, and consist mainly of andesite, basalt, and andesitic basalt; also very subordinate dacite, rhyolite, intermediate to acid tuff, volcanic breccia, and agglomerate. They belong to the calc-alkaline series, and are occasionally oversaturated with aluminium. The associated intrusives are quartz-diorite and diorite bodies, and peridotite, gabbro, diabase, diorite, diorite-porphyrite, monzonite porphyry, granite, and granite porphyry dykes. Elsewhere there are batholithic and smaller granitic intrusions of Indosinian and Yanshanian age. The elongated shape of many intrusions, and the linear distribution of some volcanic belts parallel or sub-parallel to the numerous fractures in the region, seem to indicate tectonic control of the location of magmatic activity. In fact Zhao has suggested that these fractures were genetically related to a late Triassic northward to north-eastward subduction zone. Exposures of Himalayan basalt are also found in the extreme north-western part of this region of dominant Indosinian influence.

In the district roughly south of 28 °N, well within the Transverse Ranges, the characteristics of igneous activity are more varied and complicated. Besides the Proterozoic igneous rocks mentioned above, those formed during the major periods of both Palaeozoic and Mesozoic magmatic activity are found. There are some occurrences of Caledonian or Caledonian-Hercynian granites in south-western Yunnan. The products and influence of Hercynian magmatism are much more widespread. In addition to the Upper Carboniferous to Lower Upper Permian submarine spilitic rocks of the extreme north-western part of Yunnan at the western limit of this sub-region, the Upper Permian tholeiitic plateau basalt, locally accompanied by pyroclastic rocks and even minor marine facies in the basal part, occurs in many places in West-central Yunnan and South-west Sichuan. There are also rather small ultrabasic–basic and basic intrusions, including gabbroic, diabasic, olivine-gabbroic, pyroxenitic, and olivine-pyroxenitic types. Some of these are genetically related to the formation of vanadiferous, titaniferous magnetite deposits and others to nickel mineralization. Their emplacement is closely followed by the intrusion of aegirine-nepheline-syenite in the Xichang district, South-west Sichuan, where the late Palaeozoic magmatism is usually considered to be closely related to the N–S fracture zone. Indosinian granitoids and basic and ultrabasic intrusions are present at many places, and most of them are of very small size. According to a paper by the Yunnan Institute of Geological Science in 1982, whole rock samples from the southernmost part of the largest Indosinian batholith Lincang–Menghai body in South-west Yunnan, which is composed chiefly of biotite-granite and biotite-granodiorite, give a Rb–Sr isochron age of 715 ± 40 Ma. This batholith may thus have a rather complicated petrogenetic history and must contain remnants of older granitoids. To the south of it, in Menyang, there is an Upper Triassic spilito-keratophyric volcano-sedimentary series which contains ferriferous spilite (Plate XIV.8) and layers of amygdaloidal magnetite rock of unusual, indeed unique, composition, structure, and texture (Plate XIV.7, 9), evidently recrystallized from an 'iron magma' of spilitic affinity. Small Yanshanian intrusions of granitic and alkaline granitoidal composition are found in many localities in Yunnan, and some larger ultrabasic ones occur along certain fracture belts to the south-west of the Red River fracture zone.

The second sub-region (Fig. 17.1, IV_2), roughly corresponding to the geotectonic regions III_B, III_F, IV_B, IV_C and IV_D in Fig. 19.3, includes the greater part of Xizang north of the Yarlung Zangbo fracture zone, and small areas of south-west Qinghai and north-west Yunnan. It is characterized by the dominance of Yanshanian and Himalayan magmatism in many places and of Hercynian and Indosinian at certain localities (Liu *et al.* 1980). Among the volcanic rocks, which include some occurrences of Permo-Carboniferous andesite, basalt porphyrite, and basalt, Indosinian andesite and rhyolite, and Himalayan and Quaternary basalt, the outstanding feature is the presence of the extensive belt of late Yanshanian–Late Cretaceous to Himalayan andesite, dacite, rhyolite, etc. This belt is 3 000 m thick or more, and is found in the Gandise (Kangdese) Mountains and extends eastward to the Nyainqentanglha Mountains in the environs of Lhasa (Plate XIV.4). As to intrusives, there is a spectacular discontinuous zone of dominant late Mesozoic magnesian (Mg/Fe_{tot} 9–11) ultrabasic rocks composed mainly of strongly serpentinized harzburgite, dunite, olivine-pyroxenite, etc., with occasional associations of basic intrusives and extrusives. It attains an aggregated area of well over 1 100 km^2, and extends from the Nujiang fracture (the upper reaches of the Salwan River) approximately westward to the Bangong Lake at the border along a distinct fracture belt which has been interpreted as a suture zone by several geologists, including Chang Chengfa, Xiao Xuchang, Wang Xibin *et al.*, who worked in this region recently. There are also two granitoid zones to the south of the ultrabasics just mentioned (Liu *et al.* 1980). The first, northern one, extending from Selincuo, north-west of Lhasa, east-south-eastward to Chayu in South-east Xizang, is composed of over sixty intrusions, many of composite nature, including a few elongate batholiths. They consist mostly of biotite-granite, porphyritic biotite-granite and granite, and subordinately of quartz-diorite, diorite, and granodiorite. Most of them are unconformably overlain by the Tertiary, with isotopic ages of 80–145 Ma and occasionally of 16–35 Ma, and hence are dominantly late Yanshanian and partly Himalayan. The second, southern zone containing batholithic complexes, appears immediately to the north of the Yarlung Zanbo fracture, stretching from Gadac in the west, through Lhasa to the Linzhi district in the east. The western section is composed chiefly of diorite, quartz-diorite, and granodiorite, and subordinately of biotite-granite; the middle section from Ngamring to Quxu mainly consists of a diorite-granodiorite-granite series; and the eastern section is chiefly of biotite-granite and secondarily of two-mica granite, diorite, and granodiorite. A part of the granite to the west of Lhasa is porphyritic and contains abundant xenoliths, showing evidence of being intensively contaminated, and probably of a genetic relationship between the metasomatic potash feldspar 'phenocrysts' and the nearby relict assimilated xenoliths (Plate XV.5, 6). The middle and western segments of this intrusive belt and the Gandise–Nyainqentanglha volcanic belt mentioned above actually constitute a composite magmatic zone. The general intrusive sequence is from diorites through granodiorites to granites and petrochemically it varies from Si- and Al-oversaturated in the west, to normal calc-alkaline in the east. The intrusions are mostly late Yanshanian with isotopic values of 70–120 Ma for biotite and zircon, and also Himalayan with isotopic dates of 20–30 Ma for recrystallized biotite, according to a paper by Zhang Yuquan *et al.* in 1981. There are Cainozoic basalt and Yanshanian as well as Himalayan granites in westernmost Yunnan. The Yunnan Institute of Geological Science reported in 1982 a Rb–Sr age of 51–52 Ma and a K–Ar age of 52–54 Ma for biotite collected from the biotite-granite in the vicinity of Tengchong, Yunnan.

The third sub-region (Fig. 17.1, IV_3), geotectonically corresponding to IV_A and IV_E of Fig. 19.3, lies to the south of the Yarlung Zangbo fracture zone which was first suggested as a suture zone between Gondwana and Eurasia by Chang Chengfa (Chang *et al.* 1973) and the nature of which has been investigated in detail since 1979 by many geologists. It is characterized by the strong influence of late Yanshanian to Himalayan magmatism genetically related to the formation of the suture zone just mentioned. There are, however, some signs of older igneous activity. The intrusives are mainly found in three belts (Liu *et al.* 1980). From north to south, the first one is a magnesian (Mg/Fe, 8–13) ultrabasic belt composed of intrusions covering a total area about 2 000 km^2 and occurring mainly along the southern side of the Yarlung Zanbo fracture zone, attaining a length of over 1 500 km. The rocks are chiefly different types of harzburgite accompanied by dunite to the east and peridotite in the west. Together with associated intermediate to basic volcanic rocks, gabbroic rocks, and Cretaceous radiolarian siliceous rocks, they constitute the Yarlung Zangbo Ophiolite Suite formed during late Yanshanian to early Himalayan (see below). Then comes an indistinct belt along the low watersheds of South Xizang, composed of scattered small intrusions. These are mostly gneissic, two-mica granites, with subordinate gneissic biotite-granite and muscovite-granite, and partly accompanied by marginal migmatization. While most of them are of Himalayan age with isotopic dates of 22–40 Ma (Liu *et al.* 1980), some belong to the Caledonian (Geol. Bureau and Soc., Xizang 1982), with the Kangmar intrusion having a Sr–Rb date of 484 ± 7 Ma. The third, and southernmost belt appears along the Himalaya Mountains and consists of scattered small intrusions of tourmaline-bearing muscovite-granite and two-mica granite, and also pegmatite with isotopic dates of 10–20 Ma (Geoch. Inst. 1981), and is thus of late

Himalayan age. Volcanic rocks are less developed, with ages ranging from Triassic (basalt and andesite) to Jurassic (andesite).

South China region (South-central and South-east China region)

Situated to the south and south-east of the other major regions, this area may be provisionally called the South China region. It is characterized by an abundance of Palaeozoic and Mesozoic igneous rocks in the south-eastern part. Igneous rocks are scarce or absent in other districts. It may be divided into two sub-regions.

The first (Fig. 17.1, V_1) includes the greater part of the Yangzi Platform (Fig. 19.3, III_A), with the exception of the Kham–Yunnan Uplift in the west, and of the Ankang–Tongbo (Caledonian) Fold Zone and the Dabie Uplift in the east (Fig. 19.3, III_{A3}, III_D and III_{A2}), and also of the Youjiang (Indosinian) Fold Zone (Fig. 19.3, V_C). This is one of the few regions where magmatism was almost absent in most places for a very long span of its geological history and which had remained 'stable' probably since late Proterozoic times. Middle or Upper Proterozoic volcanic rocks are found in north-western Hubei near its northern boundary, certain places in Hunan, Guangxi, Guizhou and Jiangxi (i.e. Sibao and Banxi groups), in South-west Sichuan and also to the north-west and south-west of Kunming along the western boundary. Late Proterozoic granitic rocks, often showing signs of relatively strong contamination, and locally associated with migmatization, are present along its western limit, to the west of Nanchang (Plate XVI.8) and in a few other localities. There are also many outcrops of Upper Permian basalt, near and to the north-east of Kunming and in north-western Guizhou, and of Middle Cretaceous alkaline trachyte (Plate XIV.3) near Yao'an, less than 200 km WNW of Kunming. Yanshanian volcanic rocks, which include andesite (Plate XIV.2) and rhyolite with associated granitic intrusions, and Tertiary basalt flows occur in the Lower Yangzi depression (Fig. 19.3, III_{A5}).

Geotectonically corresponding to the South-east China (Cathaysian) (Caledonian) Fold Zone Belt, the South-east (Hercynian) Fold Belt and the Taiwan (Himalayan) Fold Zone (Fig. 19.3, V_A, including V_{A1}–V_{A4}, V_B and V_E), the second sub-region is marked by the abundance of igneous rocks, especially in the eastern and southern parts. In general, it seems that many granitoids of older ages often occur on the northern or northwestern side of the younger ones, and the Late Mesozoic Yanshanian granites become more abundant eastward or south-eastward. This indicates the tendency of the granite regions to migrate from the interior of the continent towards the coastal region over time. But more detailed investigations reveal that there is partial or local inward shifting of granite formation in later tectonogenetic cycles or later phases of the same cycle. The north-western part lies approximately within the limits of the Xiang–Gui Fold Zone (Fig. 19.3, A_1) and the south-eastern part of the Youjiang (Indosinian) Fold Zone (Fig. 19.3, V_C), and contains the greater part of Hunan (Xiang), and part of Guanxi (Gui) and Jiangxi. There are within it some outcrops of Middle to Upper Proterozoic volcanics, including Sinian tuffaceous rocks, near the northern and western boundary. There are numerous Caledonian granitic bodies partly accompanied by marginal migmatization (see below) and in some places by regional migmatization, and some Hercynian, Indosinian, or Hercynian to Indosinian granitoids, and a few granitic intrusions of Yanshanian age. While the southern and eastern limit of the exposures of Jinningian granitics, and the northern and western limit of the Caledonian granites may be taken as the boundary between this sub-region (V_2) and the first sub-region (V_1), it is difficult to distinguish accurately the north-western and the south-eastern part of this sub-region.

The south-eastern part, including the Jian'ou Uplift, the Min-yue Fold Zone, the Hainan (Hercynian) Fold Zone, the Fuding–Xiamen Fold Zone, and the Taiwan (Himalayan) Fold Zone (Fig. 19.3, V_{A3}, V_{A4}, V_{B1}, V_{B2}, V_E), is dominated on the mainland by Yanshanian granites and volcanic rocks (Plate XIV.6). There is a subordinate zone of scattered exposures of Caledonian granite and gneissic granite in western Fujian Province, and just east of them some outcrops of Hercynian to Indosinian granitoids, and a few outcrops of Tertiary or Quaternary basalt. The Yanshanian granites may be of either Middle to Late Jurassic or Cretaceous age. The igneous rocks of Hainan Island are granites of Hercynian, Indosinian, and Yanshanian ages and Quaternary basalt, and those of Taiwan Island are Yanshanian granite and related rocks, Himalayan andesitic rocks and ultrabasic intrusions, and Tertiary and Quaternary basalts.

THE YARLUNG ZANGBO OPHIOLITE THRUST BELT—AN EXAMPLE OF THE OPHIOLITE SUITE

This is the eastward continuation of the Indus River Ophiolite Zone and has a width of 10–20 km. It extends in an approximately E–W direction for more than 1500 km along the Yarlung ZangBo River valley and its southern flank in Xizang, from Mapangyongcuo (north of Pulan) and Zhongba in the west, through

Sage, Angren, Shigatze, Renbu and Zedong, to the east of Lang County in the east. It then turns southeastward to join the ophiolite zones in Burma and Thailand. This belt is an integral tectonic unit of the Yarlung Zangbo Suture Zone; to the north of it there is a succession of three other units of this zone, viz. the flysch wedge of the forearc basin, the molasse of the epicontinental mountain chain and the Gangdise–Nyainquentanglha volcano-intrusive arc (see above) and to the south the flysch *mélange* belt of the subduction zone. In the Renbu–Shigatze–Angren segment of this zone, ophiolite outcrops are distributed over an area more than 300 km long and 10–20 km wide. Though it occurs mainly in upthrusted bodies and masses that are often complicated further by later ruptures, the original ophiolite suite still preserves in part a more or less complete and continuous succession of the original oceanic crust and the uppermost mantle. The rock sequence of the Shigatze district, which is typical of its kind, is listed in ascending order as follows:

(1) The ultramafic rocks (uppermost mantle). Distributed in a zone with a general width of 2–8 km, and a maximum of 14 km, these are mainly serpentinized harzburgite, diopside-harzburgite (Plate XVI.1), and lherzolite with minor dunite. The rocks are characterized by metamorphic fabrics and subsolidus deformation, indicated by the twisted zonal structure, undulose extinction (Plate XVI.1, 2), and recrystallization of the olivine. The enstatite in harzburgite contains exsolved clinopyroxene occurring in aligned narrow stripes (Plate XVI.3). The upper part of the ultramafic rocks contains many gabbro-diabase and rodingite dykes and inclusions of deformed dykes. The diopside-harzburgite and lherzolite, representing the residue of partial melting of the upper mantle, tend to occur at a depth of 1–2 km below the top of this unit. Auto-clastic *mélange* is found locally at the base of the ultramafics. It consists of blocks of metagabbro, garnet-amphibolite, diabase, and rodingite, and a crushed serpentinitic matrix, and occasionally occurs on a base of thin garnet-amphibolite and schistose quartzite beds. Petrochemically the rocks of this unit show great resemblance to Alpine ultramafic types (Table 17.1).

(2) The cumulate. Showing a sharp contact with the underlying ultramafic rocks, this only occurs in some places, such as Yelong in Jiding, and in the Dazhu and Baigang regions, and it can be divided into three parts.

(a) The lower part consists of alternating layers of dunite, plagioclase-bearing dunite (Plate XVI.7), troctolite, allivalite or wehrlite, and olivine-pyroxenite, often showing layered and banded structures.

Table 17.1. *Average chemical composition of the rocks from the Yarlung Zangbo Ophiolite Zone (after Geol. Bur. Xizang and Geol. Soc. Xizang 1982).*

	Ocean crust					Upper mantle	
	Mafic lavas (basalt)	Sill-dyke complex (dolerite diabase)	Cumulate			Ultramafic rocks	
			Layered gabbro	Diallage-allivalite	Dunite	Harzburgite	Dunite
SiO_2	50.94	50.98	45.50	40.93	37.97	40.79	38.91
TiO_2	1.08	1.13	0.12	0.07	0.09	0.03	0.03
Al_2O_3	14.70	15.29	17.96	15.53	0.77	1.37	1.04
Fe_2O_3	4.56	3.16	1.72	1.92	9.02	4.60	3.18
FeO	4.39	6.09	3.78	4.04	1.87	3.42	3.38
MnO	0.15	0.15	0.09	0.10	0.10	0.12	0.11
MgO	6.10	6.97	11.40	19.37	36.43	37.56	40.94
CaO	8.06	7.82	12.87	9.97	0.19	1.08	0.34
Na_2O	4.18	4.27	1.73	0.76	0.01	0.04	0.13
K_2O	0.35	0.32	0.23	0.04	0.03	0.01	0.03
Cr_2O_3	—	—	—	—	—	0.45	0.26
NiO	—	—	—	—	—	0.31	0.24
H_2O^+	3.18	2.76	3.93	6.01	11.79	9.97	10.15
H_2O^-	0.45	0.43	0.32	0.54	1.19	0.25	0.28
CO_2	1.26	0.40	0.20	0.28	0.39	0.17	0.34
P_2O_5	0.11	0.11	0.03	0.00	0.10	0.07	—
Total	99.51	99.88	99.88	99.56	99.95	100.24	99.99

(b) The middle part is composed of layered olivine-gabbro cumulate (Plate XV.1) with alternating melanocratic and leucocratic bands, locally giving rise to a gneissic structure.
(c) The upper part is essentially a massive gabbro or gabbro-diabase grading into the rocks below and above.

The total thickness of this unit is less than 1000 m, and the layered cumulate is 150–600 m thick. The basicity of the cumulate decreases upward gradually. Trondhjemite (Plate XVI.9) or quartz-diorite is rare in both the homogeneous gabbro and in the sheet-like sills, such as occur in the Dazhu district, and these rocks are the final crystallization product of the parent magma.

(3) Sheet-like sills and clusters of dykes. The occurrence of such intrusives reflects a tensile tectonic environment and provides evidence of sea-floor spreading. Many of them are found in a zone usually several hundred metres to one kilometre wide. They crop out chiefly to the north of the ultramafic rocks with an approximate E–W or ENE–WSW trend, essentially parallel to that of the ophiolite suite. The sills and dykes are frequently parallel to each other and closely spaced (Plate XV.4), usually 1–2 m thick with indistinct chilled edges, but they are sometimes thinner, or as thick as 7–8 m. The rocks so far observed include diabase (Plate XVI.4), diabase-gabbro, and dolerite (Plate XVI.5). Some of them have cut the pillows of the overlying pillow lavas.

(4) Mafic lavas. These lavas may appear as massive flows or show typical pillow structure (Plate XV.3). Some pillow bodies have preserved a well developed zonal structure characterized by longitudinal 'growth lines' (Plate XV.2). The rocks may be basalt or spilite, with or without variolitic or amygdaloidal structures, and doleritic or porphyritic textures (Plate XVI.6). Infrequently there may occur some intercalated volcanic breccia and tuff beds.

Petrochemically the lavas are of the ocean-ridge tholeiite type (Table 17.1). They are further characterized by low LREE, with La/Sm = 0.52 and an average δ_{Eu} value of 0.96.

MESOZOIC CONTINENTAL VOLCANIC ROCKS OF THE EASTERN PART OF CHINA

Mesozoic continental volcanic rocks of dominant intermediate to acid composition are widespread in the eastern part of China, extending all the way from the northern part of North-east China to the southern coastal regions of the country. The approximate western boundary of their outcrops is shown in Figure 17.1. It has been suggested recently that the formation and evolution of these rocks and related granitoids are closely related to the plate tectonics of Eastern Asia during the Mesozoic, especially in Jurassic–Cretaceous times, mainly in the time-span of the Yanshanian tectonic stage. (Li, Chunyu 1980; Wu Liren et al. 1982; Chapter 19.)

It seems that the distribution of such rocks is related to NE to NNE-trending fractures and tectonic zones, and also to the nearly E–W fractures to the north of Beijing and to the west of Nanjing. Some of these had begun to appear in the Late Jurassic, and others of more ancient date were reactivated in Jurassic or Cretaceous times. They may be divided into three NNE–SSW belts arranged en echelon. The western one includes the rocks of the western part of North-east China, Nei Mongol, and the Yanshan region. They are mainly Late Jurassic intermediate to basic or intermediate to acid lava, and volcano-clastics, with subordinate Early Cretaceous acid and intermediate to acid volcanics, separated by coal- or oil shale-bearing formations. The middle one, extending from the eastern part of North-east China southward to Shandong, contains scattered areas of volcanic rocks which are mostly of Cretaceous age and locally of Late Triassic to Late Jurassic age. The rocks are frequently intermediate to acid in composition and partly intermediate to basic, containing many intercalations of sedimentary rocks. The eastern belt covers extensive areas of the south-eastern provinces, including southern Jiangsu, Zhejiang, Fujian and part of Guandong, and also part of Anhui, Jiangxi, and Hubei further inland. The most intensive eruption occurred in the Late Jurassic, consisting of an early phase of acid material with intercalated sandy and argillaceous rocks, and a later phase of intermediate to acid. However, there are also Late Cretaceous acid volcanics. The volcanic series increases in thickness from 3000 to 5000 m (locally 8000m) along the coast, to around 1000 m or less inland. Both fissure and central types of eruptions occurred.

It is apparent that the age of violent eruption decreases from west to east. This is especially evident in North-east and North China. For instance, while strong effusive activity took place in Late Jurassic to Early Cretaceous times in the Da Hing'an Mountains and the Yanshan regions, that in the eastern flank of the Songliao Plain and Shandong occurred in the Early Cretaceous, and that in the eastern part of North-east China, in the Late Cretaceous. In North China, and especially in North-east China, the tendency is for the intensity of eruption to increase towards the Pacific. The volcanics are often associated with, and closely followed by, granitoid intrusions in many regions, the latter being more abundant and of greater extent in the outermost belt.

In addition to the coal, oil shale, oil, gas, and gypsum that occur in the intercalated sedimentary formations in some of the volcanic basins, there are also many kinds of mineral deposits, such as bentonite, alunite, kaolin, pearlite, fluorite, Fe, Cu, Pb, Zn, Ag, Au, and U, found in the volcanic or other rocks genetically related to volcanism.

Wu Liran et al. (1982) have divided the Mesozoic volcanics into three petrological regions. The northern region, which includes North-east China, Nei Mongol, Beijing, northern Hebei (north of 40 °N), and Central and West Shandong (east of the Tancheng–Lujiang Deep Fracture), is in general characterized by basalt–andesite–dacite–rhyolite association, and also by the frequent presence of quartz-trachybasalt, quartz-trachyandesite, quartz-latite, and rhyolite of a transitional type from alkali to calc-alkaline. The volcanics of the central region occur in the down-faulted Mesozoic basins in south-eastern Hubei, Anhui, and Jiangsu along the Lower Yangzi River, and as scattered outcrops in western Shandong (west of the Tancheng–Lujiang Deep Fracture), south-eastern Shanxi, and south-eastern Henan. They are well represented by a potash-rich trachyandesite–latite–trachyte–(phonolite) association of alkaline affinity in many localities. Locally, in Lishui in southern Jiangsu and Fanchang in southern Anhui they are characterized by the basalt–andesite–dacite–rhyolite association of the calc-alkaline series. The southern region extends from the Shanghai district southward and south-westward to the southern limit of the Mesozoic volcanics, and is dominated by the andesite (dacite, quartz-trachyandesite, quartz-latite)–rhyolite association, especially by rhyolite, including its alkaline variety. Basalt and trachybasalt, however, occur locally. These rocks are commonly high in alkalis content with K_2O/Na_2O greater than one, and low in titanium with $TiO_2 < 1$ per cent. The postulated depth of magma-formation is greatest in the northern region, and decreases through the central region to the southern region.

The trend of evolution and degree of differentiation of the magma differ considerably in different places. For instance, the volcanic rocks of the Yanshan region to the east of Beijing are only slightly differentiated, while those of the Lower Yangzi faulted basins are often the products of well-marked magmatic differentiation. The latter is exemplified by a volcanic series around 2500 m in thickness in the Nanjing–Wuhu region of the Jiangsu–Anhui borderland, which shows a distinct evolution of the magma from slightly basic intermediate (K- to Na-rich), through intermediate, to alkaline (Iron research group of Middle–Lower Changjiang Valley 1977). The rocks belong to four volcanic cycles of Late Jurassic to Early Cretaceous age. The lowermost, the Longwangshan Cycle, consists chiefly of rocks of hornblende-bearing basaltic andesite–andesite and basaltic trachyte–trachyte associations (of the calc-alkaline series). These are more basic and richer in potash than other cycles. The second, the Dawangshan Cycle, is composed mainly of biotite- and augite-bearing basaltic andesite–andesite and basaltic trachyandesite–andesite associations of the calc-alkaline series rich in soda. The rocks of the third, the Gushan Cycle, belong to the andesite–dacite association of the calc-alkaline series, and are often hornblende- and/or biotite-bearing and infrequently quartz- or augite-bearing. Those of the uppermost cycle, the Niangniangshan Cycle, are alkaline intermediate types and are of the alkaline trachyte–phonolite association containing feldspathoids such as leucite, nosean, and sodalite, and alkaline mafic minerals such as aegirine. Each cycle begins violently and ends with relatively quiet lava effusion and related subvolcanic intrusion. It may then be followed by local subaqueous volcano-sedimentation prior to the volcanism of the subsequent cycle. The first two cycles are provisionally taken as J_3–K_1 in age, the third belongs to the Early Cretaceous with a K–Ar age of 113–116 Ma, and the fourth is of the Late Cretaceous with a K–Ar age of 94–106 Ma. The K–Ar age of the associated granitoids is c. 107 Ma and that of the augite-basaltic andesite–gabbroic diorite porphyrite 120–125 Ma. This is followed by gabbro with a K–Ar age of 82–115 Ma, genetically related to a series of iron, and other mineral, deposits.

GRANITIC ROCKS OF THE EASTERN PART OF SOUTH CHINA

It is evident from the descriptions in the section on 'major regions of igneous rocks' that granitic rocks are of frequent occurrence in the second sub-region (Fig. 17.1, V_2) of the South China Region and also in the eastern part of the first sub-region (Fig. 17.1, V_1), and the Lower Yangzi Depression (Fig. 19.3, $IIIA_5$), which may be loosely termed the eastern part of South China. Intensive investigations on such rocks have already been carried out and some of the results have been published (Dept. Geol., Nanjing University 1972, 1980; Guiyang Inst. Geochemistry 1979; Mo et al. 1980). Many important mineral deposits, including W, Sn, Bi, Mo, Pb, Zn, Cu, Fe, Nb, Ta, etc. in this region are genetically related to the granitic intrusions, and warrant a special section on their geological features.

Geochronological groups

So far as present knowledge goes, there are six geochronological groups of granitics in terms of tectonic stage and orogenic movements. They are Sibaoan, Jinningian (Xuefengian), Caledonian, Hercynian, Indosinian, and Yanshanian. The first and the second may together be termed Late Proterozoic. The Hercynian and Indosinian granites are difficult to distinguish in some regions (Fujian, Guandong, etc.), but those of the Caledonian and Yanshanian are often further subdivided by age.

Typical Sibaoan granitoids (Rb–Sr isochron age of 1063 Ma and U–Pb age of 1100 Ma) are only found in northern Guangxi, appearing as small intrusions of granodiorite having an initial $^{87}Sr/^{86}Sr$ ratio of 0.7001. They also occasionally occur as plagioclase-granite, often with marginal quartz-dioritic rocks. They are unconformably overlain by the Upper Proterozoic Banxi Group and intimately associated with a suite of ultrabasic and intermediate (including diorite and quartz-diorite) rocks, all probably being end products of differentiation of a basic magma from the mantle that gave rise earlier to the spilito-keratophyric volcanics of the Sibao Group.

The Jinningian rocks (K–Ar and U–Pb ages of c. 840–900 Ma) are unconformably covered by the uppermost Proterozoic Sinian or younger formations, and are represented by a number of granitic bodies, often accompanied by marginal migmatization, in south-eastern Anhui, north-western Jiangxi and northern Guangxi, within the Yangzi Platform (Fig. 18.3, III$_A$). The Motianling granite of Guangxi has an initial $^{87}Sr/^{86}Sr$ ratio of 0.735 (Mo et al. 1980), probably indicative of its being derived from the remelting of crustal material. The Jiuling Granite of north-western Jiangxi may contain up to 17–18 per cent of cordierite, thus suggesting strong contamination.

The Caledonian granitoids are of much wider distribution, and can be subdivided. There is an earlier phase, mostly of Late Cambrian to Early Ordovician age (e.g. U–Pb age of 522 Ma and K–Ar or U–Pb ages of c. 400–500 Ma), represented by the granite, migmatitic granites, and migmatites of the Yunkai Mountains and Wuyi Mountains, and the granodiorite and granite intrusions of other regions. The later phase, mostly with K–Ar and U–Pb ages of c. 380–400 Ma, is exemplified by some granitic bodies in Hunan, Jiangxi, Guangdong, and Guangxi, some of which are overlain unconformably by Devonian sedimentaries.

Hercynian granites are mostly biotite-granite and partly cordierite-bearing (southern Guangxi), and locally granodiorite, and are found in quite a number of localities in Guangxi, Jiangxi, Fujian, and Hainan in Guangdong, with isotopic ages chiefly within the range of 230–295 Ma (K–Ar, U–Pb and Rb–Sr ages). Most of the intrusions are seen to have invaded either Carboniferous or Permian rocks.

Indosinian granites, which are frequently biotite-granite, occur in Hainan and some localities in Anhui, Jiangxi, and Hunan, with K–Ar or U–Pb isotopic ages of 200–220 Ma. Some of them occur as intrusions in Permian or older rocks, and are uncomformably overlain by either Upper Triassic to Lower Jurassic, or Middle to Upper Jurassic, formations.

Yanshanian granitoids are very widely distributed, especially in the regions near the coast (Fig. 17.1), and can be subdivided into at least two geochronological groups. Those of the early phase are of Jurassic age and mostly of biotite-granitic composition and partly a late stage two-mica granite. They may be further divided: an earlier group (K–Ar or U–Pb age, 175–185 Ma) intrudes into the Lower Jurassic or older formations, but is unconformably overlain by Middle or Upper Jurassic volcano-sedimentaries; and a later group (K–Ar or U–Pb age, 135–160 Ma) occurs as intrusions in Lower to Upper Jurassic rocks, but shows an unconformable contact with the overlying Cretaceous, or even Lower Tertiary, formations. Rocks of the later phase are more varied in composition, and include granite, miarolitic potash-granite, granodiorite, and also diorite. They may intrude into Cretaceous, or older, formations and early Yanshanian granitoids, with isotopic ages (K–Ar, U–Pb and also Rb–Sr determinations) of 100–135 Ma and 69–106 Ma. Some are unconformably overlain by Tertiary formations. At many localities, composite granitoid bodies composed of different stages of both early and late Yanshanian intrusions are found. For instance, the Jiuyishan intrusion along the Guangdong–Hunan border is composed of separate bodies belonging to three stages. There is an early phase consisting of granite dated 186 Ma, granodiorite of 188 Ma and fine-grained biotite-granite of 172 Ma; a later granite has an isotopic age of 155 Ma, and the latest is a small granite prophyry stock of 138–140 Ma. All of these are further invaded by aplite. The Jiulianshan intrusion on the Guangdong–Jiangxi border consists of medium- to coarse-grained biotite-granite, fine-grained two mica-granite and granophyre, dated 138–146, 112, and 88 Ma respectively. It is noteworthy that both the Jiuyishan and Jiulianshan rocks are intruded into the Caledonian granodiorites to form composite intrusions of even greater dimensions. Composite intrusions consisting of rocks ranging from Caledonian, through Hercynian and/or Indosinian, to Yanshanian ages are also found. In fact, single-phase late Yanshanian bodies are not very frequent.

With the exception of Precambrian rocks, the evolution of the granitoids may be divided into two stages. The first one includes Caledonian, Hercynian, and Indosinian rock associations which are composite in nature. For instance, those of the Early Caledonian are mostly migmatitic types; the Late Caledonian are chiefly granite and granodiorite, with subordinate adamellite, and the Hercynian–Indosinian granites often show contamination characteristics, being frequently the regional products of the further evolution of the Early Palaeozoic migmatites and migmatitic granites (Mo *et al.* 1980). The second Yanshanian stage is much simpler, being predominantly biotite-granite, supplemented by a unique miarolitic potash granite of Late Cretaceous age.

Structural control of granitic activity

Granitic activity, which was often associated with slightly earlier or later volcanism, was in general closely related to the tectonic evolution of this region, especially to the regional rifting and fracturing beginning in the Late–Middle Proterozoic. While the Late Proterozoic intrusions are mostly intimately related to the regional fractures situated within the limit of the Yangzi Platform, especially near its southern and eastern boundaries, the Caledonian granitic activity was genetically connected to both the folding and faulting of the Palaeo-Caledonian Geosynclinal Belt located further south-east. The associated intrusions often occur at the intersections of NE–SW and E–W fracture systems. The Hercynian granitoids seem to be of wider distribution than those of the Caledonian, coinciding with the greater extent of the Hercynian fracturing up to the coastal islands of China, including Hainan and probably also Taiwan. The extent of the Indosinian fracturing, which controls the emplacement of Indosinian granites, is somewhat ambiguous and may be slightly greater than that of the Hercynian, but the depth and the intensity of its influence seem to be less. The Yanshanian granitic activity coincided with the Middle to Late Mesozoic fracturing in space and time. It was very widely distributed and very intense in relation to both magmatism and tectonic activity, with the intensity increasing eastward and south-eastward. The easternmost fracture belt along the Fujian coast on the mainland is marked not only by granitic and migmatitic activity but also by slightly earlier metamorphism, both related to the subduction of the Pacific plate under the Asiatic plate (Chapters 18 and 19). The migration of the zones of granitoidal activity towards the coast (especially its cumulative effect) is related to tectonic development and the accretion of the continental crust (Mo *et al.* 1980). It is also related to the evolution of metamorphic belts from Middle Proterozoic to Mesozoic and Tertiary times discussed elsewhere in this book (Chapters 18 and 19).

Petrochemical and evolutional features

Following an intense, collaborative study of the data on granites of different ages of the Nanling region (east of 108 °E, and between 21 °40′–26 °40′ N), obtained chiefly by geological surveying on a scale of 1:200 000, Mo and his colleagues (Mo *et al.* 1980) classified them into five rock-types, namely granodiorite, adamellite, biotite-granite, mica-granite and miarolitic potash-granite. Of these, the biotite-granite is of the most widely distributed spatially (Table 17.2) and temporally, ranging from Jinningian to Late Yanshanian in age. It is followed by granodiorite and adamellite, the former being characteristic of the Sibaoan and also of some Caledonian and Hercynian intrusives. The miarolitic potash-granite is of Late Cretaceous age and of alkaline affinity, and is only found in the Gushan district near Fuzhou, Fujian. The mica-granite is of very local occurrence, and is found in some granites of

Table 17.2. *Area and percentage coverage of different types of granitoids of the Nanling region (after Mo Zhusun* et al. *1980).*

	Total area (km²)	%
Granodiorite	3895	5.1
Adamellite	5172	6.8
Biotite-granite	65141	85.7
Mica-granite[a]	590	0.8
Miarolitic potash-granite	1200	1.6
Total	75998	100.00

[a] Including muscovite-granite.

Yanshanian and other ages. The overall chemical composition of these five types weighted on the basis of their areal coverage (Table 17.2) is listed in Table 17.3. This weighted average composition is slightly lower in Al_2O_3, Fe_2O_3, MgO, CaO, and Na_2O and slightly higher in SiO_2, FeO and K_2O than the arithmetic average composition of the granites of China which Li et al. calculated in 1963 (Table 18.3, 5). Mo later suggested (1983) that the biotite-granite type was probably derived from anatexis or granitization of the sedimentary type described by B. W. Chappel and A. J. R. White (1974).

Mo et al. also calculated the areally weighted average chemical composition of the biotite-granite of five age-groups of the Palaeozoic and Mesozoic of the Nanling region (Table 17.4) to study the petrochemical evolu-

Table 17.3. *Average chemical composition of granitoids of various geological ages of the Nanling region, in recalculated weighted percentage based on areal coverage (after Mo Zhusun et al. 1980).*

	Granodiorite	Adamellite	Biotite-granite	Mica-granite[a]	Miarolitic potash-granite	Average
SiO_2	67.16	69.68	73.12	74.19	76.80	72.65
TiO_2	0.49	0.43	0.25	0.14	0.14	0.27
Al_2O_3	15.04	14.42	13.43	13.77	12.12	13.56
Fe_2O_3	1.74	1.24	0.84	1.01	0.88	0.92
FeO	2.97	2.54	1.88	1.16	0.72	1.97
MnO	0.06	0.08	0.07	0.06	0.06	0.07
MgO	1.51	1.07	0.52	0.47	0.16	0.60
CaO	3.09	2.21	1.22	0.52	0.36	1.36
Na_2O	3.22	3.19	3.08	3.35	4.12	3.11
K_2O	3.80	4.25	4.87	4.62	4.46	4.76
P_2O_5	0.24	0.13	0.11	0.23	0.01	0.11
Loss on ignition	0.68	0.76	0.61	0.57	0.17	0.62
Total	100.00	100.00	100.00	100.00	100.00	100.00
Number of samples	54	78	444	12	49	—

[a] Including muscovite-granite.

Table 17.4. *Average chemical composition of biotite-granite of various geological ages of the Nanling region, in recalculated weighted percentage based on areal coverage (after Mo Zhusun et al. 1980).*

	Early Palaeozoic	Late Palaeozoic to Triassic	Early to Middle Jurassic	Late Jurassic	Early Cretaceous
SiO_2	71.79	72.28	72.46	73.46	74.09
TiO_2	0.33	0.28	0.28	0.22	0.20
Al_2O_3	13.54	13.72	13.59	13.35	13.20
Fe_2O_3	1.42	1.05	0.94	0.72	1.02
FeO	2.47	1.71	1.99	1.88	1.35
MnO	0.07	0.06	0.07	0.07	0.06
MgO	1.00	0.65	0.48	0.40	0.39
CaO	1.42	0.84	1.16	1.23	1.06
Na_2O	2.58	2.80	3.13	3.10	3.39
K_2O	4.65	5.12	5.06	4.91	4.65
P_2O_5	0.16	0.14	0.16	0.08	0.04
Loss on ignition	0.57	1.35	0.68	0.58	0.55
Total	100.00	100.00	100.00	100.00	100.00
Number of samples	45	39	88	166	46

tion of these rocks during this time-span. It is interesting to note that with the progress of time, both SiO_2 and Na_2O contents show a gradual increase, and the percentages of TiO_2, FeO, MgO, and CaO decrease, while the amounts of Al_2O_3, MnO, Fe_2O_3, and K_2O remain almost constant or only fluctuate within a small range. This may illustrate some of the factors which account for the prolific tungsten, tin, and related mineralization of these Yanshanian granitoids. There may also be some geological significance in the sudden decrease in the Fe_2O_3 content from the Early Palaeozoic granitoids to those of the Late Palaeozoic and Triassic, and in the great drop in the P_2O_5 percentage from the Hercynian–Indosinian to the late Yanshanian. At about the same time (1979), the Guiyang Institute of Geochemistry computed the arithmetic average composition of the granitoids of different ages from the Sibaoan and Xuefengian to the Late Yanshanian of the Nanling region together with the district further north up to the Yangzi River in Jiangxi, Anhui, and Jiangsu (Table 17.5). The results obtained show an overall petrochemical evolution within similar to that above. The only differences are the decrease in the Al_2O_3 content, especially from the Proterozoic to the Caledonian, and the continuous increase in percentage of K_2O from the Caledonian through the Hercynian to the Yanshanian, as compared with the relative constancy of these values of the biotite-granites of the Nanling region during Palaeozoic and Mesozoic times.

Granite series

Weng Wenhao (Wong Wen-hao or W. H. Wong) was the first (1920) to classify the Yanshanian granitoids of this region into two types: a granodioritic type of deeper origin genetically related to iron and copper mineralization, exemplified by the intrusions of the middle to lower Yangzi Valley; and another type of granitic composition genetically related to tin and tungsten mineralization, represented by the intrusions of the Nanling region. These two were named as Yangtze (Yangzi) and Hongkong types respectively by Xie Jiaron (Hsieh Chia-yong or C. Y. Hsieh) in 1936. The related mineralogenetic problems have been discussed further by various geologists in China since the 1950s (Wang Liankui et al. 1982), especially since the middle 1970s by geologists of the Geology Department, Nanjing University, Guiyang Institute of Geochemistry, Jiangxi Geological Institute, and Yichang Institute of Geology and Mineral Resources, and by

Table 17.5. *Chemical composition of granitoids of various geological ages of South China (after Guiyang Inst. 1979).*

	Sibaoan and Xuefengian	Caledonian	Hercynian	Early Yanshanian and Indosinian	Late Yanshanian
SiO_2	69.31	70.53	71.23	72.76	73.80
TiO_2	0.43	0.42	0.43	0.23	0.20
Al_2O_3	14.87	14.04	13.38	13.51	13.15
Fe_2O_3	0.82	0.86	1.20	0.90	0.89
FeO	3.26	3.14	3.18	1.77	1.48
MnO	0.13	0.07	0.06	0.07	0.06
MgO	1.34	1.15	1.04	0.51	0.46
CaO	1.24	1.81	1.60	1.30	0.97
Na_2O	2.83	2.78	2.23	3.27	3.36
K_2O	3.92	3.86	4.36	4.71	4.69
H_2O	—	—	0.93	0.55	0.59
P_2O_5	0.20	0.14	0.14	0.09	0.05
Loss on ignition	1.45	1.19	1.26	0.88[a]	0.63
Total	99.80	99.99	99.98	100.00	99.74
Number of samples	61 (47[b])	143 (58[b])	62 (13[b], 45[c])	272 (72[c], 36[d])	91 (15[c], 60[b])

[a] Volatile matter.
[b] For loss on ignition.
[c] For H_2O.
[d] For volatile matter.

Mo Zhusun and Zhang Chongzun. On the basis of the probable source materials from which the rocks might have been derived and the geological setting in which they occur, Xu Keqin and his colleagues (Geol. Dept. Nanjing Univ. 1980, Xu *et al.* 1982) have subdivided the Mesozoic granites in this region into two genetic series, namely the transformation series and the syntexis series. The former occurs mainly to the west of the Zhenghe–Dapu Fracture (extending from 121 °E, 30 °N to 115.5 °E, 22.6 °N), 'in a well-developed crust inside the continental plate', and is derived chiefly from crustal material through transformations, including metasomatism (migmatization and granitization), partial anatexis, etc. The syntexis series appears chiefly to the south-east of the same fracture, 'in the active continental margin and along some deep fault zones and fault depressions inside the continental plate', and is derived chiefly from syntexis of crustal materials admixed with others originating from the mantle that invaded the geosynclines. The transformation series consists mainly of normal granites and is genetically related to mineral deposits of tungsten, tin, tantalum, niobium, beryllium, etc. The syntexis series is mainly intermediate in composition, which usually varies from diorites→quartz-diorites→granodiorites→quartz-monzonites→potash-granites, and is noted for its close connection with iron, copper, and molybdenum mineralization. Some of the other differences between them are briefly tabulated as follows:

Syntexis series	Transformation series
K_2O-SiO_2: exhibiting a linear relation	no linear relation
K_2O/Na_2O: small, increasing inland (0.77–1.70; 0.90–1.77)	larger (0.82–2.41; 0.85–2.55)
Accessories: mt., 5214 p.p.m.; ilm., 171 p.p.m.; sph., 966 p.p.m. (average)	mt., 345 p.p.m.; ilm., 25 p.p.m.; sph., 84.7 p.p.m.
Evolution: Na-rich→K-rich; high in Ca, low in Al	K-rich→Na-rich; high in Al, low in Ca
$^{87}Sr/^{86}Sr$: usually 0.705–0.710	c. 0.710 or more

Geologists of the Guiyang Institute of Geochemistry and related institutions, especially Wang Liankui and his colleagues (Guiyang Institute 1979; Wang Liankui *et al.* 1982), proposed two petrogenetic and mineralized series for the granitoids in South China which are similar to those proposed by Nanjing University. Series I, or the Nanling Series, is formed chiefly through anatexis of the crustal materials and Series II, or the Yangzi Series, is derived chiefly from the mantle or lower part of the crust. The former shows the following sequence of petrogenesis and mineralization: monzonitic granite or granodiorite→biotite-granite→leucocratic granite→granoporphyry, or quartz porphyry→intermediate–basic dykes; REE→Nb, Ta (Li, Rb, Cs), Be, Sn, W, Mo, Bi, As→Cu, Zn, Pb→Sb, Hg, U. The Yangzi Series exhibits a different sequence: pyroxene-diorite (or gabbro)→diorite or quartz-diorite→granodiorite (quartz-monzonite) or monzonitic granite→granite→K-felspar granite→granoporphyry or quartz-porphyry (syenitic porphyry or quartz-syenitic porphyry)→intermediate–basic dykes; Fe→Cu (Au)→Mo (W)→Zn, Pb→Pb (Ag). Other differences between them may be listed briefly as follows:

	Series I	Series II
Bulk composition	granitic, high in SiO_2 and K_2O	andesitic, higher in Mg and Ca
Formation T	600 °–680 °C	980 °–1140 °C
Biotite	high in Mg	high in Fe
Accessories	mt.-sph.-apat., or mt.-il.-REE-rich zir.	complicated, mt.-il.-zir., or monaz.-xenot.-zir.
trace elements	predom. by F, Li, Rb (Cs), and Be, with $\varepsilon_Y > \varepsilon_{Ce}$, relatively low δEu with distinct depletion	predom. by Cl and Sr, with $\varepsilon_{Ce} > \varepsilon_Y$ in rock and accessories, relatively high δ_{Eu} (0.74–0.99) without depletion
Other features	$\delta^{18}O > 10‰$, greatly varying $\delta^{34}S$, low in Pt-group elem., high $^{87}Sr/^{86}Sr$ (0.7112–0.7360)	$\delta^{18}O < 10‰$, $\delta^{34}S$ close to meteoritic sulphur, relatively high in Pt-group elem., very low $^{87}Sr/^{86}Sr$ (0.7036–0.7085), intermediate-basic dykes in the final evolutionary stage.

Wang *et al.* have pointed out that the granitoids of these two series show considerable dissimilarities from those of Australia (Chappell and White 1974) and Japan (Ishihara 1977) in $^{87}Sr/^{86}Sr$, $\delta^{18}O$ and $\delta^{34}S$ values.

MAGMATIC SERIES OF THE ULTRABASIC INTRUSIONS

The ultrabasic intrusions of China have been classified into three magma series by Song Shuhe and his colleagues (Hou Shihchun et al. 1977, 1979; Song 1984), on the basis of the different evolutionary series of the peridotite magma. The magnesian magma series is poor in aluminium and alkali, rich in magnesium, relatively low in iron, and very low in calcium, with some variation in the relative proportions of magnesium and iron. With the progressive increase of iron there occur in succession Mg-rich dunite, harzburgite, and orthopyroxene-olivinite, while in the case of higher amounts of calcium and aluminium, lherzolite-clinopyroxenite may be formed in small quantities. Harzburgite often forms the main part of the intrusions. Rocks of this series are frequently found in the ultrabasics of the mobile belts in north-eastern, north-western and south-western China, of Hercynian, Caledonian, and other ages. The associated mineralizations are chiefly of chromite* and, to a lesser extent, of copper-nickel, apatite and asbestos.

The second, ferruginous magma series is rich in iron, high in aluminium and alkali, and relatively low in magnesium. The variation in petrographic types depends on the content of magnesium and calcium. With the gradual increase of calcium, Fe-rich dunite, harzburgitic and olivinitic rocks, and pyroxenite are formed consecutively. The rocks usually contain plagioclase and clinopyroxene, and thus often pass progressively into troctolite, norite, gabbro, etc. They occur more often in the more stable regions and are accompanied by relatively insignificant mineralization. Apatite–magnetite deposits have been found associated with the more alkaline rocks in North and North-west China.

The third, ferromagnesian-calcic magma series is relatively low in iron (within a narrow range of variation) and high in aluminium and alkali. The petrographic types vary with the content of calcium and magnesium. With decreasing calcium, there appear Mg-rich dunite, harzbgurite, and orthopyroxene-olivinite. As calcium gradually increases, there appear successively lherzolite, websterite, and clinopyroxenite, with a subsequent transition to basic rocks. Members of this series are of frequent occurrence in both mobile and stable regions. In the uplifted stable districts, the emplacement of the rocks is evidently controlled by tectonic movements, especially by deep fractures. Copper–nickel and platinum deposits are more commonly associated with the calcium- and iron-rich rocks (Gansu), and vanadium-bearing titaniferous magnetite deposits with the iron-rich types ($Mg/Fe_{tot} < 2$; south-western Sichuan), such as pyroxenite or gabbro. Simple platinum mineralization has been found in some more basic differentiates, such as clinopyroxene-olivinite, olivine-clinopyroxenite, and clinopyroxenite–gabbro masses. Chromite deposits are usually related to calcium-low and magnesium-high varieties. The emplacement of the rocks found in the mobile geosynclinal belts of Caledonian (Qilian Mountains), Hercynian (Xinjiang, Nei Mongol, etc.), or mainly late Mesozoic (Xizang) ages are also controlled by tectonics in various ways. As with the magnesian magma series, the related chromite deposits also occur either in the dunite or Mg-rich harzburgite zones.

*Rocks with Mg/Fe_{tot} ratio of 8–13.

18. METAMORPHIC SERIES AND METAMORPHIC BELTS OF CHINA

Cheng Yuqi

INTRODUCTION

Except in the vast plains of the eastern part of China and the various large and small inland drainage basins further west, metamorphic series and related metamorphic rocks are quite widely distributed (Fig. 18.1). They are exposed in regions extending from the Karakorum Mountains in the west to the mountains and hills of the south-east and Taiwan Island, and stretching from the Da Hing'an and Xiao Hing'an Mountains, the Mongolian Plateau, and the Altay Mountains in the north to Hainan Island in the south. Core samples show them to be present at depths under the superficial sediments or sedimentary rocks of some islands in the South China Sea. There are metamorphic series and metamorphic rocks representing all the main metamorphic epochs known. Some of the metamorphic formations, and especially the older ones, have often gone through two or more epochs of metamorphism. There are also examples of nearly all protolithic types and of all degrees and types of metamorphism. They may constitute very complicated metamorphic or migmatitic complexes. They also appear as host formations or host rocks for many mineral deposits, the more important ones being iron, copper, gold, mica, magnesite, talc, graphite, apatite, and certain rare and dispersed elements. These serve as part of the evidence showing that the upper crust in China is not only very complicated in its constitution, but has also undergone a prolonged, complex, geological history involving multiple geological transformations and tectonogenesis, and so exhibits a variety of geological and structural features in different areas.

Our knowledge of the metamorphic rocks and metamorphic series was very meagre at the beginning of this century. The metamorphic rocks, even some of the less metamorphosed ones, were usually considered to be Precambrian. Thus the geological map compiled in the late 1940s still contained large geologically blank areas and indicated a much greater extent of Precambrian metamorphic terrain than that in Fig. 18.1.

Extensive and comprehensive geological surveys, including geological mapping (mainly at a scale of 1:200 000) and regional geochemical and geophysical investigations, have been carried out since the middle 1950s. We are therefore now in a much better position to give a concise picture of the metamorphic geology of China and to present a general review and discussion of related problems. The most detailed metamorphic studies relate to North China and to the southern part of North-east China.

It should be pointed out here that for the metamorphic series and belts of a polymetamorphic nature, only their dominant metamorphic age is shown in Fig. 18.1.

METAMORPHIC SERIES AND METAMORPHIC BELTS—THEIR DISTRIBUTION AND CHARACTERISTICS

Metamorphic series chiefly of Archaean metamorphic age

Although many Archaean metamorphic rocks often exhibit signs of being affected by later metamorphic events, the salient characters of nearly all such rocks are those of the Archaean metamorphic epoch or epochs. Hence the distribution and characteristics of the metamorphic series of dominant Archaean metamorphic age (Fig. 18.1, I) are those of the Archaean formations (Fig. 4.1), which have already been described in some detail in a previous chapter (Chapter 4). It will therefore be sufficient to give a short supplementary account of the conditions and nature of Archaean metamorphism and its accompanying migmatism in selected regions.

As has already been mentioned in the chapter on the Archaean (Chapter 4), the Qianxi Group of the Yanshan region about 180 km to the east of Beijing has suffered two periods of metamorphism, the first probably before 3000 Ma and the second around 2500 Ma. According to Gao Jifeng (Kao 1981) and others, the common mineral assemblages of the earlier granulite facies metamorphism are: (1) hypersthene–diopside–plagioclase–(quartz); (2) plagioclase–quartz–biotite–(hornblende); (3) plagioclase–hypersthene–quartz–(garnet); (4) plagioclase–diopside–quartz–(hornblende). There is also a plagioclase–biotite–quartz–(garnet–

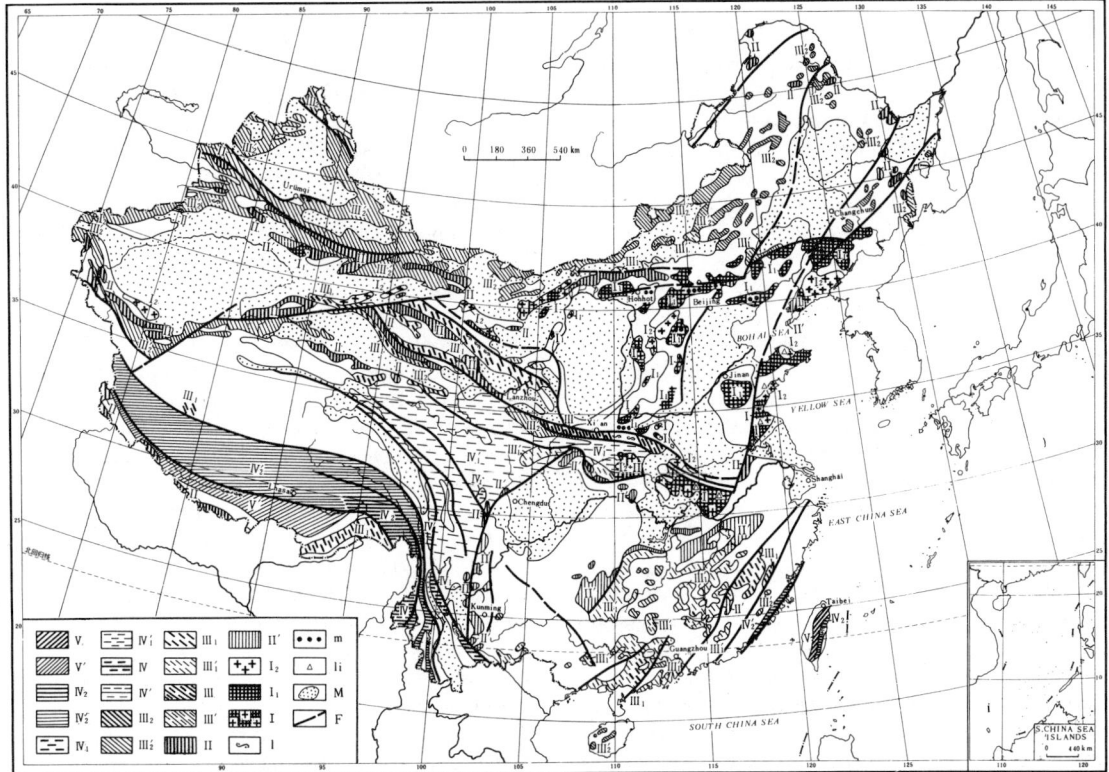

Fig. 18.1. Sketch map showing the distribution of metamorphic series and belts of the various metamorphic epochs of China. (Modified after the original map of Cheng Yuqi and Zhang Shouguang 1979.) V, chiefly of Cainozoic (Himalayan) metamorphic age, greenschist to amphibolite facies; V', as V, generally of greenschist facies or of lower grade. IV$_2$, chiefly of Late Mesozoic (Yanshanian) metamorphic age, greenschist to amphibolite facies; IV'$_2$, as IV$_2$, generally of greenschist facies or of lower grade. IV$_1$, chiefly of Early Mesozoic (Indosinian) metamorphic age, greenschist to amphibolite facies; IV'$_1$, as IV$_1$, generally of greenschist facies or of lower grade. IV, chiefly of Mesozoic metamorphic age, undifferentiated, greenschist to amphibolite facies; IV', as IV, generally of greenschist facies or of lower grade. III$_2$, chiefly of Late Palaeozoic (Hercynian) metamorphic age, greenschist to amphibolite facies; III'$_2$, as III$_2$, generally of greenschist facies or of lower grade. III$_1$, chiefly of Early Palaeozoic (Caledonian) metamorphic age, greenschist to amphibolite facies; III'$_1$, as III$_1$, generally of greenschist facies or of lower grade. III, chiefly of Palaeozoic metamorphic age, undifferentiated, greenschist to amphibolite facies; III', as III, generally of greenschist facies or of lower grade. II, chiefly of Middle to Late Proterozoic metamorphic age, greenschist to amphibolite facies; II', as II, generally of greenschist facies or of lower grade. I$_2$, chiefly of Early Proterozoic metamorphic age, of greenschist to amphibolite facies. I$_1$, chiefly of Archaean metamorphic age, mostly of amphibolite facies and partly of granulite or greenschist facies. I, chiefly of Archaean and/or early Proterozoic metamorphic age, mostly of amphibolite facies and partly of granulite or greenschist facies.

1, Localities of glaucophane-schist and related rocks; m, Localities of granulite facies rocks; li, Localities of eclogite; M, Mesozoic and Cainozoic basins; F, Fractures, observed or inferred.

silimanite-potash feldspar) association for the plagioclase–gneiss type. The plagioclase has an anorthite percentage which increases with the Ca content of the rock concerned up to a value of An$_{52}$; it has a degree of order of 0.8–1.0 and occasionally shows characteristic antiperthitic texture. The garnet is basically a Mg-almandine characterized by an MgO/FeO+MgO ratio up to 0.5 and an MnO content increasing with the degree of metamorphism. The temperature of metamorphism of these rocks can be determined by using the two-pyroxene geothermometers devised by various authors (Zhang *et al.* 1982; Wang and Chen 1982). Values obtained are about 750 °C and 817–845 °C depending on which geothermometer is used. Zhang *et al.* (1982) also obtained a value of about 13 kb for the pressure of metamorphism using an orthopyroxene–

garnet geobarometer, and Liu obtained an estimate close to 10 kb by comparing the value of $K_D^{Mg/Fe}$ for the pyroxene pair in the Qianxi granulite with corresponding values for rocks worldwide. Hence the metamorphism is mainly of medium-pressure type. The associated migmatization is marked by the formation of various migmatites and an anatectic migmatitic granite of charnockitic affinity containing more Na_2O than K_2O, and showing either transitional or cross-cutting relationships with the metamorphic rocks. When the diopside is either transformed into granules of a second-generation diopside (Plate XVII.1) or occurs as relics in a later hornblende (Plate XVII.2), the rocks concerned must definitely have been affected by a second metamorphic epoch and hence are of polymetamorphic nature. The frequent mineral association of the second metamorphic epoch of dominant amphibolite facies of the upper succession of the Qianxi Group and the amphibolite facies rocks of the Badaohe Group, and the corresponding Luanxian Group are: (1) plagioclase–quartz–biotite(–garnet); (2) plagioclase–quartz–hornblende–epidote; (3) microcline–quartz–biotite(–magnetite). Using the garnet-biotite geothermometer, Gao suggested 475–480 °C as the temperature of metamorphism of the amphibolite facies and of the retrogressive metamorphism which the rocks had undergone during the second metamorphic epoch. This temperature range seems to be rather too low for such a mineral facies, and Wang and Chen gave a much higher temperature, 57–694 °C. This metamorphism is probably also of medium-pressure character, and is accompanied by the generation of pale grey migmatitic granite and migmatites, mainly by permeation-metasomatism, and enriched in potassium rather than sodium. There is a slightly later pinkish to purplish-red migmatitic granite, and related rocks, formed chiefly by injection metasomatism, and of dominant potassic composition, under pressure and temperature conditions similar to those of the amphibolite facies rocks.

The Archaean Miyun Group (roughly correlated with the Qianxi Group in the broad sense), outcropping in the western section of the Yanshan Mountains and about 80 km to the north-east of Beijing, has also undergone two epochs of metamorphism, accompanied by migmatism, to various extents during the Archaean (Lu Liangzhao et al. 1981). The rocks of the first epoch have mineral associations of the granulite facies similar to those of the Qianxi Group, including clinopyroxene–hypersthene-granulite with characteristic antiperthitic plagioclase (Plate XVII.8), and have been considered to be metamorphosed at about 800 °C and 8–10 kb, as suggested by the two-pyroxene and garnet-clinopyroxene pairs. The second epoch of dominant amphibolite facies metamorphism, with its earlier phase still exhibiting certain features of the low granulite facies, has a metamorphic temperature calculated as 650 °–700 °C.

From data reported by Cui Wenyuan in 1981, Jin Shiqin and Hu Zhaohua recalculated in 1982 the metamorphic temperature and pressure of the granulite facies rocks of the Jining Group near Jining in the eastern Yinshan region, obtaining 980 °C and 5–7 kb. Hence these rocks were metamorphosed under higher temperature and lower pressure conditions than those of the Miyun and Qianxi Groups which occur in the same granulite belt further to the east.

As pointed out by You Zhendong and Han Yuqing in 1981, the Late Archaean regional metamorphism suffered by the Dengfeng Group (see above) of Mount Songshan, Dengfeng, Henan Province may be divided into two stages. The earlier one is characterized by amphibolite facies mineral assemblages of the moderate pressure series formed under medium temperatures above 500 °C, according to the garnet-biotite geothermometer. The temperature of the accompanying migmatism, probably of alkaline metasomatic nature and attaining its maximum in the axial part of an anticlinorium, was about 550–580 °C as indicated by the two-feldspar geothermometer, and at a pressure of about 5 kb. The later one, restricted to the NNE-shear zones, belongs to the greenschist facies and was superimposed on the earlier metamorphic stage.

Metamorphic series of dominant Early Proterozoic metamorphic age

In the regions where the Early Proterozoic formations are divisible into two stratigraphic units, that is the Wutai and Hutuo Groups and their equivalents, they are usually characterized by two periods of regional metamorphism. In their type localities in the Wutai region, the ends of the metamorphism of these two groups are respectively c. 2250 Ma (Cheng et al. 1982a) and c. 1850 Ma, the upper age limit of the Hutuoan (see above). The Wutaian was metamorphosed at moderate pressure and moderate temperature, and contains mineral associations of the amphibolite facies (Plate XVIII.7) and high greenschist facies. The rocks derived from the protolithic volcano-sedimentaries are often associated with chalcopyrite and pyrite mineralizations. The metamorphism of the Hutuo Group, known as the Luliangian metamorphism, is usually characterized by rocks of the lower greenschist facies of low pressure–low temperature type. Hence the amygdaloidal structure of the protolithic andesite is well-preserved (Plate XVIII.3). Within the North China Platform there are some metamorphic series whose metamorphism has an age and nature similar to that of the Wutai. These are: the Luliang Group of the Luliang Mountains; the locally cupriferous Jiangxian Group of the Zhongtiao Mountain; the Sanheming and Erdowa Groups of the Yinshan region; the Shuangshanzi Group and probably also the Qinglonghe Group of the Yanshan Mountains; probably part of the Liaohe Group of Liaoning, and the Jiaodong Group of eastern Shandong; and the Helan Group to the south-west of Yinshan. The Luliang Group is however of somewhat lower grade, and contains some phyllite and hornblende-schist with protolithic amygdales still quite well preserved. Indistinct zones of almandine, staurolite–kyanite and sillimanite have been observed in the restricted terrain of the Jiangxian Group. Part of the Dabie Group of the Huaiyang region within the northern boundary of the Yangzi Platform is probably of Wutaian metamorphic age.

Of the various metamorphic groups occurring in the North China Platform with an age and nature of metamorphism similar to those of the Hutuo Group, the groups widely distributed and best investigated are: the Zhongtiao Group of the Zhongtiao Mountain; the Majiadian Group of the Yinshan Mountains; the partly magnesite-bearing Liaohe Group and its equivalents in Liaoning and Jilin; the Fenzishan Group of eastern Shandong; the Songshan Group of Mount Songshan; and the Longshoushan Group to the northeast of the Qilian Mountains. However, many of these have been metamorphosed to a higher degree than the Hutuo Group and are characterized by metamorphism that is partly of moderate pressure and moderate temperature. For instance, the mineral association of the protolithic pelitic type of the Liaohe Group, besides partly belonging to the low greenschist facies (Plate XVIII.9), may also be of the amphibolite facies. Metamorphic series of other regions which are also of Luliangian metamorphic age, and worthy of mention here, are the locally phosphatic Susong and Hongan Groups of the Dabie Mountains and neighbouring regions, part of the Wudang Group in north-western Hubei Province, the Dahongshan Group to the south-west of Kunming, and the Kangding Group in the northern section of the Transverse ranges. Some of these have also been metamorphosed to a higher degree than those of the Hutuo Group. Metamorphic formations of Wutaian and/or Luliangian metamorphic age are also found in the north-west and the northern part of North-east China, such as: the Xinditag Group of the Quruktagh Mountains; the Aksu Group and probably also the Mouzat Group of the western Tianshan Mountains; part of the metamorphic series along the northern slope of the Altyn Mountains; the Hualong Group of the south-eastern end of Central Qilian, and probably part of the Heilongjiang and Mashan Groups of Heilongjiang Province. The regional metamorphism of the metamorphic series just mentioned is variable in degree and nature, and the rocks belong to either the low or high greenschist facies, or even the amphibolite facies, and are partly affected by migmatization.

In addition to these series of Early Proterozoic metamorphic age, there occur in many regions rocks of dominant Archaean metamorphic age which have been overprinted by Wutaian and/or Luliangian metamorphism, accompanied by migmatization and related igneous activities. They show appropriate K–Ar and Rb–Sr isochron age values: $c.$ 1850 Ma for the Anshan Group; $c.$ 2230 Ma for the Taishan Group of Xintai, Shandong; $c.$ 1900 Ma for the Jining, Qianxi and Fuping Groups; and $c.$ 2200–2300 Ma and 1800–2000 Ma for the Taihua Group and its equivalents on the northern slope of the eastern Qinling Mountains and neighbouring regions.

Metamorphic series and belts of dominant Middle to Late Proterozoic metamorphic age

Middle to Late Proterozoic formations that have been metamorphosed during Precambrian times are practically absent within the limits of the North China Platform, except the limited exposures of the low greenschist facies rocks of the Penglai Group of East Shandong and its equivalent near Dalian, at the tip of the Liaodong Peninsula. Over the vast area to the north of this platform, there are scattered outcrops of rock formations of dominant Middle to Late Proterozoic metamorphic age (Fig.18.1, II) in the northern part of North-east China, including the Xinghuadukou Group and part of the Heilongjiang and Mashan Groups and their equivalents. The rocks range from the greenschist to the amphibolite facies, being partly migmatized. It is noteworthy that the Mashan Group occurring around Mashan, Heilongjiang may locally consist of granulite rocks which are often strongly migmatized, while rocks found in other localities may contain andalusite and cordierite, and are thus of the moderate to low pressure and moderate to high temperature type. To the south-west near the northern boundary of the North China Platform, there is a roughly E–W belt of Proterozoic metamorphic age, the Bayan Obo Group, the Langshan Group, and their equivalents to the west, containing greenschist and amphibolite facies rocks. Further west there exists another E–W to WNW–ESE belt of similar metamorphic age extending from the western section of the Beishan region in north-western Gansu, i.e. west of 100 °W, discontinuously to the axial zone of the Central Tianshan and the Quruktagh Mountains on the southern flank of the Tianshan Mountains. It includes the Kawabulak Group, the Yangjibulak Group, the Pargontagh Group, and their equivalents, comprising both low- and medium-grade metamorphic rocks. To the south of the Tarim Platform (Fig. 19.3, II$_A$), there are rocks of probably similar metamorphic age and character exposed on the northern slope both of the western Kunlun and of the Altyn Mountains.

On the south-west of the extreme western portion of the North China Platform, there are three subparallel WNW–ESE zones of essentially similar metamorphic age. These lie in the Qilian Mountains, i.e. within the Qilian Caledonian Fold Belt (Fig. 19.3, II$_G$), and are composed of the Zhulongguan, Jintieshan, Daliugou, Huangyuan, Danghe, and Dakendaban Groups and their equivalents. Beyond the south-west of the Qaidam Massif, there is also a belt of metamorphic rocks of dominant Middle to Late Proterozoic metamorphic age. While many rocks of these groups are of the greenschist facies, there are also rocks of the amphibolite facies (Fig. 18.1, II) some of which have been somewhat migmatized.

A Precambrian metamorphic age is attributed to nearly all these stratigraphic units chiefly on the basis of geological reasoning, correlation, and their unconformable contacts with the overlying Sinian or Lower Palaeozoic formations, because isotopic data are very scarce. With the exception of a K–Ar age of 1580 Ma (Tainjin Inst. 1979) for hornblende from the schist of the lower Dakendaban Group, the few available values (such as the K–Ar age of 935 Ma (Geol. Inst.-C.A.G.S. 1964) for phlogopite from some of the Mashan Group rocks, and that of 740 Ma (Geol. Inst.-C.A.G.S. 1963) for biotite from a Huangyuan Group schist) probably correspond respectively to the age of the Jinning Movement and the Chengjiang Movement prevailing in many places in South China.

Weakly metamorphosed Middle to Late Proterozoic rocks are quite widely distributed in the Transverse Ranges along the western border of the Yangzi Platform. They are mostly low greenschist facies rocks, including phyllitic and partly recrystallized types (Fig. 18.1, II'), and belong to the Kunyang Group and its equivalents. It seems that among the various isotopic age data, including Rb–Sr isochron and K–Ar values ranging from $c.$ 1750 Ma to $c.$ 850 Ma, only those between $c.$ 850 and $c.$ 900 Ma are the ages of metamorphism, approximately corresponding to the age of the Jinning or Xuefeng Movement. Further east within the boundary of the Yangzi Platform, that is in North Guangxi, South-east and East Guizhou, West and North Hunan, the northern part of Jiangxi, and the southernmost part of Anhui, usually known as the Jiangnan Oldland, there exists an almost continuous terrain of low-metamorphosed rocks of the same age as those just mentioned. These are often composed of the lower Sibao Group and the unconformably overlying Banxi Group and their equivalents. The Sibao rocks have a more limited extent and an approximate metamorphic age of 1150 Ma at the type locality in Guangxi, and the Banxi and their corresponding rocks have metamorphic ages of $c.$ 850–900 Ma or 700–750 Ma. There are also high greenschist or even low amphibolite facies rocks locally influenced by migmatization, such as those found to the west of Yichang, Hubei, in Baoxing, Sichuan, and Yuanmou, Yunnan.

There are two Middle to Late Proterozoic stratigraphic units, very probably of polymetamorphic origin, in Xizang. One of them is the amphibolite facies Qomolangma (Jolmo Lungma) Group and its equivalent Nielamu (Nyalam) Group of the Himalaya Range, exhibiting garnet, staurolite, kyanite, and sillimanite zones (Plate XVII.5) and having Rb–Sr and U–Pb isochron metamorphic ages of 640–660 Ma and a superimposed K–Ar metamorphic age of 10–20 Ma (Guiyang Inst. 1972). It is still questionable whether the Late Precambrian metamorphic event is dominant. The other unit is the Nyainquentanglha Group composed of gneisses and schists of greenschist to amphibolite facies, the age of which has long been debatable. It might have been metamorphosed first in the Late Precambrian, though Xu Ronghua in 1982* suggested that the main metamorphic imprint on the rocks was of Himalayan age.

In 1973–7 Zhang Quisheng and his colleagues investigated (Zhang *et al.* 1977) the polymetamorphic belt of prominently Caledonian metamorphic imprint occurring in the Qinling Mountains in Shaanxi, and the Funiu Mountains in western Henan (see below). They found that both the Middle Proterozoic Kuanping and Taowan Groups still retain the petrological features of the moderate pressure greenschist facies prior to $c.$ 1300 Ma. In the Ailaoshan polymetamorphic belt of dominantly Mesozoic metamorphic age (see below) lying to the west of the Red River Fault Zone, Yunnan, the Cangshan Group and part of the Ailaoshan Group might have been metamorphosed in Late Proterozoic times. The Lincang Group of the Lancangjiang meta-

---

*Xu Ronghua (1982). *Isotopic geochronological experimental methods and the geochronological study of the Lhasa Massif* (in Chinese). M.Sc. thesis, Institute of Geology, Academia Sinica.

morphic belt further west could also have been metamorphosed towards the end of Proterozoic.

Metamorphic series and belts chiefly of Early Palaeozoic (Caledonian) metamorphic age

The distribution of the metamorphic rocks of dominantly Early Palaeozoic metamorphic age generally coincides with Caledonian fold zones, and very locally with the Xingkaian. They occur in scattered, rather small areas usually of low or very low grade metamorphic facies (Fig. 18.1, III'_1) in a region roughly to the east of 110 °E and north of 42.2 °N, in eastern Nei Mongol, Heilongjiang, part of Jilin, and the extreme northern part of Liaoning. Those found in the north are the Ergunahe Group (ϵ, or D–S)* and the Suhuhe Group (O_1–O_2) composed of schist and phyllite of the low greenschist facies and other rock types of submetamorphic nature; and also part of the Mashan Group which is often metamorphosed to a higher grade. Those present in the south close to the northern limit of the North China Platform, are the Hulan Group (ϵ–S) in the east composed of kyanite-bearing and garnetiferous schists characteristic of greenschist facies, and also low amphibolite facies rocks of the moderate pressure series; and the Wendu'ermiao (Wendurmiao) Group (ϵ–S) in the west in Nei Mongol consisting mainly of greenschist rocks, and locally of glaucophane-schist of the high pressure series (see below). It is possible that relics of Caledonian metamorphic rocks are present in the metamorphic belt of dominantly Hercynian metamorphic age in the Altay Mountains. Further south in the western part of the Altyn Mountains there are also rocks of probable Caledonian metamorphic age.

In the south-west of the western part of the North China Platform, there are three metamorphic belts chiefly of Caledonian metamorphic age: two in the Qilian Mountains (Fig. 18.1, III_1) and another further south-west along the north-eastern border of the Qaidam Basin, all trending WNW–ESE. The first, occurring in the northern Qilian Mountains, is mostly composed of either low or high greenschist facies rocks derived from sedimentary rocks of Middle Cambrian to Silurian ages, containing in its middle section a zone of various types of glaucophane-bearing schists (see below). It extends discontinuously along strike for over 100 km (Fig. 18.1, symbol for glaucophane schist), and is very probably of the high pressure series. There are scattered outcrops to the south-west, for instance in the vicinity of Lanzhou and Datong, Qinghai Province, where Upper Precambrian as well as Lower Palaeozoic rocks have been metamorphosed simultaneously during the early episode of the Caledonian (492–544 Ma, Geol. Inst.-C.A.G.S. 1963) probably under moderate temperature and moderate pressure to form amphibolite and greenschist facies. The second one in the southern Qilian Mountains, and the third one along the Qaidam Basin, are mostly of greenschist facies derived from Cambro-Ordovician rocks, which are probably of moderate to low pressure types. Low amphibolite facies rocks containing andalusite and cordierite have also been found. The metamorphic rocks to the south-west of the Qaidam Basin (Fig. 18.1) may have been metamorphosed in both the Caledonian and Hercynian stages.

Further south is the Qinling metamorphic belt, extending from South-east Gansu through South Shaanxi to West Henan. It is composed of a northern sub-zone characterized by the superposition of Caledonian metamorphism, chiefly of the moderate pressure amphibolite facies, on Middle Proterozoic rocks already weakly metamorphosed during the Middle Proterozoic (see above), and a southern subzone consisting of metamorphosed Lower Palaeozoic formations, mostly of the greenschist facies and locally of low amphibolite facies, indicated by plagioclase–staurolite–almandine and other associations. Not far from the Qinling belt is the slightly metamorphosed to submetamorphic Bikou Group of the Gansu–Shaanxi–Sichuan borderland, which has undergone regional dynamic metamorphism at some time in the Palaeozoic.

Still further south is the broad zone of dominantly Caledonian metamorphism covering many places in Jiangxi and Hunan roughly to the south of 28 °N, western Fujian, western and northern parts of Guandong, and the eastern part of Guanxi. This zone approximates geotectonically the Xiang-Gui Fold, the Yunkai Uplift, and the Jian-ou Uplift (Fig. 19.3, VA_1, VA_2 and VA_3), and part of the Min–Yue Fold Zone (Fig. 19.3, VA_4). In the northern part of this zone the stratigraphic units involved include both Sinian and Lower Palaeozoic rocks, and those in the other regions are mainly Cambro-Ordovician and even Silurian, mostly with metamorphic ages of c. 350–400 Ma. Most of the rocks belong to the greenschist facies, often to its lower part and sometimes passing imperceptibly into submetamorphic types, probably under regionally rather weak stress conditions. However, there are also places, such as the Wugong Mountain in West Jianxi, the Wuyi Mountains along the Fujian-Jiangxi border, and the Yunkai Mountain along the southern section of the Guanxi–Guandong borderland, where high

---

*Geological ages in parentheses are those of the protolithic rocks.

greenschist to amphibolite facies rocks occur. These have low to moderate pressure and moderate to high temperature mineral associations, including plagioclase–staurolite–almandine–cordierite/andalusite, plagioclase–kyanite–biotite, and plagioclase–andalusite–sillimanite. It is also at these places that the metamorphism is often associated with, and usually immediately followed by migmatization (Plate XIX. 2), most of which seems to be genetically related to certain regional fractures.

There are also exposures of very low-grade metamorphic rocks chiefly of Caledonian metamorphic age in South-east Yunnan, and of metamorphic rocks mostly of amphibolite facies probably of similar dominant metamorphic age in South-east Xizang and in the region to the north of Gaize, Xizang.

Metamorphic series and belts chiefly of Late Palaeozoic (Hercynian) metamorphic age

Metamorphic series chiefly of Late Palaeozoic metamorphic age are usually found in the Hercynian fold belts of China, where protolithic rocks are mostly of Late Palaeozoic age and occasionally also include those of older ages. Some of them may be further metamorphosed during the Indosinian, and even later, orogenies.

The metamorphic series are widely distributed in the vast areas north of 41°–43°N up to the northern border. Those of Nei Mongol and North-east China are mostly low greenschist facies rocks or of submetamorphic nature, but often strongly deformed. Glaucophanic schist has been found at Heige'aola, Nei Mongol in 1981 (Bei Wenji, personal communication). In addition to the low-grade types spread extensively in Xinjiang, there are many places in this region where rocks of higher metamorphic grades are found to occur. For instance, according to the Regional Geological Survey Team of the Geological Bureau of Xinjiang, the metamorphic belt of dominant Hercynian metamorphic age of the Altay Mountains (Cheng et al., in press) is composed of a central anticlinorium containing chlorite, biotite, andalusite, staurolite, and sillimanite zones formed under moderate to high temperature and moderate to low pressure conditions; a northern zone of greenschist facies rocks with an association of sericite, chlorite and embryonic biotite; and a southern one characterized by a mineral association similar to that of the northern. The rocks of the sillimanite zone are often migmatized and closely related to the frequent migmatites and migmatitic granitoids; these are often associated with pegmatites that are workable for muscovite or rare elements.

Rocks of these three zones are strongly sheared and dynamically metamorphosed along a number of fracture zones; their metamorphosis was occasionally accompanied by metasomatism of alkaline nature, probably towards the end of the Palaeozoic. The Hercynian metamorphic belt of the Tianshan Mountains also consists of three zones (Lou et al. 1982). The northern one of protolithic Lower and Middle Carboniferous rocks, and the southern one derived from Silurian and Devonian formations are mostly of low greenschist facies and are submetamorphic rocks. The central zone, which originated from Silurian to Lower Carboniferous formations, is of low greenschist to low amphibolite facies rocks of the moderate pressure series, locally containing kyanite–staurolite–mica–schist and even glaucophane–schist of moderate to high pressure character derived from the Silurian formation south of Tekesi in the western section (Cheng et al. 1982c).

South of the broad metamorphic terrain just described, there are weakly metamorphosed rocks of Hercynian metamorphic age in most parts of the Kunlun Mountains, and low greenschist facies rocks of Early Hercynian metamorphic age derived from Carboniferous beds in the eastern continuation of the southern Qilian Mountains close to the eastern periphery of the Qaidam Basin. However, almandine-sillimanite-staurolite-bearing biotite-schist and andalusite-schist and andalusite-phyllite of the high greenschist to low amphibolite facies of moderate to low pressure transitional nature are locally found in the latter region. Almost coinciding with, and partly parallel to, the southern Caledonian metamorphic belt of the Qinling, the Early Hercynian metamorphic belt is represented by the Devonian Liuling Group and its equivalents, and is composed of either garnetiferous mica-schist or actinolite-schist, or of much less metamorphosed sedimentary to volcano-sedimentary rocks. These are shown as an undifferentiated Palaeozoic metamorphic belt in Fig. 18.1. They may be partly influenced by further weak Late Hercynian or Indosinian metamorphism, as indicated by the presence of glaucophane-schist in the eastern section (i.e. in Shanyang, Shaanxi and Neixiang, western Henan) of Triassic metamorphic age (Cheng et al. 1982c), and by the presence of submetamorphic Carboniferous and Permian rocks which had probably undergone transformations under regional stress and deformation in the region further south, causing weak regional dynamic metamorphism.

Further south-east and lying on the south-eastern flank of the Caledonian metamorphic belt in Jiangxi, Hunan, and western Fujian, there occurs a NE–SW belt of Hercynian, partly overprinted by Indosinian, metamorphism extending from North–Central Fujian, through the north-western portion of the Fujian-–Guandong boundary, to Hainan Island. Since part of it is superimposed on the Caledonian belt and its influence is weaker than that of the Early Palaeozoic metamorphism, the area of dominantly Hercynian metamorphism is limited (Fig. 18.1). In South-west Fujian (Li, Zhang et al. 1983), the Upper Devonian and Lower Carboniferous beds have been metamorphosed to low greenschist facies rocks to the northwest, locally containing minute biotite flakes, and to rocks of a higher grade of metamorphism to the south-east, such as garnetiferous biotite-schist and biotite-granulitite, and even kyanite-bearing schist. It is also in this south-eastern sub-zone that the fossiliferous Lower Permian rocks have been metamorphosed to chloritoid–chlorite-schist and related types, probably under fairly high stress conditions especially along certain fracture zones. The rocks of Hainan Island and a few localities on the Guandong mainland are mostly of submetamorphic to low greenschist facies.

Evidence for the influence of Hercynian metamorphism has been found in western Yunnan and Xizang but it is usually camouflaged by stronger later metamorphic transformations.

Metamorphic series and belts chiefly of Early Mesozoic (Indosinian) metamorphic age

Metamorphic series chiefly of Indosinian metamorphic age are mainly exposed in the eastern part of the Qinghai–Xizang Plateau, part of the western and eastern Qinling Mountains, and the northern part of the Transverse Ranges. Formerly they have been designated by various Triassic stratigraphic names, including Xikang Group, Dege Group, and Caodi Group. They are usually represented by submetamorphic slaty rocks or phyllitic types of the low greenschist facies (Fig. 18.1, IV_1'), including chiastolite-slate in certain regions, probably genetically related to regional heat flow rather than to intrusions. But in the vicinity of Danba (30°52'N, 101°47'E) in West Sichuan, near the south-eastern margin of this broad metamorphic terrain, there occurs an anticlinorium, including a few dome-anticlines, which are composed of high greenschist to amphibolite facies rocks of dominantly Indosinian metamorphic age (Fig. 18.1, IV_1). These are probably derived from Sinian, Lower and Upper Palaeozoic, and Triassic sedimentary and volcano-sedimentary rocks. Progressive metamorphic zonation at moderate pressure is well displayed (Cheng 1944; Cheng et al. 1963; Cheng et al. 1984), the various successive zones observed being characterized by index minerals such as biotite, almandine, kyanite and staur-olite, kyanite and sillimanite, and other relevant mineral assemblages. It is interesting to note that while the occurrence of the sillimanite zone is in general located between the kyanite zone and the zone of partial migmatization, the latter is exposed at the axial parts of the anticlines, and the appearance of sillimanite may also (in some places) be more directly related to the abundant muscovite-bearing pegmatites in the mica-rich gneisses. Since the protolithic Upper Palaeozoic formations are believed to have been disconformably overlain by Triassic rocks, it is possible that they had already been somewhat metamorphosed during the Hercynian Stage.

In the western part of Yunnan there are three major polymetamorphic belts* probably of dominantly Mesozoic metamorphic age. The Ailaoshan metamorphic belt in the east occurs in a fracture zone the nature of which is still in dispute. It is composed partly of andalusite- or sillimanite-bearing rocks associated with migmatites in the eastern sub-zone and greenschist facies types in the western sub-zone, separated by a strongly sheared and mylonitized zone, locally associated with rocks containing intercalated minerals of glaucophanic affiinity. They are probably derived from Upper Precambrian to Upper Palaeozoic rocks, having a metamorphic age of 230 Ma, belonging to the Hercynian to Indosinian stages, and with an overprinting age of 135–171 Ma and 30–60 Ma (Fan 1982). It is also possible that some of the older rocks had already been metamorphosed in Late Precambrian times. The Lancangjiang metamorphic belt in the middle has a geological and metamorphic history essentially similar to that of the Ailaoshan. Its southern segment is composed of a western high pressure, low temperature sub-zone with index minerals of glaucophane, 3T-phengite (showing a Rb–Sr age of 240–260 Ma), barroisite, riebeckite, and stilpnomelane; and an eastern low to moderate pressure and high to moderate temperature zone characterized by andalusite, staurolite, cordierite, sillimanite, and green-brown hornblende, with associated granitic rocks, 193–235 Ma in age (Peng et al. 1983). These have been interpreted as paired metamorphic zones related to Indosinian plate tectonics by Wang Kaiyuan et al. (1979), and recently by Peng and others. The geological and metamorphic history of the western Gaoligong metamorphic belt is not very different, but it has also been metamorphosed in the Early Hercynian as shown by tectonic evidence and by a biotite K–Ar age of 394 Ma from an associated granite; and the effect on it of Himalayan metamorphism (K–Ar age of 23–33 Ma on mica from the migmatite) may

---

*The names of these three belts were proposed by Wang Kaiyuan based on studies by Wang and his colleagues in 1975.

be somewhat stronger than on the other two (Fan 1982).

A metamorphic influence probably of Indosinian age can be traced in Fujian and even in Guandong, but it is usually weaker than that of Hercynian or Yanshanian age in the same district. Hence a belt of dominantly Indosinian metamorphic age cannot be delineated, and in fact such metamorphism is often a continuation of Hercynian metamorphism in quite a number of localities.

Metamorphic series and belts chiefly of Late Mesozoic (Yanshanian) metamorphic age

Metamorphic belts and metamorphic formations of dominantly Yanshanian age are well displayed along the coast of Fujian and in eastern Taiwan. The former extends from Mazu Island in the north-east, via Quanzhou and Dongshan up to the Nan'ao Deep Fracture Zone (Li et al. 1983). It consists of two nearly parallel sub-zones. The western one, characterized by rather open folds, consists of slightly metamorphosed Upper Triassic to Upper Jurassic sandstone, volcanic rocks, and phyllite, partly of low greenschist facies character with incipient biotite flakes, which pass westward into unmetamorphosed rocks. The eastern one, marked by tight and complicated folds, is composed of various kinds of granulitite, mica-schist and leuco-granulitite, derived from rocks of similar ages. There are high greenschist facies types occurring in the axial part of the syncline, but the rocks found in the axial part of the anticline are mostly of low amphibolite facies, associated with migmatites and even migmatitic granite, and containing mineral associations such as biotite-muscovite-andalusite-quartz-(sillimanite), hornblende-potash feldspar(plagioclase)-diopside-quartz, and hornblende-almandine-plagioclase-quartz, indicating high to moderate temperature and moderate to low pressure conditions. According to a paper published in 1982 by the Regional Geological Survey Team of the Geological Bureau of Fujian, this metamorphism ended at the close of the Late Jurassic; this is shown by the unconformable contact between the Lower Cretaceous agglomeratic lava and the underlying metamorphosed rock series, the K–Ar isochron ages of 99.2 and 103 Ma, and the Rb–Sr isochron age of 97 Ma for the rocks and minerals. The formation of the metamorphic and migmatitic rocks was related to the subduction of the Pacific Plate under the Asiatic one. The second belt is located in the eastern part of Taiwan Province, extending over 200 km in a NNE–SSW direction. It is composed of an eastern Yuli Belt characterized by high pressure and low temperature glaucophane-schist, and a western Tailuko Belt, composed of schists, crystalline limestone, and gneiss, partly of the greenschist facies and partly of the low amphibolite facies, of low pressure and high temperature, derived from Permian and probably also Carboniferous and Triassic rocks (Ho 1979). T. P. Yen proposed in 1963 that they formed paired metamorphic belts which are now in direct fault contact. The age of the main stage of regional metamorphism is probably Middle Mesozoic (Yen 1981), but according to J. G. Liou (1981) the blueschist facies recrystallization took place only 8–14 Ma ago, as suggested by Rb–Sr age determinations. It seems that the rocks are of polymetamorphic character, the main metamorphic age of which may vary from place to place in the same belt.

A broad metamorphic belt of dominantly (probably Late) Yanshanian metamorphic age extends from the western borders eastward through the central part of Xizang, to the western limit of the Transverse Ranges, with protolithic formations of various geological ages. Most of the rocks are of low or higher greenschist facies, but those of the south-eastern portion and of some localities in the western section may contain amphibolite facies types partly containing andalusite and/or sillimanite associated with migmatitic rocks, and so must have been formed under moderate to high temperature conditions. Some of them have also been influenced by the Himalayan metamorphism, as indicated by the K–Ar ages of 10–110 Ma for the associated granites and of 21 and 119 Ma for a migmatite and an amphibolite respectively (Zhang and Li 1981).

Most regions affected by the Yanshanian Movement only display thermal metamorphism around the intrusions. However, some Mesozoic and older formations in a few Mesozoic depressions, such as part of the western hills of Beijing, had actually been metamorphosed to low greenschist facies rocks at an earlier stage of the Yanshanian at low temperature and moderate stress, on which was superimposed a later metamorphic stage of high to moderate temperature and moderate stress (see below).

The isotopic age data for the Yanshanian metamorphism so far obtained fall into two groups, namely, $c.$ 90–120 and $c.$ 165 Ma (Cheng et al., in press). Modified K–Ar ages ranging from $c.$ 85 to 160 Ma have been determined for the gneisses of the Early Precambrian terrains, such as the Dabie region.

Metamorphic series and belts chiefly of Cainozoic (Himalayan) metamorphic age

Himalayan metamorphic series and belts are of very limited distribution. The Early Tertiary slate formation, composed of submetamorphic and slightly metamorphosed rocks, of the Central Range and the Hsuehshan Range of Taiwan (Fig. 18.1, V) were metamorphosed during Middle to Late Tertiary times, with isotopic metamorphic ages of 9.7–52.5 Ma (Ho 1975). The Nielamu Group and its equivalent in the Himalaya Range show a superimposed Himalayan metamorphic age (see above). The region further north up to the Yarlung Zangbo River is also a belt of dominantly Himalayan metamorphic age, as indicated by the K–Ar age of 30–50 Ma (Geoch. Inst. 1981) for the associated granites, and also by geological considerations. The protolithic formations are Permian to Cretaceous, and the rocks are mainly of the greenschist facies (Plate XVII.6) but may pass into low amphibolite facies rocks (Plate XVII.4) on the one hand and submetamorphic types, and even locally unmetamorphosed ones on the other. Examples of imprinting of Himalayan metamorphic ages on earlier metamorphic rocks have been encountered frequently in West Yunnan.

Summary (Figs. 18.1 and 19.3)

There are certain regularities with regard to the distribution of the metamorphic series and metamorphic belts. The metamorphic series of Archaean and Early Proterozoic metamorphic age are distributed mainly on the North China Platform, being unconformably overlain by unmetamorphosed, or largely unmetamorphosed, Middle to Upper Proterozoic formations. This indicates that the basement of the Platform was formed at the end of the Early Proterozoic. Metamorphic rocks of Middle to Late Proterozoic metamorphic age are found in the western and eastern parts of the Yangzi Platform, in some regions around the Tarim Platform and in the Qinling Mountains, in the districts extending from the Yinshan to Tianshan, and at some other localities. In most of these places they are usually unconformably covered by the unmetamorphosed Sinian formations, thus denoting the appearance of another period of platform formation in the tectonic history of China towards the close of the Proterozoic, approximately corresponding to the Jinningian. Except for those which originated between the platforms, the later metamorphic series and metamorphic belts have tended to form successively further away from these platforms. For instance, from the western part of the North China Platform and the eastern part of the Tarim Platform southward, and partly south-westward up to the Himalaya Ranges, metamorphic series and belts of Caledonian, Hercynian, Indosinian, Yanshanian, and Himalayan metamorphic epochs appear almost consecutively, with local gaps or overlaps. From the Yangzi Platform south-eastward to Taiwan Island, metamorphic series and belts of Caledonian, Hercynian or undifferentiated Hercynian and Indosinian, Yanshanian, and Himalayan metamorphic epochs are also exposed almost consecutively. The recent discovery (Cheng et al. 1982c) in the central part of Nei Mongol and northern Liaoning of a narrow and discontinuous Caledonian metamorphic belt between the dominantly Middle to Late Proterozoic metamorphic belt to the south, and the extensive Hercynian metamorphic terrain to the north, serves as an example of the consecutive 'outward growth' of metamorphic belts of different ages with the progress of time.

The phenomenon of orderly migration of metamorphic belts is not at all incidental, but has to be considered as a logical consequence of the migration of the mobile (folding) belts. In fact, this has been noticed by a number of geologists, especially structural geologists, and could be explained by the continued accretion of the platforms and the extinction of geosynclines of various geological ages. More recently it has been partly accounted for by Li Chunyu and others as an outcome of the successive growth of the continental plate and a series of subductions of the oceanic plates underneath the continental one. On the other hand, temporary and local reversal of the direction of migration of the metamorphic belts does exist, probably due to the complexity of the evolution of the Earth's crust.

EVOLUTIONAL FEATURES OF METAMORPHISM

Metamorphism is a physico-chemical process, but should be more properly regarded as an important geological one. Its birth, development and possible further transformations are all closely related to, and influenced by, other geological processes. Hence the metamorphism of each metamorphic epoch in China has its own peculiarities as well as some common characteristics.

Although there are terrains of greenstone belt affinity in the regions of Archaean metamorphic rocks, they are not 'typical' when compared with most of those of other countries, being of higher or much higher metamorphic grade (Plate XVIII.8; Plate XIX.8). This fact, together with the widespread occurrence of amphibolite, and even granulite, facies rocks, often with associated migmatites, in the Archaean terrains of China, implies that the geological setting for the Archaean metamorphism in China was quite uniform over an extensive area. This was probably due to the dominant influence of ascending heat flow from the upper mantle at this early stage, together with the additional cumulative metamorphic effect of the great thickness of Archaean rocks that had already formed towards the later stages of the Archaean, so that a relative 'homogenization' of high-grade metamorphic conditions, probably of predominantly moderate pressure, was reached.

The conditions for the Early Proterozoic metamorphism were quite different from those of the Archaean. For example, the Wutai Group is composed of eugeosynclinal volcano-sedimentary and sedimentary rocks accumulated and metamorphosed in elongated marine basins. These are more or less trough-shaped, and of rather limited extent, developed on the Archaean sialic crust. They were metamorphosed at moderate, or moderate to lower, pressure with the partial formation of progressive metamorphic zones of greenschist to amphibolite facies, and of associated migmatites in some places. The Hutuo rocks are chiefly miogeosynclinal clastic sediments deposited in even smaller sea basins, metamorphosed to, greenschist facies, and basically of mono-zonal character.

Middle to Late Proterozoic metamorphism in South China is mainly developed within the Yangzi Platform, with the metamorphosed rocks, derived from miogeosynclinal flysch and flyschoid formations, usually only attaining to areal, mono-zonal, low greenschist facies or lower grade, and of a low to medium pressure nature. Metamorphic conditions in other regions are more varied, and the rocks thus formed often belong to greenschist to low amphibolite facies, but locally to an even higher grade, and are associated with migmatitic rocks.

The Palaeozoic metamorphic conditions were even more varied. In general, in the less mobile miogeosynclinal belts where igneous activity was rather weak or practically absent, the effect of heat flow not prominent and the metamorphism took place at comparatively shallow depths. The resulting regional dynamic metamorphism is usually characterized by areal, mono-zonal, low greenschist facies, partly with submetamorphic rocks. Typical examples are the metamorphic rocks of dominantly Caledonian metamorphic age in Hunan and Jiangxi, and a large tract of Hercynian metamorphic rocks in many places in northern Northeast China and Nei Mongol. In the more mobile eugeosynclinal belts where magmatism, often accompanied by migmatism, was intensive, the effect of heat flow prominent, and deformation strong, related physico-chemical transformations usually took place at greater depths. The metamorphism is then often characterized by well-marked progressive metamorphic zonation up to amphibolite facies of the moderate to low pressure series. The character of metamorphism is further controlled by local geological structures; for instance, the centre of metamorphism within a region is often located in the axial part of an anticline or anticlinorium or along a particular fracture zone. This is exemplified by the Hercynian metamorphic series and belts of the Altay Mountains. Paired metamorphic belts are found in a few places.

There are still more types of metamorphism and of related metamorphic rocks in the dominantly Mesozoic or Cainozoic metamorphic belts. The first type is of regional metamorphic affinity with a prominently linear distribution genetically related to fracture zones, or even to sutures between plates. Examples occur in the Central Range of Taiwan and in Xizang, along the coast of Fujian, and in the Lancangjiang metamorphic belt, some of which are accompanied by the presence of paired metamorphic zones. The second type is regional dynamic metamorphism of areal extension and essentially of mono-zonal, low greenschist facies, represented by the Indosinian metamorphism and related rocks of the Bayan Har Mountains of Qinghai, and the partly dissected extensive plateau of western Sichuan. The third type is the well-developed thermal metamorphism associated with distinct marginal migmatization, sometimes of considerable extent. It is present in the vicinity of quite a number of granitoidal bodies in South-east and South China. The fourth type is the dynamic metamorphism associated with migmatization along a number of fractures such as those found in south-eastern and south-western China. However it is difficult sometimes to distinguish the fourth from the first kind of metamorphism.

To sum up, all the Archaean and Early Proterozoic rocks are metamorphosed in Archaean and Early Proterozoic times respectively, while those of the later ages are only metamorphosed in appropriate geological environments, notably in the mobile belts and fracture zones of various origins. It seems that in general the temperature and areal extent of regional metamorphism tend to decrease from the Archaean, through the Proterozoic and Palaeozoic, to later geological ages, with a concomitant increase in the effect of stress and in the number of greenschist facies regions and even of laumontite–prehnite–pumpellyite facies regions. Distinct metamorphic belts with clear metamorphic zoning appear from the Early Palaeozoic onwards, and their linear character is most prominent in those of Cainozoic age. While Archaean migmatization is usually of the regional type, that of the Mesozoic and Cainozoic is generally on the margins of granitoidal bodies and in fracture zones (Cheng et al. 1963; Cheng et al. 1982c) (see below).

POLYMETAMORPHISM AND POLYSTAGE METAMORPHISM

Polymetamorphism and polystage metamorphism are metamorphic processes of different orders of magnitude, yet the former term is often used loosely. Dong Shenbao pointed out in 1981 that polymetamorphism is not concerned with different metamorphic events of a particular metamorphism, but deals with more than two epochs of metamorphism with distinctly different characters and belonging to different (geological) cycles. Cheng et al. (1982c) have recently redefined 'polymetamorphism' as follows.

Polymetamorphism is defined as the superposition of metamorphic events of different metamorphic cycles. A metamorphic cycle includes the entire course of evolution of metamorphism, usually characterized by an early stage of dominantly increasing temperature and/or pressure and a late stage of dominantly decreasing temperature and/or pressure, suggestive of a cyclic change in $P-T$ conditions during a period corresponding roughly to a major tectonic stage or cycle, such as the Caledonian, Hercynian, etc. It is evident that from the beginning to the end of such a cycle, a series of complicated changes takes place, including the increase and decrease of temperature and pressure, so that two or more genetically related metamorphic stages may appear. The occurrence of different metamorphic events (stages) within a particular metamorphic cycle is called poly-stage metamorphism of the same metamorphic epoch. It is however not necessary that the mineral assemblage, or the relics, of each stage should be clearly preserved, since the minerals of the earlier stage are usually intensively 'remoulded' or modified under new physico-chemical conditions at a later stage. However it is more often the case that the mineral associations of different stages may be preserved to a certain extent. When the fall of temperature in the last stage is very rapid the minerals of the early stage may be almost completely preserved. Thus the effects are varied and complicated.

The rocks of the Qianxi Group are a good example of polymetamorphism. The superposition of an amphibolite facies metamorphism of Late Archaean age, at or before 2500–2600 Ma, on the granulite facies type of an earlier Archaean regional metamorphism, probably prior to 3000 Ma, occurred in the rocks of the Qianxi Group of the Yanshan region, has already been mentioned. Other examples are the three polymetamorphic zones of the dominantly Mesozoic metamorphism of West Yunnan, and the northern sub-zone of the Qinling metamorphic belt of dominantly Caledonian metamorphic age already referred to. In the latter district (Zhang et al. 1977), the Middle Proterozoic Kuanping and Taowan Groups were first subjected to regional open folding, and metamorphosed prior to $c.$ 1300 Ma into greenschist facies rocks of the moderate pressure series, that is characterized by mineral associations such as quartz–chlorite–(albite), chlorite–epidote–(zoisite)–actinolite–albite, chlorite-quartz-albite, and calcite–quartz–sericite. These then underwent regional fan-shaped folding and greenschist to amphibolite facies metamorphism, also of the moderate pressure series, during the Caledonian tectonogenesis at $c.$ 420 Ma, with mineral associations of quartz–oligoclase–biotite–almandine–kyanite; quartz–oligoclase–garnet–hornblende, and quartz–biotite–tremolite–calcite, etc. in the vertical axial part, and of quartz–albite–sericite(muscovite)–biotite, quartz–albite–actinolite–epidote–(calcite–hornblende), and quartz–epidote–calcite–(actinolite), etc. in different parts of both limbs. At almost all the localities where these two groups are exposed, there are unevenly unoriented, deep brown ($Ng \cong Nm$) porphyroblastic biotite flakes, either single or aggregated, which may be associated with some porphyroblastic actinolite in some places. They are considered as products of the Yanshanian ($c.$ 150 Ma) metamorphism by Zhang, and seem to be related to the typical Yanshanian contact metamorphic rocks that contain mineral assemblages such as diopside–tremolite, or biotite–andalusite–(corundum), or (potash feldspar)–wollastonite–vesuvianite–andradite, around some granitic intrusions.

Two examples of polystage metamorphism may be noted. The first is the two-stage Late Archaean metamorphism exhibited by the Dengfeng Group of Mount Songshan, Henan (You and Han 1981), already mentioned in the previous section on metamorphic series chiefly of Archaean metamorphic age. The mineral associations of the first stage are quartz–oligoclase–biotite–almandine–(staurolite), quartz–oligoclase–biotite–muscovite–(potashfeldspar), and plagioclase(An_{30})–quartz–hornblende–(epidote), etc. Those of the second stage are quartz–biotite–phengite–chlorite–epidote, and quartz–chlorite–epidote–biotite–actinolitic hornblende, formed under retrogressive conditions.

The second example is the two-stage metamorphism of Early Yanshanian age in the Western Hills of Beijing, especially shown by the country rocks of the Fangshan granodiorite body to the south-west of Beijing (Liu and Wu 1977). Stratigraphical units, ranging from the upper part of the Middle to Upper Proterozoic, through the Cambro-Ordovician and Permo-Carboniferous up to the Jurassic, exposed on the eastern and south-eastern slopes of the region, were involved in the warping and folding, and influenced by the ensuing intrusive activity of the early stage of the Yanshanian

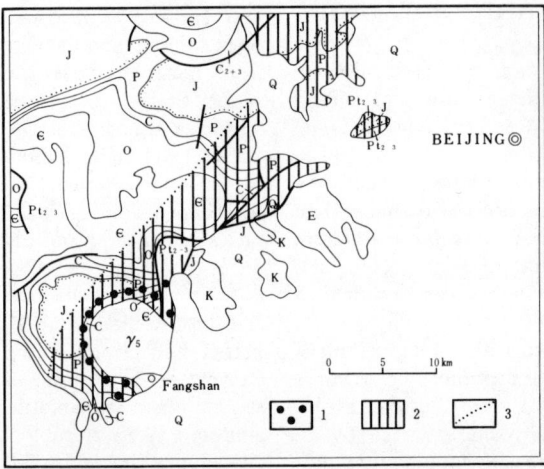

Fig. 18.2. Sketch map showing the distribution of two successive metamorphic phases of the Yanshanian metamorphic age of the Western Hills of Beijing (after Cheng Yuqi and Zhang Shouguang 1979). 1, second phase, moderate to low pressure and moderate to high temperature; 2, first phase, low to moderate pressure and low temperature; 3, boundary of first metamorphic phase; γ_5, Yanshanian granite; Q, Quaternary; K, Cretaceous; J, Jurassic; P, Permian; C, Carboniferous; O, Ordovician; E, Cambrian; Pt_{2-3}, Middle to Upper Proterozoic.

Movement, probably between 145–170 Ma. They were transformed into a SW–NE to SSW–NNE belt of greenschist facies metamorphism (Fig. 18.2) during the first phase (or stage), and then further metamorphosed during the second phase around some intrusive bodies, concealed or exposed, such as the Fangshan intrusion. The first phase is characterized by sericite–chloritoid–quartz, sericite–chloritoid–chlorite–quartz (Plate XVIII.6), and sericite–chlorite–epidote assemblages in pelitic to semipelitic types, partly with some calcareous materials; and by a chloritoid–chlorite–sericite–(plagioclase) association, and nearly monomineralic chloritoid rock (Plate XVIII.1) in basic volcanic rocks of slightly different composition, and also in calcite-bearing rocks in the impure carbonate types. Thus they belong mainly to the quartz–albite–muscovite–chlorite sub-facies of the low temperature and moderate to low pressure series, formed under regional stress conditions. The second phase metamorphism is of much more limited distribution and is superimposed on the first. It is the direct outcome of the ascent of the acidic magma under structural control, and of the final emplacement of the related intrusions, probably under continued regional stress and with gradually increasing heat flow until a high maximum temperature was reached. In the regions around the Fangshan granodiorite body, four successive metamorphic zones have been distinguished by Liu and Wu. The rather broad, outermost biotite zone was formed first and is of the greatest extent. It is marked by transformation of the chlorite and chloritoid in the slate and phyllite (of the first phase), to greenish to brownish yellow scaly biotite at the periphery, and to yellowish to reddish brown 'hornfels' biotite towards the inner boundary. However, no distinct mineral changes have been observed in either pure or impure carbonate rocks. The next, and second in chronological order, is the narrower andalusite zone. The index mineral, porphyroblastic andalusite, in the pelitic or semipelitic slate or phyllite is colourless in thin section. It is either not well crystallized and maculated with minute quartz, some biotite, and occasional chloritoid, or chiastolite of better crystal form (Plate XVIII.2); in either case it is embedded in a groundmass of chlorite and/or chloritoid, sericite, larger biotite, etc. The carbonate types are recrystallized rocks or marbles containing scanty metamorphic muscovite. The third zone, which was formed still later, is the even narrower almandine–staurolite zone. It contains mineral associations such as andalusite (reddish)–biotite(muscovite)–almandine, (andalusite)–biotite–staurolite(chloritoid)–almandine, and biotite–staurolite(chloritoid), in pelitics and semipelitics, and tremolite–anorthite–(calcite), hornblende–andesine (oligoclase)–epidote–(biotite)–quartz, and hornblende–biotite–andesine–(quartz), in calcareous and basic tuffaceous rocks. It is interesting to note that the almandine contains small grains of colourless andalusite, the staurolite contains inclusions of chloritoid, and both may be accompanied by perfect crystals of reddish andalusite containing Al^{3+} replaced by Fe^{3+} at higher formation temperatures than those of the earlier colourless variety. The innermost sillimanite zone, the latest formed, situated right at the intrusive contact, is usually only 20–30 m wide, characterized by sillimanite–muscovite(biotite)–almandine, sillimanite–muscovite–(andalusite), and sillimanite–biotite–muscovite in pelitic schist or hornfels and olivine–diopside–(tremolite), diopside–anorthite–(vesuvianite), diopside–wollasite–hedenbergite, etc. in carbonate rocks of varied composition. These four zones are considered to be the products of four successive episodes of the second phase of the Early Yanshanian metamorphic epoch under continuously increasing temperature conditions; taken together, they seem to have formed under moderate to high temperature and low to moderate pressure conditions. The first and second zones have been named chloritoid–andalusite–phyllite facies by Liu and Wu, the third as almandine–staurolite–hornfels (schist) facies, and the fourth as muscovite–sillimanite–schist (hornfels) facies. They are roughly comparable to albite–epidote–hornfels facies, hornblende–hornfels facies, and pyroxene–hornfels facies of the high temperature and low pressure series. It is apparent that the metamorphic phenomena just described serve as excellent examples of quite complicated polystage metamorphism but they serve even better as an illustration of the dialectical relationship between the time and space problems of the formation of metamorphic zones during particular stages of metamorphic evolution. Hence it may be inferred that the evolution of the Precambrian metamorphism in China may be much more complicated than at first thought.

EXAMPLES OF ARCHAEAN AND EARLY PROTEROZOIC METAMORPHIC TERRAINS

The Wutai Mountain and part of the Taihang Mountains*

Being situated in the northern part of the borderland of the Shanxi and Hobei Provinces, this is one of the few regions in China where not only both the Archaean and Lower Proterozoic metamorphic formations are well developed, but their stratigraphic relationships and their individual characteristics can also be clearly observed (see Chapters 4 and 5). A short description and synthesis of the metamorphic geology of this region may serve as a key to the understanding of the evolution of the crust in North China during the Archaean and Early Proterozoic times. As shown in Fig. 18.3, the Archaean Fuping Group (ArF_1 and ArF_2) appears in general as domes in the eastern and south-eastern part of the region, and the Archaean Longquanguan Group, which for simplicity is labelled as 'ArF_3', occurs on the western flank of the Fuping, unconformably overlain by the Wutai Group or still younger rock formations. The Lower Proterozoic Wutai Group crops out in an elongated depression (synclinorium) to the north-west of the dome region, while the unconformably overlying Lower Proterozoic Hutuo Group is found in certain minor structural units within the Wutai region, and also in a NE–SW belt on the south-eastern flank of the dome. A study of the basal part of the Wutai Group shows that it is composed of protolithic, conglomeratic, and coarse sandy rocks deposited when the underlying erosion surface was a depression, and of finer arkosic types deposited when the latter was a protruding area. It seems that most of the detritus came directly from rocks identical to those below the unconformity surface; some of the feldspathic materials were well preserved and the basal rocks were still fresh, indicative of strong denudation but rather weak chemical weathering.

Metamorphism of the Archaean rocks

Although it is quite evident that in general the degree of metamorphism of Archaean rocks increases with their original depths, and that their foliation is parallel to the protolithic bedding of the sediments, with very few exceptions, this does not mean that the lowest rocks of the stratigraphic sequence all attain the maximum metamorphic grade of the region. In fact, granulite facies rocks, such as granulite containing both hypersthene and clinopyroxene (Plate XVII.7), hypersthene–almandine-taconite, and olivine-marble, are usually found in the undulatory central part of the dome, whereas the often strongly folded peripheries are characterized by rocks of a lower degree of metamorphism, including various types of the amphibolite facies. This may indicate that the metamorphism took place under conditions of high temperature mainly caused by heat flow from the mantle and the deep protocrust, and high hydrostatic pressure produced by the great depth of the Archaean sequence. It seems that the stress related to the peripheral folding did not play an important part in the metamorphic process. It is also near the central part of the dome, where both pressure and temperature were probably the highest in the region, that the rocks are often intensively migmatized, mainly under anatectic conditions, resulting in the formation of an areal,

---

* A great part of the geological data on which this account is based is kindly supplied by Wu Jiashan.

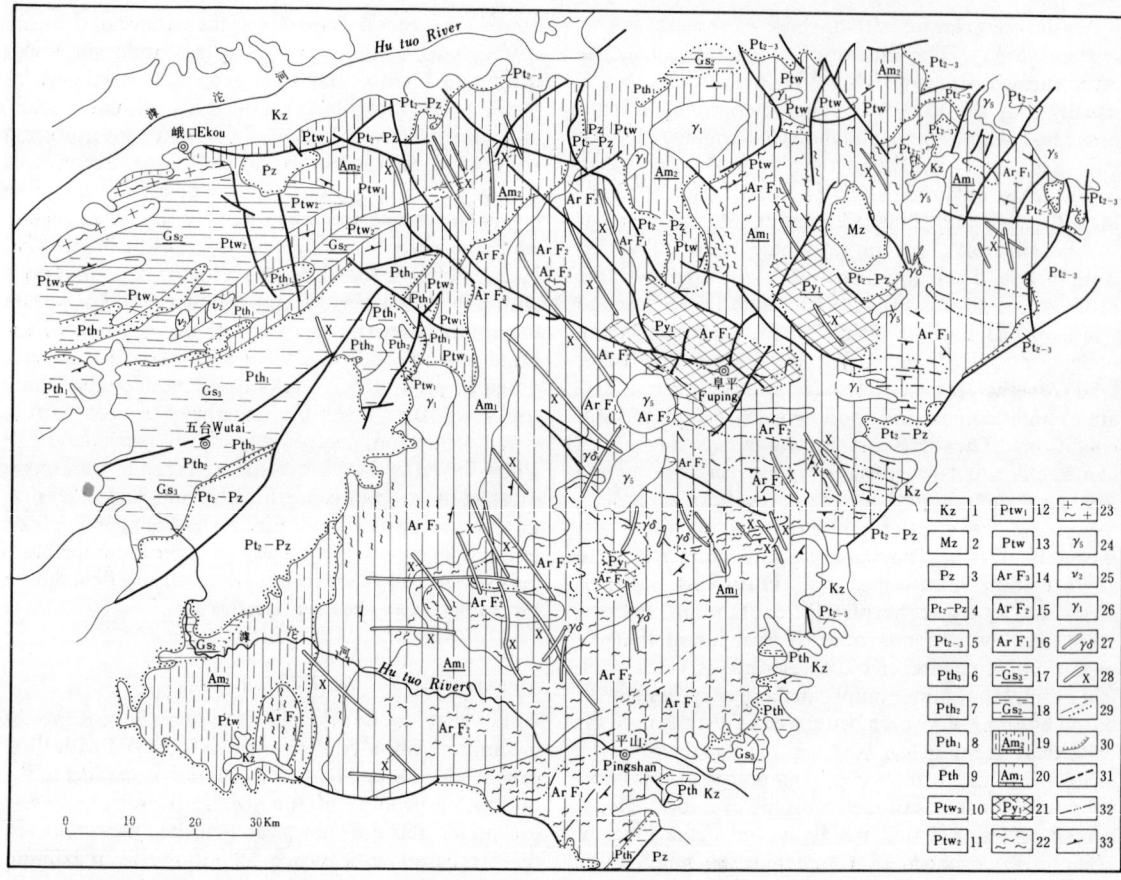

Fig. 18.3. Geological sketch map of the metamorphic series of the Wutai Mountain and part of the Taihang Mountains (modified after Wu and Zhang 1979). 1, Cainozoic; 2, Mesozoic; 3, Palaeozoic; 4, Middle Proterozoic to Palaeozoic; 5, Middle to Upper Proterozoic; 6, Upper Hutuo Group; 7, Middle Hutuo Group; 8, Lower Hutuo Group; 9, Hutuo Group, undifferentiated; 10, Upper Wutai Group; 11, Middle Wutai Group; 12, Lower Wutai Group; 13, Wutai Group, undifferentiated; 14, Upper Fuping Group; 15, Middle Fuping Group; 16, Lower Fuping Group; 17, Luliangian greenschist facies; 18, chiefly Wutain greenschist facies; 19, chiefly Wutain amphibolite facies; 20, chiefly Fupingian amphibolite facies; 21, chiefly Fupingian pyroxene-granulite facies; 22, migmatite; 23, migmatitic granite; 24, Yanshanian granite and/or granodiorite; 25, Luliangian basic rocks; 26, Wutain and Fupingian granites; 27, intermediate to acid dykes; 28, basic dykes; 29, geological boundary, observed or inferred, boundary between metamorphic facies; 30, unconformity; 31, faults, observed or inferred; 32, index beds (marble or amphibolite) in the lower Fuping Group; 33, strike and dip of foliation.

though somewhat irregular, distribution of migmatitic rocks without distinct migmatitic zonation. At the peripheries, the metamorphic formations were intensely folded and fractured, thus offering favourable conditions for the circulation of fluids and the emplacement of intrusives. Here the rocks are migmatized in one of two ways; either along fractures and fracture zones by alkaline solutions probably derived from the remelting of the centre of the dome region, giving rise to belts of dominant pegmatization; or around gneissic granitic bodies to form zones of marginal migmatization of metasomatic nature (see below).

Many metamorphic rocks were retrogressively metamorphosed to a certain extent. For instance, the pyroxenes were transformed into hornblende and/or biotite, and olivine into serpentine, but the exact age of the retrogressive metamorphism is still an unsolved problem.

It seems that metamorphism started somewhat earlier than the deformation which was almost coeval with

Metamorphic and migmatitic rocks of the Taishan Group

Basic features of the Taishan Group

The Yanglingguan region is situated to the south-east of Jinan, the capital of Shandong, and north-west of Xintai (Fig. 18.4). In this vicinity the Taishan Group is widely distributed and consists of three successive stratigraphic units, the Renjiazhuang, Yanlingguan and Shancaoyu Formations. They are unconformably overlain by the non-metamorphosed Cambrian formations.

The lower Renjiazhuang Formation is exposed in the north-eastern part of the region, occupying the axial part of a small anticlinorium. It is composed chiefly of a north-eastern lower belt of biotite-bearing, oligoclase-rich, migmatitic 'granite' of trondhjemitic affinity (Table 18.3, 2), and a south-western upper one of slightly streaky, biotite–oligoclase-homogeneous migmatite.* Both belts are probably derived mainly from biotite-granulitite and biotite–plagioclase-gneiss through migmatization of a dominantly Na-metasomatic nature, and show a mutual transitional relationship. Relict bands of the metamorphic rocks are found both in the migmatite and migmatitic granite. The homogeneous migmatite shows a Rb–Sr isochron date of 2586 Ma (Geol. Inst.-C.A.G.S. 1976), probably indicative of the age of migmatization, and a K–Ar age of c. 2000 Ma.

The middle Yanlingguan Formation outcrops to the south-west of the Renjiazhuang Formation and within the inner belt of the south-western limb of the anticlinorium. The latter on the whole shows a roughly parallel (concordant) selective migmatization contact with the former, and the topmost part of the Renjiazhuang Formation is only very weakly migmatized. They are distinctly conformable with each other. The Yanlingguan Formation was originally a dominantly basic volcano-sedimentary series 1000–1400 m thick, containing subordinate basic intrusions and ultrabasic rocks. Only the rocks of the lowermost part have been clearly influenced by migmatization to a certain extent. The available K–Ar and Rb–Sr isotopic data for both mineral and rock samples from the Yanlingguan all suggest a metamorphic age similar to that of the

Metamorphism of the Wutain rocks

The lower part of the Wutai Group (PtW$_1$), which occurs in general in the outer belt of the west–southwestward pitching synclinorium in the north-west of this region, is usually composed of amphibolite facies rocks of the moderate pressure series, exhibiting kyanite-, straurolite-, almandine- and sillimanite-zones at certain places. This could have been caused by a not unusual spatial variation of temperature during metamorphism in a geosynclinal environment that was probably trough-shaped. The situation is further complicated by the emplacement of parautochthonous migmatitic granitic bodies both in the lower and lower middle parts of the Wutai, often partly associated with marginal migmatization as in the vicinity of Okou, Wangjiahui. The middle (PtW$_2$) and middle to upper parts of the Wutai, which are mostly found at or near the central part of the synclinorium and of the ancient geosyncline, often consist of widely distributed greenschist facies, and subordinately of low amphibolite facies rocks. It is noteworthy that they are intensely folded yet practically free from the influence of magmatism. It seems that the increase in the grade of metamorphism of the Wutai rocks was more related to the intensity of migmatic and magmatic activity than to the stress set up during the deformation. While the boundaries between the metamorphic facies and subfacies often agree well, and are concordant with the structural features and elements of the rock formations, they may also crosscut them at the eastern end of the Wutai region. This indicates the post-tectonic date of the end stage of metamorphism.

Metamorphism of the Hutuo Group

Being accumulated and deformed in local marine basins, the rocks of the Hutuo Group are often strongly folded and highly cleaved and puckered, but practically free from the influence of igneous activity except for a few small basic intrusions. They have mostly been metamorphosed to phyllitic rocks of the low greenschist facies under regional stress and low temperature conditions.

It seems that the difference in the metamorphic features between the three stratigraphic groups described above not only reflects the difference in their exact geological settings and the related P–T conditions of their formation, but also serves to trace the basic characteristics of the evolution of the Wutai–Taihang region during Archaean and Proterozoic times.

---

*This term was proposed by Cheng Yuqi during field-work near Anshan, China in 1950 and published in Chinese (jun-zhi-hun-he-yan) (Cheng et al. 1963) for a fairly massive migmatite of relatively homogeneous structure and composition, formed at an advanced stage of migmatization, and roughly corresponding to part of the 'diatexite' of K. R. Mehnert.

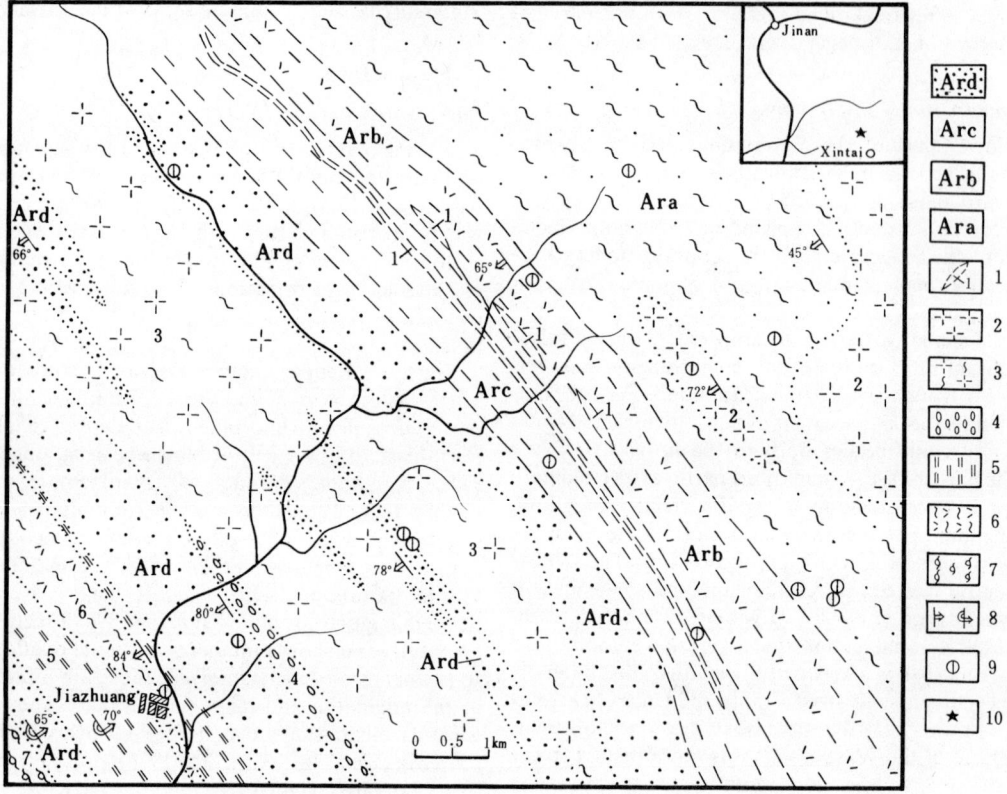

Fig. 18.4. Map showing the distribution of metamorphic and migmatitic rocks of the Archaean Taishan Group of the Yanlingguan region, Xintai, Shandong (modified after Cheng Yuqi et al. 1964). Ard, Shancaoyu Formation, chiefly biotite-granulitite and biotite-bearing granulitite; Arc, Upper Yanlingguan Formation, plagioclase-amphiobolite with hornblende-granulitite intercalations and basal pebble-bearing hornblende-granulitite; Arb, Lower Yanlingguan Formation, chiefly plagioclase-amphibolite, transitional to chlorite-actinolite-schist in the south-eastern-most part; Ara, Renjiazhuang Formation, biotite-oligoclase-homogeneous migmatite, with screens of mica-schist and mica-granulitite and scanty amphibolite. 1, serpentinite, actinolite-tremolite-schist and talc-tremolite-schist; 2, Xinfushan biotite-bearing, oligoclase-rich migmatitic granite; 3, Cishan porphyroblastic biotite-microcline-migmatitic granite; 4, Moshishan fine-granular biotite-feldspar-homogeneous migmatite with quartzo-feldspathic aggregates; 5, partly migmatized belt of biotite-plagioclase-migmatite and biotite-granulitite; 6, belt of biotite-plagioclase-migmatite and biotite-hornblende-plagioclase-migmatite; 7, partly migmatized belt of biotite-microcline-augen-migmatite and biotite-granulitite; 8, strike and dip of foliation; 9, localities of isotopic age specimens; 10, Yanlingguan region.

Sancaoyu, and this is supported by geological evidence.

Based upon the stratigraphic succession of the metamorphic rocks and the characters of the protolithic volcano-sedimentaries, the Yanlingguan Formation is further subdivided into two sub-formations, each containing five beds and constituting a 'major' eruption-sedimentational cycle. The lower 'major' cycle consists of four minor ones and the upper comprises two. Each part of such a cycle usually begins with protolithic tuff or lava (plagioclase-amphibolite), which is succeeded by lava (Plate XVII.3) of the main eruption phase, and ends with tuffaceous sandy rocks (hornblende-granulitites) often with intercalations of semipelitic types (biotite-granulitites) of the late dominant sedimentation phase. Calc-silicate rocks are present locally.

The metamorphosed basic lava and tuff and related intrusions are of common magmatic origin. The lava usually corresponds to the high-ferromagnesian, and especially high-magnesian, type of tholeiitic basalt (Table 4–3, III), but lavas of komatiitic composition in association with ultrabasic intrusions are also found locally (Table 18.1). Judging from the petrochemical

Table 18.1. *Partial chemical composition (weight %, recalculated on the dry basis) of protolithic komatiitic rocks (chlorite-actinolite-schist) of the Yanlingguan Formation (after Cheng et al. 1982b).*

	Serial No.	
	1 (Ay 1036)	2 (Ay 1517)
SiO_2	45.31	50.78
TiO_2	0.48	0.35
Al_2O_3	10.58	6.03
CaO	9.81	8.60
MgO	22.14	23.69
K_2O	0.09	0.04
CaO/Al_2O_3	0.93	1.40
$FeO/FeO+MgO$	0.31	0.28

data, interpreted according to current views, the eruption could have taken place for part of the volcano-sedimentary period in a transitional belt between the island arc area and the deep marine basin. All other geological characteristics, including the original structures of the lava, tuff, and other rock types, as well as their facies changes and petrographic features, reveal the frequent variation of their conditions of formation under the influence of the instability of the crust during this period. In fact, parts of the volcano-sedimentaries were formed under subaqueous, subaerial, and even lagoonal conditions. The top and bottom portions of the subaerial lava flows exhibit characteristic slender tubular vesicle-filling structures, and many thin subaqueous lava flows show elegant 'fusuline-shaped' vesicle-filling structures (Plate XIX, 6) throughout the whole thickness of the flow, and both are uniquely and beautifully preserved.

The upper Shancaoyu Formation, which lies comfortably on the middle Yanlingguan Formation, is up to 4000–5000 m thick and occupies the outer part of the south-western limb of the Xinfushan anticlinorium. It is composed mainly of biotite-granulitite and biotite-bearing granulitite, derived chiefly from silty rocks with a certain greywacke affinity (Table 4.2, III), locally mixed with basic volcanic clastics. It also contains plagioclase-amphibolite of basic igneous origin. Most of the rocks were formed in rather stable sedimentary conditions, and some of them have been influenced by migmatization and granitization in places, thus resulting in the formation of four roughly parallel rock belts (Fig. 18.4, 3, 4, 5–6, and 7). The biotite-granulitite practically free from the influence of migmatization shows a biotite K-Ar age of 2460 Ma.

The regional metamorphism affecting the three formations described above terminated at the end of the Archaean. They were intruded by microcline-pegmatites at an earlier date of 2500 Ma (K-Ar age of muscovite) and a later one of 2250 Ma (also K-Ar age of muscovite), accompanied by muscovitization and tourmalinization of their country rocks.

While the rocks of the Yanlingguan Formation were chiefly the products of an unstable basic volcano-sedimentary period, those of the Sancaoyu were formed during a more stable, predominantly sedimentary period, with only weak basic volcanism at the beginning. Judging by the rock types and their petrological features and their geological setting, the rocks of these two formations show some resemblance to those of the greenstone belt sequences of certain Archaean regions in other countries, yet differ from these by their higher degree of metamorphism and the absence of a part of the 'hypothetical stratigraphic succession' of Windley (Windley 1977, p. 25).

Characteristics of metamorphism

Most of the metamorphic rocks of the Yanlingguan and Shancaoyu Formations belong to the amphibolite facies of the low to moderate pressure series, but some have been retrograded to the greenschist. As in other Archaean metamorphic terrains, the regional metamorphism of this area does not exhibit evident zonation. The mineral associations of the different facies of the main rock types are shown in Table 18.2.

The association of metamorphic minerals under similar $P-T$ conditions are obviously controlled by the chemical and mineral compositions of the original rocks. This is especially true for clastic rocks of partly volcano-sedimentary and partly sedimentary parentage, and it is natural that certain particular rock types are only formed in certain stratigraphic (lithological) horizons, such as the laminated to thin-banded cum-

Table 18.2. Mineral associations of the main rock types of the Yanlingguan and Shancaoyu formations in different low–moderate pressure mineral facies series.

Rock types	Mineral facies		
	Greenschist facies (partly of retrogressive and partly of alteration nature)	Low amphibolite facies[a]	High amphibolite facies
Ultrabasic rocks: talc-tremolite-schist, actinolite-schist, tremolite-schist, etc.		1. Tremolite + serpentine ± chlorite + talc. 2. Tremolite ± chlorite + actinolite. 3. Actinolite ± tremolite.	
Basic and related rocks: plagioclase-amphibolite, amphibolite, hornblende-schist, almandine-amphibolite, cummingtonite-granulitite, hornblende-granulitite, etc.	1. Actinolite + tremolite ± chlorite. 2. Actinolite + albite. 3. Actinolite + epidote (zoisite)	Hornblende + plagioclase (An < 35) + epidote + quartz.	1. Hornblende + plagioclase (An < 35) ± diopside (scanty) + almandine 2. Cummingtonite (anthophyllite) + plagioclase + quartz.
Semipelitic rocks (mostly of greywacke character): biotite-granulitite and mica-schist.		1. Biotite + plagioclase + quartz. 2. Biotite + staurolite + almandine + quartz. 3. Biotite + plagioclase + epidote + quartz.	3. Biotite + plagioclase + almandine + quartz.
(Pelitic rocks: (andalusite)–mica–quartz-schist)	(Sericite pseudomorph after andalusite + mica + quartz)	(1. Biotite + andalusite + quartz. 2. Muscovite + andalusite + quartz + staurolite)	(1. Biotite + quartz + plagioclase.)
Calc-silicate rocks: diopside-, zoisite-, garnet-, hornblende- and vesuvianite-bearing rocks.		Zoisite + epidote + hornblende + calcite.	Diopside + vesuvianite + calcite.

[a] Including rocks of the epidote–amphibolite facies (after Cheng et al. 1977, 1982b).

mingtonite-granulitite derived from a certain type of tuffaceous sandstone, the laminated to thin-banded almandine-quartz-amphibolite, probably derived from tuffaceous sandstone with much basic crystalline tuff, and staurolite-bearing andalusite–muscovite–quartz-schist of dominantly pelitic parentage. Almandine-bearing and almandine-free biotite-granulitite (and plagioclase-amphibolite) are usually directly in contact and interbedded.

While some of these rocks had suffered thermally metamorphic transformation adjacent to basic and ultrabasic intrusions prior to regional metamorphism, others were subjected to later migmatization, especially during the hydrothermal phase. Some of the regional metamorphic rocks were further influenced by dynamic metamorphism along certain north-western fractures and close folding belts, and retrograded. For instance, some plagioclase-amphibolite may thus be transformed into chlorite-, actinolite-, epidote- or tremolite-bearing types.

Migmatization zones and migmatization

Both the Renjiazhuang and Shancaoyu Formations have suffered various degrees of moderate to intensive migmatization, resulting in the formation of five distinct parallel zones of migmatitic rocks of different widths, with or without intervening zones of metamorphic rocks that are practically free from migmatization or are only slightly migmatized. From the north-eastern part of the district south-westward, they occur successively as follows:

1. Zone of migmatitic granite and homogeneous migmatite. It consists of the following two rock types:
   (1) Xinfushan biotite-bearing and microcline-bearing, oligoclase-rich migmatitic granite (Fig. 18.4, 2), occurring mainly at the axial part of the anticlinorium. Its Na_2O content is much greater than K_2O (Table 18.3, 2).
   (2) Biotite-oligoclase homogeneous migmatite (Fig. 18.4, Ara), containing more Na_2O and less K_2O (Table 18.3, 1) than the migmatitic granite.
2. Zone of Cishan porphyroblastic biotite-microcline migmatitic granite (Fig. 18.4, 3). Its north-eastern border shows a distinct contact with the neighbouring biotite-granulitite of the Shancaoyu Formation, probably due to very local and partial 'rheomorphic' migration of the magma. It contains belts of partly to highly migmatized Shancaoyu granulitite (Fig. 18.4, Ard) showing transitional relationships in the interior, and many thin parallel layers of migmatites of biotite-granulitite parentage and screens of biotite-granulitite of various degrees of migmatization

Table 18.3. *Chemical composition of migmatitic granite and homogeneous migmatite of the Yanlingguan region, Xintai (weight %) (after Cheng* et al. *1977).*

	Serial No.				
	1	2	3	4	5
SiO_2	70.61	71.56	71.84	71.34	71.27
TiO_2	0.14	0.20	0.20	0.33	0.25
Al_2O_3	15.03	14.91	14.63	14.32	14.25
Fe_2O_3	0.36	0.95	0.42	0.82	1.24
FeO	1.66	1.49	1.71	1.71	1.62
MnO	0.04	0.02	0.04	0.01	0.08
MgO	1.15	0.17	0.67	0.71	0.80
CaO	2.29	1.62	1.54	1.50	1.60
Na_2O	5.50	5.00	3.76	3.52	3.79
K_2O	2.17	3.40	4.86	4.89	4.03
H_2O^+	0.31	0.26	0.48	0.38	0.56
H_2O^-	0.05	0.10	0.11	0.14	—
P_2O_5	0.15	—	0.079	0.085	0.16
CO_2	0.26	0.18	—	0.135	0.33
Total	99.72	99.86	100.339	99.89	—

1, Biotite-oligoclase homogeneous migmatite (Renjiazhuang Formation), average of two samples; 2, Biotite-bearing, oligoclase-rich migmatitic granite (Renjiazhuang Formation); 3, Cishan biotite-microcline-migmatitic granite, average of eight samples; 4, Moshishan fine granular biotite-feldspar-homogeneous migmatite with quartzo-feldspathic aggregates, average of two samples; 5, Average of 221 granite samples of China (Li Tong *et al.* 1963).

in the south-western marginal zone. These are indicative of the essentially autochthonous nature of the migmatitic granite. Petrochemically it differs from the Xinfushan migmatitic granite chiefly by the dominance of K_2O over Na_2O (Table 18.3, 3), but shows great resemblance to the average composition of granites of various ages in China (Table 18.3, 5; Li Tong et al. 1963).

3. Moshshan fine-granular biotite-feldspar homogeneous migmatite with quartz-feldspathic aggregates (Fig. 18.4, 4). It usually passes gradationally across the strike either to the Cishan granite on the north-east or to the Shancaoyu granulitite on the south-west, with transitional rock types. Its chemical composition (Table 18.3, 4) is very similar to that of the Cishan migmatitic granite.

4. Zone of homogeneous migmatite and partly migmatized rocks. It consists of the following two subbelts:
   (1) Partly migmatized belt of fine- to medium-grained biotite-granulitite (Fig. 18.4, 5).
   (2) Belt of medium- to coarse-grained biotite-plagioclase-migmatite and biotite-hornblende-plagioclase-homogeneous migmatite (Fig. 18.4, 6).

5. Partly migmatized belt of biotite-microcline-augen migmatite and biotite-granulitite (Fig. 18.4, 7).

Although all of the above five zones have their individual characteristics of migmatization, they seem to belong to two types. The first is represented by zones 1 and 4, formed by Na-rich alkaline metasomatism of the Renjiazhuang Formation and the Shancaoyu Formation respectively, with a K–Ar age of c. 2000 Ma and Rb–Sr age of c. 2586 Ma. The other three zones belong to the second type which is characterized by K-rich alkaline metasomatism showing a K–Ar age of c. 2250 Ma (Cheng et al. 1972, 1973; Cheng et al. 1977). In the formation of homogeneous migmatite and migmatitic granite, there is in general an influx of Si, Na and K, and outflow of Fe, Mg, Ca, and Al, accompanied by a series of alternating Na- and K-metasomatisms. Furthermore, the process of partial or very slight migmatization is usually marked by local internal migration of substances within the rock, through diffusion or circulation of fluids, although trifling additions of Si and Na, or K, and slight outflow of Fe, Mg, and Al may also take place.

LOCALITIES OF GLAUCOPHANE-SCHIST AND RELATED ROCKS AND PAIRED METAMORPHIC BELTS

Miyashiro's concept of metamorphic facies series and paired metamorphic belts (Miyashiro 1961, 1972) has aroused great interest among Chinese geologists. During the last decade or so there have been more discoveries of metamorphic rocks of the high-pressure type, chiefly glaucophane-schist and related rocks, and much discussion of the subject of the occurrence of paired metamorphic belts and its bearing on the problem of plate tectonics.

Present knowledge shows that the localities of the high pressure glaucophane-schist and related rocks lie mainly in four belts.

(1) Tianshan–Nei Mongol belt. Here the metamorphic rocks are mostly of Palaeozoic age. For instance, according to the recent investigation by Wang Suying and the Geological Survey Team of Xinjiang, the glaucophane-schist localities discovered in 1959 south of Tekesi in the western part of the Central Tianshan Mountains, are found in a roughly W–E zone of dominantly Upper Silurian formations metamorphosed during the Late Palaeozoic, extending discontinuously for about 180 km. The mineral associations are phengite–epidote–glaucophane–(albite), almandine–epidote–glaucophane–(albite), chlorite–albite–glaucophane, etc. The glaucophane-schist found in the Lower Palaeozoic Wendu'ermiao Group (see above) is probably of dominantly Caledonian metamorphic age. It is composed mostly of spilitic volcano-sedimentary protolithic rocks. It lies in Central Nei Mongol along the railway between Beijing and Ulan Batu. The mineral associations so far identified and mentioned in Hu Rao's paper (1981) are glaucophane(crossite)–chloritoid–epidote–chlorite–albite–stilpnomelane, glaucophane (or crossite)–sericite–quartz–epidote(chlorite)–calcite, crossite–calcite–(quartz–dolomite), crossite–quartz–magnetite(hematite)–sericite–chlorite, etc. According to Yan Zhuyun and Teng Kedong (1982), lawsonite and phengite are also found in some rocks. The Heige'aola glaucophanic schist (see above), which contains magnesio-riebeckite and crossite, is situated more than 300 km to the north-east, being derived from Late Palaeozoic ultrabasic and basic rocks, chiefly of Early Hercynian metamorphic age. The occurrence of pumpellyite–glaucophane–schist in the Suhuhe Group (O_1–O_2) in the Da Hing'an Mountains, as reported by Mo Youzhen in 1980, is of Caledonian metamorphic age. However, the rocks discovered by Wang Wuyan, Xiong Jibin, and others, south-west of Aksu on the southern slope of the western Tianshan Mountains, more than 200 km south-west of the Tekesi locality, belong to the Pre-Sinian, Upper Proterozoic Aksu Group. They are of Late Proterozoic metamorphic age, and the oldest of their kind in China. The mineral associations as worked out by Xiong and Wang are albite–epidote–glaucophane(crossite), stilpnomelane–glaucophane–epidote–chlorite, chlorite–epidote–glau-

cophane, and stilpnomelane–albite–muscovite (2Ml type)–quartz.

(2) Qilian–Qinling–Dabie belt. The metamorphic rocks are of different ages. For example, the glaucophane-schist of the middle segment of the northern Qilian Mountains (see above) is of Caledonian metamorphic age (Xiao et al. 1978; Wu 1980, 1982). Two tectonic sheets (Wu 1982) of ancient oceanic crust are found in the Qingshuigou locality. Serpentinites and glaucophane-bearing eclogites (C type) occur in the lower sheet, and glaucophane-schists and greenschists in the upper. The mineral associations identified in the schist are glaucophane–epidote–chlorite–phengite–chlorite–(quartz), glaucophane–chlorite–phengite–quartz–(stilpnomelane–albite), garnet–epidote–glaucophane and glaucophane–quartz. The common mineral assemblage for the eclogite is glaucophane-garnet-omphacite, often associated with chloritoid, phengite, etc. Crossite, barroisite or magnesio-riebeckite may also occur. The garnet formed at the early stage is Fe^{2+}- and Mn-rich, while at the late stage it is Mg- and Ca-rich (Wu 1982).

The rocks of the eastern Qinling occur in two belts over 100 km long in the eastern part of an Indosinian fold zone covering the borderland of Shaanxi, Henan, and Hubei (Wu 1980). The minerals derived from different types of rocks, including basic volcanics, volcano-sedimentaries, acid volcanics and iron-rich carbonate rocks, are albite, crossite, magnesio-riebeckite, and also glaucophane, phengite, and aegirine. They were probably formed in the Early Mesozoic under lower pressure and higher temperature than the Qilian rocks. The occurrence of phengite (both 2M and 3T types) in the mica-schists and other rocks, and of eclogite (C type) near Xinyang, south-east of the eastern Qinling locality, has been regarded by Ye Danian et al. (1980) as an example of high pressure, low temperature metamorphism, probably of Hercynian age and genetically related to plate tectonics.

There are also reported findings of Precambrian rocks of glaucophane-schist affinity in the Dabie Mountains.

(3) Yarlung Zangbo–West Yunnan belt. The rocks are of either Mesozoic or Tertiary ages. In the broad metamorphic terrain of dominantly Himalayan (and subordinately Yanshanian) metamorphic age, there is a high pressure, low temperature metamorphic belt, nearly 350 km long and about 1000 m wide, immediately south of the western Xigaze ophiolite suite (see above) of the southern Yarlung Zangbo Suture Zone (see below). It extends from the Daji Mountain in the west, via Zangsang, to Zisong (south-west of Xigaze city) in the east. The constituent rock types are mainly stilpnomelane-greenschist, glaucophanic amphibole-stilpnomelane-greenschist, schists containing amphiboles of glaucophanic composition, etc. (Xiao and Gao 1982). The mineral associations recognized are stilpnomelane–chlorite–epidote–aragonite(?), glaucophanic amphibole–stilpnomelane–chlorite–sericite–quartz and glaucophanic amphibole–chlorite–actinolite–albite. The glaucophanic amphiboles include the type intermediate between crossite and tremolite-glaucophane (winchite), that between crossite and magnesio-riebeckite and that between glaucophane and tremolite, probably barroisite. Pumpellyite has been found in certain rocks. The metamorphic rocks of Mesozoic-Cainozoic metamorphic age, derived mostly from the volcanic rocks of the Sangri Group—located to the north of the ophiolite belt just noted and actually in the northern Yarlong Zangbo Suture Zone—occur chiefly within part of the Gangdise magmatic zone (Chapter 17). They are characterized by a high- to moderate-temperature, low-pressure mineral assemblage including andalusite, cordierite, biotite, etc. These rocks and the high pressure belt mentioned above are considered to be integral parts of paired metamorphic belts of Late Mesozoic to Tertiary age genetically related to the Yarlung Zangbo Suture Zone (*Guide to Yarlong Zangbo geological excursion* 1982).

As has already been stated previously, the glaucophane-schist and its high temperature partner of the paired metamorphic belts of the Lancangjiang (Yunnan) metamorphic terrain, are of essentially Indosinian metamorphic age. The glaucophanic schist of the Ailaoshan metamorphic belt to the east, also in Yunnan, is of similar metamorphic age and is of very limited distribution. The glaucophanic mineral is an intermediate type between crossite and magnesio-riebeckite.

(4) Yuli belt of Taiwan. A short description of the Yuli glaucophane-schist belt has already been given in a previous section on 'metamorphic series and belts chiefly of Yanshanian metamorphic age'. While T. P. Yen (1963) proposed that the Yuli and the high temperature Tailuko belts formed the paired metamorphic belts of Middle Mesozoic age (Yen 1981), Bor-ming Jahn (Jahn 1974) and the Regional Geological Survey Team of the Geological Bureau of Fujian (1978, 1982) suggested pairing the Yuli belt with the batholiths of south-eastern China (90–109 Ma), and the high temperature, low pressure Yanshanian metamorphic belt (97–103 Ma) along the Fujian coast respectively.

TYPES OF MIGMATIZATION AND EXAMPLES

The Chinese term for migmatite, 'hun-he-yan', which

literally means 'mixed rock', was introduced into Chinese geology in 1937 by Cheng Yuqi in a note entitled 'Comments on a new classification of rocks' in Volume 2 of the *Geological Review*, published by the Geological Society of China. The related terms, 'regional migmatization' and 'marginal migmatization', correspond essentially to the migmatizing processes for the formation of H. H. Read's 'regional migmatites' and 'contact migmatites' (Read and Watson 1962). They were also proposed by Cheng, and have been widely used in China since 1963 (Cheng et al. 1963). The granitization related to regional migmatization was further classified into 'geosynclinal' and 'basement' types by Cheng et al. (1963), the latter being restricted chiefly to Early Precambrian terrains. A third type of migmatization, which occurs along fracture zones of various geological ages, has been named 'fracture zone migmatization' (Cheng and Zhang 1982c).

The regional migmatization, caused by anatexis and/or alkaline metasomatism, is well developed in the Precambrian basement, especially in Archaean metamorphic terrains (see above), such as the Anshan–Benxi region of Liaoning (Anshan Group), Central and West Shandong (Taishan Group), the Yanshan region of eastern Hebei (Qianxi and Badaohe Groups), the Wutai–Taihang region (Fuping Group), and the northern slope of the eastern Qinling Mountains (Dengfeng and Taihua Groups). They are characterized by the formation of a great variety of migmatites, including agmatite (Plate XIX.1), banded migmatite (Plate XIX.4), streaky migmatite (Plate XIX.5), augen-migmatite, dictyonite, nebulite (Plate XIX.3), etc., and migmatitic granites. Some of the rocks of these regions are strongly sheared and strained and further migmatized (Plate XVIII.5). There appear altogether four stages of regional migmatization in one or other of the above regions during the Archaean and Early Proterozoic, the approximate dates being c. 3000, 2500, 2230, and 1800–1900 Ma (Cheng et al. 1973, 1982a). Migmatization also occurs in the medium- to high-grade metamorphic axial part of some of the geosynclinal fold zones in the mobile belts of Palaeozoic and later ages, such as the Wugong Mountain in West Jiangxi (Caledonian) and the Altay Mountains in northern Xingjiang (Hercynian).

Marginal migmatization appears on a local scale with a width of tens to hundreds of metres just outside the margin of a variety of granitoid bodies with different evolutional histories and geological ages. These bodies are generally absent in the stable area, and usually occur in the peripheral parts of the mobile belts or in the relatively mobile regions. While this phenomenon is more often seen around Palaeozoic and younger granitoids in appropriate geological settings, there are also some examples with Proterozoic and even Archaean ages, such as that occurring in the Fuping Group of the Wutai Mountain (see above). Migmatites of the marginal migmatization zones are produced from granitoids, as a result of the metasomatic activity of related fluids, mostly of alkaline nature. An example is the 'marginal migmatite' developed along the northern and southern flanks of the Early Yanshanian Fangshan granodiorite,* and the two-stage Yanshanian metamorphism of its country rocks has already been described in a previous section of this chapter. Guo Huqi's observations on this marginal migmatite are worth quoting as follows.†

Such gneissic and related rocks are 'marginal migmatite' and related rocks formed along the periphery of the Late Mesozoic Fangshan intrusion, shortly after being regionally metamorphosed by the Mesozoic Yanshanian Orogeny, from the partly ferriferous, calcareous pelitic rocks of the Xiamaling Formation of the Late Proterozoic Qingbaikou System.

The marginal quartz-diorite zone of the main porphyritic granodiorite with large plagioclase 'phenocrysts' passes gradually outward into a belt of assimilation-contamination rocks 30–150 m wide, which in turn is transitional to a zone of marginal migmatite 100–200 m in width. This last zone also shows a gradational contact with the further outward spreading Xiamaling Formation composed of hornblende- and/or biotite-bearing granulitite and even phyllite, which have already been partly affected by very weak migmatization.

The assimilation–contamination zone consists mainly of a partly gneissic biotite–quartz–dioritic contaminated rock which contains up to about 50 per cent slightly transformed biotite-hornblende-granulitite enclaves showing variable shapes [Plate XIX.7]. The marginal migmatite zone may be subdivided according to its degree of migmatization into two sub-zones, i.e. an inner zone of migmatite and an outer one of partly and slightly migmatized granulitite. The former consists chiefly of partly slightly streaky homogeneous migmatite and streaky migmatite, with a variable amount of roughly interbedded (biotite)–plagioclase–amphibolite and biotite–hornblende-granulitite relict bands or masses with their strikes and foliation almost parallel to those of the migmatite, and also in the main parallel to those of the granulitite and phyllite of the outward lying Xiamaling Formation, the boundary of the granodiorite intrusion, and the longer axes of the rock inclusions of the assimilation–contamination and quartz-diorite zones. In addition to some plagioclase-amphibolite, these relict bands and enclaves are composed mostly of (biotite)-hornblende-granulititic types petrographically similar to those of the outward lying zone.

... Rock inclusions in different portions and zones of this

---

*Based chiefly on Guo Huqi's paper 'Petrological characteristics and origin of the gneissic rocks along the northern flank of the Fangshan granodiorite, Beijing' (in Chinese, with English summary), M.Sc. thesis, Chinese Academy of Geological Sciences 1981, and also on field observations of Cheng, Liu Guohui and Wu Jiashan.

†Quotations from the English summary by kind permission of Guo Huqi.

intrusion were basically xenoliths of metamorphosed country rocks further assimilated by the ascending magma, and metasomatism played an important role in the mechanism of assimilation. Hence the transformation process which these xenoliths had undergone was very similar in many respects to that of the relict bands and masses of the migmatite zone. It is to be further noted that hornblende in the slices of quartz-diorite is frequently found as xenomorphic grains many of which have already developed into rather irregular 'phenocrysts' containing minute grains of plagioclase and quartz, and exhibiting a sieved texture [Plate XIX.9]. They could not have crystallized directly from the cooling magma but were formed by the further development of the highly assimilated calc-ferromagnesian substance of the relics of the 'captured' country rock, being the products of contamination of the magma by such assimilated rocks. In other words, they may be interpreted as products of a fairly advanced stage of assimilation–contamination yet still prior to the final homogeneity, as a result of the interraction of the magma and the country rocks under the conditions transitional from metamorphism to magmatism.

... Different types of migmatite and partly migmatized granulitite exhibit in thin sections certain phenomena of K- and Na-metasomatism related to the activity of certain postmagmatic fluids. In general, there was outflow of Ca, Mg, Fe, Ti and influx of K and Si with inconsistent amount of interchange of Na and Al during the metasomatism.

Plagiocase porphyroblasts of metasomatic origin found in the rocks of the marginal migmatite zone often show different types of zoned texture which varies with the degree of migmatization. These, in general, reflect the shorter duration and more rapid decrease in temperature of the Mesozoic marginal migmatization of this region, which thus differs markedly from the regional migmatization of ancient metamorphic terrains characterized by a much longer duration of higher P–T conditions so as to render the homogenization of the newly formed plagioclase devoid of zoned texture.

It is evident from the details of their field occurrence and mineralogical, petrographical, and petrological characters that the migmatized rocks and migmatites are the products of the corresponding protolithic rock types through a series of episodes of metasomatism. So far no phenomenon of re-fusion has been observed.

The brief history of geological evolution of this region during the Yanshanian Orogeny may be roughly subdivided into three stages. At the beginning, the rock formations were subjected to the Yanshanian tectonism and the accompanying regional metamorphism of rather limited areal distribution which began in the Jurassic, with the production of greenschist facies rocks and the gradual birth of the Fangshan dome-anticline and Nandazhai Fault and other fractures. This was succeeded by the ascending of granodioritic magma to form the proto-Fangshan intrusion, resulting in its contact with the upper succession of the Late Precambrian formations, which was upgraded to epidote-amphibolite and amphibolite facies metamorphic rocks under rather high P–T conditions, and was partly assimilated, accompanied by contamination of the magma by the assimilated substance. The belt of assimilation–contamination now preserved near the margin of the intrusion is part of the relic products of such complicated processes. Still later at the declining stage of intrusive activity, when the already metamorphosed country rocks just outside the assimilation–contamination belt were replaced by the alkaline fluids derived from the intrusion through intergranular percolation under the regional stress almost persistent ever since the beginning of the intrusion, the zone of marginal migmatite with planary structures inherited from the original metamorphic rocks was finally formed after a series of metasomatic transformations.

It seems that Guo has not only studied in some detail the characteristics of the migmatites and related rocks of the marginal migmatization zone of the Fangshan granodiorite bodies and the geological setting for their formation, but has also traced back to a certain extent the genetic relationship between the assimilation–contamination process related to the emplacement of the intrusion and the marginal migmatization. This may help us to understand further the cause, background, and mechanism of marginal migmatization in general, though there are still unresolved problems.

The occurrence of migmatization along certain fractures or fracture zones in China is in the main an example of distinct directional control of migmatization and thus of definite linear extension. It may be related to the ascending heat flow along such weak zones in the crust causing local re-fusion of the pre-existing rocks at particular depths and/or of magma (migma) and related fluids of various origins. It is often accompanied by strong dynamic metamorphism and may be of polymetamorphic nature, for some of the fracture zones were continuously active for a considerable time. Examples of this type already noted in previous sections of this chapter and in Chapter 17, are the pegmatization zones at the periphery of the Archaean Fuping dome region, most of the Caledonian migmatization zones of the Yunkai Mountain of the Guangxi–Guangdong border, and of the Wuyi Mountains along the Fujian–Jiangxi border, those occurring along the long fracture zones at the south-western flank of the Hercynian metamorphic terrain of the Altay Mountains, Xinjiang, the polymetamorphic zones of dominantly Mesozoic metamorphic age of western Yunnan, and the eastern sub-zone of the metamorphic belt along the Fujian coast of dominantly Yanshanian age. Some of these are considered to be genetically related to plate tectonics. It is apparent that the exact conditions for the formation of fracture zone migmatization differ to some extent in different places. Actually there are more examples of this category of Palaeozoic and/or Mesozoic age in the southern part of China, as in the complicated metamorphic phenomena observed along the Shiwandashan–Qinjiang Fracture in Guangxi and the Wuchuan–Sihui Fracture in Guangdong, that were first studied by Mo Zhusun and others in 1962.

PART IV
GEOTECTONIC DEVELOPMENT OF CHINA

PART IV

OUR ENERGETIC VIOLENT UNIVERSE

19. THE TECTONIC FRAMEWORK AND THE GEOTECTONIC UNITS

Wang Hongzhen

INTRODUCTION

The first comprehensive treatise on the geology and tectonics of China was given by the late Professor Li Siguang (J. S. Lee), whose *Geology of China* was published in 1939. An excellent analysis and summary of the tectonics of China by Professor Huang Jiqing (T. K. Huang) appeared in 1945. After the foundation of the Peoples' Republic, Li developed his concept of tectonic systems and established the well-known school of geomechanics. Huang continued to investigate his idea of the polycyclic tectonic evolution of China. Meanwhile, Zhang Wenyou (1958, 1974) has sought to interpret the tectonics of China in terms of fault blocks. Both Huang and Zhang have compiled geotectonic maps of China based on their views (1958, 1960).

In recent years Yin Zanxum, Li Chunyu and other Chinese geologists have paid considerable attention to the theories of plate tectonics and sea floor spreading, and have adopted these ideas to elucidate the tectonic characteristics of China and of Asia. The main ideas and methods of analysis discussed here involve the concept of 'Mobilism' with respect to global tectonics and of 'Development Stages' with respect to crustal evolution (Wang 1981, 1982a). The concept of mobilism states that, as a result of horizontal movement of the lithosphere on the underlying asthenosphere, the position of continents and seas relative to the earth's axis, and that of the continents relative to each other, have undergone important changes during geological time. The idea of 'Development Stages' proposes that crustal development constitutes a continuous and progressive process, but that the main changes are accomplished by stages through rapid, concentrated alterations within comparatively short time intervals. The tectonic characteristics may show a relative stability within a particular stage, but they differ from each other in different stages. There is no single tectonic model that could apply to all geological periods.

Various kinds of evidence are available for analysis. Stratigraphy provides information on sedimentation type, which is a reflection of sedimentary environment as well as of tectonic setting. Palaeobiogeographic and palaeogeomagnetic evidence of the juxtaposition of widely different biota and diverse pole positions of an area in different periods is an indication of continental separation and convergence through time. Geochronology and magmatic events provide data to mark the time intervals between different tectonic stages. Indeed, the idea of mobilism may be seen as an important improvement on the classical geosynclinal theory, and the concept of development stages may prove to be a useful supplement to the theory of plate tectonics. The following discussion attempts to analyse the tectonic make-up of China in the light of these concepts, and on the basis of data already presented.

It has been pointed out (Wang 1981) that the continental margin is not a single boundary line, but encompasses a complicated tract of transitional crustal type, of variable width, and may include different kinds of island arcs and marginal seas (Fig. 19.1). Two types of island arcs may be discerned, the mature type containing ancient sialic massifs that may represent detached parts of a continent, and the juvenile type probably having its origin in subduction and superposition of oceanic crustal parts (Fig. 19.1). The former is represented by the present Japanese islands, and the latter exemplified by the Tonga and Fiji arcs. The island arcs may occupy wide regions and be separated by inter-arc basins, as in the West Pacific today and probably also in ancient times, e.g. in South China in the Late Proterozoic (see below).

The inner marginal seas are often devoid of volcanic activity and receive land-derived sediments, and are thus 'miogeosynclinal' in nature. In the inter-arc basins and on the arc-front slopes, the sediments deposited are autochthonous and volcanism is common. They are thus of 'eugeosynclinal' type. The relationship between different crustal types, and the disposition of different kinds of island arcs and marginal seas, are illustrated schematically in Fig. 19.1.

GEOTECTONIC UNITS OF CHINA

Terminology

The heterogeneity and regional differentiation of the

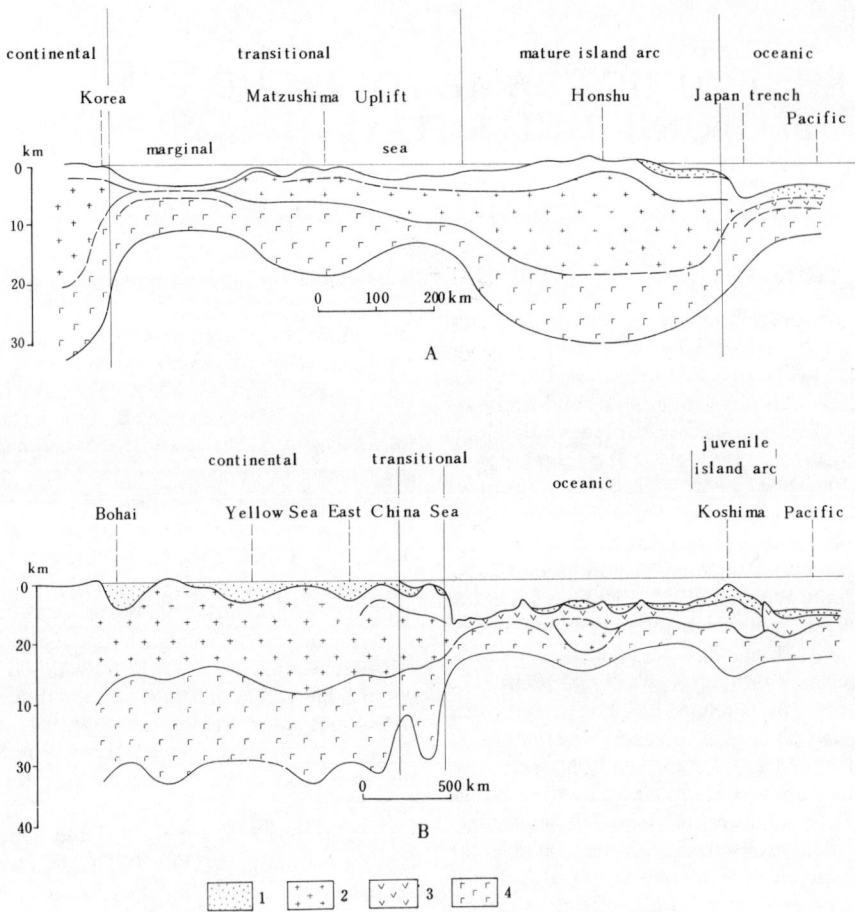

Fig. 19.1. Ideal sections through the Sea of Japan (A) and the East China Sea (B) showing crustal types, island arcs and marginal seas (simplified after The Second Team of Marine Geology, Ministry of Geology and Mineral Resources 1980). 1, Cainozoic sediments; 2, upper (granitic) layer; 3, volcanics; 4, lower (gabbroic) layer.

earth's crust is universal, and the first rank of differentiation is manifested by the difference between the continental and the oceanic type of crust. The transitional type is mainly distributed in a belt of variable width situated between the continental and the oceanic crustal sectors, and roughly corresponds to the continental margin tracts. It includes the island arcs and marginal seas, the essential characteristics of which consist in their high tectonic activity and the variable thickness of the crust.

The present continents are composed of extensive stable regions which are separated or surrounded by mobile fold belts. The stable regions correspond to platforms consolidated in the Precambrian, while the fold belts may include median massifs of Precambrian basement and fold zones that were consolidated in different stages in the Phanerozoic. If we call the Precambrian platforms cratonic or continental, then the fold belts subsequently formed between the continental massifs may be denominated intercontinental. The mobile regions are in fact composed of two, or more than two, ancient continental margins. Thus within the scope of the present continents there may be distinguished two kinds of tectonic regions, the continental and the continental marginal.

In view of the concept of mobilism in geological history, one of the principal aims of palaeotectonic analysis is to locate the ancient continental margins, the original sites of the island arcs, and the crustal consumption zones from which marginal seas, and even open seas, of different ages have disappeared. The positive vestiges of extinct sea areas are the ophiolite suites that represent the oceanic sedimentary associations preserved along the crustal consumption zones. The transformation of continental margins into fold belts actually represents the very process of continental accretion. If we combine the ideas of crustal consumption and continental accretion, we may distinguish several kinds of principal boundary lines within the scope of a continental margin tract. Thus the lines of subduction may be called subduction zones before the oceanic crust is completely consumed and the marginal seas entirely filled up and folded (Fig. 19.2, SZ). When the marginal seas are entirely elevated and converted into fold zones, the original subduction zones will become a vestige of crustal consumption within the continental margin tract. Such boundary lines may be formed in successive zones as the continental margin extends towards the ocean side (Fig. 19.2, AC 1–2, AC 1–2), or may be superposed upon each other over roughly the same site. As they are the products of crustal consumption and of continental accretion within a continental margin, they are here named accretional consumption zones. On the other hand, the open seas between two opposite and remote continental margins may also be entirely consumed, and the final extinction of the marine realm will lead to the coalescence and convergence of the two opposite continents. As the line of convergence and collision is the vestige of crustal consumption between two converging continents, it is here called a convergent crustal consumption zone (Fig. 19.2 CC). Naturally accretional zones of different ages may be superimposed upon each other, and as a result of continental collision may also have the final convergent zone superimposed upon them. In any case, the distinction between convergent and accretional crustal consumption zones is of special interest and importance in understanding the palaeogeographic and palaeotectonic evolution of the continental margin. A diagrammatic sketch illustrating the development of continental margins and the formation of crustal consumption zones is given in Fig. 19.2.

Under conditions of predominantly tensional stress, faulted troughs may be developed in the interior of the platforms, as well as on the inner side of the platform margin uplifts. Such troughs are usually called aulacogens or rift zones, and the sedimentary type and thickness of strata formed in them often differ remarkably from the platform cover sequence. It has been pointed out that further rifting and opening will lead to the appearance of oceanic crust and the entire separation of the continental massifs. Therefore the troughs may be distinguished as miogeosynclinal or eugeosynclinal depending on whether magmatic activity is developed or not, but if development of the sea troughs is impeded no sediments will be found in surrounding regions, either of marginal or open sea type. When the troughs are closed up through compression, the result will be the formation of intra-platform fold zones.

The sedimentary development on the platforms is as a rule controlled by syndepositional faults, which often cut through the basement complex. When transcurrent faults of large displacement occur, they often form transcurrent fracture zones cutting across the whole

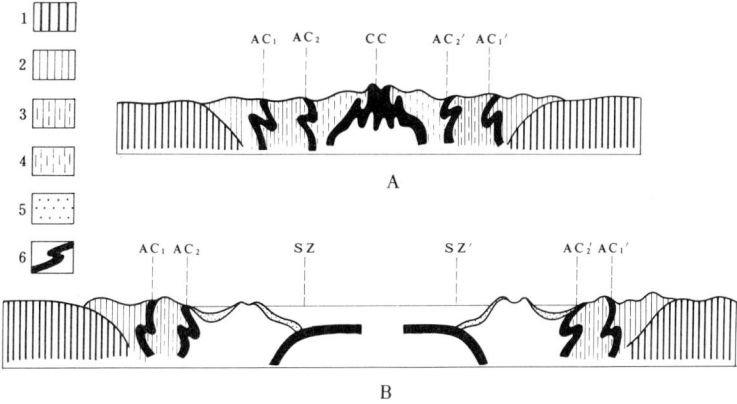

Fig. 19.2. Ideal profiles showing the development of continental margins and formation of crustal consumption zones. A, after convergence; B, before convergence. 1, ancient basement of platform; 2–4, fold zones of different stages; 5, marginal sea and island arc deposits; 6, oceanic crust. CC, convergent crustal consumption zone, AC1, AC2, AC1', AC2', accretional crustal consumption zones, SZ, SZ', subduction zones.

region. The formation of these fracture zones is the result of huge shearing stresses, and they usually cut through the crust and even the lithophere, presumably in episodes of profound tectonic change. Naturally they exert a strong influence on the formation and distribution of contemporary sedimentary types, and cause conspicuous displacement of all geological bodies existing prior to their formation.

The principal tectonic domains of China

After the general discussion of the continental margins and the crustal consumption zones, there follows a brief description of the basic tectonic frame and the main geotectonic units of China. The largest geotectonic unit on the earth's surface may be designated a tectonic domain. This may either be composed of one, or more than one, continental terrain with surrounding marginal tracts, or of an extensive and complicated continental margin terrain by itself. Five tectonic domains are recognized in China, three of which are continental in nature, North China, South China, and part of Gondwana; and two have the nature of continental margins, one connected with the Siberian–Mongolian in the north, and the other opening to the Western Pacific in the east. Within the territories of China, there are two outstanding convergent crustal consumption zones, each divided into two segments, that separate the three northern tectonic domains. From north to south, the first is the convergent zone between the Siberian–Mongolian continental margin in the north and the North China continental margin in the south. It is divisible into two segments, the western Aibi (Ebinur)–Juyan zone (Fig. 19.3,1) running roughly along the northern slope of North Tianshan to Beishan, and the eastern Suolun–Xilamulun (Har Moron) zone (Fig. 19.3,2) starting from Suolun Hill of Nei Mongol and crossing the border line on the Tumen River in eastern Jilin. The second is the convergent zone between the North China and South China continental domains, running mainly along the Kunlun–Qinling line. Its western segment, known as the Xiugou–Maxin convergent zone (Fig. 19.3,3), extends further westward, being truncated by the Altyn transcurrent fault near the Jilia Pass and crossing the border line on the northern slope of the Karakhorum. The eastern segment, the Shangyang–Tongcheng convergent zone (Fig. 19.3, 4), starts from southern Shaanxi and enters the Yellow Sea near Lianyun Harbour, being divided by the Tancheng–Lujiang transcurrent fault in eastern Anhui and northern Jiangsu.

In addition to the four tectonic domains that are separated by the convergent crustal consumption zones described above, there is the fifth, the East China Continental Margin Domain, which is a part of the giant circum-Pacific Mesozoic volcanic belt. The main part of it comprises the South China Caledonides and the Hercynian to Mesozoic fold belts stretching from the south-eastern coastal region to the Nadanhada Mountains in North-east China. In the Palaeozoic the boundary between it and the South China Continental Domain was located at the southern boundary of the Yangzi Platform. After the Indosinian Movement, when the East Asian continental margin became active through its interaction with the West Pacific one, this line migrated inland to the western slope of the Hingan, Taihang, and Xuefeng Mountains.

To sum up, the tectonic units of China consist of four laterally extended tectonic domains separated by convergent crustal consumption zones, the northern continental marginal, and the three southern continental ones, which are flanked by and partly superimposed by the fifth, the eastern continental marginal. The main geotectonic units and tectonic elements of China are illustrated in Fig. 19.3.

THE NORTH (SIBERIAN–MONGOLIAN) CONTINENTAL MARGIN DOMAIN (Fig. 19.3, I)

This is the northernmost tectonic domain of China and constitutes a part of the complicated continental margin tract on the southern side of the Siberian–Mongolian continent. As stated previously, the southern boundary of this domain is the convergent crustal consumption zone stretching from the Aibi Lake in North Tianshan, via Juyan Lake and Suolun Hill on the Sino–Mongolian border, to the Jingbo Lake in eastern Jilin. It is fundamentally a terrain of Palaeozoic folding and includes two Precambrian median massifs, the Junggar in the north-west and the Songliao in the north-east. Palaeozoic fold belts comprise the Ergun and Jiamusi of Xingkaian age, both in the north-east; the Nenjiang of Caledonian age; and the Transjunggar, Hingan–Nei Mongol and Zhangguangcailing of Hercynian age. They will be treated separately as follows.

The North-west region

The Junggar Massif is triangular in shape and is entirely covered by Mesozoic and Cainozoic terrestrial sediments. There are different opinions about the actual nature of the basement. Recent aeromagnetic surveying revealed the presence of a rigid foundation underneath the basin cover, which is consistent with the tectonic lineament of the close Lower Palaeozoic folds

THE TECTONIC FRAMEWORK AND THE GEOTECTONIC UNITS 241

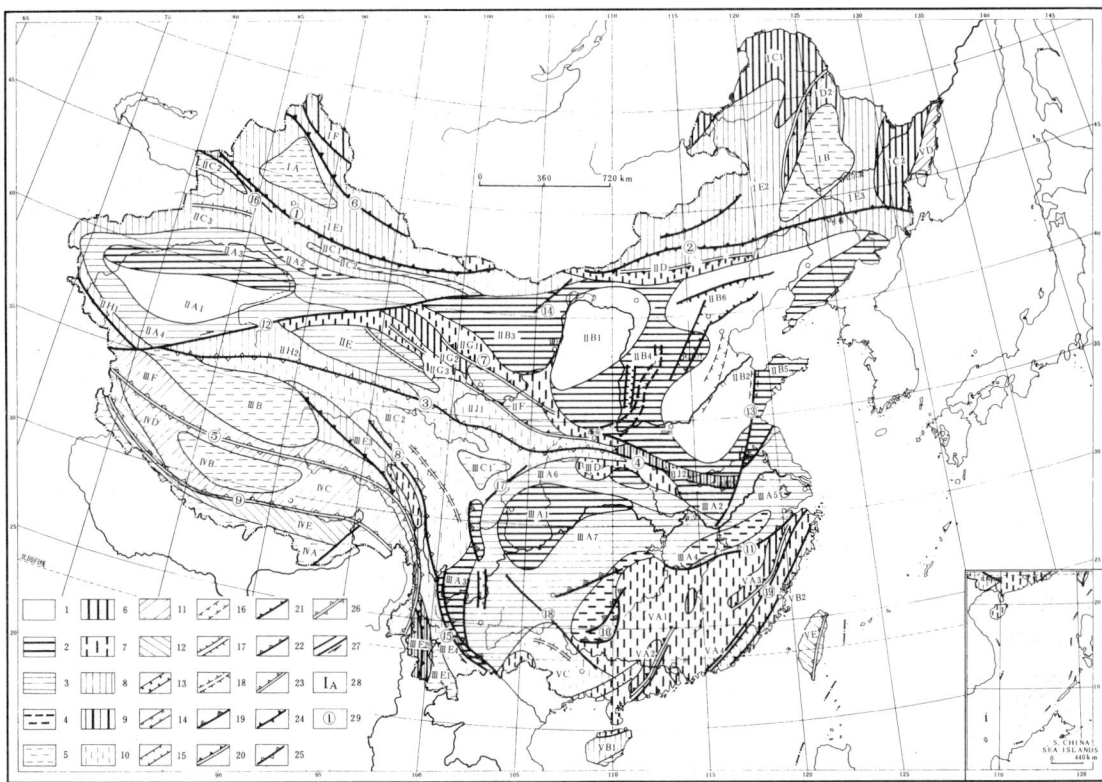

Fig. 19.3. A geotectonic outline map of China (modified after Wang 1981). 1, continental nuclei, consolidated 2500–2600 Ma; 2, consolidated 1850 Ma; 3, platforms, consolidated 850 Ma; 4, uplifted regions, consolidated 850 Ma; 5, median massifs and uplifts in fold regions, consolidated 800–600 Ma; 6, early Caledonian (Xingkaian) fold zones; 7, Middle and Late Caledonian fold zones; 8, Hercynian fold zones; 9, Palaeozoic fold zones not differentiated; 10, Indosinian fold zones; 11, Yanshanian fold zones; 12, Himalayan fold zones; 13, aulacogen, mainly Middle Proterozoic; 14, aulacogen, mainly Sinian or Ordovician–Silurian; 15, aulacogen, mainly Permian–Triassic; 16, aulacogen, mainly Cainozoic; 17, sea spreading zones, mainly Caledonian; 18, sea spreading zones, mainly Hercynian and Indosinian; 19, accretional crustal consumption zones, Jinningian and older; 20, accretional crustal consumption zones, mainly Caledonian; 21, accretional crustal consumption zones, mainly Hercynian; 22, accretional crustal consumption zones, mainly Indosinian; 23, accretional crustal consumption zones, Yanshanian and Himalayan; 24, convergent crustal consumption zones, Hercynian; 25, convergent crustal consumption zones, Indosinian; 26, convergent crustal consumption zones, Yanshanian; 27, depth faults and transcurrent faults; 28, number of geotectonic units; 29, number of crustal consumption zones and main transcurrent faults.

I, North (Siberian–Mongolian) Continental Margin Domain: IA, Junggar Massif; IB, Songliao Massif; IC, Xingkaian fold zones: IC1 Ergun, IC2 Jiamusi; ID, Caledonian fold zones: ID1 Altay, ID2 Nenjiang; IE, Hercynian fold zones: IE1 Transjunggar, IE2 Xingan–Nei Mongol, IE3 Zhangguangcailing. II, North China Continent and Continental Margin Domain: IIA, Tarim Platform: IIA1, South Tarim Nucleus; IIA2, Quruktagh Aulacogen; IIA3, North Tarim Depression; IIA4, Tieklik–Altyn Uplift. IIB, North China Platform: IIB1, Ordos Nucleus; IIB2, Ji-Lu Nucleus; IIB3, Alxa (Alashan) Massif; IIB4, Wutai–Zhongtiao Uplift; IIB5, Jiaoliao Uplift; IIB6, Yanshan Aulacogen. IIC–D, Northern continental margin tracts: IIC, Tianshan (Hercynian) Fold Belt: IIC1, North Tianshan Fold Zone; IIC2, Central Tianshan Uplift; IIC3, South Tianshan Fold Zone; IID, Southern Nei Mongol (Palaeozoic) Fold Zone. IIE–J, Southern continental margin tracts: IIE, Qaidam Massif; IIF, Lanzhou–Xining Massif; IIG, Qilian (Caledonian) Fold Belt: IIG1, North Qilian Fold Zone; IIG2, Central Qilian Uplift; IIG3, South Qilian Fold Zone; IIH, Kunlun (Hercynian) Fold Belt: IIH1, West Kunlun Fold Zone; IIH2, East Kunlun Fold Zone; IIJ, Qinling Fold Belt: IIJ1, West Qinling (Indosinian) Fold Zone; IIJ2, East Qinling (Palaeozoic) Fold Zone. III, South China Continent and Continental Margin Domain: IIIA, Yangzi platform: IIIA1, Central Sichuan Massif; IIIA2, Dabie Uplift; IIIA3, Kham–Yunnan Uplift; IIIA4, Jiangnan Uplift; IIIA5, Lower Yangzi Depression; IIIA6, Northern Border Depression; IIIA7, Upper Yangzi Depression; IIIB, Qiangtang Massif; IIIC–F, Continental margin tracts:

(*see over*)

IIIC, Songpan–Garze Fold Belt: IIIC1, Songpan Massif; IIIC2, Bayanhar–Shaluli (Indosinian) Fold Zone; IIID, Ankang–Tongbo (Caledonian) Fold Zone: IIIE, Qamdo–Simao Fold Belt: IIIE1, Lincang Massif; IIIE2, Ximeng (Palaeozoic) Fold Zone; IIIE3, Kaixinling (Hercynian) Fold Zone; IIIE4, Simao (Indosinian) Fold Zone; IIIF, Karakhorum (Yanshanian) Fold Zone. IV, South (Gondwana) Continent and Continental Margin Domain: IVA, Himalaya Massif; IVB, Gandise Massif; IVC, Lhasa–Nagqu (Yanshanian) Fold Zone; IVD, Ngali (Yanshanian) Fold Zone; IVE, Yarlung Zangbo (Himalayan) Fold Zone. V, East China (Circum-Pacific) Continental Margin Domain: VA, South-east China (Cathaysian) (Caledonian) Fold Belt: VA1, Xiang–Gui Fold Zone; VA2, Yunkai Uplift; VA3, Jian'ou Uplift; VA4, Min–Yue Fold Zone; VB, South-east Hercynian Fold Belt: VB1, Hainan Fold Zone; VB2, Fuding–Xiamen Fold Zone; VC, Youjiang (Indosinian) Fold Zone; VD, Nadanhada (Yanshanian) Fold Zone; VE, Taiwan (Himalayan) Fold Zone.

Convergent crustal consumption zones: 1, Aibi–Juyan; 2, Suolun–Xilamulun; 3, Xiugou–Maxin; 4, Shanyang–Tongcheng; 5, Bangong–Nujiang. Accretional crustal consumption zones: 6, Bayitik; 7, North Qilian; 8, Jinshajiang; 9, Yarlung Zangbo; 10, Sibao; 11, Yichun–Shaoxing. Transcurrent fault zones: 12, Altyn; 13, Tancheng–Lujiang; 14, Langshan; 15, Honghe (Red River). Depth faults: 16, Central Tianshan (North); 17, Longmen; 18, Zhaotong–Lingshan; 19, Lishui–Haifeng.

which surround the massif. Devonian and Lower Carboniferous geosynclinal deposits show a clear change to stable facies and a decrease in thickness towards the massif, especially in the northern part near the Ulungur River. Subsidence and terrestrial deposition began in the Permian when the surrounding regions were folded and elevated in the late Hercynian. In the Mesozoic and Cainozoic, the extent of the basin changed from time to time. The periods of most intensive subsidence and extensive deposition are Early to Middle Jurassic and Late Tertiary, while the Late Jurassic to Early Cretaceous and the Early Tertiary are marked by coarse red deposits of limited extent. The slopes of the basin floor are steep in the south and gentle in the north, and the centre of subsidence shows a westward migration along the southern border, from Qitai via Mannas to Usu, in the period from Triassic to Late Tertiary. The Mesozoic and Cainozoic deposits amount to 12 000 m near Mannas, including the prominent Pliocene and Pleistocene molasse deposits along the piedmont zone. It is also along the southern border of the basin that E–W trending folds and southward-dipping thrust faults are developed.

The Hercynian fold zones around the Junggar Massif between the Aibi–Juyan Convergent Crustal Consumption Zone and the Baytik (Ertix) Accretional Crustal Consumption Zone (Fig. 19.3 6) together occupy what may be called Transjunggar, which includes also the Bogda Mountain and part of Beishan. In the western part of this region, the Lower Palaeozoic begins with Lower Ordovician ophiolites, and the whole of the Ordovician and Silurian is mostly eugeosynclinal in nature. The overlying Devonian and Lower Carboniferous are marine and rich in volcanics, chiefly tuffs, andesite-porphyrites, and rhyolites. A flagrant unconformity separates the Middle Carboniferous and younger rocks from the underlying strata, and terrestrial deposits are predominant in the later part of the Palaeozoic. Land plants are common in the Middle Devonian (*Protolepidodendron scharyanum*) and an Angara flora, represented by *Angaropteridium* and *Noeggerrathiopsis*, became dominant in the Middle Carboniferous. Another unconformity surface was found below the Upper Permian, which is a thick purple and red molasse deposit. From Late Permian onwards, the basin deposits became inland and intermontane in nature.

It is evident that this region was dominated by an oceanic regime of deposition in the Early Palaeozoic. The first important orogenic movement took place after the Early Carboniferous (Early Hercynian or Tianshanian, see Table 20.1), and after the Late Hercynian (Yiningian or Nilkaan) orogeny it was converted into land with prevalent continental regime. The folding trends tend to follow the western border of the Junggar Massif in a NE direction and turn to WNW in conformity with the tectonic grain of eastern Khazachstan in Central Asia.

The eastern part includes the Bogda Mountain and its eastern extension to the north of the Turfan Basin. Here the stratigraphic development is peculiar in the absence of the Lower Silurian and the pronounced Caledonian Movement. Devonian and Carboniferous volcano-clastics amount locally to 10 000 m, and the Early Hercynian Movement is also outstanding in the northern part, but tends to be weakened in the Bogda region, where marine conditions lingered as late as the Early Permian. It is noticeable that the Silurian *Tuvaella* fauna was found in the Hongliujia Formation. The presence of the Early Palaeozoic *Tuvaella* fauna and the Late Palaeozoic Angara flora in this region is a strong indication of its intimate relation with the Siberian–Mongolian continent before the final closure of the marine realm between Junggar and north Tianshan in the Early Permian.

The eastern part is characterized by a prevalent

WNW–ESE folding trend and forms the eastern limb of the Transjunggarian arc with its apex situated approximately at the Ulungur Lake.

To the north of the Baytik (Irtix) Accretional Crustal Consumption Zone (Fig. 19.3, 6) are the Altai Palaeozoic and the Baytik Hercynian fold zones. The Altai is a part of the extensive fold zones bearing the same name in the Soviet Union and the Mongolian People's Republic. In the Chinese part, the Ordovician and Silurian are represented by a miogeosynclinal type, generally slightly metamorphosed and locally migmatized. The Devonian is chiefly volcano-clastic, and the Middle Devonian Altai Formation may reach 4800 m in thickness. In more than one place the Middle Devonian was observed to overlie the Silurian unconformably, although Caledonian orogeny is much less pronounced than in the Soviet Altai. The Lower Carboniferous is comparatively thin and probably represents a cover strata, as pointed out by Ren, J. et al. (1980).

In the Baytik Mountains the Caledonian Movement is manifested by the conspicuous unconformable contact between the Ordovician Huangcaopo Group and the Middle to Upper Silurian Kekexiongkuduk Formation, and also by a break between the latter and the Devonian. The Devonian is variable in facies and contains littoral and paralic volcano-clastics as well as deeper sea basic eruptives and radiolarian cherts. Similar submarine basic eruptives were also reported at the same horizons near the Ulungur Lake. The more or less continuous zone of ultrabasic and ophiolitic rocks, extending to the Irtix Zone beyond the border line near the Zaison Lake, is a crustal consumption zone of early Hercynian age, probably with a northward subduction.

The North-east region

This region encompasses the northern part of Northeast China excluding the Nadanhada Mesozoic Fold Zone, and the northern part of Nei Mongol excluding the part to the south of the Suolun–Xilamulun Convergent Crustal Consumption Zone. The Songliao Massif situated in the middle part is entirely covered with Mesozoic and Cainozoic deposits, and is probably underlain by a Precambrian basement, although its actual extent, especially its south-eastern boundary, is uncertain. An indication of the presence of an ancient foundation below the basin cover is shown by the outcrop of Precambrian rocks along the Nenjiang fault defining the western basin border, and by the recent discovery by Ma Jiajun and others of Early Palaeozoic subduction zones toward the basin in the Xiao Hingan Mountains and Nenjiang regions. The inland basin development probably began in the Jurassic, when isolated fault basins of limited extent were formed. The main bulk of the basin deposits is Lower Cretaceous and contains red clastics in the lower part and dark coloured fine sediments of deep lacustrine facies (the main oil-producer) in the upper part. The extent of the lake began to shrink in the Late Cretaceous, and the Lower Tertiary is represented only by small isolated basin deposits. The structure within the basin is characterized by a series of *en echelon* short-axis folds.

The Songliao Massif is surrounded by fold zones of different ages. Two Xingkaian or early Caledonian fold zones, the Ergun and Jiamusi, occupy the northernmost part and are continuous with fold zones of the same age in the Soviet Far-East and the Mongolian People's Republic. In the Jiamusi region, Late Proterozoic granites are common and a Late Sinian *Ediacara* type of fossil was found near Jixi. On the western and northern sides of the Songliao Massif there occurs a Palaeozoic (possibly Caledonian) fold zone, in which the Lower Palaeozoic is best developed near Aihui at the northern end of the Xiao Hingan Mountains and in a belt on the west of the Nenjiang fault. Here the Middle and Upper Ordovician Duobaoshan Group contains andesitic and dacitic porphyrites and spilites, evidently of island arc type, and is followed by Silurian greywackes, tuffaceous sandstones, and slates bearing the conspicuous *Tuvaella* fauna (Tang, Su, and Wang 1982). There is evidence that these rocks represent the island arc deposits off the Songliao Oldland to the east. The Lower Palaeozoic is truncated by an unconformity, upon which rests the marine Devonian, which also contains spilites in the lower (Handaqi Formation) and intermediate to acid volcanics in the upper part (Daminshan Formation). Another discontinuity separates the Devonian from the Carboniferous and Permian, the last two being generally incomplete and scattered in distribution.

The Hercynian fold zones include two parts, the midsouth Hingan and northern Nei Mongol in the west, and the Zhangguangcailing in the east. In the western part, Lower Palaeozoic rocks are well exposed in the Sonid Quoqi–Irsh zone. To the north of this zone, in the Erenhot area, the Devonian and the Lower Carboniferous contain an immense amount of submarine basic eruptives. Paralic deposits bearing an Angara flora began to appear in the Middle Carboniferous. In the Ujimqin Qi, the Middle and Upper Carboniferous Benbatu and Amushan Formations are characterized by profuse medium to acid volcanic rocks. The Lower Permian Jesu Formation marks the last marine transgression in the region, and the sea ways became entirely closed along the convergent crustal consumption zone in Late Permian. The demarcation between the

Angara type and the Cathaysian type of flora is clear and sharp and provides an excellent example of distinct biogeographic provincialism resulting from continental collision.

The Mesozoic of this region, excepting the Songliao basin, is composed chiefly of Late Jurassic small semi-graben coal basins and basic to intermediate lavas and volcano-clastics. The continental volcanic belt extends from the Da Hingan Mountains to Nei Mongol, and is probably continuous with the Zhangguangcai Mountain belt across the Songliao Basin. These volcanics belong in general to the calc-alkali series, but show some variation in different parts. A notable phenomenon of this region is the especially widespread occurrence of Jurassic volcanics and Hercynian to Yanshanian granites.

THE NORTH CHINA CONTINENT AND CONTINENTAL MARGIN DOMAIN (Fig. 19.3, II)

In this domain are included the two large platforms, the North China (Sino–Korean) in the east and the Tarim in the west. The continental margin tracts consist of a northern Tianshan Fold Zone and its eastern extension, the narrow southern Nei Mongol zone, and a southern extensive and complicated terrain, including the Qilian Fold Belt, the Qaidam Massif, and the Kunlun–Qinling Belt to the north of the Xiugou–Maxin and Shangyang–Tongcheng Convergent Crustal Consumption Zones. For convenience the platform and its adjoining continental margins will be discussed together, and the extensive and complicated southern marginal tract will be treated separately.

The North China Platform and adjoining continental margins

The ancient foundation of the North China Platform (Fig. 19.3, II B) was consolidated at the end of the Early Proterozoic through the Luliangian Movement. The Middle and Upper Proterozoic (including the Sinian) may be called paracover strata and are intermediate between the basement and the Palaeozoic genuine cover sequence (Wang 1978). Upon all of these are the superimposed basin deposits that started to form after the Indosinian Movement. Thus the subdivision of the platform into different geotectonic units is based mainly on the pre-Indosinian development, especially on the nature of the basement.

The basement structure of the platform comprises two continental nuclei, consolidated at the end of the Archaean, Ordos in the west and Ji–Lu in the east. These nuclear massifs show a magnetic anomaly without polarity, in contrast to the clear directions characteristic of the Early Proterozoic fold zones. The basement has subsided deep under the Ordos basin, but the granulite gneiss exposed in Helanshan and southern Inshan may represent its up-turned parts. The Ji–Lu nucleus is more complicated; its basement is represented by the Taishan Complex in western Shandong, and by the Qianxi and Anshan groups in northern Hebei (Ma et al. 1981). It is notable that these oldest terrains are often deeply subsided and covered by the thickest Mesozoic and Cainozoic sediments.

The Lower Proterozoic forms the basement of the remaining parts of the platform. The Wutai–Zhongtiao Uplift (Fig. 19.3, IIB4) between the two nuclear massifs has a conspicuous NE–SW trend and include the type regions of the Lower Proterozoic Wutaian and Hutuoan, which are separated by a pronounced unconformity. The Alxa (Alashan) Massif situated to the west of Ordos also has a pre-Luliangian basement, as the Longshoushan Group in the southern border part is intruded by granites 1789 Ma in age, and the Alashan Complex exposed in the north may be even older. Both these basement rocks are unconformably overlain by stable-type Middle and Upper Proterozoic strata. The easternmost geotectonic unit is the Jiaoliao Uplift, where the cover sequence begins with the Qingbaikouan, although Sinian subsidence was strong in the Liaodong Peninsula. Finally, the Yanshan Aulacogen is developed on the Pre-Luliangian basement, and is partly manifested by the Middle Proterozoic flyschoid deposits, including the Wumishan calcareo-siliceous paraflysch, and the Dahongyu potassium-rich alkali-calcic volcanics.

The character and distribution of the Middle and Upper Proterozoic reflect clearly the difference between the various units. The thick accumulation of sediments in the Yanshan Aulacogen continues into the northern Taihang, but it quickly thins out in surrounding regions. Rifting and volcanism were prominent in southwestern Shanxi and western Henan, where another type of partially terrestrial aulacogen was developed. The widest transgression occurred in the Late Changhengian as evidenced by the overlap of the Gaoyuzhuang Formation. The Upper Proterozoic Qingbaikouan is confined to the Yanshan and the Jiaoliao regions and shows a genuine cover sequence character.

During the Palaeozoic and Early Triassic, the platform was divisible into three parts. The western part, the regions west of Ordos, is peculiar in the intermediate sedimentary type of the Ordovician, and in the local presence of the Upper Ordovician and Silurian, which are entirely lacking in other regions. The Upper

Palaeozoic shows also a more mobile character, sometimes ten times as thick as in the eastern regions. This may suggest that the Alxa Massif in the west should not be included in the North China Platform. The major or the central part of the platform had remained stable and covered by epicontinental seas in the Cambrian and Ordovician. Early Cambrian transgression coming from the Qinling seas first inundated the eastern part of the platform, and extended to the Yanshan and Liaoning regions probably by way of Shandong. The Early Cambrian sea transgressed westward and northward, until the Middle Cambrian when inundation was almost complete. The Ordovician seas were more extensive and probably deeper than in the Cambrian, but in late Early Ordovician, the southern border of the platform began to rise and the littoral and shallow sea of western Henan was obviously cut off from the Qinling marine realm. Except for the Alxa Massif, the whole platform was raised above sea level after the Middle Ordovician (Fig. 19.4).

Paralic conditions prevailed from the Middle Carboniferous to Early Permian. While the Middle Carboniferous seas came mainly from the east and north-east, the Late Carboniferous sea waters transgressed mainly from the east and north-west, the Late Carboniferous and Early Permian seas being probably connected with the Qinling marine realm in the south. From the Late Permian onwards large inland basins began to take shape in North China. After the Indosinian disturbance, the Ordos basin underwent further subsidence, but the eastern part of the platform began to be activated by volcanism and faulting. In the Yanshan and northern Taihang regions, the Indosinian Movement had caused local overfolding and metamorphism. Volcano-clastic basins began to form in the Early Jurassic in the Yanshan region. Since the Middle Jurassic inland basin development has been the main trend in crustal evolution, and large-scale rifting and subsidence came to a climax in the Palaeogene. These will be dealt with in connection with the East China circum-Pacific Domain.

The structure and development of the northern and the southern marginal tracts of the platform is interesting, but many problems still remain unsolved. The narrow belt between the Suolun–Xilamulun convergent zone and the Inner Mongolian axis of Huang is an E–W-trending Palaeozoic fold zone. In the western segment, the Proterozoic Bayan Obo Group seems to pass upwards into the Middle Proterozoic and merge, from the Yinshan Range northward, in the fashion of a passive continental margin. The Windurmiao Group exposed near the convergent zone, some 4000 m thick, including spilites in the lower and silicolites and basic lavas in the upper part, is overlain by fossiliferous Middle to Upper Silurian, probably with an unconformity. The main bulk of the group includes oceanic ophiolite suites, and has an age from Late Proterozoic to Ordovician, as Sinian and Cambro–Ordovician fossils (radiolarians etc.) were recently reported by An Taixiang. The Hulan Group, of similar lithology, distributed in the eastern segment in Jilin, has yielded the Silurian coral *Circophyllum* and is intruded by a Caledonian granite 420 Ma in age. Thus a narrow Caledonian zone seems to have been formed immediately to the north of the platform border, probably with a southward subduction. The Upper Palaeozoic is variable in different parts of this zone, but generally incomplete and miogeosynclinal in nature, except in the Yanbian area, where the Carboniferous and Permian reach a thickness of 10 000 m and contain medium to acid volcanic eruptives. It is noteworthy that a Cathaysian flora and Tethyan benthonic fauna characterize this region and the demarcation from the northern provinces is sharp and clear. The final closure of the sea-ways was effected in the Late Permian, and the emplacement of granites lasted from the Late Hercynian to Early Indosinian, as a result of continental collision after crustal convergence.

The southern border of the platform is flanked by the East Qinling, sometimes also known as North Qinling, which is geologically continuous with the North Qilian and extends from southern Shaanxi to northern Anhui.

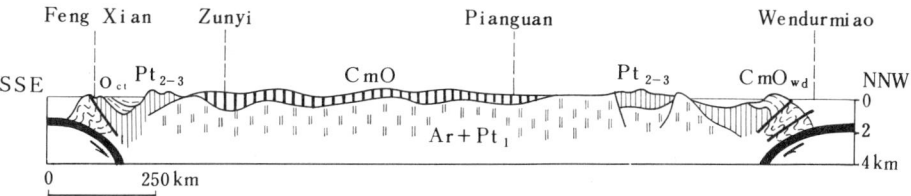

Fig. 19.4. Schematic tectonic profile across the North China Platform from Qinling to Nei Mongol in the Caledonian stage (end of Ordovician). O_{ct}, Caotangou Group; CmO_{wd}, Wendurmiao Group.

In southern Shaanxi the oldest rocks consist of the Qinling Group which may be partly Lower Proterozoic, and the Kuanping and Taowan Groups, mainly Middle and Upper Proterozoic (Wang et al. 1982). The earliest orogenic movement recorded is the unconformable contact between the Upper Proterozoic Taowan Group and the underlying strata observed from Luonan to Fangcheng. This orogeny, as corroborated by a granite (dated 999 Ma) intruded in the Jixianian dolomites near Lingbao, western Henan, was probably responsible for the general uplifting of the platform after the Middle Proterozoic. The second orogenic event occurred in the Middle to Late Ordovician and resulted in the formation of the Taibai–Danfeng subduction zone along which Caledonian ophiolites and ultrabasic rocks are distributed. This subduction is reflected in the interior of the platform by a universal uplifting in the late Middle Ordovician (Fig. 19.4). The later history of this region consists of the formation of Devonian and Triassic flysch, its final convergence with the northern continental margin of the Yangzi Oldland along the Shanyang–Tongcheng Crustal Consumption Zone in the Indosinian Stage, and granite intrusion and intermontane basin development after collision.

The Tarim Platform and adjoining continental margins

The Tarim Platform (Fig. 19.3, IIA) and its northern margin the Tianshan Fold Belt (Fig. 19.3, IIC) are intimately connected with each other. The present platform is limited by the South Tianshan in the north and by the West Kunlun in the south. The Tarim Basin is covered by Cainozoic deposits, but Palaeozoic and occasionally Precambrian rocks are exposed in narrow latitudinal stripes in the south-western part of the basin. Basement rocks are, however, well exposed in the bordering highlands, notably in Quruktagh and Kalpin in the north, and in Tieklik in the south. Aeromagnetic survey has revealed an ancient massif in central Tarim, the southern part of which may well represent an Archaean nucleus (Wang 1981). As the Middle and Upper Proterozoic constitute a part of the folded basement in the Quruktagh Aulacogen (Fig. 19.3, IIA$_2$), the final consolidation of the whole platform may be referred to the Jinningian Movement. The Quruktagh Aulacogen was a strongly subsiding belt in the Sinian and Early Palaeozoic, when andesite porphyrites and felsite-keratophyres were formed. The Upper Palaeozoic is incomplete, and the whole belt was uplifted in the Permian as a result of Hercynian orogeny in the adjoining Tianshan region. The Palaeozoic platform sequence is best developed in the Kalpin area, and extends southwards beneath the Mesozoic and Cainozoic cover into the North Tarim Depression and grows thinner in the interior of the basin, according to data from bore holes. The Tieklik Range in front of the West Kunlun is characterized by a stable type of the Middle and Upper Proterozoic, an incomplete development of the Lower Palaeozoic and a thicker sequence of the Upper Palaeozoic. On the whole the platform remained comparatively stable and was widely inundated in the Palaeozoic.

After the Hercynian Movement, the Tarim entered into a new era of basin development. Three stages may be recognized, the Triassic–Jurassic down-faulting stage, the Cretaceous–Palaeogene down-warping stage, and the Neogene general subsiding stage. The general configuration of the basin consists of a central uplift which separates a northern and a southern depression distinct from each other in stratigraphical development. The Triassic and Jurassic are well developed in the northern depression, especially in the Kocha area, but are lacking in the interior of the basin and incomplete in the southern depression. On the other hand, the Cretaceous seems to be confined to faulted basins only in the north, but is widely overlapping in the south. Late Cretaceous and Palaeogene marine ingression from Central Asia reached as far as 84 °E in the peripheral depressions. Cretaceous and Palaeogene purple to red, coarse clastics and gypsiferous mudstones and limestones may amount to over 5000 m thick in the southern depression. The mid-Tertiary Himalayan Movement brought about strong uplifting of the surrounding mountain ranges and very widespread subsidence of the basin. Pliocene and Early Quaternary piedmont molasse over 4000 m thick, the Atoush Group, and the Xiyu gravels in the Kash region, show a clear upward coarsening (Huang and Chen 1980).

In the border areas of the basin, the Palaeozoic is sometimes gently folded owing to the Hercynian orogeny. Triassic and Jurassic rocks occur, and are truncated by Cretaceous and younger strata as a result of the Yanshanian Movement. The Himalayan Movement was responsible for the strong uplifting, and formation of piedmont deposits, and thrust faults dipping steeply to the north, 400 km in length, can be observed near Kalpin. The WNW–ESE-trending broad- and short-axis folds in the western part of the basin are also products of the Himalayan Movement.

The Tianshan region is a Palaeozoic fold belt consisting of a median uplift bounded on both sides by Hercynian fold zones. The median uplift includes the Yining Massif (Fig. 19.3, IIC$_2$) at the western end, and a narrow zone of Middle and Upper Proterozoic rocks in the eastern segment, continuing into the Beishan.

This marks the northern margin of the ancient Tarim Platform, while South Tianshan seems to be an intraplatform aulacogen or geosyncline developed in the Palaeozoic. Subsidence of the South Tianshan trough probably began in the Cambrian, and it was not until the Middle Silurian that volcanism began to appear. Marine Devonian and Carboniferous carbonates and clastics bearing typical Tethyan forms such as the Upper Devonian *Yunnanellina* and the Lower Carboniferous *Striatifera* constitute a continuous sequence, and the Lower Permian contains volcanics ranging from quartz porphyries to olivine basalt, often unconformable upon the Silurian and Devonian. The Lower Permian is in turn overlain unconformably by Upper Permian conglomerates on the southern slope of Harliqtau.

Central Tianshan is characterized by the presence of Proterozoic rocks and was intermittently transgressed by Palaeozoic seas. The Palaeozoic is incomplete and contains several breaks. The final Yiningian (Nilkaan) orogeny, bringing about the folding of the geosyncline, occurred near Nilka, and is recorded by the unconformable contact between the Upper Permian conglomerates and the underlying Lower Permian acid eruptives and coal measures (see Table 20.1).

North Tianshan, a narrow Hercynian zone on the northern side of the ancient Tarim Oldland, had remained under the sea until the mid-Permian, after which time the whole region was uplifted. It is noteworthy that Tethyan forms *Striatifera* and *Yuanophyllum* occur in the Choltagh zone, but at the same time the Angaran form *Noeggerathiopsis* is found in association with Cathaysian forms in North Tianshan, and even in the western segment of Central Tianshan. Similar conditions occurred in Beishan, where *Noeggeratiopsis* is mixed with the Cathaysian *Protoblechnum* and *Sphenophyllum* cf. *thonii*.

West Kunlun probably represents the southern continental margin of the ancient Tarim continent, and the extensive proto-Tethys ocean was probably further to the south.

The south-western continental margin tract

This triangular region encompassing Qilian, Qaidam, East Kunlun, and West Qinling may be regarded as a complicated continental margin tract to the south of the eastern part of Tarim and the western part of the North China Platforms (Fig. 19.3, IIE–IIF).

The Qilian Fold Belt is a Caledonian terrain composed of a median uplift bounded by fold zones on both sides, the North Qilian and the South Qilian. The median uplift is mainly composed of Precambrian rocks and is divisible into the Lower Proterozoic basement and the Middle and Upper Proterozoic epimetamorphic sedimentaries. Palaeozoic cover is incomplete. In North Qilian, the Middle and Upper Proterozoic are mobile in nature with abundant basic eruptives, thus differing from Alxa and Qaidam. In North Qilian WNW–ESE-trending Lower Palaeozoic tectono-sedimentary zones are predominant. The Middle Cambrian to Lower Ordovician rocks are thick continental slope miogeosynclinal deposits in the northern zone, but contain typical island arc deposits in the southern zone (Fig. 19.5). The Ordovician Gulangian Movement is prominent, and the Upper Ordovician and Silurian consist of volcano-clastic and flyschoid deposits marking the late development stage of the geosyncline. A prominent WNW–ESE-trending zone containing numerous ultrabasic rock bodies of Caledonian age has been observed along the southern slope of the North Qilian (Nanshan) Fold Zone. Considered in connection with the island arc nature of the Lower Palaeozoic, it is most probable that this zone represents residual oceanic crustal patches after subduction (Li Chunyu *et al.* 1978).

The South Qilian trough is situated between the Central Qilian Uplift and the Oulongbruk Uplift, the latter being regarded as the northern border of the Qaidam Massif. The stratigraphic record within the trough begins with Lower Ordovician volcano-sedimentaries and consists of an immense thickness of volcanic eruptives, greywackes, and silicolites up to Late Silurian in age. This strongly suggests an aulacogen sequence as pointed out by Wang Zewen *et al.* in 1981. It is therefore very probable that the Qaidam Massif and the Central Qilian Uplift were originally united and later severed by the South Qilian geosyncline.

The Caledonian Movement is most conspicuous in North Qilian. After this a large scale depression was formed along the northern side of the corridor region and received Devonian molasse, paralic Carboniferous, terrestrial Permian, and younger deposits. It is also in North Qilian and the border region of the Alxa Massif that granitic rocks ranging from 530 to 360 Ma in age were emplaced. Hercynian intrusives comprise small granite stocks distributed in South Qilian, and are related to Hercynian folding.

The basement of the Qaidam Massif was mainly consolidated in the Jinningian Stage, as evidenced by the Sinian cover sequence formed by the Quanji Group in Oulongbruk. The Sinian is succeeded by a stable type Lower Palaeozoic which is unconformably overlain by the Upper Palaeozoic, all of limited thickness. A narrow aulacogen filled with Ordovician and Silurian volcano-clastics, the Aszha Group (Fig. 19.5, O_{as}),

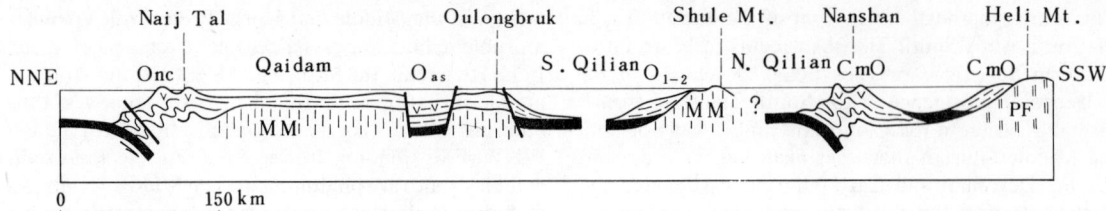

Fig. 19.5. Schematic tectonic profile from Burhan Buda to Heli Mountain, northern Gansu in the Caledonian stage (end of Ordovician). O_{nc}, Naij Tal Group; O_{as}, Asizha Group (lower part); PF, platform basement; MM, median massif basement.

extends along the northern border of the Qaidam Basin. The overlying Upper Palaeozoic consists of Devonian coarse clastics and Permo-Carboniferous marine deposits, as in Oulongbruk. The basement rocks are extensively exposed on the northern slope of Burhan Buda, and are directly covered by the Upper Palaeozoic. It seems that the aulacogen on the northern border of the basin parallels that of South Qilian, and both were developed on the originally united pre-Ordovician basement. An alternate interpretation would be that the Qaidam Basin is a depression within the Palaeozoic folded terrain, the basement rocks being only exposed along fold axes (Ren, J. et al. 1980).

The Qimantagh and the Burhan Buda mountains on the southern border of the Qaidam Massif form a complicated Hercynian fold belt. On the southern slope of Burhan Buda the Lower Palaeozoic Naij Tal Group and the Devonian volcano-clastics were folded together and unconformably covered by Early Carboniferous red molasse. In the Qimantagh the Upper Palaeozoic is still entirely of a mobile type, and the whole belt was probably folded up in the Late Hercynian, as evidenced by the huge granite zone, 280–260 Ma in age, all along the southern boundary of the basin.

The last unit to be mentioned in this connection is the West Qinling Early Indosinian Fold Zone, which is bounded in the south by the Xiugou–Maxin Convergent Crustal Consumption Zone, and in the north by the Lanzhou–Xining Massif (Fig. 19.3, IIF). The oldest rocks exposed are the Devonian Xihanshui Group and Dacaotan Group, representing respectively shallow sea and paralic thick rhythmic deposits, with detritus supplied by the northern oldland. The Upper Palaeozoic and Triassic sequence is on the whole marine. The Lower and Middle Triassic consist of some 9000 m of clastic and calcareous paraflysch deposits, the Early Indosinian Movement being manifested by an unconformity between them and the overlying Upper Triassic coal-bearing paramolasse. A narrow branch of geosynclinal Triassic penetrates westward along the northern border of Qaidam, where Early to Middle Triassic seas inundated the South Qilian. The marine realm finally disappeared after the Early Indosinian Movement.

THE SOUTH CHINA CONTINENT AND CONTINENTAL MARGIN DOMAIN

This domain is bounded in the north by the Xiugou–Maxin and the Shangyang–Tongcheng Crustal Consumption Zones and in the south by the Bangong–Nujiang Crustal Consumption Zone. The main continental massifs in this domain are the Yangzi Platform and the Qiangtang Massif. The Indosinides are by far the most extensive among the variously dated fold zones. Four regions may be recognized, the Yangzi Platform (IIIA), the Qiangtang Massif (IIIB) and adjacent parts, the Songpan–Ganzi Fold Belt (IIIC), and the Qamdo–Simao Fold Belt (IIIE), and in addition a narrow Caledonian belt to the north of the platform (IIID).

The Yangzi Platform

The ancient foundation of the Yangzi Platform is heterogenous and was finally consolidated through the Jinningian orogeny at 850–1050 Ma. Huang (1973) has established the Yangzi orogenic cycle which contains two phases, the Jinningian and Chengjiangian, and Ren et al. (1980) have given an excellent synopsis of the geological development of the Yangzi region. The oldest part of the basement lies under the Sichuan Basin and is at least Early Proterozoic in age, as represented by the Kongling Group in the Yangzi Gorge. The Dabie Massif, which is temporarily included in the Yangzi Platform, may include the Archaean. The most mobile part of the basement occupies the south-eastern margin along the Jiangnan Uplift, which is chiefly Upper Proterozoic. Intermediate between the two is the western tract, the Kham–Yunnan Uplift, with a Proterozoic (including Lower Proterozoic) basement. The platform cover strata begin with the Sinian, the Lower Sinian being unstable in lithology and distribution, and the Upper Sinian widely transgressive on the platform. The whole cover strata may be divided into three sequences, the lower from Sinian to Silurian, the middle from Devonian to Triassic, both mainly marine, and the upper Jurassic and younger, mostly terrestrial basin deposits. They are different in sedimentary types as well as in distribution and type of tectonic deformation. On the basis of these differences, seven units may be recognized within the platform as shown in Fig. 19.3. The development of the platform will be discussed briefly in three stages.

The pre-Sinian

The oldest part of the basement is probably the Central Sichuan Massif which is roughly congruous in area with the present red basin, but may extend eastward to western Hubei and westward to the Kham–Yunnan Uplift. According to data from bore holes in the basin, the Sinian rests directly on the Lower Proterozoic crystalline basement as in the Huangling area in the gorge region. The basement of the rest of the platform is composed of Middle and Upper Proterozoic epimetamorphics. The innermost outcrop of basement rocks in the Upper Yangzi Depression is the Middle Proterozoic Fanjingshan Group of north-eastern Guizhou, which probably represents the earliest island arc to the south of the Central Sichuan Massif. A second series of island arcs was situated near Luocheng, northern Guangxi, as indicated by the Sibao Group containing ophiolites. At both localities the basement rocks are overlain unconformably by the Late Proterozoic Banxi Group, which was mainly deposited in the interarc basins and also contains ophiolites in the Yuanbao Mountain area near Luocheng (Fig. 19.6). It is obvious that eastern Guizhou and northern Guangxi represent a complicated continental margin that bordered the ancient massif to the north, although the land massif and its wide marginal parts seem to be out of proportion when compared with recent examples (Fig. 19.6). Along the Jiangnan Uplift, the Proterozoic basement rocks are extensively exposed, and an unconformity is usually observed between the Middle and the Upper Proterozoic. This relationship of the island arcs to the northern mainland may be observed all along the southern margin of the platform.

The nature of the northern margin of the Yangzi Platform is more complex, especially as the tectonic setting of the Dabie Massif remains problematic. The Dabie Massif was originally regarded as a constituent part of the North China Platform owing to the deep metamorphism and seemingly ancient age of the Dabie Group. Later discovery of Palaeozoic metamorphic rocks on the northern side of the Shanyang–Tongcheng Crustal Consumption Zone has led to transference of the Dabie Massif to the Yangzi Platform, i.e. to the south of the convergent, collision zone between North China and the Yangzi Platform. Recent research by Yang Sennan *et al.* (1983) has however raised the possibility that it was originally an element of the north, and the Qinling Geosyncline was an aulacogen developed in the Middle and Late Proterozoic that had separated it from the northern mainland. If so, the Dabie Massif was probably incorporated with the Yangzi Platform through the Caledonian orogeny, as a narrow Caledonian zone occurs to the south of it.

At the northern protuberance of the platform near Hanzhong, the island arc volcano-clastic rocks are represented by the Huodiya Group, which overlies unconformably the Hannan complex of Early Proterozoic age. Similar relations are observed all along the western border of the platform. The Lower Proterozoic basement rocks include the Kangding complex of western Sichuan, the Dahongshan Group and the Hekou Group of south-western Sichuan and central Yunnan. The overlying Middle and Upper Proterozoic comprise, roughly from north to south, the Dengxiangying, Yanbian, Huili, and Kunyang Groups, all of which are characterized in various degrees by island arc volcanism. An outstanding feature is the sharp truncation of structural lineaments between the two sets of strata, the lower trending east–west and the upper filling the

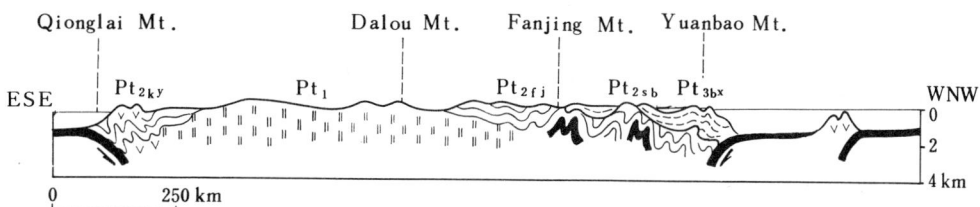

Fig. 19.6. Schematic tectonic profile from Qionglai Mountain, western Sichuan to Yunkai Mountain, western Guangdong in the Jinningian stage (Late Proterozoic). Pt_{2ky} Kunyang Group; Pt_{2fj} Fanjingshan Group; Pt_{2sb} Sibao Group; Pt_{3bx} Banxi Group; PF, platform basement; CA, Caledonian basement; C, Carboniferous.

longitudinal troughs developed on the basement. As recently pointed out by Feng Benzhi (1983), this may indicate that the ancient basement was more extensive, and was cut off by faulting in later orogenic movements. The unconformity mentioned above is obviously referrable to the Luliangian orogeny, but more important is the Jinningian Movement, which is evidenced by a universal unconformity below the Sinian, and by a continuous longitudinal zone of granitoid intrusives, ranging from 867 to 808 Ma in date and intimately associated with andesitic eruptives. All these suggest the presence of a huge subduction zone in the Late Proterozoic along the western margin of the platform (Luo 1979).

Palaeozoic and Triassic marine transgressions

The marine sequence from the Sinian to the Middle Triassic records the stage of stable development of the platform. The Lower Sinian clastics are confined to the peripheral parts of the platform, while the Upper Sinian is comparatively uniform in lithology and universal in distribution. Differentiation became pronounced in the Early Palaeozoic. The Kham–Yunnan Uplift was elevated and supplied sediments to the eastern depressions. The Lower Palaeozoic is generally of moderate thickness, but Cambrian evaporitic dolomites may be very thick in local depressions of south-western Sichuan and north-western Guizhou. Within the scope of the platform, the Silurian is more clastic and limited in distribution, the thickest deposits being found on the inner side of the Jiangnan Uplift in north-western Hunan. The upheaval of the platform culminated in the Early Devonian, when terrestrial deposits were dominant but restricted in distribution. The second extensive transgression, comparable with the Late Sinian, occurred in the Early Permian, when central Sichuan was once more entirely flooded. A NW–SE-trending fault zone, engendering an unusual kind of sedimentation in south-western Guizhou, began in the Devonian and continued throughout the Late Palaeozoic. It was also along this zone and in eastern Yunnan that an enormous amount of basalt, mainly terrestrial but partly submarine, was poured out in Late Permian times. Carboniferous and Permian basic eruptives and intrusives were also formed on the Kham–Yunnan Uplift as a result of intensive rifting. Differentiation of sedimentary facies was conspicuous in the Late Permian. Apart from the basalt flow in the southwest, coal-bearing deposits are very widespread and are divisible into a terrestrial facies confined to the neighbourhood of the Kham–Yunnan Uplift, and a much more extensive paralic facies in the southern part of the platform. Topographic relief was even more pronounced in the Triassic. The elevated terrains in the west supplied detritus to the eastern marine basin in much the same way as in the Cambrian and Ordovician. The Lower Triassic may be divided into a western Feixianguan clastic facies, and an eastern Daye carbonate facies, with the dividing line passing through central Guizhou. Deep depressions or lagoons receiving Early to Middle Triassic gypsiferous dolomite deposits also occurred in south-western Sichuan as in Sinian and Cambrian times. This marks the last marine invasion, after which the whole region entered into a new stage of inland basin development.

Mesozoic and Cainozoic inland basin development

After the Indosinian Movement, the demarcation between the western and the eastern part of the platform became more conspicuous as a result of the increasing interaction between the East Asian continent and the West Pacific ocean. The upwarping of Mid-Triassic upland along the Hunan–Quinzhou border and its prolongations became an effective barrier between a western paralic basin with Tethyan *Burmesia* fauna, and an eastern part (including the lower Yangzi) characterized by Pacific brackish water *Bakewelloides* fauna (Wang 1981). The lingering seas retreated completely from the platform before the end of the Triassic, after which the basins became entirely terrestrial. The large Jurassic basin covering Sichuan in the north and central Yunnan in the south was probably continuous through southern Sichuan. Late Jurassic normal marine fossils occur in the Longmen Mountain along the western border of the basin. After the Late Indosinian Movement at the end of the Triassic, the sea-ways in connection with the Tethys were largely cut off, but Jurassic marine ingressions carrying brackish water faunas were still found in Central Yunnan. As a matter of fact, the Central Yunnan region had remained an on-shore or shoreline basin throughout the later part of the Mesozoic. The terrestrial sequence in the Sichuan basin starts with Lower Jurassic lacustrine argillaceous and marly deposits, and these are followed by Middle and Upper Jurassic red clastics of arid climate origin. The basin was greatly reduced in area during the Cretaceous, when thick and coarse molasse was deposited along the western piedmont belt. In the meantime Central Yunnan underwent strong subsidence, and an enormous deposit of purple and red sands and silts was accumulated. The basin ceased to exist at the end of the Cretaceous, although Palaeocene clastics are known in some residual subsiding zones.

Conditions were different in the eastern part of the platform. The Jianghan Basin in central Hubei started to develop in the Cretaceous and strong subsidence took place in the Early Tertiary. Further to the east, in the lower Yangzi, medium to small volcano-sedimentary basins were developed in the Late Jurassic and Early Cretaceous. Late Cretaceous to Early Tertiary red basins are also found on the platform, although they are more common to the south of the Jiangnan Uplift.

Magmatic activity is not as a rule frequent on the platform. Granitic intrusives exposed on the uplifts are connected with the Jinningian (c. 850 Ma), and partly with the Chengjiangian Movement (700 Ma), in the Kham–Yunnan region, where Hercynian intrusives associated with basalt flow are also reported. The long interval from the Cambrian to the Triassic is in general a time of magmatic quietness in the interior of the platform, and it is the Indosinian orogeny that first affected the western border of the platform and the lower Yangzi. Granites were emplaced on the Kham–Yunnan Uplift, and intensive folding and thrusting took place in the Longmen Mountain region. However, the most important folding affecting the whole platform resulted from the Yanshanian Movement. In the surroundings of the Sichuan Basin, especially in the Upper Yangzi Depression in the south, quite strong folding, involving the whole Palaeozoic and Triassic, makes up a continuous giant, S-shaped, folded terrain. This was developed between, and controlled by, the ancient central Sichuan massif and the Kham–Yunnan and Jiangnan border uplifts. On the western border of the Kham–Yunnan Uplift, a narrow strip of relatively close folding of Yanshanian age occurs in the Yanyuan–Lijiang region, which had undergone strong subsidence in the Palaeozoic. Faults trending N–S to NNE–SSW are prominent, although E–W-trending faults are also present. Yanshanian folding and small-scale granitic intrusives are also prominent in the lower Yangzi. In addition, Jura type folding is developed in the inner belts immediately surrounding the red basin.

A Caledonian fold zone composed of pre-Sinian volcanics (Yaolinghe Group), and Sinian to Lower Palaeozoic slightly metamorphosed strata, occupying the middle and upper reaches of the Hanshui River, flanks the platform in the north from Ankang in southern Shaanxi to south of Yingshan, north-eastern Hubei. An incomplete stable sequence from Devonian to Triassic in age overlies the Lower Palaeozoic, but without pronounced unconformity. It is probable that this Caledonian zone was formed on the northern passive continental margin of the Yangzi Platform.

*The authors are grateful for personal communications from Guo Tieying and Liang Dingyi, Wuhan College of Geology, and from Fan Yingnian, Chengdu Institute of Geology and Mineral Resources.

The western elements of the domain

The western portion of the domain consists of various elements of complicated structure which remain relatively uninvestigated. The Qiangtang Massif has been regarded as Precambrian (Wang 1978), because an Ordovician to Devonian platformal sequence occurs near Lazhulong in the northern Karakhorum, and stable-type Devonian has been found to be underlain by metamorphic rocks near Mayigang (about 34 °N, 97 °E). The Silurian of the Lazhulong area yields *Hindella xizangensis*, and the Upper Devonian contains a fairly rich coral fauna consisting of *Phillipsastraea* and *Macgeea*, which is followed by Permo-Carboniferous carbonates and clastic deposits bearing the Visean *Avonia*, the Late Carboniferous *Triticites*, and the Early Permian *Spinomarginifera* and *Neoschwagerina*. In the eastern part of Qiangtang, no Lower Palaeozoic has been reported, but the Upper Palaeozoic is entirely of stable type; the Middle Devonian Chasang Formation, the Lower Permian Xiaocaka Formation, and the Upper Permian Raggyorcaka Formation bearing a southern Cathaysian flora are all mainly stable-type deposits of limited thickness. Moreover, the Triassic and Jurassic in the whole Qiangtang region show a much decreased thickness in comparison with the northern Tangula region and the southern Rutog–Nagqu region, the Cretaceous being generally lacking. The Jurassic Yanshiping Group has a thickness only one third that of deposits in the Tangula region, and is widespread and comparatively less disturbed. But to the south of the line linking Chabu north of Garze to Myggargang north of Nyima, from 34 °N in the west to 33 °N in the east, both the Permo-Carboniferous and the Jurassic are of a mobile type. The Upper Carboniferous Cameng diamictite Formation and the Lower Permian *Eurydesma*-bearing Zhanjin Formation are very thick, and abundant in volcanics. The Jurassic is likewise of mobile type and is intensely deformed.* A tholeiitic basalt belt containing ultrabasic rocks was recently found along the boundary line noted above, and this may well be the demarcation between the Cathaysian and the Gondwanan realms in the Late Palaeozoic. Between this belt and the Bangong–Nujiang Crustal Consumption Zone (Fig. 19.3, 5) is the Yanshanian fold zone (Fig. 19.3, IIIF) composed of folded Upper Palaeozoic and Jurassic deposits intruded by Yanshanian granites. Ophiolite suites and ultrabasic rocks are well developed in the Dengqen and Amdo region, and near Rutog and Garze, where they are overlain by the Upper Jurassic. The occurrence of Yanshanian granites and the general uplifting of the Qiangtang Massif in the Cretaceous indicate that the convergence and collision between this massif and the southern continental elements were effected mainly in the Late Jurassic (Wang 1981).

The triangular region between the Qiangtang Massif and the Yangzi Platform consists mainly of Indosinides to the east of the Jinsha River, with a narrow western rim of Palaeozoic fold zone along the Transverse Mountains. The Palaeozoic zone (IIIE) may be subdivided into three segments. The Kaixinling Fold Zone is situated to the north of Qamdo and is of Hercynian age, where two unconformities occur both above and below the Upper Triassic Jiezha Group. Near Yushu the Upper Palaeozoic contains more than 5000 m of Lower Carboniferous greywackes and bioherm carbonates, 4000 m of Mid–Late Carboniferous coal-bearing clastics and carbonates, 4000–5000 m of Lower Permian (Kaixinling Group) clastics and carbonates with abundant volcanics of a calc-alkali suite (Fig. 19.7), followed by the Wuli coal beds yielding a Cathaysian flora closely comparable with that found in the Raggyarcaka and Tuoba Formations (see below). Lower Palaeozoic strata were reported recently from near Nangqen and the whole belt may represent a Late Palaeozoic island arc, but the subduction relation is uncertain (Wang 1983). The Median segment, the Qamdo Caledonian zone, includes an Upper Palaeozoic platformal sequence which overlies the Ordovician flysch unconformably, but the metamorphic Jiayuqiao Group in the Nujiang valley may contain Lower Carboniferous strata, and Early Hercynian folding may therefore be involved (Chen 1982).

Conditions are different in western Yunnan between the western border of the Yangzi Platform and the Bangong–Nujiang Crustal Consumption Zone. A Precambrian median massif (Lincang Massif) is present in the west of the Indosinian Simao Fold Zone, and Precambrian rocks, the Lower Proterozoic Ailao Group, and the Middle to Upper Proterozoic Cangshan and Lancang Groups are in general comparable with the Precambrian of Central Yunnan, and therefore may have been separated from the latter in Palaeozoic and later times. A Palaeozoic fold zone stretching from Ximeng to north of Changning consists of the metamorphic Ximeng Group which is covered by a Lower Carboniferous ophiolitic suite near Changning. It seems probable that the Qamdo–Ximeng Palaeozoic zone may have been formed on the western border of the platform but moved away in a later period.

The extensive Indosinides occupying the triangular region of western Sichuan are mostly covered by Triassic, especially Upper Triassic sediments. The presence of the Songpan Massif is chiefly inferred from stratigraphical analysis, as the Palaeozoic is in general thinner and more stable in character, and the basement may be the south-western continuation of the Precambrian Bikou Group. Permo-Carboniferous carbonates including basic eruptives are occasionally exposed within this region, but Devonian and older rocks are chiefly confined to the border parts. An Indosinian accretional crustal consumption zone is inferred to exist along the Jinsha River based on the presence of Triassic *mélanges* and olistostromes, but opinions differ as to the direction of subduction. In view of the spreading nature of the Late Palaeozoic seas indicated by the extensive basic effusives, the main direction may be eastward, as shown in Figure 19.3 8.

The Himalaya and Gangdise regions to the south of the Bangong–Nujiang convergent zone evidently belong to Gondwanaland. The principal structural features and tectonic evolution have recently been ably analysed by Gansser (1964) and Chang Chengfa (Chang et al. 1981), and Wen has summarized the stratigraphy and palaeogeography of the whole Xizang region (Wen 1981).

In the south of the Yarlung Zangbo, the stratigraphic sequence is fairly well established. A Middle Proterozoic age for the Precambrian crystalline rocks has recently been confirmed. A complete platformal marine sequence from Ordovician to Palaeogene is to be found in the northern Himalaya (Tibetan) belt, the

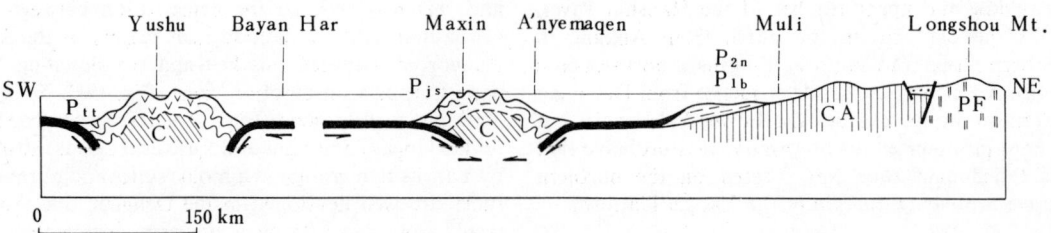

Fig. 19.7. Schematic tectonic profile from Yushu, southern Qinghai to Longshou Mountain, northern Gansu in the Hercynian stage (end of Permian). P_{tt}, Tongtianhe Group; P_{js}, Jishishan Group; P_{lb}, Bayinhe Formation; P_{2n}, Nuoyinhe Formation.

Palaeozoic sometimes containing fossils of Gondwanan affinity that are generally of low diversity and probably come from high latitudes. Only Upper Palaeozoic and Mesozoic strata are exposed in the Yarlung Zangbo valley, the mobile type beginning with the Upper Triassic and containing ophiolites and radiolarian cherts.

On the northern side of the Yarlung Zangbo, the Gandise Massif is limited in extent and suffered intense change during the Yanshanian and Himalayan orogenies. The basement rocks composed of the Middle Proterozoic Nyaiqentanglha Group are exposed in the Damxung region and extend at least to the west of Hainza, where platformal Lower Palaeozoic much like that in the north Himalaya is developed. The Upper Palaeozoic is also of stable type, but the Lower Carboniferous contains some warm water faunas, although Upper Carboniferous to Lower Permian glacigene pebbly slates are widespread as far north as the Karakhorum region (Liu, B. et al. 1983, Liang et al. 1983). The Triassic is not developed; the Jurassic and Cretaceous are very thick and contain abundant volcanic rocks, the Late Cretaceous to Early Palaeogene acidic eruptives extending all along the northern side of the Yarlung Zangbo for a distance of over 1000 km in association with contemporaneous granitic intrusives. Himalayan plutonic activity was principally confined to the south of the Yarlung Zangbo, but marine and paralic Palaeogene mobile type deposits are found in the Ngali region to the west of the Ngali Yanshanian Fold Zone (Fig. 19.3, IVD).

It seems that the Gandise Massif was originally united with, or close to, the northern margin of ancient Gondwanaland. Their separation began in the Late Palaeozoic and oceanic crust started to form in the Late Triassic, after which wide oceans may have existed between them. The final convergence and collision of the Himalaya region and the Gandise Massifs took place from the end of the Cretaceous to the Eocene, after which the main process of development has been further collision and final uplifting.

THE CIRCUM-PACIFIC CONTINENTAL MARGIN DOMAIN

The major part of this domain consists of the extensive Caledonides in South China and the bordering fold zones of various ages. The Cathaysian Caledonian belt (Fig. 19.3, VA) includes two fold zones, the inner or western Xianggui Fold Zone, and the outer or eastern Min-Yue Fold Zone, demarcated in the middle by the Jien'ou Uplift. Bordering the Caledonides in the east are the coastal fold zones, including Hainan Island (Fig. 19.3, VB), which are of Hercynian to Early Indosinian age. An Indosinian fold zone is found in the Youjiang region of western Guangxi (Fig. 19.3, VC), which probably represents a secondary geosyncline developed on the Caledonian basement. Yanshanian and Himalayan fold zones are found on the eastern border of North-east China and in Taiwan Province. These units will be briefly discussed as follows.

The Cathaysian Caledonides

As mentioned in Chapter 5, the extensive terrain to the south-east of the Yangzi Platform has been a complex continental margin tract since the Middle Proterozoic. This kind of geotectonic setting was most conspicuous in the Early Palaeozoic (Fig. 19.8). In pre-Sinian times the crust was mainly oceanic, but islands that supplied detritus to nearby regions seem to have existed along the Yunkai-Jian'ou Uplift, especially in north-western Fujian, where Early Sinian and even older clastic deposits have been identified. These island groups or arcs were probably not continuous, but show a long-term tendency to uplifting and repeated magmatism and metamorphism in various epochs, as pointed out by Mo Zhusun in 1980. In the western Xiang-Gui region, the Caledonian basement consists of slightly metamorphosed, folded and faulted, arenaceous and argillaceous strata. These are covered by the Upper Palaeozoic, with obvious facies differentiation, especially in the Middle Devonian and Lower Permian. The cover sequence includes the Upper Palaeozoic and the marine Triassic, and the main folding seems to be post-Triassic, which resulted in a series of short-axis folds in Central Hunan. Granite intrusives are of Indosinian and Yanshanian age, Caledonian granites also being known in the Zhuguang Mountains on the Hunan-Jiangxi border.

The eastern Min-Yue Fold Zone is more extensive and also covers parts of Jiangxi and Zhejiang provinces. Its distinction from the western zone consists first in the wide distribution of Mesozoic granites and the well-developed Jurassic and Early Cretaceous volcanics, thus obscuring the Caledonian basement. The Sinian and Lower Palaeozoic are composed of clastics with silicolitic flysch and paraflysch deposits, containing variable amounts of volcanic material, that were formed in trough-like marine basins of oceanic crustal nature. These are unconformably overlain by Devonian terrestrial deposits which were followed by paralic and shallow marine Permo-Carboniferous clastics and carbonates, with marked lateral facies changes. A more remarkable feature is the strong Indosinian Movement, the Anyuan Movement, which brought about a basic

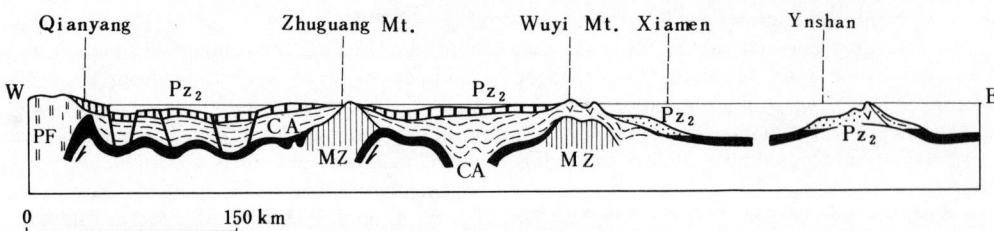

Fig. 19.8. Schematic tectonic profile from western Hunan to Taiwan Province in the Hercynian stage (end of Permian). PF, platform basement; CA, Caledonian basement; MZ, Sinian massifs.

change of the tectonic frame of South China. Late Triassic to Early Jurassic para-molasse deposits were formed upon the folded Middle Triassic and older strata. Intensive magmatic activity took place in the Late Jurassic, when large-scale eruptions occurred all over the region and formed an intracontinental volcanic zone along the East Asian continental margin. Profuse granitic intrusions of Yanshanian age were emplaced, especially along the Lishui–Haifeng (20.5–8) and Wuchuan–Sihui depth faults.

Hercynian and Indosinian fold zones

These include the south-east maritime belt of Hercynian to Indosinian age and the Youjiang Fold Zone of Indosinian age. Since the discovery of epimetamorphic strata bearing Middle Carboniferous fossils below the Mesozoic volcanic series near Fuding in north-eastern Fujian, the presence of a geosynclinal zone younger than the Caledonian is generally accepted, but the actual boundary of this zone has remained undefined. The traditional view of a stable maritime Cathaysian Oldland is not adequate, but there did occur subaerial areas that supplied detritus to the western sedimentary basins, notably in Cambrian, Devonian, and Early Permian times. It seems improbable that a broad geosyncline had developed to the east of the Lishui–Haifeng Fault, but the fold zone was probably confined to the coastal zone (Fig. 20.1, VB2) and was connected with the contemporaneous geosyncline of Taiwan and adjacent regions (Fig. 19.8). A narrow NE–SW-trending subsiding zone filled by Lower to Middle Triassic silicolitic argillaceous deposits was formed near Hua'an to the west of Zhangzhou, which may be connected with the strongly subsiding trough in north-eastern Guangdong, in which very thick Late Triassic to Early Jurassic deposits (the Lantang Group), partly of marine origin, were formed. Metamorphic beds in Hainan Island, formerly assigned to the Tuolie Group of Cambrian age, have yielded Carboniferous fossils, and it is probable that the maritime geosyncline extended to the Hainan region and was continuous with the Hercynian Changshan Geosyncline of northern Vietnam.

The Youjiang Fold Zone is an Indosinian arcuate zone, convex to the north, and roughly parallel to the local southern border of the Yangzi Platform. The Caledonian basement consists of a northern carbonate belt and a southern flysch belt of Lower Palaeozoic age, both very thick and slightly metamorphosed. The Upper Palaeozoic cover sequence is also characterized by its great thickness and its volcanic content at many horizons, especially the medium to acidic eruptives in the Permian. The main constituent rocks of the fold zone are Triassic, metamorphosed and closely folded greywackes, volcanics, and flysch beds including a typical ophiolite suite in the southern belt. The geosyncline seems to have originated through rifting and opening of the basement, when oceanic crust was formed, as evidenced by the presence of ophiolites. As the whole history of geosynclinal development was confined within the Triassic, the oceanic basin could not have been very wide. It might have been a retro-arc marginal sea in connection with a more important subduction of Indosinian age effected in the south, but the exact conditions are not clear. The Youjiang Indosinian zone is apparently truncated by the NE–SW-trending Qinfang trough, which represents an aulacogen which originated in the Ordovician to the west of the Yunkai Uplift and continued to receive large amounts of clastic sediments of olistostrome and para-molasse type during the Middle and Upper Palaeozoic. An unconformity separates the Lower and the Upper Permian, and a westward shift of the subsiding centre occurred in the Triassic. It is also along this zone that Hercynian granites are developed (Fig. 17.1).

Yanshanian and Himalayan fold zones

The Nadanhada Fold Zone is a part of the Pacific

Sikota Alin Fold Belt of the Soviet Far East, which was developed on the Hercynian basement. A Hercynian geosyncline, receiving Carboniferous and Permian eugeosynclinal deposits, probably constituted a part of the circum-Pacific continental margin in Palaeozoic times, which stretched from North-east Asia via the Japanese Islands to Taiwan and the Phillipines. It is characterized by a Late Palaeozoic marine fauna of the Pacific type found in the Yanbian region, which is quite distinct from that found in Nei Mongol. Geosynclinal deposits in the Nadanhada Fold Zone begin with the Upper Triassic Zhenjiang Formation composed of silicolites, tuffs, and basalt flows bearing *Entomonotis ochotica*. The Lower and Middle Jurassic include silicolitic and tuffaceous rocks of immense thickness and are overlain by the Middle–Upper Jurassic Longzhaogou Group of paralic to continental facies. The Qihulin Formation in the lower part of the group yields *Arctocephalites* of Callovian age. As the Longzhaogou Group is unconformably overlain by Lower Cretaceous continental volcanics, the Nadanhada Fold Zone is attributed to the Yanshanian or Pacific age.

The history of the Taiwan Fold Zone may be traced back to the Late Palaeozoic, as Lower Permian eugeosynclinal deposits carrying *Neoschwagerina* are found in the Danan'ao Group in the Central Range Geanticline. The folding and metamorphism of the Danan'ao and Yuli Groups are dated by the age of 86 Ma of an intrusive quartz-diorite, and may be compared with the Ryoke and Sanbagawa coupled metamorphic belt of Japan of the Yanshanian or Pacific stage. The western geosyncline which occupied the major part of Taiwan contains Early Tertiary metamorphic flysch and is followed by Miocene and Pliocene marine argillaceous rocks with a pronounced unconformity. This unconformity resulted from the Early Himalayan Puli Movement dated to 33 Ma. The Central Range is demarcated from the eastern Late Himalayan geosyncline by the Great Longitudinal Valley Fracture, which is a transcurrent fracture of sinistral character. The eastern geosyncline begins with Miocene marine volcanics and is characterized by the presence of ophiolites and radiolarian cherts in the sediment wedge spreading directly to the Pacific oceanic basin. The eastern geosyncline was finally folded and uplifted in the Late Pliocene to early Quaternary (Juan 1975). Evidently the Taiwan Fold Zone is a part of the West Pacific island arc system, but is peculiar in the reversed westward convexity of the arc and the apparent absence of a marginal sea between it and the mainland.

Finally it is probably appropriate here to make a few remarks on the effects upon the eastern part of China resulting from the interaction of East Asia and the West Pacific in the Mesozoic and Cainozoic. This influence is to a large extent reflected by the development of different kinds of basins and the distribution of various types of magmatism. The nature and development of Mesozoic and Cainozoic basins are mainly controlled by the stress conditions prevailing at the time (Zhu and Chen 1980). The Indosinian Movement brought about a basic transformation of the general tectonic frame. While Palaeozoic tectonic directrices are mainly E–W, tectonic zones with prevalent NE–SW to NNE–SSW trends began to appear in the Late Jurassic, which marks probably the first large-scale westward subduction of the West Pacific under the East Asian continent. Concomitant with this is the profuse magmatic activity along the East Asian coastal belt in the Late Jurassic. The volcanic rocks may be divided into three belts, the inner Hingan–Jiaoliao intracontinental type, the median Min–Zhe–Lingnan continental margin type and the outer Sakhalin–Japan island arc type (Cong *et al.* 1977). The inner belt volcanics are a rhyolite–andesite assemblage of the alkali–calcic series; the median belt, consisting of quartz-andesite and rhyolite, includes the maritime provinces of South-east China and belongs to the calc-alkali series; while the island arc type extends from off the south-east coast via Japan to Sakhalin Island. Yanshanian granites are abundant along the Lishui–Haifeng and other depth faults, probably related to the Late Jurassic westward subduction of the West Pacific.

Within the time-span from the Late Triassic to the Early–Middle Jurassic after the Indosinian Movement, the basins in South-east China were mostly of downwarped type and the principal stress condition seems to have been E–W compression. Compressional conditions continued in the Late Jurassic to Early Cretaceous. From the Late Cretaceous, stress conditions in East China underwent important changes. There developed a giant NE–SW- to NNE–SSW-trending downfaulted belt including the Xialiaohe, North China and Jianghan basins of extension type (Wang *et al.* 1983). Cretaceous to Palaeogene small, semi-graben, red basins were formed on the upland of South-east China, probably also under tensile conditions. Meanwhile marginal basins formed of sea-floor-spreading-retroarc type began to appear in the Palaeogene, e.g. the Japan Sea and the South China Sea. In short, tensile force seems to have prevailed, and the East Asian continental margin began to dissociate from the Palaeogene onwards. This tensile fracturing grew weaker in the Neogene and consequent subsidence formed draping deposits in many basins. However, rift valleys such as the Fenwei Graben in North China, were still active in the Neogene.

20. GEOTECTONIC DEVELOPMENT
Wang Hongzhen

INTRODUCTION

In this chapter emphasis will be laid on the temporal or historical aspect of the crustal development of China. Some tectonic and geotectonic terms used in the following discussion will first be defined. As pointed out above, one of the main doctrines is the concept of 'Development Stages' in regard to crustal evolution. The idea of tectonic stages is intimately connected with that of tectonic or geosynclinal cycles and that of folding phases. The idea of folding cycles was first defined by T. C. Chamberlain in 1909 and was greatly elaborated in 1935 by Stille, who put forward his concept of the tectono-magmatic cycle in connection with the idea of geosynclinal development. This concept was still largely followed by Aubouin in 1975. The belief that geosynclinal development exerts important effects on neighbouring platforms has led to the application of the geosynclinal cycle in the subdivision of natural stages in regional, and even global, geological history. Furthermore, in the 1920s, Stille established his famous scheme of orogenic epochs and phases on the basis of the synchronism of folding phases, in which he distinguished three ranks of classification in terms of time; the megacycles (Progaikum and Neogaikum), the periods (Assyntian, Caledonian, etc.), and folding phases of different intensities. This well-known system has exerted a profound influence on geological thought and has been widely applied all over the world. The idea of synchronous folding phases all over the world was opposed by some geologists from the very beginning, and probably few would believe in it seriously (*sensu stricto*) today. However, two facts stressed by Stille seem to stand firm. First, orogenic movements are not evenly distributed in geological history but are concentrated within comparatively short intervals separated by long quiet periods. Second, the short durations (usually with a time-span of about 30 Ma) of intensified crustal movements may be roughly compared with each other in different continents, and some of them may be correlated with profound changes in the nature of sedimentation, magmatism, and tectonic deformation, as well as in the organic world and the overall geotectonic regime and palaeogeographical frame. Indeed, using these ideas S. V. Bubnoff (1951) and J. H. F. Umgrove (1947) sought to interpret the earth's history in terms of cyclicity and periodicity. In China, T. K. Huang (1978) has advanced a complete subdivision of the orogenic cycles and geological events of China, and Yin *et al.* (1978) have given a comprehensive review of the folding phases of China. Using these concepts attempts have been made to distinguish different ranks of development stages in crustal evolution (Wang 1981, 1982*a*). Three ranks are discerned. The first is called a 'Megastage'. It is distinguished by basic changes in the thickness, nature, and mode of movement of crustal sectors, and consequent overall resetting of the geotectonic pattern (Table 20.1). The second rank is called a 'Stage'. It corresponds roughly to the geosynclinal cycle or orogenic cycle, which is characterized by pronounced changes in the palaeogeographical frame, but the time-span of a Stage usually covers several periods and includes long quiet intervals separated by epochs of intensified orogenic movement, through which conspicuous changes are brought about. The third rank is called a 'Movement'. It marks an intensified epoch or episode of tectonic deformation, sometimes corresponding to a 'folding phase', but often containing more than one of them.

The acquisition of radiometric data has greatly increased in China in recent years. Sun and Cui (1980) have given a very good summary of the information available. Figure 20.1 is mainly based on their work, with supplementary data collected since 1980. There is a good correlation between the peaks indicating geothermal events and the orogenic movements. Indeed, the concentrated temporal distribution of the data seems to provide a good basis for the subdivision of tectonic stages, and also to reflect the characteristics of different tectonic regions.

MEGASTAGES AND STAGES IN THE CRUSTAL DEVELOPMENT OF CHINA

The criteria for the subdivision of megastages are mainly of a global nature. Among the profound changes in fundamental conditions may be mentioned

Table 20.1. *Tectonic stages and orogenic movements of China.*

Geologic time scale		Tectonic stages		Orogenic movements	Main geologic events	Orogenic movements of Laurasia		
							Europe	N. America
Cainozoic	Quaternary	Megastage of Pangaea (Gondwana) disintegration	Himalayan Stage	~Himalayan 2~	Upheaval of Qinghai Tibet Plateau	Young Alpedic	~Rodanian~	
	—2.0—							
	Neogene			~Himalayan 1~	Collision of Himalaya and Gangdise		~Savian~	
	—24.6—							
	Eogene			~Yanshanian 3~	Opening of South China sea		~Pyrinean~	~Laramidian~
	—65—							
Mesozoic	Cretaceous		Yanshanian Stage	~Yanshanian 2~	Collision of Gangdise and Qiangtang		~Late Cimmerian~	~Nevadian~
	—144—			~Yanshanian 1~				
	Jurassic				Activation of East China continental margin			
	—213—					Old Alpedic	~Early Cimmerian~	
	Triassic		Indosinian Stage	~Indonesian 2~	Convergence of Yangzi and N. China to form Palasia			
	—248—			~Indonesian 1~			~Pfalzian~	
Late Palaeozoic	Permian	Megastage of Pangaea (Laurasia) formation		~Yiningian*~	Convergence of N. China and Siberian-Mongolia	Hercynian	~Sudetian~	~Alleghenian~
	—286—							
	Carboniferous		Hercynian (Variscan) Stage	~Tianshanian~	Disruption of W. border of Yangzi Platform		~Bretonian~	~Acadian~
	—360—							
	Devonian			~Qilianian~	Formation of S. China Caledonides and close of Qilian troughs		~Erian~	
	—408—							
Early Palaeozoic	Silurian		Caledonian Stage			Caledonian		~Taconian~
	—438—							
	Ordovician			~Gulangian~			~Salairian~	
	—505—							
	Cambrian			~Xingkaian~	Formation of Junggar and other median massifs		~Assyntian~	
	—600—							
Late Proterozoic	Sinian	Megastage of platform formation	Jinningian Stage	~Chengjiangian~				
	—850—			~Jinningian~	Formation of Yangzi Platform and Qaidam Massif etc.		~Gothian~	
	Qingbaikouan							
	—1050—							~Grenvillian~
Middle Proterozoic	Jixianian			~Sibaoan~				
	—1400—						~Karelian~	~Hudsonian~
	Changchengian			~Zhongyuean~				
	—1850—		Luliangian (Zhongtianoan) Stage	~Luliangian~	Formation of N. China and Tarim Protoplatforms			
Early Proterozoic	Hutuoan							
	—2200–2300—			~Wutaian~				
	Wutaian			~Fupingian~	Formation of Ordos and Jilu Nuclei		~Belomorian~	~Kenoran~
	—2500–2600—							
Archaena	Fupingian	Megastage of continental nuclei formation	Fupingian Stage					
	—2900–3000—							
	—3800—							
	Hadean							
	—4500—							

* or Nilkaan.

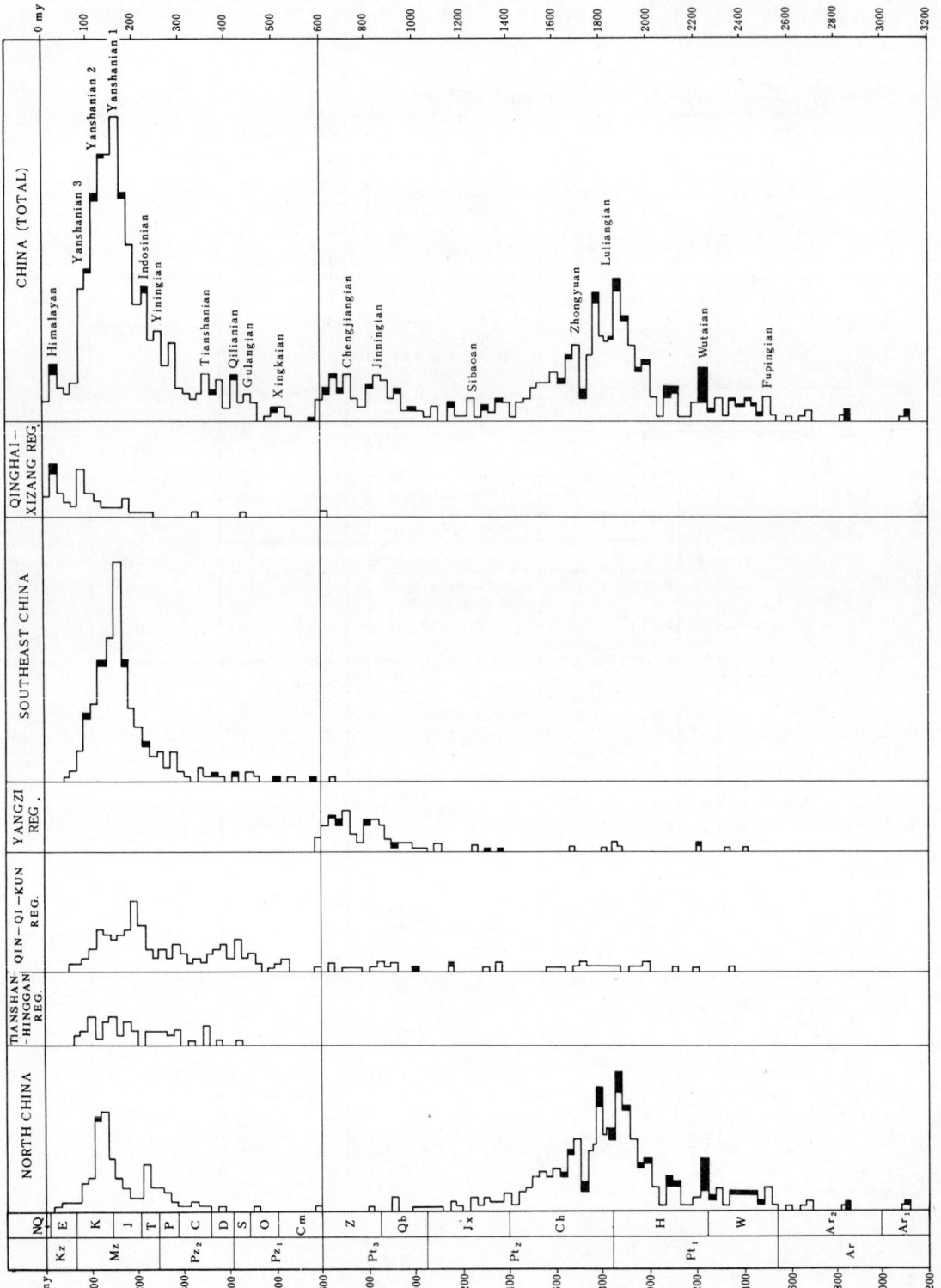

Fig. 20.1. Radiometric age data distribution in relation to orogenic movements and geological events of China.

the variation in composition of basic media such as the hydrosphere and atmosphere, the variation in thickness and scope of sialic massifs, changes in relative relief of the earth's surface and in the palaeogeographic pattern, i.e. the distribution of continents and oceans. Some of the main changes are progressive and irreversible, but the relief and the distribution of the continents may change repeatedly. Using the data collected in China and taking into consideration results obtained in other parts of the world, we have tentatively subdivided the history of the crustal development of China, and probably of the whole earth, into four megastages.

The first megastage comprises the formation of continental nuclei. By continental nuclei are meant the ancient sialic massifs of considerable size that had developed by the end of the Archaean. These are composed mainly of high-grade granite terrains and greenstone belts. The main characteristics of the first megastage were probably the highly mobile state of the crust and a high geothermal gradient. As organic life was still primitive and had no conspicuous geological effects, and as the compositions of the atmosphere and hydrosphere were different from the present ones, sedimentation processes and products were probably rather different from those of today. The process of formation of continental nuclei seems to be the progressive aggregation of small massifs. Once a critical size is reached, they tend to build up the cores or nuclear parts of the accreted primitive platforms. Under these conditions probably no subduction zones comparable with modern ones could be formed. Tectonically the Archaean ended with the Fupingian Movement at about 2500–2600 Ma (Table 20.1).

The megastage of formation of the platforms started in the Proterozoic and ended before the Sinian (2500–2600 to 850 Ma). It may be subdivided into two stages: the Early Proterozoic transition stage (2500–2600 to 1850 Ma) and the Middle and Late Proterozoic stabilizing stage (1850–850 Ma). The most significant change, according to P. E. Cloud (1972), was probably the transformation of hydrosphere and atmosphere from a reducing to an oxidizing state, which brought about profound changes in sedimentation and organic life. The transitional stage ended with the formation of protoplatforms through the Wutaian and Luliangian Movements in China. The stabilizing stage leading to the formation of platforms was characterized by stable carbonate sedimentation and flourishing lower plant life. This stage was terminated by the Jinningian Movement, when a number of platforms and median massifs were consolidated, and stable terrains were probably very extensive both in China and in the rest of the world, thus allowing continental margins of different types to develop. In all probability the formation of island arcs and subduction zones, and also the average thickness of the crust in this stage, had become comparable with Palaeozoic and later conditions. Many Chinese authors believe in the presence of an integral Chinese Platform, which was a part of the supposed Late Proterozoic Pangaea. But this requires further consideration (see below). The Sibaoan Movement of South China (Table 20.1) is tentatively dated at 1100–1200 Ma.

The time-span from the Sinian to the Middle Triassic forms the megastage of the formation of Pangaea. This includes the traditional Caledonian, Hercynian (Variscan), and Indosinian Stage or Cycle, the Xingkaian being considered as a substage of the Caledonian. We have preferred to use the term 'Qilianian' instead of 'Guangxian', because the former displays a more typical and clearer geosynclinal development. It is probably a peculiar feature in China and in East Asia that the Hercynian and the Indosinian are sometimes inseparable from each other, and the most conspicuous change occurred not at the end of the Hercynian but at the end of the Indosinian, especially in Central and South China. (The term 'Hercynian' is preferred to 'Variscan': we wish to emphasize the name of the geosyncline concerned (Hercynian) rather than the trend of the lineament that resulted (Variscan).) The closing orogeny between the Early and Late Permian was called the Dongwu Movement by Li Siguang from a section near Nanjing, but the evidence was subsequently found to be inadequate. The term Yiningian (Boluoan) is here proposed to represent the widespread glaring unconformity between the Lower Permian volcano–sedimentary Wulang Group and the Upper Permian coarse clastic Tiemulik Formation on the southern slope of the Borohoro and Yishgiri Mountains to the north of the Yining Basin in the Xinjiang Autonomous Region.

The fourth and youngest megastage, the megastage of Pangaea disintegration, probably marks the most important change in the history of crustal development, and is characterized by typical plate tectonics. The term Yanshanian is here reserved for the Late Mesozoic, although it comes from an intracratonic range overprinted by a Late Mesozoic orogeny, because subdivision of different phases is here clearly revealed. The Yanshanian is entirely comparable with the Pacific of Soviet usage. The juxtaposition of the orogenic epochs of Laurasia in the table is merely for general comparison; exact time correlation of the orogenic epochs and folding phases is not intended.

The following is a brief synopsis of the crustal evolution of China in the various megastages.

MEGASTAGE OF FORMATION OF THE CONTINENTAL NUCLEI: CRUSTAL DEVELOPMENT OF CHINA BEFORE THE PROTEROZOIC

Definite Archaean nuclear massifs are known only on the North China Platform, the western Ordos (Fig. 20.2, IIB$_1$), and the eastern Ji–Lu (Fig. 20.2, IIB$_2$), which may be further subdivided into a northern Yanliao massif and a southern Hehuai massif. The Archaean of these units is generally devoid of, or poor in, carbonate rocks, while in other regions, notably in Yinshan, Shanxi, Jiaoliao, and western Henan, the Archaean usually contains marbles in the upper part (Cheng et al. 1973). The nuclear massifs probably represent the consolidated and aggregated parts of the

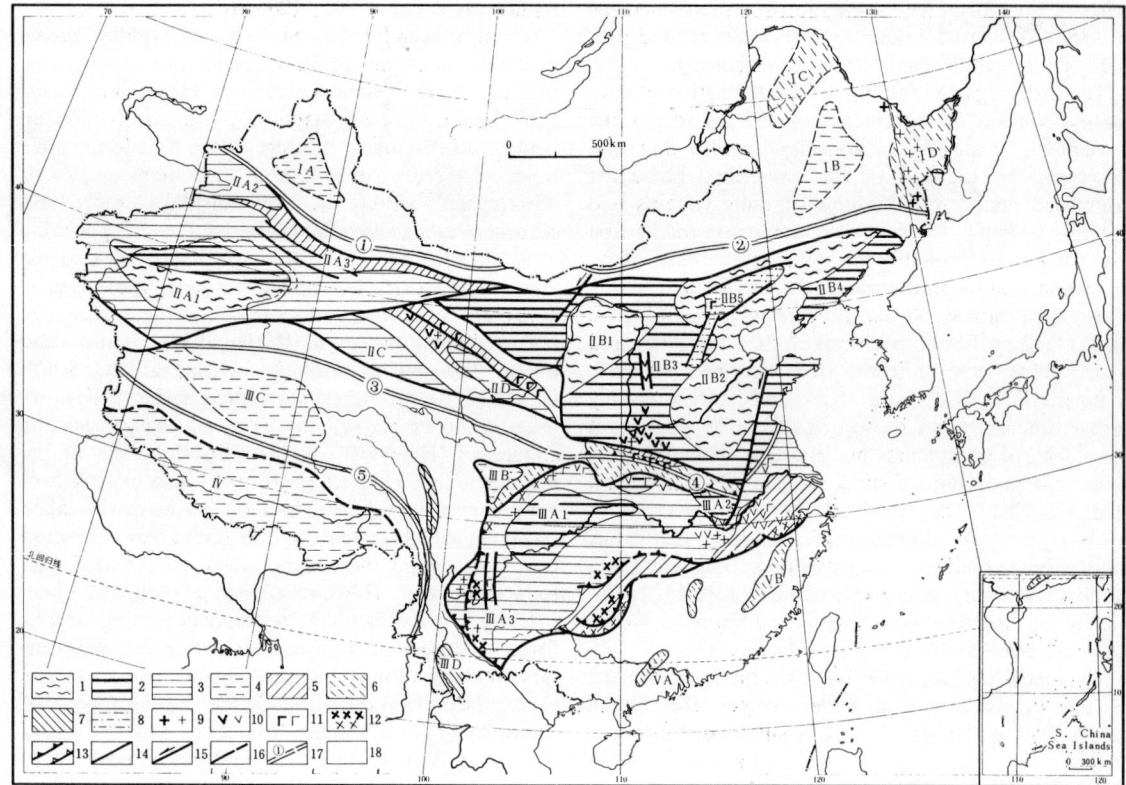

Fig. 20.2. Palaeotectonic sketch map of China prior to the Jinningian (Pre-Sinian) Movement (modified after Wang 1982b). 1–3, continental crust: 1, consolidated end of Wutaian (2200–2300 Ma); 2, consolidated end of Luliangian (1850 Ma); 3, consolidated end of Jinningian (850 Ma); 4, consolidated in Proterozoic in general. 5–9, transitional crust: 5, semi-consolidated end of Jinningian; 6, folded and uplifted end of Jinningian; 7, semi-consolidated in Proterozoic in general; 8, Pt$_2$ (1950–1050 Ma) strongly subsiding region; 9, granitic intrusives: left, Pt$_2$; right, Pt$_3$; 10, intermediate and intermediate-acidic eruptives, left Pt$_2$; right, Pt$_3$; 11, basic eruptives, left, Pt$_2$; right, Pt$_3$; 12, ophiolite suite and island arc volcano-sedimentary associations; 13, subduction zones: left, Pt$_2$; right, Pt$_3$; 14, syn-depositional faults; 15, later transcurrent (mainly Indosinian and Yanshanian) faults; 16, scope of continental massifs; 17–18, oceanic crust: 17, later crustal consumption zones; 18, oceanic realm. Crustal consumption zones 1–5 as in Fig. 19.3.
  I, Northern Continental Margin Domain: IA, Junggar Massif; IB, Song–Liao Massif; IC, Yilehuli Uplift; ID, Xingkai Uplift. II, North China Continent and Continental Margin Domain: IIA, Tarim Platform: IIA1, South Tarim Nucleus; IIA2, Yining Massif; IIA3, Central Tianshan Uplift; IIB, North China Platform: IIB1, Ordos Nucleus; IIB2, Ji–Lu Nucleus; IIB3, Hutuo–Zhongtiao Palaeo-fold zone; IIB4, Jiao–Liao Uplift; IIC, Qaidam Massif; IID, Lanzhou–Xining Uplift. III, South China Continent and Continental Margin Domain: IIIA, Yangzi Platform: IIIA1, Central Sichuan Massif; IIIA2, Dabie Massif; IIIA3, East Yunnan Depression; IIIB, Songpan Massif; IIIC, Qiangtang Massif; IIID, Lincang Uplift. IV, Southern Continent Domain (Himalaya and Gangdise). V, East China Continental Margin Domain: VA, Yunkai Uplift; VB, Jian'ou Uplift.

Archaean sialic crust, while the Archaean in other regions is represented by scattered deposits on the ancient oceanic floor. If we assume that the massifs retained their positions relative to each other within the platform, the chief mode of formation of the continental nuclei seems to be crustal aggregation *in situ*; probably no subduction *sensu stricto* occurred at that time.

The presence of a South Tarim Archaean nuclear massif (Fig. 20.2, IIA_1) is still a matter of speculation, for if it exists it is deeply buried under the basin deposits (see Chapter 19). Archaean rocks may also be included in the Daklakbulak Group, which forms the basement of the Quruktagh Aulacogen together with the folded Proterozoic. In any case, the South Tarim Massif, if present, is isolated, and forms a separate centre of sialic segregation. Furthermore, within the scope of the North China Continental Domain, scattered Archaean rocks may be included in the Dunhuang Group of eastern Tarim and in the Huangyuan Group of south-eastern Qinghuai. The general picture of the North China domain thus seems to be one of isolated nuclear massifs separated by oceanic realms with scattered mafic igneous complexes and mobile-type deposits.

So far no Archaean rocks have been identified in South China or in the intercontinental terrains of other parts of China. Individual age data over 2600 Ma have been reported or inferred for the Dabie Group in the north-eastern corner of the Yangzi Platform and in the Kham–Yunnan Uplift. The deeply buried Central Sichuan Massif has a geotectonic setting much like South Tarim and occupies the central part of the platform ($IIIA_1$). If Archaean rocks are actually included in the basement rocks, they are probably involved in the Lower Proterozoic folded terrains, and do not seem likely to have formed an integral aggregated massif. For the time being we have assumed that apart from the North China domain all other parts of China were dominated by crust of oceanic type in the Archaean.

MEGASTAGE OF FORMATION OF THE PLATFORM: CRUSTAL DEVELOPMENT OF CHINA IN THE MIDDLE AND UPPER PROTEROZOIC (PRE-SINIAN)

The megastage of formation of platforms covers the time span 2500–2600 to 850 Ma and is divisible into two stages: the earlier Luliangian stage of transitional character, which was closed by the Luliangian (Zhongtiaoan) Movement, typically developed in North China, and the later Jinningian Stage, which led to the final consolidation of most of the platforms through the widespread Jinningian Movement, typically developed on the Yangzi Platform.

In the early stage, within the area of the North China Platform, the nuclear massifs underwent consolidation and accretion through two orogenic epochs. The Wutaian Substage ended in the accretion of the Wutai greenstone belts with the nuclear massifs, mainly in Central Shanxi and southern Nei Mongol, and was terminated at around 2200–2300 Ma by the Wutaian Movement. The succeeding Hutuoan Substage is characterized by a more or less 'normal' geosynclinal sequence and the final folding and coalescence of the separate massifs to form the combined North China Protoplatform (Ma *et al*. 1981), as a result of the well-known Luliangian Movement. Well-developed molasse-like deposits are found along the folded terrains both in Shanxi and in the Liaodong Peninsula. To distinguish them from the intracratonic fold zones formed later, these ancient geosynclinal foldings have been named palaeo-fold zones (Wang 1981).

Little can be said about the ancient continental margins of the platform. The folding and metamorphism of the Longshoushan Group of Alxa and the Liaohe Group in Liaoning probably formed marginal uplifts, and the unconformity between them and overlying strata may indicate the active character of the margin. The northern margin is characterized by a metamorphic rock group (the Bayan Obo Group) straddling the Middle–Upper Proterozoic boundary, and is therefore probably passive in nature (Fig. 20.5, A).

In the Tarim region the stable terrains formed after the Luliangian Movement probably included the North Kunlun and the Tieklik Mountains, because a stable type of Middle and Upper Proterozoic was discovered in the Tieklik Range. Thus the main part of the Tarim Platform was consolidated in the Luliangian, and only the northern parts, the Quruktagh and Central Tianshan, were stabilized through the Jinningian orogeny (Fig. 19.3). P. E. Cloud (1972) has called the Aphebian (Early Proterozoic) transitional, on account of the inferred transformation from a reducing to an oxidizing atmosphere. This applies also in China, but it seems that the transitional stage may be better confined to the Wutaian (up to 2200–2300 Ma), for red beds appeared in the Hutuoan, and stromatolites began to flourish in the same period.

In the Jinningian Stage, the North China Protoplatform gradually became stabilized. Tensional conditions seem to have prevailed for most of the time, for two kinds of aulacogen appeared (Wang 1982); the Yanshan Aulacogen (Fig. 20.2, $IIB5$), inside the Nei Mongol marginal uplift (Inner Mongolian Axis of Huang), receiving 10 000 m of clastic and carbonate sediments, with potassium-rich alkali eruptives in the lower part; and the Luliang–West Henan Aulacogen, filled with partly continental sediments and with volcanic materials. Apart from these strongly subsiding belts, the Middle and Upper Proterozoic deposits are mainly of stable type. The whole region underwent uplifting (Qinyu uplifting of Qiao) at the end of the middle Proterozoic; the Qingbaikouan is limited in distribution and consists entirely of a cover sequence. In this sense the North China Platform was also finally established at the end of the Jinningian Stage; the eastern part of the platform had undergone strong subsidence in the Sinian Period.

The probably passive nature of the northern continental margin of the North China and Tarim ancient continents in the late Early Proterozoic has been alluded to above. The unconformable contact between the Jixianian Fengjiawan Formation and the Qingbaikouan strata, and the emplacement of granitic rocks 999 Ma in age (see above) in the Xiao Qinling area of western Henan, are indicative of a northward subduction from the Qinling marine realm; this is probably also responsible for the general uplifting of North China in the beginning of the Qinbaikouan.

As stated above, the Middle and Upper Proterozoic are not thick in southern Tarim or in the Yining region of Central Tianshan. Carbonates are the prevailing rocks. Thick piles of clastic and carbonate rocks were probably accumulated, and later strongly disturbed, only along the Quruktagh trough and Central Tianshan, which marks the actual northern margin of the ancient stable terrain (Fig. 20.2, IIA 3–4).

In the Jinningian Stage, the relation between Alxa, Central Qilian, and the Qaidam Massif is of special interest. The Lower Proterozoic basement does not seem to be continuous between Central Qilian and Alxa, for the Middle and Upper Proterozoic of North Qilian are of mobile volcano-sedimentary type, though they were probably not very far from each other: Upper Sinian glacial beds occur in both regions as well as in the southern part of North China.

In South China, within the area of the present Yangzi Platform, old massifs formed prior to the Middle Proterozoic include the Central Sichuan, Dabic, and Kham–Yunnan massifs, which were stabilized in the Luliangian Movement (Fig. 20.2, IIIA1, IIIA5). To the south of these old massifs, along the Jiangnan Uplift, island arcs and marginal seas of two ages, Middle Proterozoic and Late Proterozoic, were formed, folded and accreted to the Proto-Yangzi Platform (Fig. 20.2, Fig. 19.7, and Fig. 20.5, B) through the Sibaoan (Table 20.1) and Jinningian orogeny respectively. On the Kham–Yunnan Uplift, the island arc and marginal basin system is represented by the Dahongshan and Hekou Groups and forms the basement, usually with a WNW–ESE trend. The Middle and Upper Proterozoic Kunyang and Huili Groups seem to have been deposited in longitudinal rifted troughs formed on the basement. Thus it is possible that the ancient basement was of greater extent and was rifted under extension in the Luliangian orogeny. Mention has been made of the tectonic setting of the Dabie Massif (Chapters 5 and 19). It is tentatively included here. The massif was possibly a separate element in the Middle Proterozoic and was incorporated with the mainland Yangzi Platform at a later stage.

Apart from the principal continental massifs enumerated above, no Precambrian rocks older than the Middle Proterozoic have been discovered in other parts of China. Isotopic ages of 1250 Ma were obtained from metamorphic rocks of the Himalaya and Gangdise massifs (Fig. 20.2, IV) in which older strata may also be included. Except for the median massifs, the Qiangtang in the south and the Junggar and the Songliao in the north, which were composed of stabilized sialic masses in the Precambrian, all the remaining regions were then probably occupied by oceanic crust.

MEGASTAGE OF FORMATION OF PANGAEA: CRUSTAL DEVELOPMENT OF CHINA FROM SINIAN TO TRIASSIC

This megastage covers the time interval from the Sinian to the Triassic, and contains three tectonic cycles or stages; the Caledonian, the Hercynian, and the Indosinian. The term 'Pangaea' as used here refers to the generally accepted united supercontinent of Late Carboniferous to Triassic times (Pangaea A–B of Morel and Irving 1981). Piper (1976) has indicated a Proterozoic Pangaea (Pangaea E), and Seyfert and Sirkin (1979) have shown from palaeomagnetic calculations that North America and Gondwanaland united and separated several times in the Proterozoic and Palaeozoic. Although Gondwanaland had long been a united supercontinent before breaking up in the Mesozoic, the northern continents had probably remained separate until they became united in the Permian and Triassic. The congruity of apparent polar wandering curves of one northern continent with Gondwanaland does not therefore necessarily indicate the presence of Pangaea. On the other hand, many of the ancient massifs within the borders of China and Asia were probably separate in the Palaeozoic, and a united Chinese Platform does not seem to have been created until the Late Hercynian and Indosinian stages.

The Caledonian Stage

The Caledonian Stage as used here covers the time interval from the Sinian to the Silurian, and contains three orogenic epochs or movements; the Xingkaian, the Gulangian, and the Qilianian (Fig. 20.1). The main features of the palaeogeographical and palaeotectonic development comprise the differentiation of sedimentary types in the interior of the platform, and crustal rifting and opening in the continental margins and marginal rifts. The formation of the extensive Caledonides in South China was accomplished through successive accretion by the folding and uplifting of island arcs and marine troughs.

North China Platform and its marginal tracts

The internal differentiation of the North China Platform is manifested by the longitudinal sedimentary belts and the change of direction of marine transgressions. The Jiaoliao region on the eastern border of the platform underwent strong subsidence in the Sinian Period and was partly uplifted in the Cambrian (Fig. 20.3, IIB_1). The central region, or platform interior, was a peneplane of weak denudation in the Sinian, and was bordered on the south by mountainous regions with mountain glaciers. The central region had remained a stable epicontinental sea in the Cambrian and Early Ordovician and was completely elevated in the Middle Ordovician. In the western belt (Fig. 20.3, IIB_3), Late Sinian tillites have been discovered to the west of Ordos. Subsidence partly continued in the Late Ordovician and Silurian, and the western belt is thus distinct from the central and eastern belts. The direction of the transgression reflects subduction and compression in the continental margins. In the Cambrian there was a southward tilting of the whole region which caused a general northward transgression in the Ordovician. This was followed by a general upheaval of the whole region in the Middle Ordovician. The discovery of ophiolite suites and radiolarian cherts, and of basal Cambrian fossil beds, in the lower part of the Wendurmiao (Ondor Sum) Group may indicate a southern subduction of the Nei Mongol ocean floor and account for the southward tilting. The final elevation of the platform is closely related to the subduction on both the northern and southern margins. The unconformable relation of the Cambro-Ordovician 'eugeosynclinal' Wendurmiao Group with the overlying Silurian in the north, and the unconformity both below and above the Early to Middle Ordovician Caotangou Group of the Qinling region in the south (Wang, H. *et al.* 1982) are indications of subduction on the opposite margins of the platform (compare Fig. 20.6, B, C and Fig. 19.4).

Rifting and opening of the South Tianshan trough and formation of the Qilian Fold Belt

Rifting and opening leading to aulacogen formation within the continental massifs were outstanding features of North-west China in the Caledonian Stage. In the Quruktagh Mountains the Sinian System, consisting of thick piles of coarse clastics and volcanic eruptives, represents an aulacogen (Fig. 20.3, IIB_1) formed on the inner side of the Central Tianshan, the northern border uplift of the Tarim Platform. Another aulacogen-like unit, the South Tianshan faulted trough, began to appear in the Cambrian and was miogeosynclinal up to the Ordovician, but became eugeosynclinal in the Silurian and Devonian (Ren, J. *et al.* 1980). This change of sedimentary type is a reflection of the increasing depth attained by the fractures. As there is no appreciable unconformity between the Sinian and the overlying strata in Quruktagh, the aulacogen seems to be more typical, for it was filled up, not folded, at its closure. The South Tianshan has on the other hand a longer history and constitutes a normal Hercynian fold zone.

As mentioned above, the relation between the Alxa Massif and Central Qilian in the Proterozoic remains obscure. Island arc sedimentary associations appeared more than once in the North Qilian trough from the Middle Cambrian to the Late Ordovician, and oceanic crust may have been subducted in both directions, probably with northward subduction prevailing, for Caledonian granites are distributed both in North Qilian and in the Longshou Mountains to the north of the corridor (Fig. 20.3, IID_1).

As the pre-Sinian basement rocks of Central Qilian and Qaidam bear a close resemblance to one another, the South Qilian and the narrow trough on the northern Qaidam border may be better regarded as aulacogens or rifting troughs formed on a united basement in the Ordovician, but quickly closed at the end of the Silurian (Fig. 20.3, IID_3). A short-lived episode of sea-floor spreading occurred in the Ordovician, and the subduction was probably again northward.

On the southern border of the Qaidam Massif, the eugeosynclinal Early Palaeozoic Naij Tal Group is followed by the Devonian without a clear unconformity, but further northward the Devonian molasse-like Maoniushan Formation is unconformable on all underlying rocks. Northward subduction probably occurred in the late Caledonian, although the southern continental margin remained open to the Proto-Tethys ocean in the south. The unstable and solitary crustal elements then existing, or newly created in the Early Palaeozoic, were again reassembled and consolidated at the end of the Caledonian Stage, thus forming the

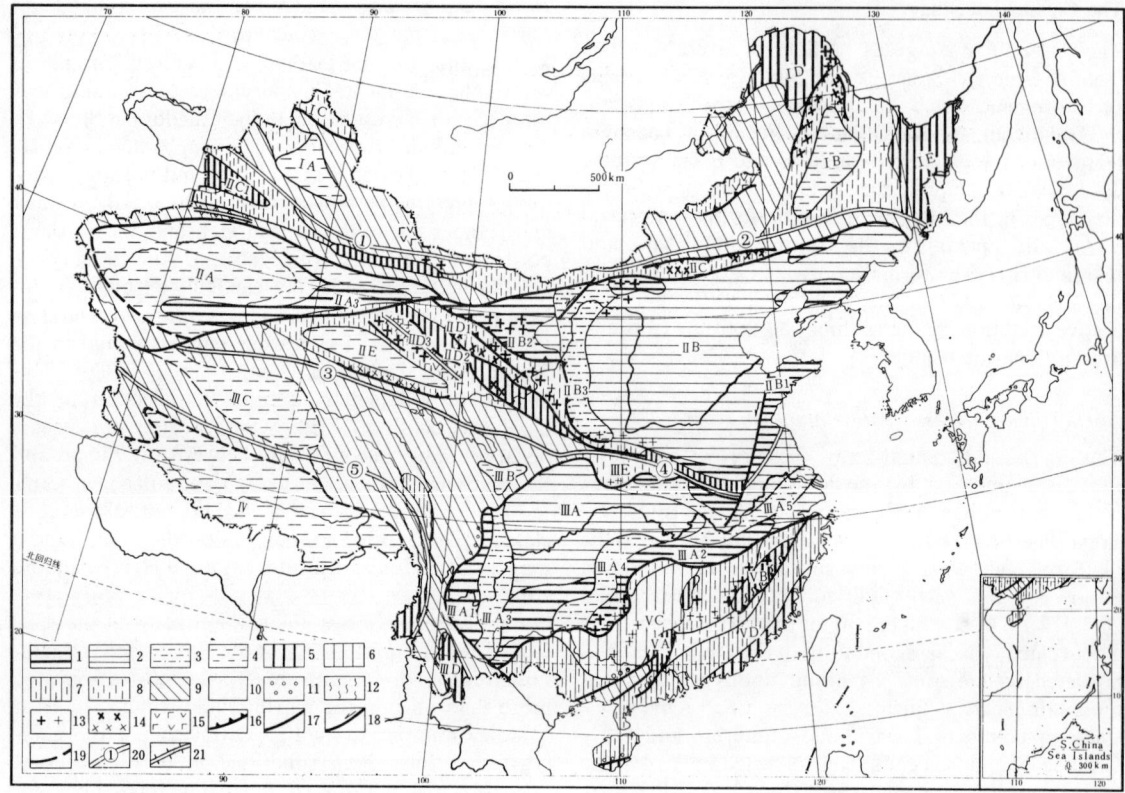

Fig. 20.3. Palaeotectonic sketch map of China in the Caledonian stage (modified after Wang 1982b). 1–8, continental crust; 1–4, stable type: 1, positive element uplifted after Cambrian; 2, epicontinental sea sedimentary regions, North China uplifted after O_2, Yangzi after S_1; 3, subsiding sedimentary regions, North China uplifted after O_3, Yangzi after S_1; 4, stable regions undifferentiated; 5–8, mobile type: 5, Xingkaian fold zone or uplifted end of Pt_3; 6, marginal seas mainly uncompensated, uplifted end of Ordovician; 7, marginal sea and trough mainly compensated, uplifted end of Ordovician; 8, sea trough, uplifted end of Silurian; 9, transitional crust: marginal sea, continental slope and sea trough, Silurian and Devonian continuous; 10, oceanic crust; 11, coarse clastic deposits, Ordovician; 12, Middle Caledonian (mainly Ordovician) metamorphic and migmatitic zones; 13, granitic intrusives: left, Early and Middle Caledonian; right, Late Caledonian; 14, ophiolite suite and island arc volcano-sedimentary associations: left, Cambro-Ordovician; right, Ordovician–Silurian; 15, submarine intermediate-basic eruptives; 16, accretional crustal consumption zones; 17, syn-depositional faults; 18, later transcurrent faults; 19, scope of continental massifs; 20, convergent crustal consumption zones; 21, sea floor spreading zones. Crustal consumption zones 1–5 as in Fig. 19.3.

I, Northern Continental Marginal Domain: IA, Junggar Massif; IB, Song–Liao Massif; IC, Yining Massif; ID, Yilehuli Uplift; IE, Xingkai Uplift; IF, South Nei Mongol Fold Zone. II, North China Continent and Continental Margin Domain: IIA, Tarim Platform: IIA1, Kuruktagh Aulacogen; IIA2, Qiemo–Ruoqiang Uplift; IIB, North China Platform: IIB1, Jiao–Liao Uplift; IIB2, Alxa Uplift; IIB3, Transordos Depression; IIC, Qaidam Massif; IID, North Qilian–North Qinling Fold Zone; IIE, South Qinling Fold Zone. III, South China Continent and Continental Margin Domain: IIIA, Yangzi Platform: IIIA1, Kham–Yunnan Uplift; IIIA2, Jiangnan Uplift; IIIA3, East Yunnan Depression; IIIA4, Bamianshan Depression; IIIA5, Lower Yangzi Depression; IIIB, Songpan Massif; IIIC, Qiangtang Massif; IIID, Lincang Uplift. IV, Southern Continent Domain. V, East China Continental Margin Domain: VA, Yunkai Uplift; VB, Wuyi Uplift; VC, Xiang–Gui Marginal Sea; VD, Min–Yue Sea Trough.

virtual southern margin of the North China Continental Domain (compare Fig. 20.8, A, B, and Fig. 19.5).

Internal differentiation of the Yangzi Platform and formation of the South China Caledonides

A conspicuous feature of the Yangzi Platform in the Caledonian Stage is the development of marginal uplifts and of depressions on their inner sides. From the Sinian up to the Middle Ordovician the Kham–Yunnan Uplift had been a constant source of detrital supply to the east (Fig. 20.3, $IIIA_1$, $IIIA_2$), but it was not until the Late Ordovician and Silurian that the Jiangnan Uplift (Fig. 20.3 $IIIA_2$) began to supply abundant detrital material, when immensely thick, Silurian variegated clastic deposits were formed in the western depressions (Fig. 20.3, $IIIA_4$). The differentiation of the platform interior is manifested by the occurrence of evaporite basins in the Sinian and Cambrian, the juxtaposition of alternating shallow seas and small restricted basins in the Ordovician, and of a large-scale marine euxenic basin of marginal sea type in the Early Silurian. The major part of the platform was elevated in the Middle Silurian. The northern and western margins of the Yangzi Platform were characterized by fragmentation, and partial detachment of minor massifs. The South Qinling region had probably undergone extension (Fig. 20.3, IIIE), the northernmost element being the Zhen'an Massif bearing a stable Sinian and Cambro-Ordovician sequence of the Yangzi type. The intervening euxenic basins receiving Sinian and Lower Palaeozoic black shales and basic eruptives probably owe their origin to crustal extension (Wang, H. *et al.* 1982) (Fig. 20.5, B). The situation is somewhat similar in western Sichuan, where the Songpan Massif (Fig. 20.3, IIIB) and the Qamdo Caledonian zone may represent detached remnants originally situated at the platform border.

The Caledonides of South China may be recognized as a further accretion and extension of the Yangzi Platform toward the south-east. The threefold subdivision of the region in the Caledonian Stage into a western marginal basin, a central island group or arc, and an eastern marine trough has been explained in Chapter 19. Suffice it to say here that the emplacement of Early and Late Caledonian granites on the western side of the Yunkai–Jian'ou belt, and the presence of migmatization and metamorphism along the same belt, indicate that small-scale westward subduction may actually have occurred (compare Fig. 17.1 and Fig. 20.3). Most parts of the region were uplifted and folded in the Middle Silurian.

Other parts of China

Three regions may be mentioned here. In North-east China, definite Early Caledonian or Xingkaian fold zones are known in Yilehuli and Jiamussi, where Sinian fossils of *Ediacara* type and Cambrian trilobites occur in the folded metamorphic rocks. In the Xiao Hingan Mountains and to the west of the Nenjiang River, Ordovician submarine metabasites and ophiolite suites of island arc type are found surrounding the Songliao Massif on the north. Similar rocks occur also in East Ujimqin Qi. Northward subduction toward the adjacent Tuotuoshan Massif near the southern boundary of the People's Republic of Mongolia may have occurred (Fig. 20.3).

In the Altai and Junggar region to the north of the Aibi–Juyan convergent zone, Lower Palaeozoic deposits of marine mobile type were formed all over the region. The Altai region was affected by the Caledonian orogeny, but was more strongly involved in the Hercynian orogeny. Mobile conditions continued from Early Palaeozoic to Devonian times in the marine realm surrounding the Junggar Massif and in the Baytik-Mountains, where the crust was evidently of a transitional type in the Caledonian Stage (Fig. 20.8, B, C).

The united Himalaya–Gandise Massif has a common stable type of Ordovician and Silurian which passes upwards into the Devonian, the Cambrian being generally poorly represented and, when present, distinct from that in the Yangzi Platform and in Australia. The Ordovician has yielded some nautiloids related to those found in Ordos and in Qilian. Silurian fossils were meagre in the Himalaya, but the Silurian near Dongqiao to the north of Baingoin bears a shelly fauna somewhat resembling the Yangzi type. It is probable that the Himalaya–Gandise region was on the northern margin of Gondwanaland and was not directly connected with the Yangzi Platform and Australia. Shallow seas may have existed between the Himalaya and the Qilian region in the Ordovician, and the Baingoin region may have occupied a lower (southern) latitude, as was the case with the Yangzi Massif in the Silurian.

The Hercynian (Variscan) and Indosinian Stages

As the Hercynian orogeny is usually not readily separable from the Indosinian, especially in Central and South China, these two stages are here treated together, and the main features in both stages are shown in the same sketch map (Fig. 20.4). The Hercynian Stage includes two orogenic epochs; the Early Tianshanian in the Early Carboniferous and the later mid-Permian Yiningian or Nilkaan, named after the Nilka District in the Yining Basin. There are two aspects to major

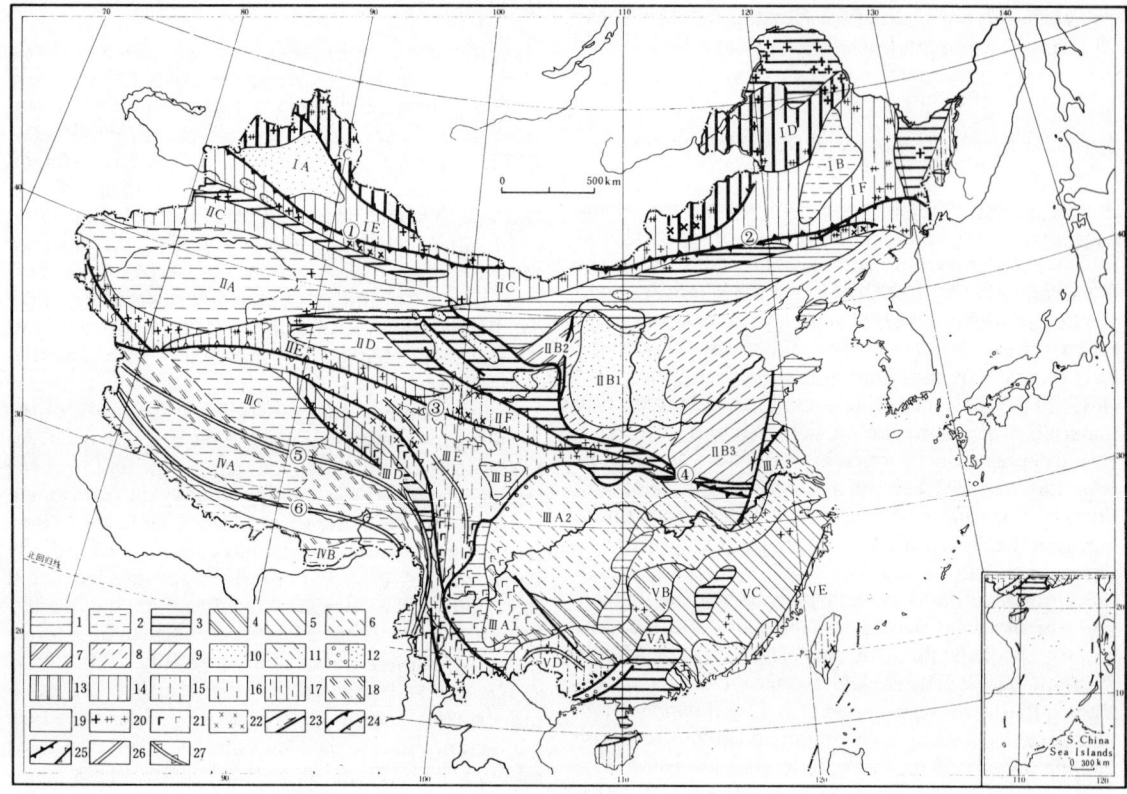

Fig. 20.4. Palaeotectonic sketch map of China in the Hercynian–Indosinian stages (modified after Wang 1982b). 1–16, continental crust; 1–12, stable type: 1, pre-Sinian basement uplifts; 2, stable regions undifferentiated; 3, Caledonian basement uplifts; 4, Devonian to Triassic strongly subsiding regions; 5, Carboniferous to Triassic epicontinental sea deposits; 6, Carboniferous to Triassic shelf sea deposits, mainly siliceous and argillaceous; 7, Carboniferous strongly subsiding regions; 8, Carboniferous paralic, Permian terrestrial deposits; 9, Carboniferous to Permian paralic deposits; 10, Triassic, mainly terrestrial deposits; 11, Triassic strongly subsiding regions; 12, coarse clastics: left, Permian; right, Triassic; 13–17, mobile type: 13, Early Hercynian fold zones; 14, Late Hercynian fold zones; 15, Early Indosinian fold zones; 16, Late Indosinian fold zones; 17, Hercynian and Indosinian fold zones undifferentiated; 18, transitional crust: Triassic and Jurassic continuous; 19, oceanic crust; 20, granitic intrusives: left, Early Hercynian; median, Late Hercynian; right, Indosinian; 21, Carboniferous to Triassic basic eruptives; left, submarine; right, continental; 22, ophiolite suite and island arc volcano-sedimentary associations; 23, faults: left, syn-depositional; right, later transcurrent; 24, accretional crustal consumption zones; left, Hercynian; right, Indosinian; 25, Convergent crustal consumption zones: left, Hercynian; right, Indosinian; 26, later crustal consumption zones; 27, sea floor spreading zones. Crustal consumption zones 1–6 as in Fig. 19.3.

I, Northern Continental Margin Domain: IA, Junggar Basin; IB, Song–Liao Massif; IC, Bayitik Early Hercynian Fold Zone; ID, Hingan–Xiliongol Early Hercynian Fold Zone; IE, Transjunggar Late Hercynian Fold Zone; IF, Zhangguangcailing Late Hercynian Fold Zone. II, North China Continent and Continental Margin Domain: IIA, Tarim Platform; IIB, North China Platform: IIB1, Ordos–Shaanxi Basin; IIB2, Trans-Ordos Depression; IIB3, He–Huai Depression; IIC, South Tianshan–Beishan Late Hercynian Fold Zone; IID, Qaidam Massif; IIE, East Kunlun Hercynian Fold Zone; IIF, West Qinling Indosinian Fold Zone. III, South China Continent and Continental Margin Domain: IIIA, Yangzi Platform: IIIA1, East Yunnan–South Guizhou Depression; IIIA2, Upper Yangzi epicontinental seas; IIIA3, Lower Yangzi Depression; IIIB, Songpan Massif; IIIC, Qiangtang Massif; IIID, Kaixinling Hercynian Fold Zone; IIIE, Bayan Har Indosinian Fold Zone. IV, Southern Continent Domain: IVA, Gangdise Massif; IVB, Himalaya Massif. V, East China Continental Margin Domain: VA, Yunkai Uplift; VB, Xiang–Gui Marginal Sea; VC, Zhe–Gan marginal sea; VD, Youjiang Indosinian Fold Zone; VE, Min–Yue Hercynian Fold Zone. Crustal consumption zones 1–6 as in Fig. 19.3.

changes in the tectonic and palaeogeographic frame. The first is the coalescence of the southern margin of the Siberian–Mongolian continent and the northern margin of the North China continent, thus bringing about the formation of extensive Hercynides and the preliminary completion of the ancient Pal-Asian continent. Secondly, rifting and fragmentation of the continental margins surrounding the Palaeo-Tethys ocean had resulted in a complicated relationship between the faunas and floras, stratigraphic types and sequences, and magmatic activities of the various massifs. Also, the Yangzi Platform had probably moved away from Australia, with vast basic eruptions in the Permian on its western border, under tensile conditions.

The Indosinian Stage corresponds approximately to the Triassic Period and marks a critical time in the crustal development of China, after which the tectonic directrices and palaeogeographic outline turned from E–W to ENE–WSW to NE–SW to NNE–SSW, and the marine realm of South China, that had continued from the Sinian up to the Triassic, had finally disappeared. The convergence of the two opposite continental margins in the north in Hercynian times further developed into continental collision in the Indosinian. More important was the convergence between the North China and Yangzi Platforms and the consolidation of the extensive Indosinides in western Sichuan and Bayan Har Mountains. On the south-eastern side of the South China Caledonides the Hercynian and Indosinian fold zones developed along the coastal belt. The main features and development of the various tectonic domains will be dealt with separately in the following pages.

Convergence of the two northern continental margins and formation of the Hercynides

The Upper Palaeozoic within the narrow Caledonian fold zone on the northern margin of the North China Platform has a limited thickness and is devoid of volcanic rocks. Here the Lower Permian is characterized by a Tethyan warm-water fauna, whereas the Upper Permian contains a Cathaysian flora. But on the northern side of the Solun–Xilamulun convergent zone substantial thicknesses of Devonian and Carboniferous basic, and intermediate to basic, volcanic rocks are abundant. Paralic deposits began to appear in Middle Carboniferous times. They contain an Angara flora. An accretional crustal consumption zone probably existed to the south of the East Ujimqin Qi, representing an island arc bordering the Tuotuoshan Massif across the border line in the north, which forms the southern limit of the Early Hercynian fold zone (Fig. 20.4, ID). Near Erenhot the Lower Permian Zhesi Formation contains a cool-water fauna of Siberian type and an Angara flora, but mixed with warm-water elements in higher horizons. Still further to the north in the Da Xingan Mountains the faunas and floras are exclusively of the northern type. A reasonable interpretation is as follows. The North China continent probably had a passive northern margin in the Devonian and Carboniferous when the intervening sea between the two opposite continental margins was still quite wide. They approached each other in the Early Permian and became joined in the Late Permian. Granites formerly designated as Hercynian may include some of Indosinian age, especially as the age limit of the Early Triassic should be considerably lowered according to Harland *et al.* (1982). The emplacement of some Late Hercynian to Early Indosinian granites is a reflection of further collision after the convergence (Fig. 20.4, Fig. 20.5, C, D).

In North-west China the northern margin of the ancient Tarim Platform is represented by Central and North Tianshan, while the regions to the north of the Aibi–Juyan Lake convergent zone represent the complicated and broad marginal tract of the Siberian Platform. This is evidenced by the occurrence of the Silurian *Tuvaella* fauna and the Permo-Carboniferous Angara flora, as in North-east China. Notable differences are the earlier convergence of the two continental margins, probably in the Late Carboniferous to earliest Permian, the earlier appearance of inland basins (Fig. 20.4, IA) in the Late Permian, and the presence of the European plant *Callipteris* mixed with the Angara forms. The Altai and Bayitik regions were strongly affected by the Early Hercynian Movement and are limited in the south by the Bayitik (Ertix) accretional crustal consumption zone (compare Fig. 19.3, 6, and Fig. 20.8, C). It is also interesting to note that near Zhusleng Hairhan in the north of the Badain Jaran desert, marine Devonian deposits bearing abundant brachiopods and corals (including *Calceola sandalina*) of obvious Tethyan affinity are found, in strong contrast to the Middle Devonian in the Central Da Hingan Mountains and East Ujimqin Qi, which contains a fauna of North American kinship. As this locality is situated in the south of the convergent zone, it is not surprising that the Devonian faunas belong to the Tethys realm and were connected with South Tianshan and Central Asian regions. Nor is the Angara flora found in the Permian and Carboniferous in this region. As a Lower Carboniferous ophiolite suite is reported from the Choltagh of North Tianshan, and as the Lower Carboniferous is lacking in Badain Jaran, a southward subduction from the northern marine realm might have occurred along the northern margin in Early Carboniferous (Fig. 20.4).

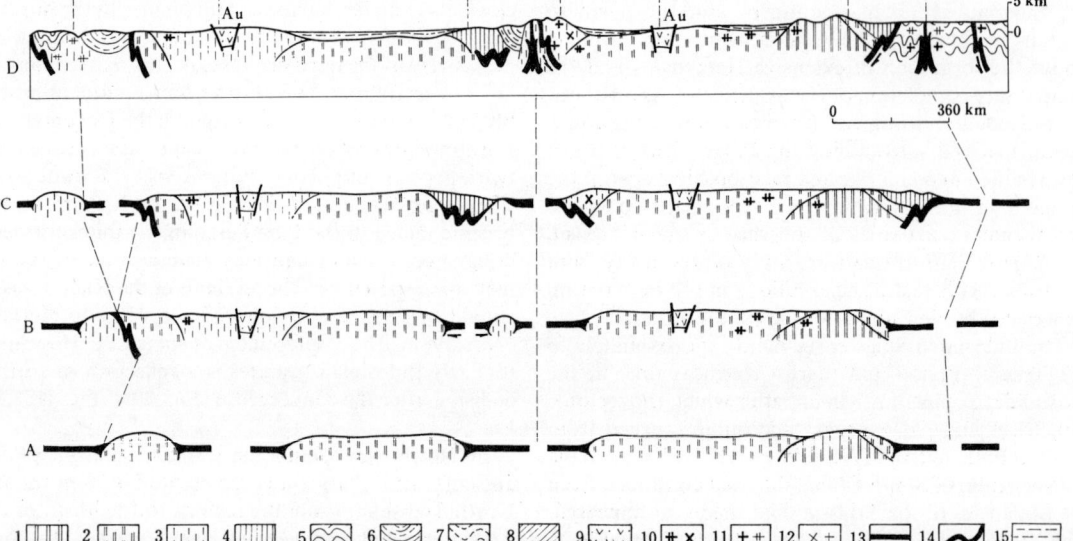

Fig. 20.5. Schematic profiles from Lincang, Yunnan to Wendurmiao, Nei Mongol, showing tectonic development in various stages (platform cover omitted, partly shown in uppermost profile). 1, Archaean nuclei; 2, Lower Proterozoic basement; 3, Middle and Upper Proterozoic basement; 4, Caledonides; 5, Hercynides; 6, Indosinides; 7, Yanshanides; 8, Himalayides; 9, aulacogen deposits; 10, Precambrian (left) and Caledonian (right) granites; 11, Hercynian (left) and Indosinian (right) granites; 12, Yanshanian (left) and Himalayan (right) granites; 13, subducted or residual oceanic crust; 14, oceanic crust; 15, platform cover. A, end of Early Proterozoic; B, Sinian; C, Devonian; D, Recent.

Convergence of the North China and Yangzi Platforms and formation of the Hercynides and Indosinides

The differentiation of the North China Platform, into a western subsiding belt, an eastern uplifted one, and a central stable belt (Fig. 20.4, IIB2, IIB1, 3), is similar to that of the Caledonian. A Middle Carboniferous to Lower Permian paralic facies was by far the most widely distributed in the platform interior. Inland basin facies appeared in Ordos and adjacent regions in the Late Permian, while marine ingressions continued to occur in the south-eastern part (Fig. 20.4, IIB3), in northern Henan, and northern Anhui. This is an indication of the presence of the Qinling Sea in the Late Palaeozoic, whence came marine ingressions toward the north. In the North Qinling of southern Shaanxi, the southern limit of the North China Platform was represented by the thick Devonian and Lower Carboniferous clastic flysch, including typical turbidites derived from a northern source area along the Shangxian–Taibai belt to the north of the Shanyang–Tongcheng convergent zone (Wang, H. *et al.* 1982). Immediately to the north of this belt is distributed the Lower to Middle Carboniferous Caoliangyi coal series of terrestrial and volcano-sedimentary facies. On the other hand, the wide marginal tract of the Yangzi Platform to the south of the convergent zone is characterized by Devonian and Carboniferous deposits of shelf type, probably at that time still far from the northern margin of the marine realm. Conditions seem to have changed in the Permian and Triassic, when the two opposite margins drew near to each other, as evidenced by the Lower and Middle Triassic Liufengguan flysch that finally filled up the trough and led to the coalescence of the opposite continental margins. The incorporation is further indicated by the emplacement of Indosinian granites followed by Yanshanian batholiths, which evidently resulted from further continental collision (Fig. 20.5, D). The narrow Indosinides pinched in between the North China and Yangzi Platforms terminate east of Zhenan and expand in the west, enclosing the eastern end of the Qaidam Massif (Fig. 20.4, IIF). As marine Triassic rocks have never been found from western Henan eastwards, it was generally assumed that the eastern segment of the Shanyang–Tongcheng convergent zone was closed in the Late Hercynian instead of the Indosinian, although some authors believe that the whole zone was closed in the Indosinian on account of the presence of Middle Triassic ammonites of North Pacific type in west Qinling, which may indicate that a sea-way leading to the Pacific still existed between the two continents in Triassic times.

In the western extension of the convergent zone in the south of Burhan Buda, some 5000 m of quartz sandstones and carbonates overlie the Naij Tal Group with a profound unconformity. The overlying Permian has a similar lithology. They may represent deposits formed on a passive continental margin, but the very extensive belt of Middle to Late Hercynian granitoid rocks along the Burhan Buda and the Arkatag (Fig. 20.4, IIE), and especially the presence of a Hercynian ophiolite suite, provides an indubitable indication of the northward subduction of the Palaeo-Tethyan oceanic realm along the Xiugou–Maxin convergent zone (Fig. 17.1).

The nature and the main active period of the Tancheng–Lujiang transcurrent fault (Fig. 19.3, 13) has been much discussed recently. There are various estimates of the magnitude of lateral displacement. Strata up to Permian in age were evidently affected by the fault, but Cretaceous basin deposits seem to have straddled the fault line. It is therefore inferred that the principal lateral motion on both sides of the fault occurred in the Indosinian and the Early Yanshanian stages.

The Yangzi Platform and its surroundings

The major part of the Yangzi Platform was above sea level in Devonian and Carboniferous times, but was submerged in the Permian and underwent strong differential subsidence in the Triassic. Basic eruptive and intrusive rocks were formed on the Kham–Yunnan Uplift in the Permian. Two sets of foundation faults, trending ENE–WSW and NW–SE respectively, were developed on the Caledonide foundation in Guangxi, and extended to western Guizhou on the platform, and this exerted an obvious control over the facies distribution throughout the Late Palaeozoic. Differentiation into a platform carbonate facies and a marine trough silicolite facies was especially conspicuous in Devonian and Permian times (Hou et al. 1982). The development of the Qinfang Aulacogen and the Late Hercynian deformation that it suffered (Fig. 20.4) were mentioned in Chapter 19, and the history of the Youjiang Indosinian geosyncline of western Guangxi has also been described.

It is noteworthy that, in the extensive Caledonides of South China, variation in facies was pronounced in Permian times and was evidently controlled by the Yunkai and Jian'ou uplifts, around which Hercynian and Indosinian granitic rocks were also intruded. The occurrence of a Hercynian–Indosinian fold belt along the south-eastern coast and in Hainan Island has already been noted. It should be mentioned that this giant arcuate fold belt is continuous with the Hercynian–Indosinian belt on the south-western side of the Honghe (Red River) of northern Vietnam and of western Yunnan.

Hercynian marginal rifting and opening were prevalent on the western border region of the Yangzi Platform. The Palaeozoic stratigraphic sequence and faunal content in the Qamdo and Batang region are very similar to those of this platform. Important crustal rifting and opening occurred in the Jinsha River and Litang belts in the Hercynian Stage, but the oceanic crust thus formed seems to be of limited extent. The Caledonian and Early Hercynian fold zones in the Qamdo region along the Lancang River valley had probably moved westwards to their present position in the Late Palaeozoic (Chen 1983), and the newly opened marine basin occupying the present sites of western Sichuan and Bayan Har Mountains was filled by Triassic, especially Upper Triassic, sediments and subsequently folded to form the extensive Late Indosinian folded terrains (Fig. 20.4, IIIE).

As mentioned above, the Palaeozoic (mainly Hercynian) geosyncline of the Qamdo region (Fig. 20.4, IIID) extended northward to the Tangula and southward to the Lancang Massif of western Yunnan, the western boundary of which was probably limited by an eastward subducting zone (Fig. 20.4). The northern section of the Hercynian zone is characterized by a conspicuous Early Permian island arc belt stretching from Jurban Ula to the south of Zhaduo, which probably represents a subduction zone toward the Qiangtang Massif. A typical Late Permian Cathaysian flora is found in the Shuanghu coal beds in eastern Qiangtang, in the Wuli coal beds in Tangula, and in the Tuoba coal beds near Qamdo. This flora can be correlated with a contemporaneous flora on the Yangzi Platform, and all three localities were thus probably situated at similar palaeolatitudes at that time. The southward subduction of this belt is probably contemporaneous with and compensatory to the sea-floor spreading in western Sichuan and Bayan Har (Fig. 20.4, IIIE).

It is evident that the incorporation and accretion of the extensive Indosinides to the northern massifs mark the final step in the formation of the Pal-Asian continent. It implies a large-scale northward movement of the Yangzi and Qiangtang massifs, thus bringing about the closure of the northern or Palaeo-Tethys and the broadening of the median or main Tethys (Hao et al.

1983; Wang 1983). Some comment may be made here on the location and extent of the median Tethys. To the south of the line from Lazhulong in the Karakhorum, passing via north of Garze, to Basu in the Nujiang valley, there occur widespread Late Carboniferous to Early Permian pebbly slates of glacial origin, which represent the northern limit of distribution of cold climate deposits on the northern margin of Gondwanaland (Liu and Cui 1983). We have used this boundary on the one hand, and the southern limit of distribution of the Early Carboniferous *Kweichouplyllum* fauna and the Late Permian Cathaysian flora on the other, to locate the median Tethys that once existed and separated the northern Cathaysian and the southern Gondwanan faunal provinces.

MEGASTAGE OF DISINTEGRATION OF PANGAEA: POST-INDOSINIAN CRUSTAL DEVELOPMENT OF CHINA

After the Indosinian Movement, East China consolidated into an integral unit, and no large-scale marine transgression has occurred since in China except in the Qinghai–Xizang region. A basic change in the tectonic framework of East China was the transformation of the E–W and ENE–WSW trend into the Neocathaysian NNE–SSW trend, and the increasing interaction between the East Asian continent and the West Pacific ocean resulting in the intensive volcanic activity of the circum-Pacific volcanic belt. Various kinds of basins were formed on the continent, and marginal seas developed on the continental margin. The Qinghai–Xizang region, on the other hand, was characterized by consecutive northward movement of massifs detached from the Gondwana mother continent and their final coalescence and collision with the northern continent. A comprehensive sketch map is given to show the main features of the whole megastage (Fig. 20.6).

The new tectonic framework of East China: basin development, volcanic activity, and formation of marginal seas

In the post-Indosinian stages East China was characterized by a Neocathaysian tectonic trend and was clearly demarcated from West China by a line joining the Helan Mountain in the north and the Kham–Yunnan Uplift in the south. It may be subdivided into three belts; a western belt of large inland basins (Fig. 20.6, IV); a central belt, mainly of onshore basins; and an eastern belt characterized by volcanic activity. A series of mountains, including the northern Hingan–Taihang and the southern Wuling–Xuefeng, marks the boundary between the western, central, and eastern belts constituting the East China Circum-Pacific Continental Margin Domain. The boundary between the central belt and the eastern belt is the Tanchang–Lujiang fracture zone. All three belts are divided into three sections by the two intervening convergent zones, and are roughly comparable with the second and the third Neocathaysian upwarped and subsidence zones of Professor Li Siguang.

In the history of the development of the inland basins and of the continental margin, two stages may be recognized; an earlier one from the Jurassic to the early Late Cretaceous corresponding approximately to the Yanshanian, and a later one from the late Late Cretaceous to the Neogene corresponding to the Himalayan. The former is dominated by a compressional type of regional stress field and the latter by an extensional type.

Tectonic history of the Yanshanian Stage

In this stage the western belt of East China was dominated by the development of large inland downwarp basins, while the eastern belt was characterized by predominant Neocathaysian (NNE–SSW) structural directrices and profuse volcanic activity. The western inland basins underwent general subsidence and reached their maximum extent in the Middle Jurassic, when conditions of low relief prevailed, and fine clastic to marly sediments were formed both in the Ordos and in the Sichuan–Yunnan basins. In the Late Jurassic and Early Cretaceous the western margin of the basins underwent strong uplifting, probably caused by lateral compression coming from the west, and the basins were much diminished in size, piedmont deluvial deposits being predominant in the western margins (Wang *et al.* 1983) (Fig. 20.6, IVB, IVC). The northern section of the western belt is somewhat different in its development of small semi-graben coal basins and conspicuous magmatic activity in the Late Jurassic (Fig. 20.5, IVA).

The central belt had remained elevated since its uplifting in the Indosinian Movement, and it was not until Cretaceous times that it began to subside, as in the Songliao Basin in the north and the Jianghan and Hengyang basins in the south. The formation of the North China basin was obviously later (Fig. 20.6, VA–C). The development of the basins usually consists of an earlier downfaulting, followed by downwarping, and finally by uplift. The Songliao Basin was developed on an old foundation and the faulting stage persisted into the mid-Early Cretaceous. General downwarping and subsidence, leading to a deep-water lacustrine environment favourable to oil formation, occurred in late Early Cretaceous to early Late Cretaceous times,

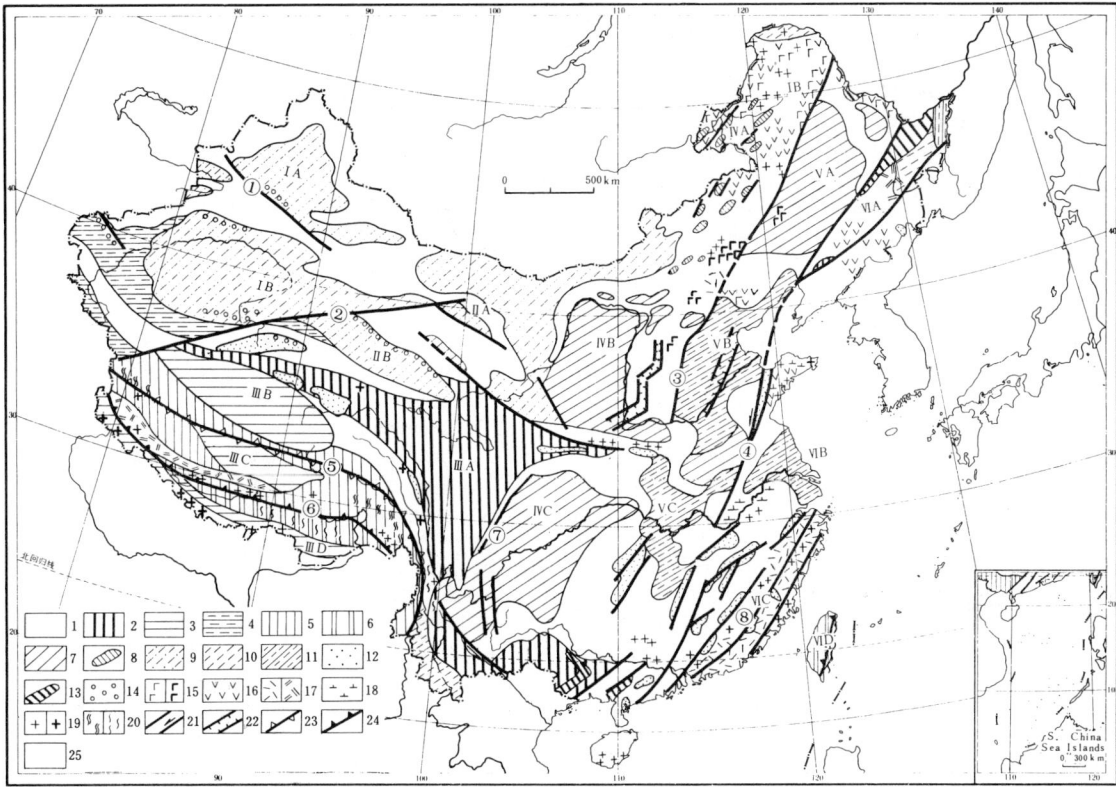

Fig. 20.6. Palaeotectonic sketch map of China in the post-Indosinian (Yanshanian and Himalayan) stages (modified after Wang 1982b). 1, Platform and pre-Indosinian folded regions; 2, Indosinian fold zones; 3, Jurassic or Jurassic to Cretaceous stable marine sedimentary regions; 4, Cretaceous to Palaeogene stable marine sedimentary regions; 5, Jurassic to Palaeogene or Jurassic to Cretaceous mobile marine sedimentary regions; 6, Jurassic or Jurassic to Cretaceous mobile paralic sedimentary regions; 7, Jurassic to Cretaceous or Cretaceous large inland basin deposits; 8, Jurassic to Cretaceous coal basins; 9, Jurassic to Palaeogene inland basins; 10, Jurassic to Cretaceous onshore basins; 11, Cretaceous to Palaeogene large basins; 12, Cretaceous to Palaeogene small faulted basins; 13, Palaeogene coal basins; 14, Late Oligocene and Neogene paramolasse deposits; 15, Basic volcanics: left, Jurassic and Cretaceous; right, Palaeogene; 16, Jurassic and Cretaceous intermediate volcanics; 17, Acidic volcanics: left, Jurassic and Cretaceous; right, Cretaceous and partly Palaeogene; 18, Jurassic and Cretaceous alkaline volcanics; 19, granitic intrusives: left, Late Yanshanian; right, Himalayan; 20, metamorphic zones: left, Yanshanian; right, Himalayan; 21, syn-depositional faults and later transcurrent faults; 22, rift valleys and basins; 23, Yanshanian convergent crustal consumption zone; 24, Himalayan accretional crustal consumption zone.

I, Tarim Tianshan terrain: IA, Junggar Basin; IB, Tarim Basin. II, Kunlun–Qilian terrain: IIA, Corridor piedmont basins; IIB, Qaidam Basin; III, Qinghai–Xizang Plateau: IIIA, Bayan Har Indosinian Fold Zone; IIIB, Qiangtang Massif; IIIC, Gangdise Massif; IIID, Himalaya Massif; IV, Western Neocathaysian belt: IVA, Hingan–Nei Mongol volcanic zone; IVB, Ordos–Shaanxi Basin; IVC, Sichuan–Yunnan Basin; V, Central Neocathaysian belt: VA, Songliao Basin; VB, North China rifted basins; VC, Jiang-Han Basin; VI, Eastern Neocathaysian belt: VIA, Changbai-Jiaodong volcanic zone; VIB, North Jiangsu–Lower Yangzi Basin; VIC, Min–Zhe volcanic zone; VID, Taiwan Himalayan Fold Zone.

Depth faults and crustal consumption zones:

1, North Tianshan; 2, Altyn; 3, Xingan–Taihang; 4, Tancheng–Lujiang; 5, Bangong–Nujiang; 6, Yarlung Zangbo; 7, Longmen; 8, Lishui–Haifeng.

and uplifting set in during the late Late Cretaceous after a local disturbance. Similar conditions obtained in the Hengyang Basin, but downwarping persisted here to the middle of the Eocene.

The eastern belt occupies the maritime provinces, from Liaoning in the north to Guangdong in the south, and is characterized by profuse magmatic activity. As already mentioned, the volcanic rocks in this region include different types arranged *en echelon* with a NE–SW trend; the Jiaoliao continental type connected with the Da Hingan Mountains, and the Min–Zhe continental margin type connected with Lingnan of southern Korea (Chapter 19). Between these two is the alkaline type distributed along the Tancheng–Lujiang fracture zone, from eastern Shandong to the Lower Yangzi (Fig. 20.6). Yanshanian granites are widespread in this belt, especially on both sides of the Lishui–Haifeng depth fault (Fig. 20.5, 8) and along the coast from Xiamen to Changlo in eastern Fujian. Upper Jurassic and Lower Cretaceous rocks underwent obvious migmatization and dynamic metamorphism, for which an Rb–Sr isotopic age of 165 Ma has been obtained. It is on this account inferred that a westward subduction zone had been active off the coast in the late Jurassic and Early Cretaceous, which may have been connected with the Ryoke–Sabagawa metamorphic belt of southwest Japan across the East China Sea.

Tectonic history of the Himalayan Stage

The main tectonic features at the beginning of this stage consist of the general uplifting of the western belt and a corresponding subsidence of the central and eastern belts. The Ordos and the Sichuan–Yunnan basins were further reduced in size and entirely disappeared at the beginning of the Palaeogene. By this time large-scale down-faulting in North China led to the formation of the on-shore Bohai and North China basins, where basalt, piedmont debris, evaporites, and deep lake turbidites with a total thickness of 5000 m were deposited. A typical continental rifting and semigraben structure is thus displayed (Fig. 20.7). Rifting and fracturing became much weakened in the Neogene and a general subsidence took place, thus bringing about the formation of draping deposits in nearly all the basins (Fig. 20.7). Uplifting and rifting were, however, still active to the west of the Taihang Mountains, where the Fenwei Graben is conspicuous and persisted to the Quaternary. The Jianghan Basin in the central belt underwent strong subsidence in the Palaeogene, but the south-eastern uplands were characterized by numerous semigraben red basins of Late Cretaceous to Palaeogene age, probably formed under tensile conditions against a background of uplift.

After the Yanshanian Movement the continental margin front of East Asia had transferred to the east of Taiwan Island and the Japanese Islands. At the same time a Palaeogene miogeosyncline appeared in western Taiwan, where more than 3000 m of flysch were deposited. Probably the most important event that occurred in East China and East Asia and the adjacent marine realms was the appearance of a series of arc-trench and marginal sea systems. This process was started in the Palaeogene, as evidenced by the initial opening of the Sea of Japan and the South China Sea, and was continued in the Neogene. Glaucophane-schists 8–14 Ma old are found in the Yuli metamorphic belt in Taiwan, which indicates a westward subduction of the West Pacific under the East Asian continent in the Late Tertiary.

Basins and ranges of North-west China

North-west China embraces the terrains lying to the west of the Helan Mountain and to the north of the Kunlun and Qinling ranges. Two regions, the western Tarim–Tianshan and the eastern Kunlun–Qilian regions (Fig. 20.6, I, II), may be recognized. They are separated by the Altyn transcurrent fault (Fig. 19.3, 12), which was probably active in the Late Mesozoic and ceased to act before the Palaeogene. The Tarim–Tianshan region was entirely elevated after the Indosinian Movement, and the Jurassic–Palaeogene sequence

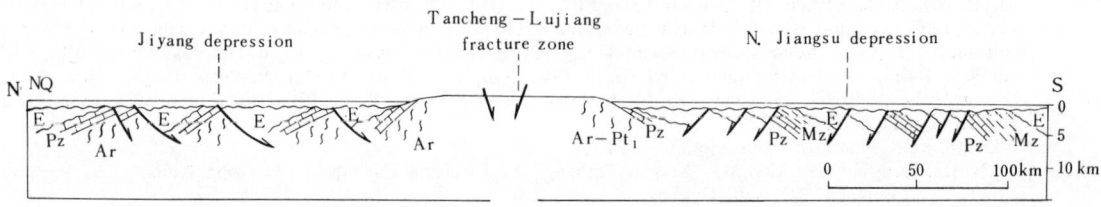

Fig. 20.7 Schematic profiles showing the semigraben structure of the North China and the northern Jiangsu basins (simplified after the Geophysical Prospecting Bureau, Ministry of Petroleum Industry 1978). Ar, Archaean, Pt, Proterozoic, Pz, Palaeozoic, Mz, Mesozoic, E, Eocene.

to the north of Tianshan is generally of stable basin type. The intermittent coarse clastics in the Bogda and Turfan basins were the result of strong differential uplifting in the Cretaceous and Palaeogene. Mesozoic subsidence in the Tarim Basin was confined to the peripheral and western parts, where intermittent shallow-sea marine and littoral deposits of Late Cretaceous to Miocene age were formed. The northern slopes of the Mustag Mountain and the Karamiran Basin are covered by coarse, thick, intermontane Jurassic and Cretaceous clastics, which are a reflection of the northward compression exerted by the north-moving Qiangtang Massif. The Qaidam Basin (Fig. 20.6, IIB) is peculiar in its strong subsidence in the Mesozoic and Palaeogene and the conspicuous deformation it suffered in the Neogene. Pronounced unconformities are present between the Lower and Middle Jurassic and between the Cretaceous and Palaeogene. Upper Cretaceous and Palaeogene red conglomerates and sandstones of piedmont facies are indicative of intensive compression coming from the south in the Late Yanshanian Stage. In the corridor region of the North Qilian and further to the north, the Jurassic to Palaeogene sediments, mainly of open basin type developed, the piedmont facies being confined to the Neogene (Fig. 20.6, IIA).

Northward movement of the northern Gondwanan massifs and formation of the Qinhai–Xizang Plateau

After the incorporation of the Indosinides with the Eurasian continent, marine waters were confined to the Xizang region, the southernmost part of Qinghai, and part of western Yunnan. Within this territory are situated three massifs: the Qiangtang, Gangtise, and Himalaya, and the mobile belts lying between them. As stated above, the convergence and junction of the Qiangtang Massif and the Qiadam Massif was accomplished in the Indosinian Movement. The Jurassic shows a distinct facies change in the region (Fig. 20.6, IIIB), a stable type being confined to the Qiangtang Massif and surrounded by a mobile type both in the south-west and the south-east (see above). The Cretaceous is incomplete and very limited in distribution. In the Ngali region, on both sides of the Bangong–Nujiang convergent zone, the Jurassic and Cretaceous are characterized by thick geosynclinal deposits. Yanshanian granitic intrusives occur immediately to the north of the convergent zone near Rutog (Fig. 20.6). As mentioned in Chapter 19, ophiolitic suites and ultrabasic rocks covered by the Upper Jurassic occur along the convergent zone from Rutog and Gerze in the west to Dengqen and Amdo in the east. This indicates that the oceanic crust was mainly subducted and consumed before the end of the Jurassic, thus leading to the formation of the Early Yanshanides in the south of the Qiangtang Massif.

In the Gangdise region the Jurassic is generally lacking. The widespread strata are the Cretaceous marine, variegated, clastic deposits and carbonates, amounting to 6000 m thick, which are unconformably overlain by uppermost Cretaceous to Palaeogene coarse red clastics and volcanics. An acidic volcanic belt of the same age, over 1000 km in length, extends along the Gandise Mountains (Fig. 20.6) (Zhou Yunsheng *et al.* 1981), and probably represents the product of the northward subduction. In the Yarlung Zangbo valley, to the south of the well-known ophiolite belt, the Jurassic includes abundant intermediate to basic eruptives; the Lower Cretaceous is silicolitic and also contains volcanic rocks: the Upper Cretaceous includes a paralic flysch facies (Jigazi Group) and a deep-sea silicolite facies (Zongchuo Formation), sometimes containing olistostromes near the Yamzho Lake. The oceanic crust in the Yarlung Zangbo marine realm was therefore largely consumed by the end of the Cretaceous, as evidenced by the Upper Cretaceous facies distribution, and coordinated by the latest Cretaceous to Palaeogene magmatism in the northern belt (Fig. 20.8, D). The highest marine horizon so far known is the Lutetian *Nummulites* limestone found in the Yarlung Zangbo valley, but the marine Palaeogene is now known to be widespread, including Lutetian corals in the Ngali region as far north as Duoma. Thus the final elevation of the region is post-Eocene, resulting from coalescence of the Gandise and Himalaya massifs, and the tremendous uplift of the Qinghai–Xizang Plateau resulted from further collision and consequent thickening of the crust, and was finally accomplished in the Late Cainozoic (Chang and Pan 1981). The idea that the plateau was formed by the successive accretion of continental fragments toward the northern Eurasian continent is justified at least in the post-Indosinian stages.

CONCLUDING REMARKS

Having reviewed the spatial differentiation and historical development of the crust of China, it is now possible to make some final comments about the crustal evolution of China in particular and of the earth in general.

(1) From the viewpoint of mobilism and development stages, the classification of geotectonic units and of tectonic stages are intimately related. Tectonic domains of the continental type took their final shape with the formation of platforms (850 Ma). The

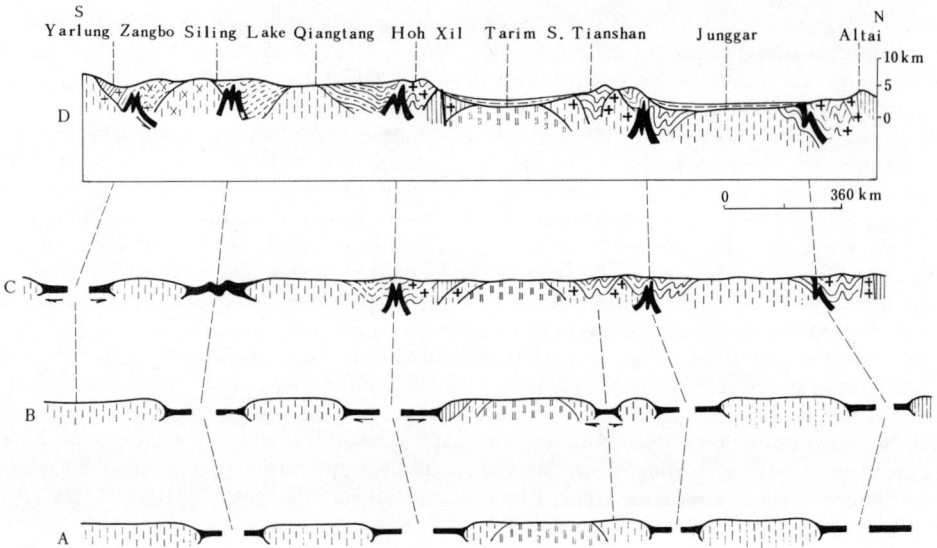

Fig. 20.8. Schematic profiles from Yarlung Zangbo valley, Xizang to Altai, Xinjiang, showing the tectonic development in various stages. Legends as in Fig. 20.5. A, Sinian; B, Devonian; C, Jurassic; D, Recent.

main process of subsequent development consists of changes in the positions of the platforms as entities, and complication of their borders, including fragmentation and detachment of minor massifs, until, in the last megastage, the platforms began to dissociate and disintegrate entirely. Tectonic domains of the continental type are united through the mutual convergence and junction of continental margin tracts at various stages. Thus the main features consist of a convergent crustal consumption zone that could be attributed to a particular tectonic stage, in addition to the accretional crustal consumption zones of various ages in the continental margin tracts. Tectonic domains of the continental margin type are peculiar in that no convergent, but only accretional, zones are developed, for no convergence or collision between opposite continents of the same rank seems to have occurred. These domains are confined to the peripheral parts of the Proto-Pacific and the Pacific, and are therefore of comparatively younger age.

(2) In accordance with the concept of development stages, the tectonic regime and structural behaviour (including the mode and extent of motion of the continental masses) differ considerably from each other in different megastages. This change with time is on the whole progressive and unidirectional. It embodies many peculiarities that are unique, as they are conditioned and determined by the changing nature of the evolving crust and essential media in the specific stages.

(3) In the megastage of formation of continental nuclei (2500–2600 Ma) the sialic nuclear massifs are composed of granite–greenstone and tonalite associations, which were mainly formed through aggregation *in situ* and were probably separated by oceanic crustal terrains. These earlier differentiates from the mantle may have attained considerable extent and thickness, as advocated by Windley (1977), but they are highly mobile in nature and are probably not rigid enough to sustain any kind of large-scale fracture and subduction.

(4) The megastage of formation of platforms (ending at 850 Ma) encompasses two stages; the earlier transitional stage (ending at 1850 Ma), still dominated by greenstone belt formation; and the later stabilizing stage, characterized by the development of paracover sequence on the protoplatform. This usually attains a size exceeding that of the subsequent platform, as shown in Central Tianshan in North-west China and in Central Yunnan in South-west China. On account of the unusual proportion between the sizes of the main massif and its surrounding marginal tracts (e.g. the Proto-Yangzi and surroundings compared with present conditions), the increase and stabilization process seem to occur chiefly through accretion in the marginal tracts without appreciable subduction

towards the main massif.

(5) The particular structural behaviour of the continental masses in the megastage of formation of Pangaea is interesting, but its destruction from the megastage of the latest 200 Ma is not altogether clear. As the concept of a Proterozoic Pangaea does not seem to be fully justified, the entire process by which Pangaea was formed can probably be designated as the coalescence or assembly of the stable massifs. It is probably to this megastage that the concepts of accretion within one continental margin tract and of the convergence of two or more opposing and adjacent continental margins are most applicable and best displayed. This implies that the median massifs or detached micro-continents probably did not move very far away from the mother continent, or that the general outlines of the main continents and their margins had more or less maintained their integrity, although their positions had changed from time to time. This inference seems to be partly supported by palaeomagnetic and palaeobiogeographic data, and also by the observation that most aulacogens are confined to the inner sides of the border uplifts and do not extend across the continental interior, and no large-scale inland basins of subsidence were formed. In other words, the chief activities seem to have been restricted to the marginal parts of the platforms.

(6) The post-Indosinian megastage of Pangaean disintegration is characterized by long-distance continental drifting and large-scale sea-floor spreading. The entire collapse and dissociation of the Gondwana supercontinent and the accelerated sea-floor spreading of the Pacific, which has itself undergone a general reduction in scope because of expansion of the Atlantic, had caused enormous crustal consumption and subduction in the circum-Pacific domains. Certain phenomena, such as the extensive and intensive magmatic activity in East China and East Asia, the appearance of different categories of Late Mesozoic and Cainozoic inland and onshore basins and marginal sea basins in East China and adjacent regions are probably unique and unprecedented (Zhu et al. 1980, 1983). The huge 'basin and range' system of alternating block mountain ranges and big intermontane basins in North-west China, and the tremendous belts of Yanshanian eruptives and granitic intrusives on the magnificent Qinghai–Xizang Plateau, are the direct result of the long-range northward motion and violent collision of the Gondwanan massifs with the Eurasian border areas. In this case the distinction between accretional and convergent crustal consumption zones becomes difficult to make, and in fact becomes less significant. This progressive change in extent and rate of continental motion may probably be partly accounted for by postulating an ever-increasing size for convection cells in the mantle.

(7) Many authors including Scotcsc (1979), Smith et al. (1980), and Ziegler et al. (1979) have constructed world maps of past ages based on palaeogeomagnetic, palaeobiogeographical, and palaeoclimatic evidence. In all the maps a clear-cut Palaeo-Tethys ocean and an integral Pangaea are generally indicated in the period from the Late Carboniferous to Triassic. There seems to be no problem about the general outline, but the actual situation might have been much more complicated. Many massifs such as the Yangzi, Qaidam, Qiangtang, Gangdise, and the Cimmerian continent of Sengor (1979, 1981) might all have been separate and might have occupied intermediate positions at different times between the Eurasian and the Gondwanan supercontinents. Indeed, the recognition of the Yangzi–Qiangtang tectonic domain as advocated here, which was close to Australia in the Sinian and Early Palaeozoic but moved northward after the Early Carboniferous, may provide a better interpretation of the complicated relationship of these massifs in the time-span concerned. Much more work remains to be done before we can confidently reconstruct the past positions of these massifs.

APPENDIX: ABBREVIATIONS OF ISOTOPIC AGE DATING INSTITUTIONS

1. Geoch. Inst.: Guiyang Institute of Geochemistry, Academia Sinica.
2. Geol. Inst.-A.S.: Institute of Geology, Academia Sinica.
3. Geol. Inst.-C.A.G.S.: Institute of Geology, Chinese Academy of Geological Sciences.
4. Geomech. Inst.: Institute of Geomechanics, Chinese Academy of Geological Sciences.
5. Kuilin Inst.: Kuilin Institute of Geology, Ministry of Metallurgy.
6. Nanjing Inst.: Nanjing Institute of Geology and Mineral Resources, Chinese Academy of Geological Sciences.
7. Shenyang Inst.: Shenyang Institute of Geology and Mineral Resources, Chinese Academy of Geological Sciences.
8. Tianjin Inst.: Tianjin Institute of Geology and Mineral Resources, Chinese Academy of Geological Sciences.
9. Ur. Geol. Inst.: Beijing Uranium Geology Research Institute.
10. Yichang Inst.: Yichang Institute of Geology and Mineral Resources, Chinese Academy of Geological Sciences.

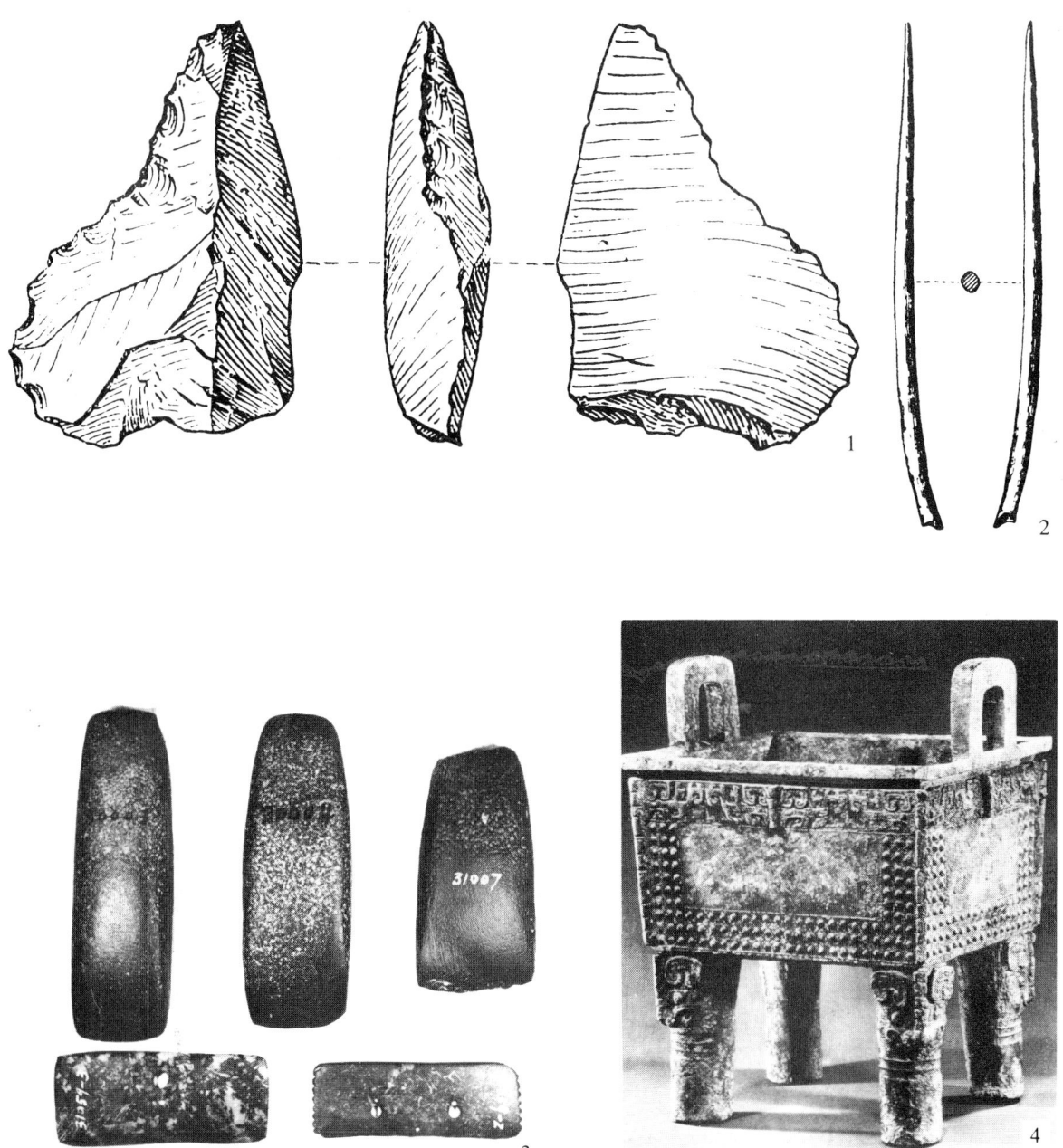

Plate I. 1, Concave side-scraper (×1), made of black flint from the Late Palaeolithic, Upper Cave industry of Zhoukoutian (Choukoutien), Beijing, China; 2, Bone needle (×1) from the same site as in I.1; 3, Neolithic Yangshao culture of Yangshao village, Henan Province, China. Upper row, three stone axes (×$\frac{1}{2}$); lower row, two stone sickles (×$\frac{2}{3}$). Courtesy of Professor Yuan Fuli (P. L. Yuan); 4, The bronze Si Mu Xin, a ceremonial vessel, unearthed from the tomb of Lady Hao, an early Yin tomb at Anyang, Henan Province weighing 128 kg (× ~$\frac{1}{7}$).

Plate II

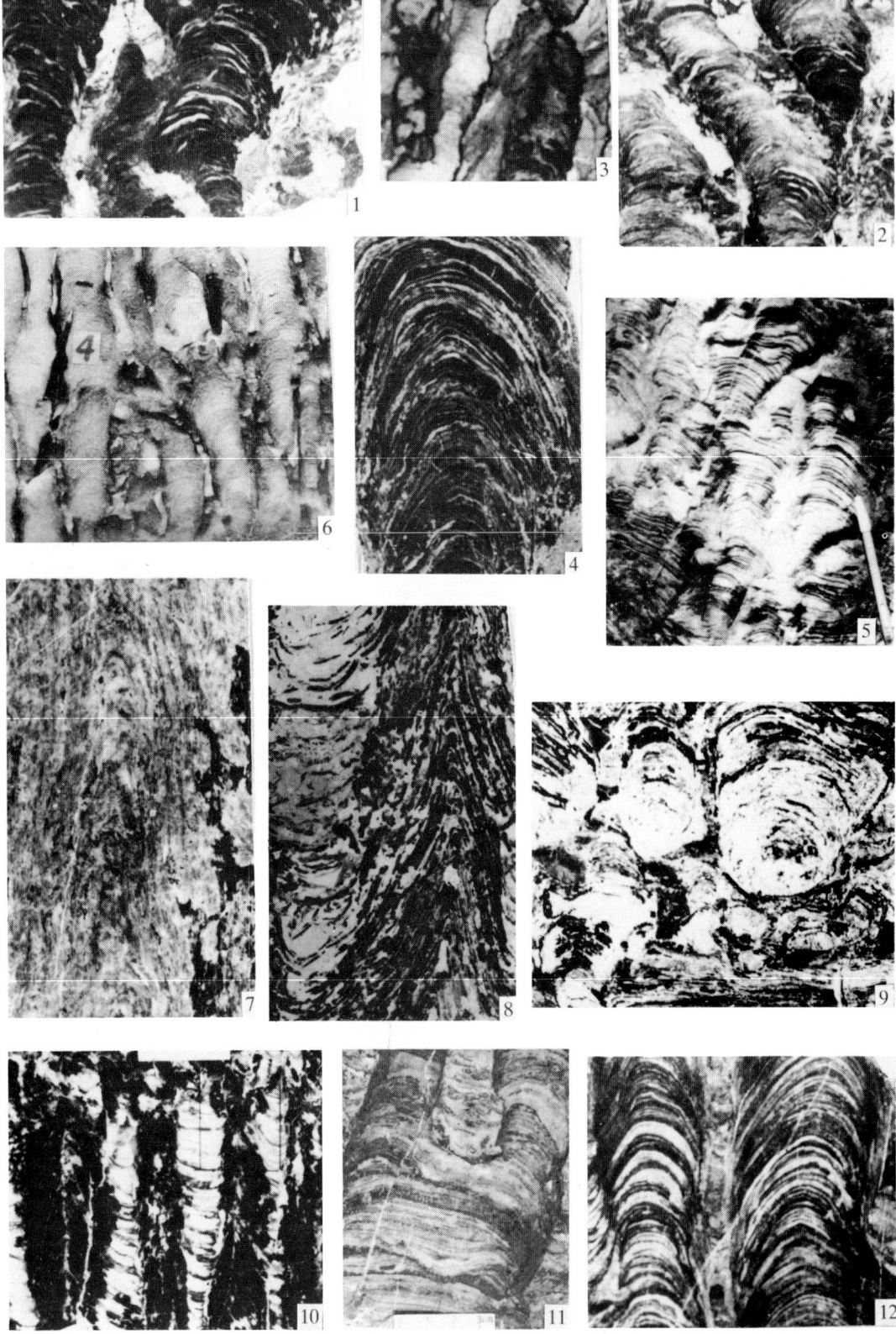

Plate II. Proterozoic stromatolites (courtesy of Drs Xing Yusheng and Zhu Shixing). 1–4, Qingbaikouan: Xiamaling Formation, Zhuolu District, Hebei Province ($\times \frac{1}{2}$): 1, *Linella sinica* Krylov, longitudinal polished surface; 2, *Inzeria tjomusi* Krylov, longitudinal polished surface; 3, *Qingbaikouia vaginata* Zhu and Du, longitudinal polished surface; 4, *Clavaphyton turbiformis* Zhu and Du, longitudinal polished surface; 5–6, Jixianian, Tieling Formation, Ji Xian, Hebei Province: 5, *Baicalia* cf. *baicalica* (Maslov) Krylov ($\times \frac{1}{5}$); 6, *Chihsienella chihsienensis* Liang densum and Cao ($\times \frac{3}{4}$); 7–9, Changchengian, lower part of Gaoyuzhuang Formation, Ji Xian, Hebei Province: 7, *Conophyton cylindricum* (Grabau). Maslov ($\times 1$); 8, *Tabuloconigera paraepiphyta*. Zhu et al. ($\times \frac{1}{2}$); 9, *Gaoyuzhuangia bulbosa* Zhu et al. $\times \frac{1}{2}$; 10–11, Hutuoan: 10, *Jacutophyton microstylum* Zhu ($\times \frac{1}{2}$), Beidaxing Formation, Wutai, Shanxi Province; 11, *Paraboxonia convexa* Zhu ($\times \frac{1}{2}$), Yaochicun Formation, Wutai, Shanxi Province; 12, *Pilbaria* cf. *perplexa* Walter ($\times \frac{1}{2}$), Yaochicun Formation, Wutai, Shanxi.

Plate III. Middle and Upper Proterozoic microplants and algae (courtesy of Drs Xing Yusheng, Zheng Wenwu and Du Rulin). 1–11, Qingbaikouan: 1, *Ellipsophysa axicula* Cheng ($\times 8$), Liulaobei Formation, Feng Xian, Anhui Province; 2–3, Longshan Formation, Huailai, Hebei Province: 2, *Vendotaenia* sp. ($\times 20$); 3, *Chuaria circularis* Walcott ($\times 20$); 4–9, Jingeryu Formation, Luan Xian, Hebei Province: 4, *Strictosphaeridium* sp.; 5, *Microptycha uniplicata* Tim.; 6, *Taeniatum simplex* Sin; 7, *Trachysphaeridium stipticum* Sin; 10–11, Xiamaling Formation, Ji Xian, Tianjin: 10, *Pseudozonosphaera sinica* Sin; 11, *Trachysphaeridium* aff. *laminaritum* Tim.; 12–18, Jixianian: 12, *Leiofusa bicornica* Sin and Liu, Hongshuizhuang Formation, Ji Xian, Tianjin; 13, *Asperatoaphaeridium partialis* Schep., Shennongjia Group, Shiqiaohe Formation, Hubei Province; 14, *Nucellosphaeridium* sp. Shennongjia Group, Taizi Formation, Hubei Province; 15, *Leiopsophosphaera pelucidus* Schep., as preceding; 16, *Favososphaeridium densum* Xing and Liu, as preceding; 17, *Quadratimorpha florentis* Sin and Liu, Hongshuizhuang Formation, Ji Xian, Tianjin; 18, *Pseudozonosphaera verrucosa* Sin and Liu, Ji Xian, Tianjin; 19, *Asperatosphaeridium umishanensis* Sin and Liu; 20, *Trachysphaeridium simplex* Sin; 21, *Leiopsophosphaera pelucidus* Schep.; 22, *Marginominuscula rugosa* Naum.; 23–24, *Leiominuscula pellucentis* Sin and Liu; 4–24 ($\times 800$).

(*see over*)

Plate IV. Sinian microplants, algae, and stromatolites (courtesy of Drs Xing Yusheng, Cao Ruiji, and Zhu Shixing). 1–2, Lower Sinian: 1, *Trachysphaeridium hyalinum* Sin and Liu ($\times 800$), Kangjia Formation, Benxi, Liaoning Province; 2, *Leiopsophosphaera apertus* Scheppen ($\times 800$), Qiaotou Formation, Benxi, Liaoning Province; 3–14, Upper Sinian: 3, *Polyedryxium polygonum* Sin and Liu ($\times 800$), Dengying Formation, Zigui, Hubei Province; 4, *Pseudozonosphaera asperalla* Sin and Liu ($\times 800$), as preceding; 5, *Pseudozonosphaera rugosa* Sin and Liu ($\times 800$), as preceding; 6, *Trachysohaeridium rude* Sin and Liu ($\times 800$), as preceding; 7, *Tr. stipticum* Sin ($\times 800$), as preceding; 8, *Nostocomorpha prisca* Sin and Liu ($\times 800$), Doushabtuo Formation, Ichang, Hubei Province; 9, *Praesolenopora formosa* Tsao and Liang ($\times 10$), Majiatun Formation, Jinxian, Liaoning Province; 10, *Multisiphonia hemicirculis* Tsao and Liang ($\times 10$), as preceding; 11, *Conophyton ocularoides* Liang ($\times \frac{1}{6}$), Shisanlitai Formation, Jinxian, Liaoning Province; 12, *Boxonia xifengensis* Zhu ($\times \frac{3}{5}$), Doushantuo Formation, Xifeng, Guizhou Province; 13, *Actinophycus ninjiangensis* Tsao and Zhou ($\times 10$), Dengying Formation, Nanjiang, Sichuan; 14, *Praesolenopora fascicularis* Tsao and Zhao ($\times 5$), Dengying Formation, Luoshan, Sichuan Province.

(*see over*)

Plate III

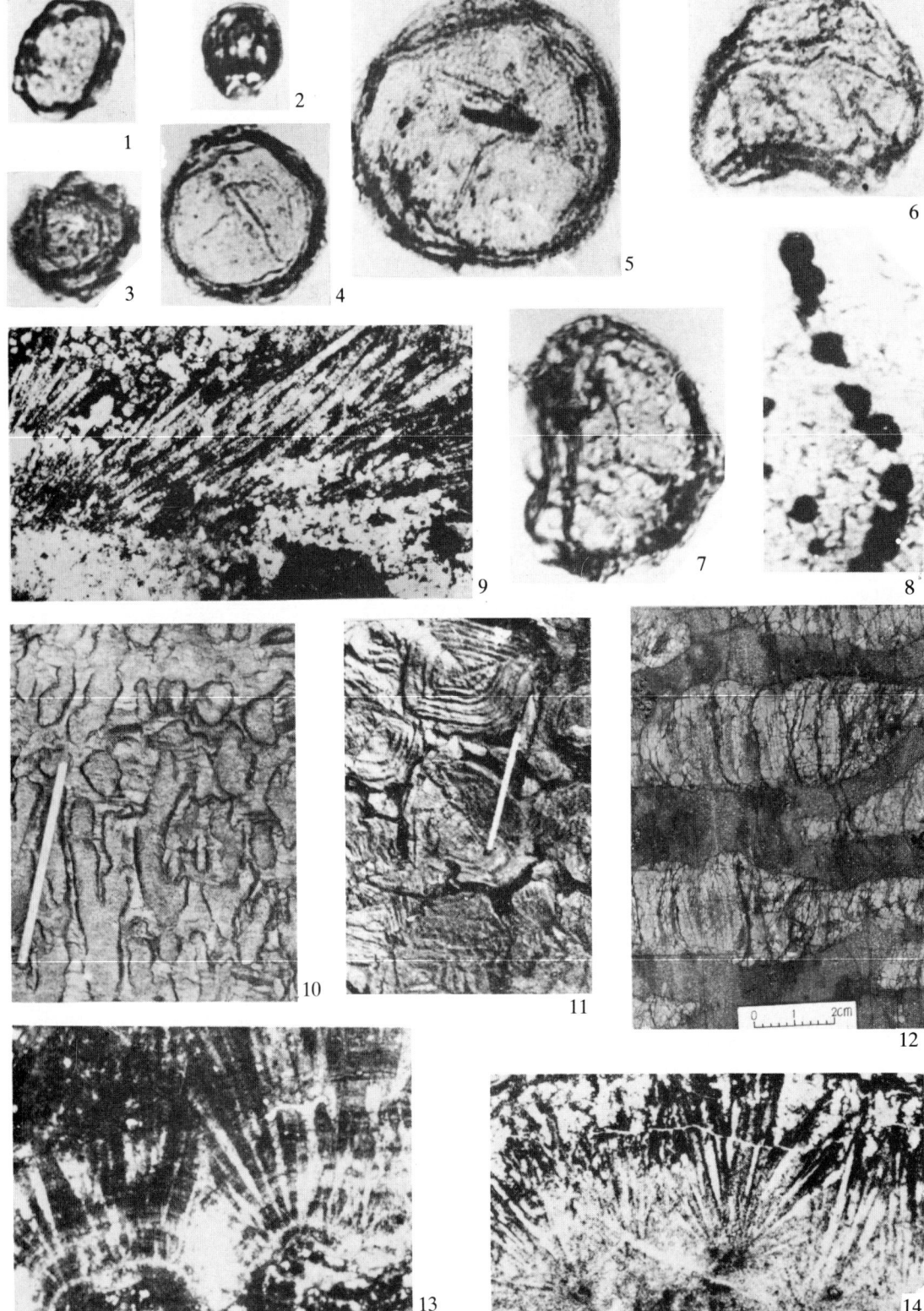

Plate IV

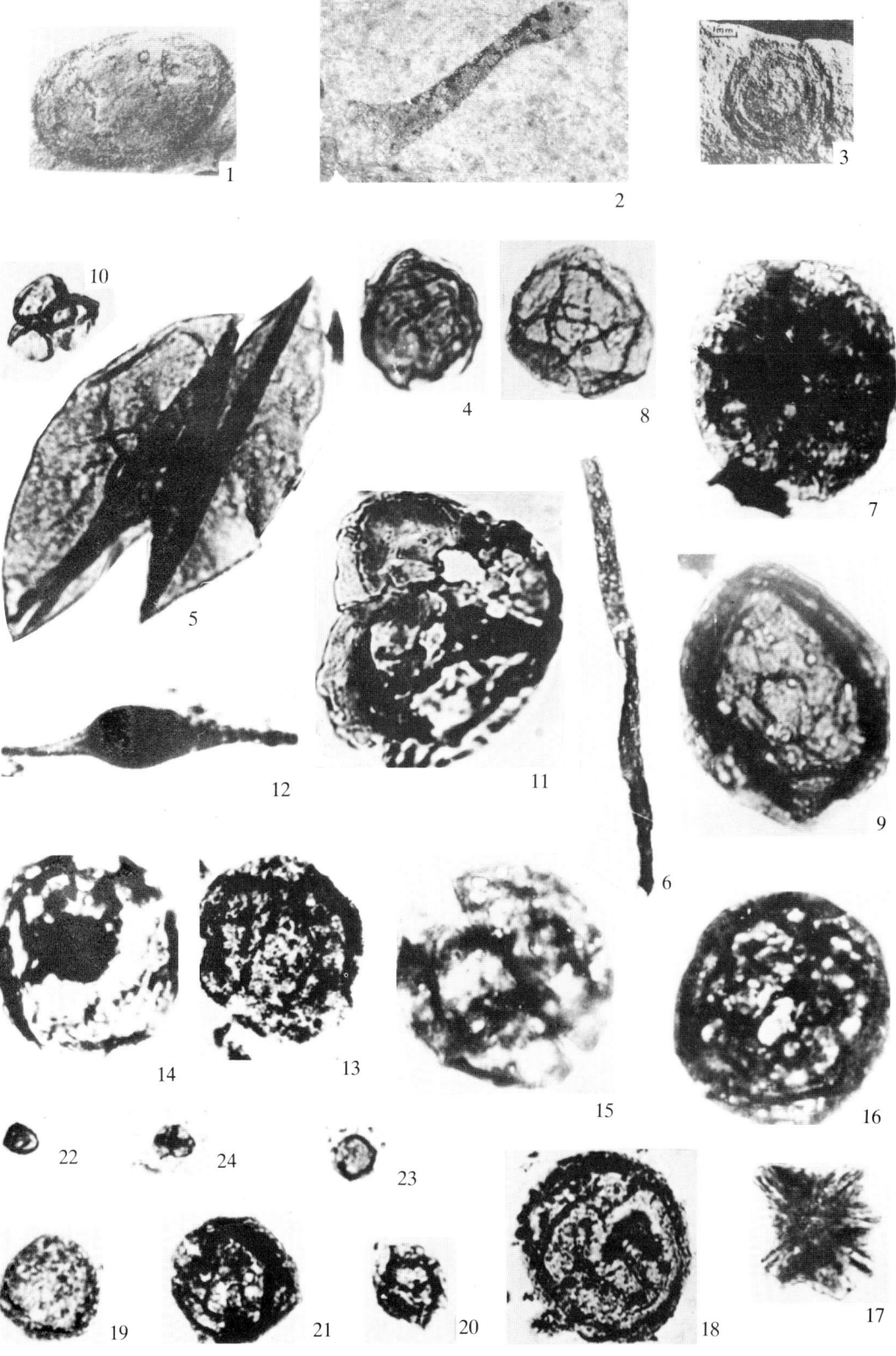

Plate V

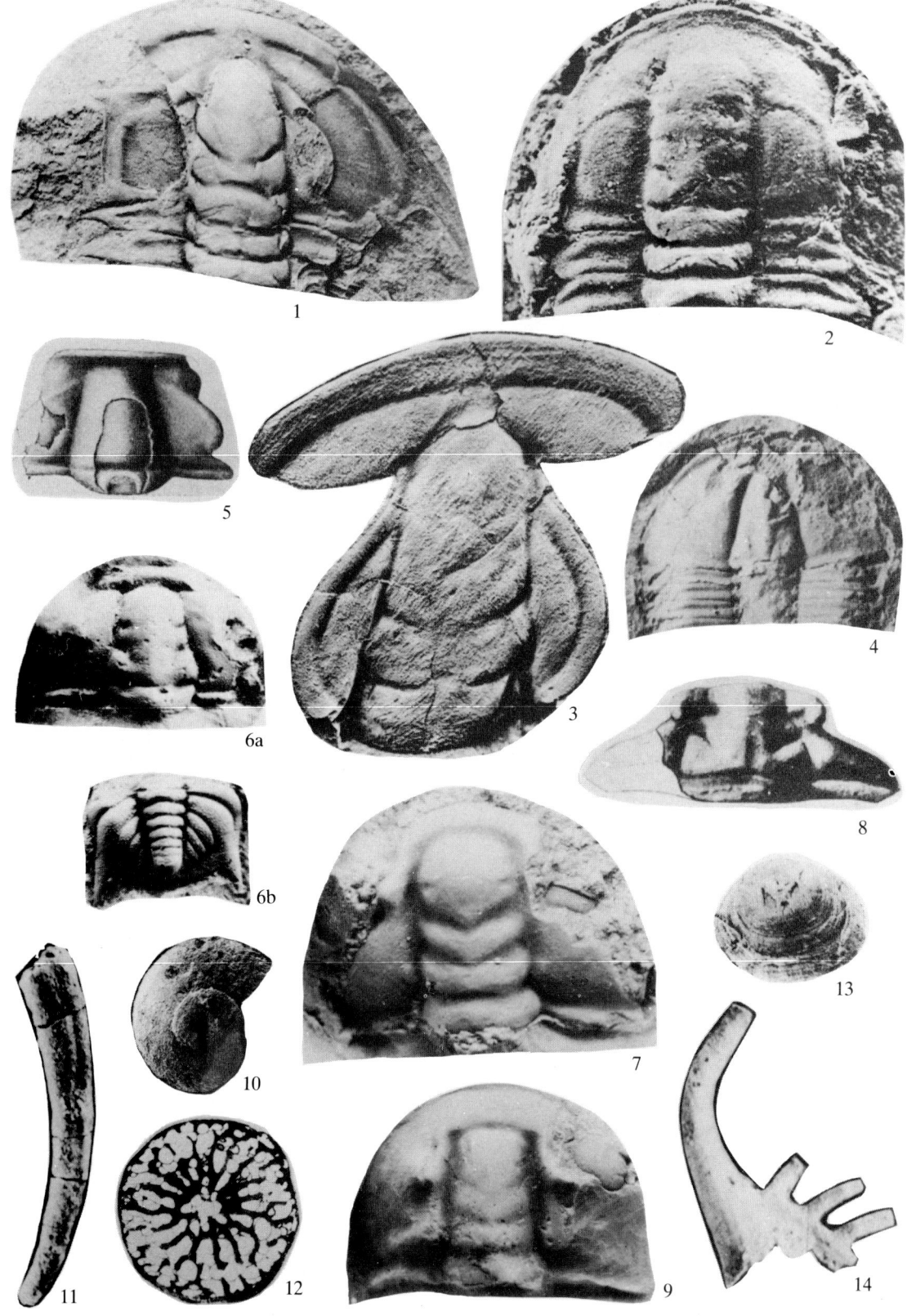

Plate V. Some important Cambrian fossils (courtesy of Dr Xiang Liwen). 1, *Eoredlichia intermedia* Lu ($\times 3$), cephalon and part of thorax, Lower Cambrian Qiongzhusi Formation, Kunming, Yunnan Province; 2, *Palaeolenus lantenoisi* Mansuy ($\times 5$), cephalon and part of thorax, Lower Cambrian Canglangpu Formation, Kunming, Yunnan Province; 3, *Redlichia murakamii* Resser and Endo ($\times 1$), glabella, Lower Cambrian Manto Formation, Sanshilipu, Liaoning Province, 4, *Bailiella lantenoisi* Mansuy ($\times 3$), cephalon and part of thorax, Middle Cambrian Xuzhuang Formation, Boshan, Shandong Province; 5, *Chuangia buchruckeri* (Lorenz) ($\times 2\frac{1}{2}$), cephalon, Upper Cambrian Changshan Formation, Shandong Province; 6, *Kaolishania pustulosa* Sun; a, cephalon; b, pygidium; ($\times 2$), Upper Cambrian Changshan Formation, North China; 7, *Ptychaspis shansiensis* Sun ($\times 3$), cephalon, Upper Cambrian Fengshan Formation, North China; 8, *Drepanura premesnili* Bergeron; 15a ($\times 1\frac{1}{2}$), cephalon; 15b, pygidium, ($\times \frac{2}{3}$). Upper Cambrian Gushan Formation, Shandong; 9, *Quadraticephalus walcotti* Sun ($\times 2$), cephalon, Upper Cambrian Fengshan Formation, North China; 10, *Aldanella* sp. apical view, ($\times 22$), Lower Cambrian Yurtus Formation (lower part), Wushi, Xinjiang Region. 11, *Anabarites trisulcatus* Missarzhevsky, lateral view, ($\times 15$), Lower Cambrian Huangshangdong Member, Shipai, Yichang, Hubei Province; 12, *Retecyathus nitidus* Yuan and Zhang ($\times 1\frac{1}{2}$), Lower Cambrian Tianheban Formation, Yichang, Hubei Province; 13, *Diangongia pista* Rong ($\times 2$), Lower Cambrian Qingzhusi Formation, Wuding, Yunnan Province; 14, *Codylodus proavus* Müller ($\times 50$), Upper Cambrian Fengshan Formation, Tangshan, Hebei Province.

Plate VI. Some important Ordovician fossils. 1, *Acanthograptus macilentus* Hsü ($\times 2$), Lower Ordovician Fengxiang Formation, West Hubei Province; 2, *Etagraptus approximatus* Nicholson ($\times 1\frac{1}{2}$), Lower Ordovician Xinertai Formation, Guozigou, Huocheng, Xinjiang Region; 3, *Didymograptus murchisoni* (Hisinger) ($\times 2$), Upper Ordovician Aijiashan Formation, Hubei Province; 4, *Cardiograptus huochengensis* Xu and Huang ($\times 2$), Upper Ordovician Talejihe Formation, Guozigou, Huocheng, Xinjiang Region; 5, *Didymograptus abnormis* Hsü ($\times 3\frac{1}{2}$), Lower Ordovician, West Zhejiang and South Anhui Provinces; 6, *Azygograptus suecicus* Moberg ($\times 2$), Lower Ordovician Meitan Formation, Zunyi, Guizhou Province; 7, *Didymograptus asperus* Harris and Thomas ($\times 2\frac{1}{2}$), Lower Ordovician, NW Qilian Mts; 8, *Orthograptus quadrimucronatus* Hall ($\times 2$), Upper Ordovician, Taojiang, Hunan Province; 9, *Tetragraptus* cf. *bigsbyi* Hall ($\times 2\frac{1}{2}$), Lower Ordovician Xinertai Formation, Guozigou, Huocheng, Xinjiang Region; 10, *Sinograptus typicalis* Mu ($\times 2$), Lower Ordovician Ninggou Formation, Zhejiang Province; 11, *Dicellograptus szechuanensis* Mu ($\times 3$), Upper Ordovician Wufeng Formation, Yuechi, Sichuan Province; 12, *Glyptograptus teretiusculus* (Hisinger) ($\times 2$), Upper Ordovician Aijiashan Formation, Guizhou and Hubei Provinces; 13, *Climacograptus diplacanthus* Bulman ($\times 1$), Upper Ordovician, Yimskan, Tawu Mt., West Xinjiang Region; 14, *Adelograptus guozigouensis* Xu and Huang ($\times 2$), Lower Ordovician Xinertai Formation, Guozigou, Huocheng, Xinjiang Region; 15, *Yangtzeella poloi* (Martelli) ($\times \frac{3}{4}$), a, dorsal view; b, lateral view; Lower Ordovician Dawan Formation; Fengxiang, Yichang, Hubei Province; 16, *Nemagraptus gracilis* Hall ($\times 1$), Upper Ordovician Miaopo Formation, West Hubei Province; 17, *Nankinolithus wanyuanensis* Cheng and Jian ($\times 3$), cephalon, Upper Ordovician, Longshan, Hunan Province; 18, *Dideroceras wahlenbergi* Foord ($\times \frac{1}{2}$), longitudinal section, Lower Ordovician Kuniutan Formation, Qijiang, Sichuan Province; 19, *Tungtzuella szechuanensis* Sheng ($\times 3$), Lower Ordovician Fengxiang Formation, Enshi, Hubei Province; 20, *Trilacinoceras aqiaense* Lai and Wang ($\times \frac{1}{2}$), Upper Ordovician Kansuo Formation, Kalpin, Xinjiang Region; 21, *Sinorthis typica* Wang ($\times 2$), a, dorsal view; b, ventral interior; Lower Ordovician Dawan Formation, Fengxiang, Yichang, Hubei Province; 22, *Sinoceras chinensis* Foord ($\times \frac{3}{4}$), longitudinal section, Upper Ordovician Mati Formation, Qijiang, Sichuan.

(*see over*)

Plate VII. Some important Silurian fossils. 1, *Glyptograptus persculptus* Salter ($\times 1\frac{1}{2}$), Lower Silurian Longmaxi Formation (basal part), West Hubei Province; 2, *Akidograptus acuminatus* Nicholson ($\times 1$), Lower Silurian Longmaxi Formation (basal part) West Hubei Province; 3, *Demirastrites convolutus* Hisinger ($\times 1$), Lower Silurian Liantan Formation, West Guangdong Province; 4, *Monograptus sedgwickii* Portlock ($\times 3$), Lower Silurian Longmaxi Formation, South China; 5, *Spirograptus turriculatus* Barrande ($\times 5$), Lower Silurian Longmaxi Formation, South China; 6, *Cyrtograptus rigidus* Tullberg ($\times 1$), Middle Silurian Upper Jenheqiao Formation, West Yunnan; 7, *Pristiograptus nilssoni* Barrande ($\times 2$), Lower Silurian, West Guangdong Province; 8, *Oktavites planus* Barrande ($\times 2$), Lower Silurian, Qilian Mts; 9, *Monograptus flexilis* Elles ($\times 3$), Middle Silurian Jenheqiao Formation, Central Yunnan; 10, *Monoclimacis griestoniensis* Nicholson ($\times 3$), Lower Silurian, Yumen, Gansu Province; 11, *Dimorphograptus zangbeiensis* Xu and Huang ($\times 3$), Lower Silurian, Xainza, Xizang Region; 12, *Monograptus scanicus* Tullberg ($\times 5$), Upper Silurian, Western Guangdong Province; 13, *Bulmanograptus macilentus* Mu *et al.* ($\times 3$), Lower Silurian, Xainza, Xizang Region; 14, *Idiophyllum multiseptatum* (Fan and He) ($\times \frac{1}{2}$), Middle Silurian Ningqiang, Shaanxi Province; 15, *Coronocephalus rex* Grabau ($\times 1\frac{1}{2}$), top of Middle Silurian, Jiangsu, Guizhou, Sichuan Provinces; 16, *Dalmanitina nanchengensis* Lu ($\times \frac{1}{2}$), Lower Silurian Nancheng Formation, Liangshan, Hanzhong, Shaanxi Province; 17, *Kailia quadrisulcata* Zhang: a, cephalon ($\times 2$); b, pygidium ($\times 1$); Middle Silurian Xiushan Formation, Yongxi, Xiushan, Sichuan Province; 18, *Aristoharpes sinensis* (Grabau) ($\times 8$), cephalon and part of thorax, Lower Silurian Pengjiayuan Formation, West Hubei Province; 19, *Ceriaster calamites* Lindström ($\times 4$), Lower Silurian Shiniulan Formation, Liangshuijing, Sinan, Guizhou Province; 20, *Pleurodium tenuiplicata* (Grabau) ($\times \frac{1}{2}$), Lower Silurian Loreping Formation, Yichang, Hubei Province; 21, *Parahelenites guizhouensis* ($\times \frac{2}{3}$) longitudinal section, Middle Silurian Xiushan Formation, Kaili, Guizhou Province.

(*see over*)

Plate VI

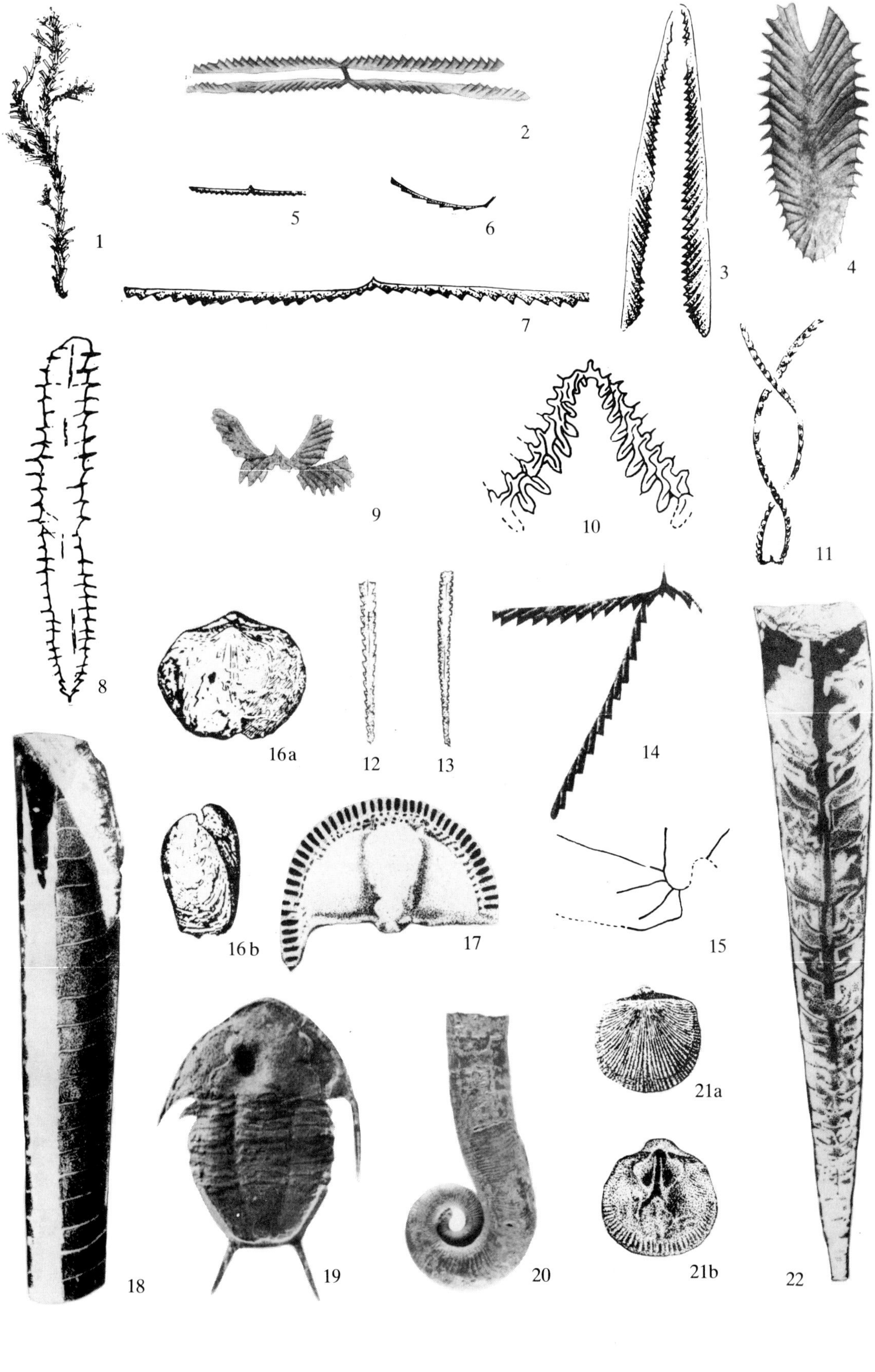

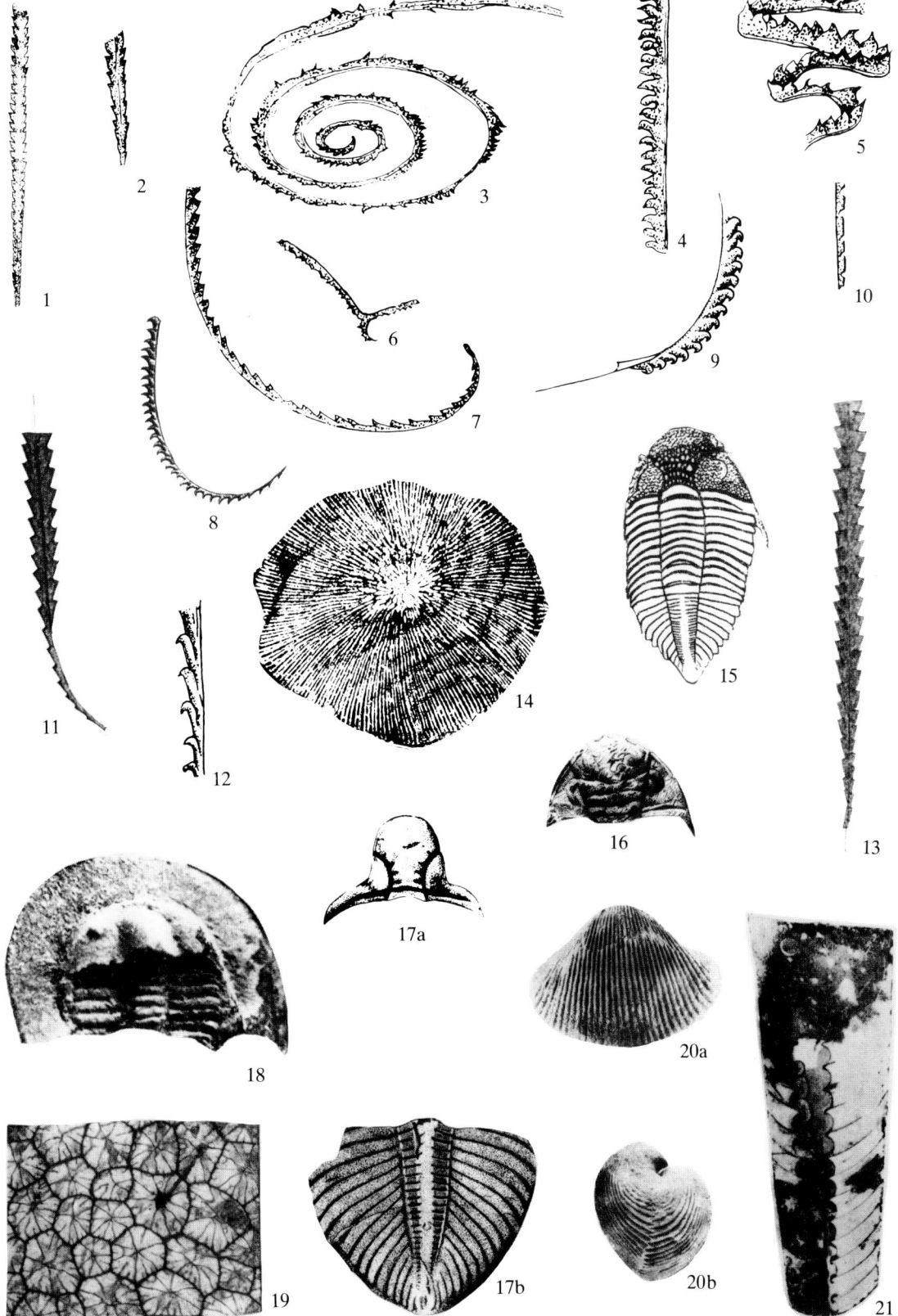

Plate VII

Plate VIII

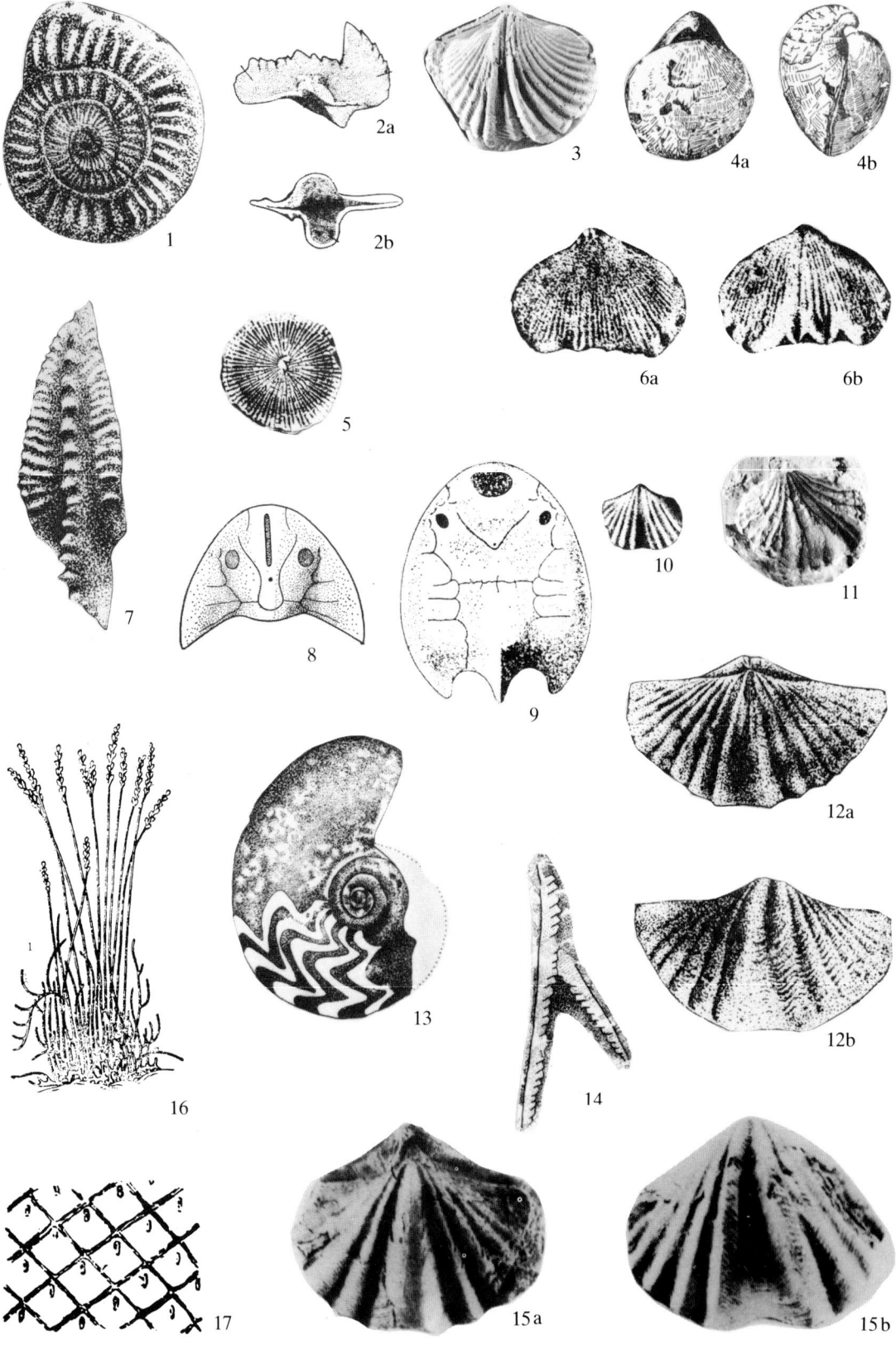

Plate VIII. Some important Devonian fossils. 1, *Erbenoceras solitarium* Barrande ($\times 1$), lateral view, Lower Devonian Chongyou Formation, Nanning, Guangxi Region; 2, *Spathognathodus exigmus kwangsiensis* Wang and Wang ($\times 27$): a, lateral; b, reverse view; Lower Devonian Yujiang Formation, Guangxi Region; 3, *Orientospirifer wangi* Hou ($\times 1$), ventral interior, Lower Devonian Jiangmu Formation, Yuling, Guangxi Region; 4, *Stringocephalus burtini* Defrance ($\times \frac{1}{3}$): a, dorsal view; b, lateral view; Middle Devonian Hualing Formation, Hualing, Yunnan Province; 5, *Temnophyllum waltheri* Yue ($\times 1$), cross-section, Middle Devonian Guanwushan Formation, Jiangyou, Sichuan Province; 6, *Yunnanella synplicata* Grabau ($\times 1$): a, ventral view; b, dorsal view; Upper Devonian Xikuangshan Formation, Xinhua, Hunan Province; 7, *Polygnathus dengleri* Bischoff and Ziegler ($\times 45$), oral view, Upper Devonian Gudling Formation, Dale, Xiangzhou, Guangxi Region; 8, *Euggaleaspis changi* Liu ($\times \frac{1}{2}$), cephalon, Lower Devonian Cuifeng Formation, Chüjing, Yunnan Province; 9, *Polybranchiaspis liaojiaoshanensis* Liu ($\times \frac{2}{3}$), dorsal view of dorsal plate, Lower Devonian Cuifeng Formation, Chüzing, Yunnan Province; 10, *Orientospirifer nakaolingensis* Hou ($\times 1$), ventral view, Lower Devonian Nagaoling Formation, Hengxian, Guangxi Region; 11, *Spiriferina supramarginalis* Khalfin ($\times 1$), ventral external cast, Lower Devonian Beijuntan Formation, Yuling, Guangxi Region; 12, *Acrospirifer ordinaris* Hou and Xian ($\times 1$): a, dorsal view; b, ventral view; Hengxian, Guangxi Region; 13, *Manticoceras wedekindi* Sun ($\times 1$), lateral, Upper Devonian Shetianqiao Formation, Changsha, Hunan Province; 14, *Monograptus uniformis* Pribyl ($\times 1\frac{1}{2}$), Lower Devonian, Fancheng, Qinzhou, Guangxi Region; 15, *Acrospirifer fongi* Grabau ($\times 1$), a, dorsal view; b, ventral view; Middle Devonian Houershan Formation, Dushan, Guizhou Province; 16, *Zosterophyllum sinensis* Lee and Zan reconstruction, Lower Devonian, Shiqiao, Guangxi Region; 17, *Leptophloeum rhombicum* Dawson ($\times 1$), Upper Devonian Huangjiadeng Formation, Changyang, Hubei Province.

Plate IX. Some important Carboniferous fossils. 1, *Fusulinella pseudobocki* Lee and Chen ($\times 7$), axial section, Upper Carboniferous Weining Formation, Weining, Guizhou Province; 2, *Eostaffella hohsienica* Cheng ($\times 25$), axial section, Lower Carboniferous Zhaojiashan Formation, Weining, Guizhou Province; 3, *Pseudostaffella nibelensis* Rauser ($\times 15$), axial section, Upper Carboniferous Weining Formation, Weining, Guizhou Province; 4, *Pseudoschwagerina miharanoensis* Akagi ($\times 6$), axial section, Upper Carboniferous Maping Formation, Weining, Guizhou Province; 5, *Yuanophyllum kansuense* Yü: a, cross-section ($\times \frac{3}{4}$); b, longitudinal section ($\times 1\frac{1}{2}$). Lower Carboniferous (upper part) in South and North-west China; 6, *Pseudouralinia gigantea* Yü ($\times 1$), cross-section, Lower Carboniferous Yanguan Formation (upper part), Yongfu, Guangxi Region; 7, *Kueichouphyllum sinense* ($\times 1$), cross-section, Lower Carboniferous, Weining, Guizhou Province; 8, *Palaeosmilia regia* Phillips ($\times 1\frac{1}{2}$), cross-section, Lower Carboniferous Zhaojiashan Formation, Weining, Guizhou Province; 9, *Kionophyllum ovatum* Wu and Chao ($\times 1$), cross-section, Upper Carboniferous Weining Formation, Shuicheng, Guizhou Province; 10, *Dibunophyllum chui* Lee and Yü ($\times 1\frac{1}{2}$), cross-section, Upper part of Lower Carboniferous, Liuzhou, Guangxi Region; 11, *Branneroceras perornatum* Yin: a, lateral; b, anterior views ($\times \frac{1}{2}$), Upper Carboniferous Wangjiakan Formation, Shuicheng, Guizhou Province; 12, *Echinoconchus liangshanensis* Wang: a, ventral, b, dorsal views, ($\times \frac{2}{3}$), Lower Carboniferous Menggongao Formation, Xiangxiang, Hunan Province; 13, *Choristites mansuyi* Chao ($\times 1$), dorsal view, Upper Carboniferous Weining Formation, Panxian, Guizhou Province; 14, *Gigantoproductus edelburgensis* Phillips ($\times \frac{1}{2}$), ventral view, Lower Carboniferous Zhaojiashan Formation, Shuicheng, Guizhou Province; 15, *Composita hunanensis* Wang ($\times 2$), dorsal view, Lower Carboniferous Menggongao Formation, Xiangxiang, Hunan Province; 16, *Vitiliproductus gröberi* Krenkel ($\times \frac{1}{2}$), ventral view, Lower Carboniferous, Weining, Guizhou Province; 17, *Dictyoclostus uralicus* Tschernyschew ($\times 1$), dorsal view, Upper Carboniferous Maping Formation, Weining, Guizhou Province; 18, *Lepidodendron szeianum* Lee ($\times 1$), Upper Carboniferous Taiyuan Formation, Shanxi, Beijing, etc.; 19, *Linopteris brongniartii* Gutb.: right, ($\times 2$); left, ($\times 3$); Upper Carboniferous Benxi Formation, Shangdong, Ningxia, etc.

(see over)

Plate X. Some important Permian fossils. 1, *Wentzellophyllum volzi* Yabe and Hayasaka ($\times 4$), cross-section, Chihsia Formation, Lower Permian, Weining, Guizhou Province; 2, *Gnathodus sichuanensis* Wang ($\times 40$): a, lateral view; b, oral view, Chihsia Formation, Lower Permian, Jiangbei, Sichuan Province; 3, *Guizhoupecten regularis* Chen ($\times 1$): a, left valve; b, right valve, Wujiaping Formation, Upper Permian Ziyun, Guizhou Province; 4, *Misellina tumida* Sheng and Sun ($\times 15$), axial section, Chihsia Formation, Lower Permian, Tianyang, Guangxi Region; 5, *Cancellina neoschwagerinoides* Deprat ($\times 8$), axial section, Lower Permian, Wanmo, Guizhou Province; 6, *Parafusulina yunnanica* Sheng ($\times 5$), axial section, Lower Permian, Hechuan, Sichuan Province; 7, *Palaeofusulina sinensis* Sheng ($\times 25$), axial section, Changxin Formation, Upper Permian, Qijiang, Sichuan Province; 8, *Pseudotirolites orientalis* Chao and Liang ($\times 1$) lateral view; b, ventral view, Dalong Formation, Upper Permian, Tianzhen, Guizhou Province; 9, *Pseudophillipsia chongqingensis* Lu ($\times 2$): a, cephalon; b, pygidium; Upper Permian, Wenxinxiang, Chongqing, Sichuan Province; 10, *Leptodus nobilis* Waagen ($\times 1$), ventral view, Heshan Formation, Upper Permian, Hechi Xian, Guangxi Region; 11, *Gigantonoclea hallei* (Asama) Gü .and Zhi ($\times 1$), Lower Shihhotse Formation, Upper Permian, Taiyuan, Shanxi Province; 12, *Lepidodendron oculus-felis* Abbado ($\times 1$), Upper Carboniferous to Permian, Daqingshan, Nei Mongol; 13, *Lobatannularia multifolia* Kon'no and Asama ($\times 1$), Upper Shihhotse Formation, Lower Permian, Wu Xian, Henan Province.

(see over)

Plate IX

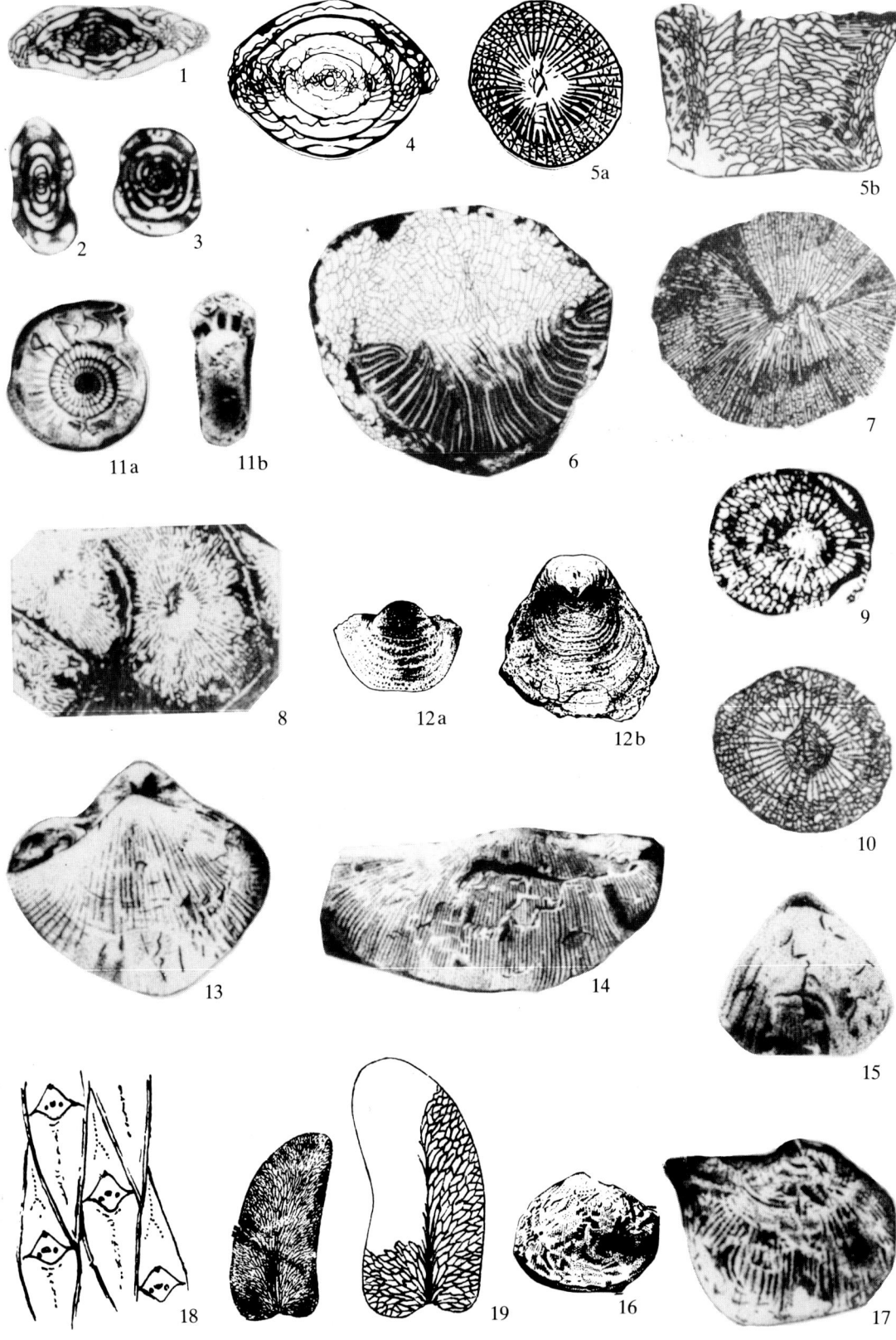

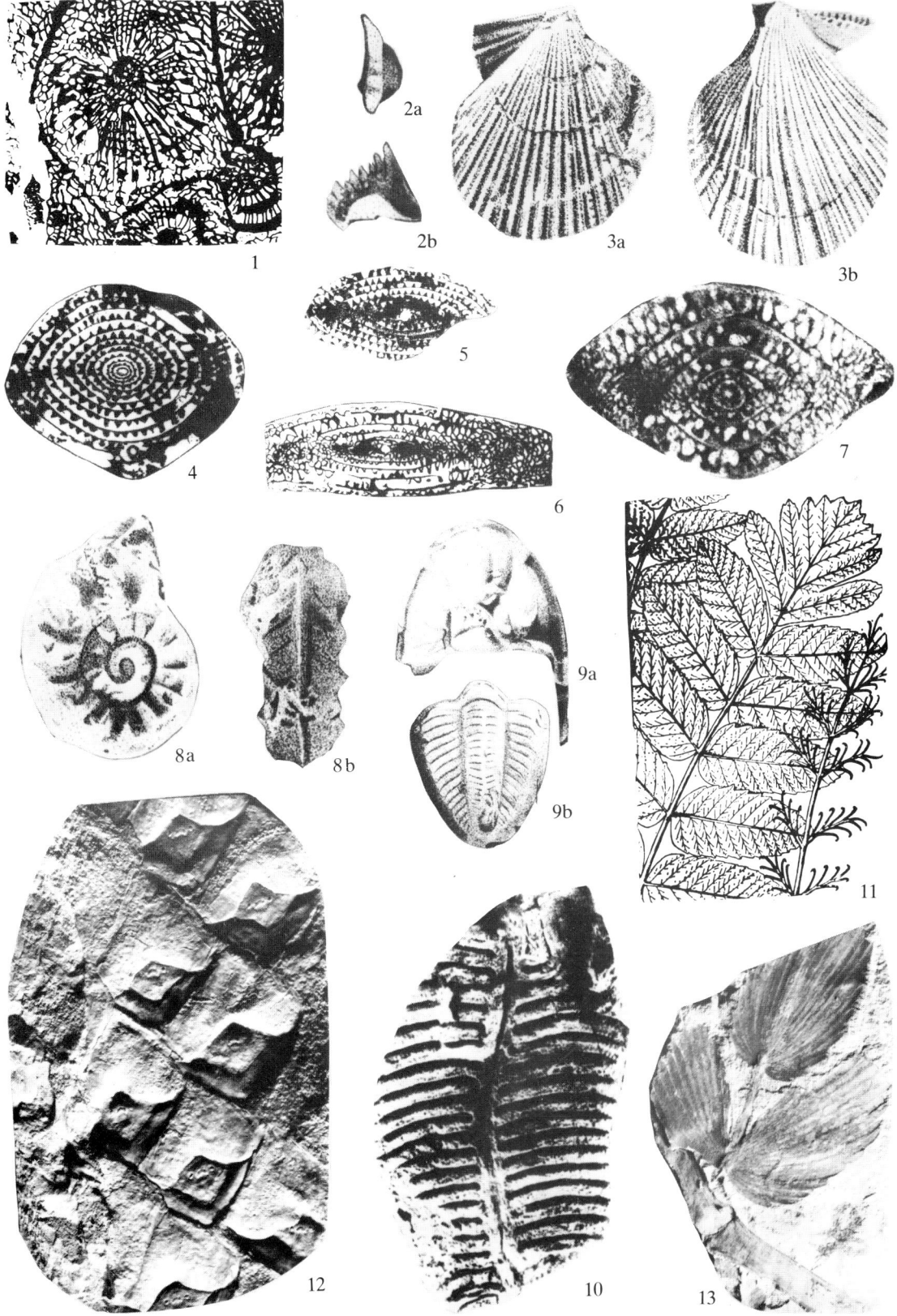

Plate X

Plate XI

Plate XI. Some important Triassic fossils. 1, *Ophiceras sinensis* Tien (× ½): a, lateral view; b, anterior view; basal part of Lower Triassic Series, Guiyang, Guizhou Province; 2, *Proptychites kwangsiensis* Chao (× ⅓), lateral view; Lower part of Lower Triassic, Tian'e Xian, Guangxi Region; 3, *Columbites costatus* Chao (× 1): a, lateral view; b, ventral view; Yongningzhen Formation, Lower Triassic, Qinglong, Guizhou Province; 6, *Trachyceras douvillei* Mansuy (× 1): a, lateral; b, ventral view; Falang Formation, Middle Triassic, Fuxi, Yunnan Province; 7, *Balatonites gracilis* Arthaber (× 1): a, lateral; b, ventral, Lower part of Middle Triassic, Tian'e Xian, Guangxi Region; 8, *Progonoceratites pulcher* Riedel (× 1), lateral, Leikoupo Formation, Middle Triassic, Weiyuan, Sichuan Province; 9, *Nodotibetites nodosus* Wang and He (× ⅔); a, lateral view; b, ventral view; Upper part of Turong Group, Upper Triassic, Nyalam, Xizang Region; 10, *Claraia wangi* Patte (× 1), right valve, Feixianguan Formation, Lower Triassic, West Guizhou Province; 11, *Eumorphotis multiformis* Bittner (× 1), left valve; Feixianguan Formation, Lower Permian, Qinglong, Guizhou Province; 12, *Leptochondria illyrica* Hsü (× 3); Leikoupo Formation, Middle Triassic, Lichuan, Hubei Province; 13, *Anchignathodus minutus* Ellison (× 50), Feixianguan Formation, Lower Triassic, Guangyuan, Sichuan Province; 14, *Burmesia lirata* Healy (× 1), right valve, Lower Member, Xujiahe Formation, Upper Triassic, Emei, Sichuan Province; 15, *Myophoria napengensis* Healey (× 1), left valve, Shizhongshan Formation, Upper Triassic, Jianchuan, Yunnan Province; 16, *Halobia neumayeri longmendongensis* Chen et al. (× ⅔), Lower part of Shujiahe Formation, Upper Triassic, Emei, Sichuan Province; 17, *Anthrophyopsis crassinervis* Nathorst (× 1), Upper Triassic, the Xiang-gan area; 18, *Ptilozamites chinensis* Hsü: left (× ⅕) showing palmate multipinnae; right (× ⅗) bifurcate pinnae; Upper part of Upper Triassic, Xiang-gan area.

Plate XII. Some important Jurassic fossils. 1, *Arietites* cf. *bonnardii* d'Orbigny (× 1), lateral view, Jingji Formation, Lower Jurassic, Dajingji, Kai'en area, Guangdong Province; 2, *Macrocephalites compressus* Quenstedt (× 1): a, lateral; b, ventral view; Middle Jurassic, Dingri, Xizang; 3, *Blanfordiceras wallichi* Gray (× ⅓): a, lateral view; b, ventral view; Upper Jurassic, Ngari Area, Xizang; 4, *Virgatosphinctes denseplicatus* Waagen (× ¾): a, lateral view; b, ventral view; Upper Jurassic, Gyirong, Xizang; 5, *Eolamprotula cremeri* Frech (× ⅔), Upper part of Ziliujing Fm, Middle Jurassic, Yunyang, Sichuan Province; 6, *Sphaerium andersoni* Grabau (× 2), right valve, Upper Jurassic, Hunyuan, Shanxi Province; 7, *Timiriasevia bella* Su and Li (× 38): a, right valve; b, left valve; Anding Formation, Middle Jurassic, Yan'an, Shaanxi Province; 8, *Belemnopsis lalonglensis* Wu (× 1): a, ventral view; b, lateral view; Lalongla Formation, Middle Jurassic; Nyalam, Xizang; 9, *Buchia spitiensis* Holdhaus (× 1), right view, Menbu Formation, Upper Jurassic, Nyalam, Xizang; 10, *Cuneopsis sichuanensis* Gu, Ma and Lan (× ⅔), right valve; Qianfuyan Formation, Middle Jurassic, Wangchang, Sichuan Province; 11, *Nestoria pissovi* Krasinetz (× 2), Baoshi Formation, Upper Jurassic, Horqin, Right Wing Front B., Nei Mongol Region; 12, *Nippononaia sinensis* Nieh (× ⅔), Haizhou Formation, Upper Jurassic, Yixian, Liaoning Province; 13, *Mesoclupea showchangensis* Ping and Yen (× ⅔), Upper Jurassic, eastern Zhejiang Province; 14, *Himalayisphaeridium nylamuensis* Xu (× 600), Pupuga Formation, Lower Jurassic, Nyalam, Xizang; 15, *Densoisporites proinatus* Couper (× 600). Baitianba Formation, Lower Jurassic, Jiangyou, Sichuan Province; 16, *Coniopteris hymenophylloides* Brongn. Seward (× 1), Lower–Middle Jurassic, Xiahuayuan, Hebei Province.

(see over)

Plate XIII. Some important Cretaceous fossils. 1, *Kilianella pexiptycha* Uhlig (× 1): a, lateral view; b, ventral view; Upper part of Lower Cretaceous, Ngari, Xizang; 2, *Neocomites indomontanus* Uhlig (× ⅔), lateral view; Lower part of Lower Cretaceous, Ngari, Xizang; 3, *Euthymiceras (Octagoniceras) hundesianum* Uhlig (× ⅓): a, umbilical view; b, lateral view; Lower part of Lower Cretaceous, Dingyongla, Ngari, Xizang; 4, *Turrilites desnoyersi* d'Orbigny (× ¾), Upper Cretaceous near Gangma, Xizang; 5, *Trigonoides sinensis* Gu and Ma (× ⅔), left valve, Guantou Formation, Lower Cretaceous, Yongkang, Zhejiang Province; 6, *Nippononaia ? jilinensis* Gu and Yü (× 1), external model, Jijialing Formation–Qingshankou Formation, Lower Cretaceous, Changling, Kilin Province; 7, *Cypridea unicostata* Galeeva (× 38): a, right valve; b, dorsal view; Zhidang Group, Lower Cretaceous, Wugamiao, Hanggin B, Nei Mongol; 8, *Estherites (Euestherites) stellasis* Zhang and Chen (× 10), Upper Cretaceous, Songliao Area, Liaoning Province; 9, *Orbitoides media* d'Archiac (× 16), axial section, Zongshan Formation, Upper Cretaceous, Gamba, South Xizang; 10, *Rotolipora cushmani* Morrow (× 52), dorsal view; Lengqingre Formation, Upper Cretaceous, Gamba, South Xizang; 11, *Yuanaia xinjiangensis* Hao and Zeng: a, ventral view; b, dorsal view (both × 70); Kukebai Formation, Upper Cretaceous, Kuzkongsu, Xinjiang; 12, *Migros hectori* Nauss: a, lateral view; b, shell margin view; Upper Cretaceous, Tarim Basin, Xinjiang; 13, *Chuhsiungichthys tsanglingensis* Liu (× ½), Lower part of Jiangdihe Formation, Lower Cretaceous, Chuxiong, Yunnan Province; 14, *Hibolites subfusiformis* Raspail (× ½), Lower Zongzhou Formation, Lower Cretaceous, Gyangzê Co., Xizang; 15, *Acanthopteris gothani* Sze (× 1), Fuxin Formation, Lower Cretaceous, Fuxin Basin, Liaoning Province.

(see over)

Plate XII

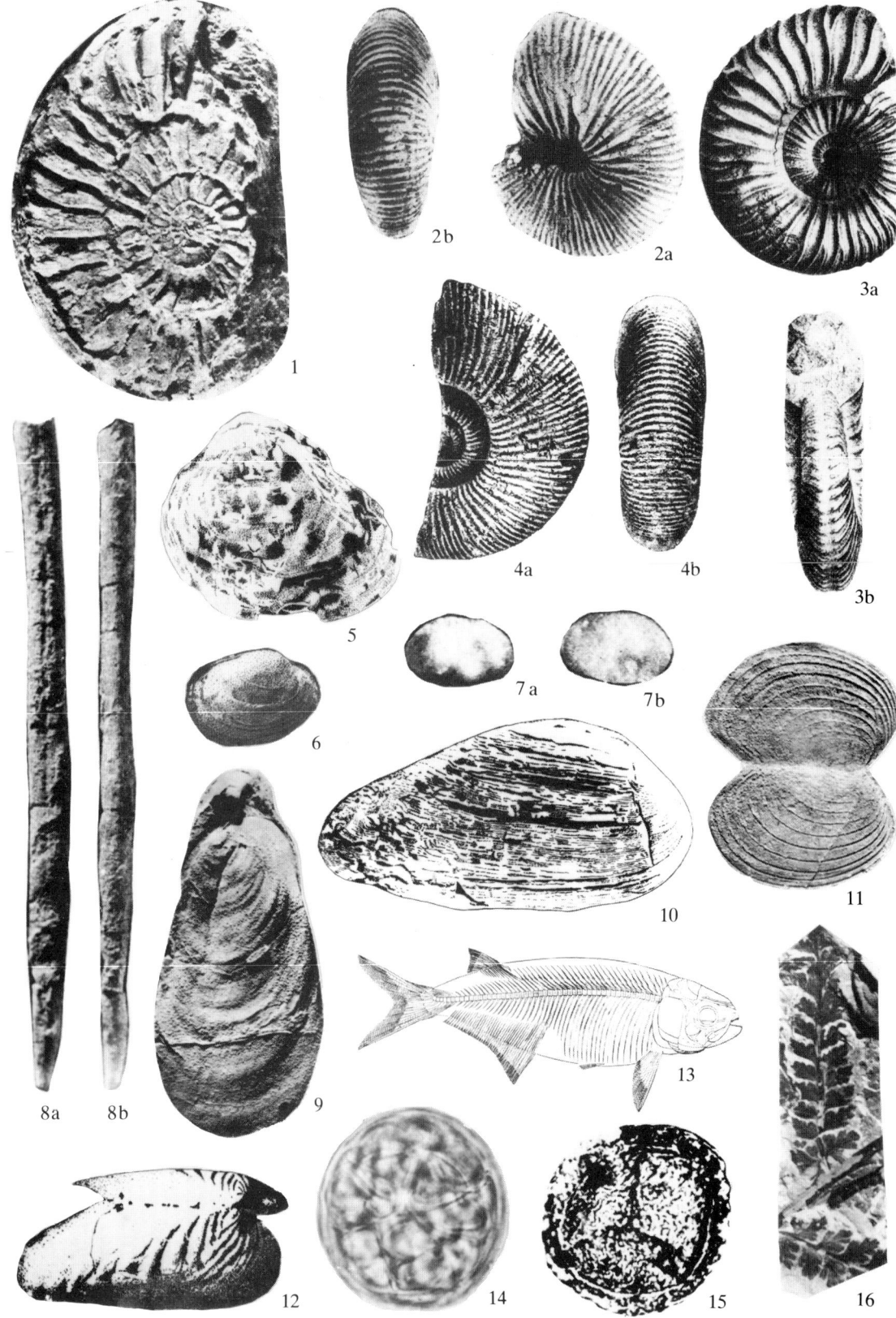

Plate XIII

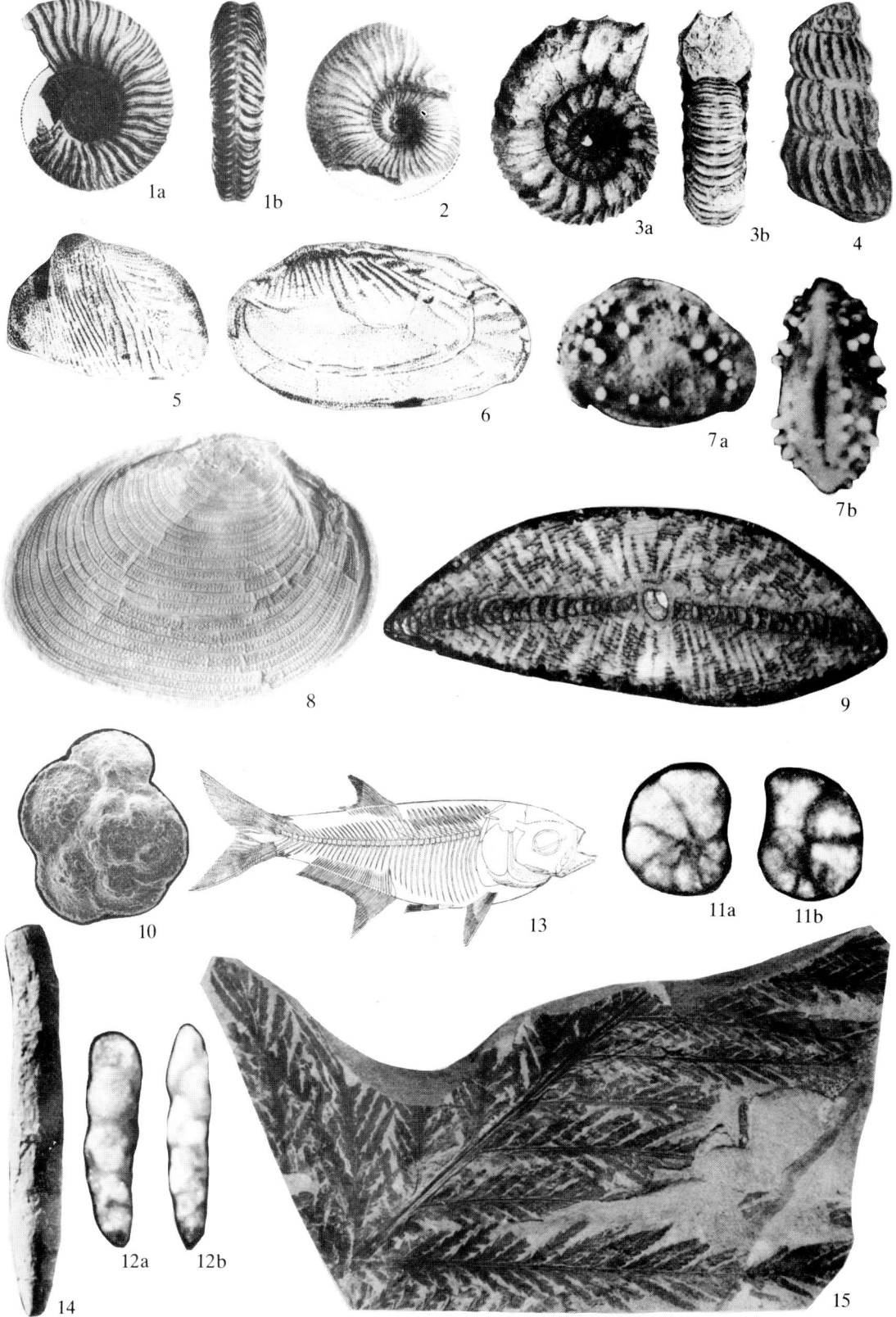

Plate XIV

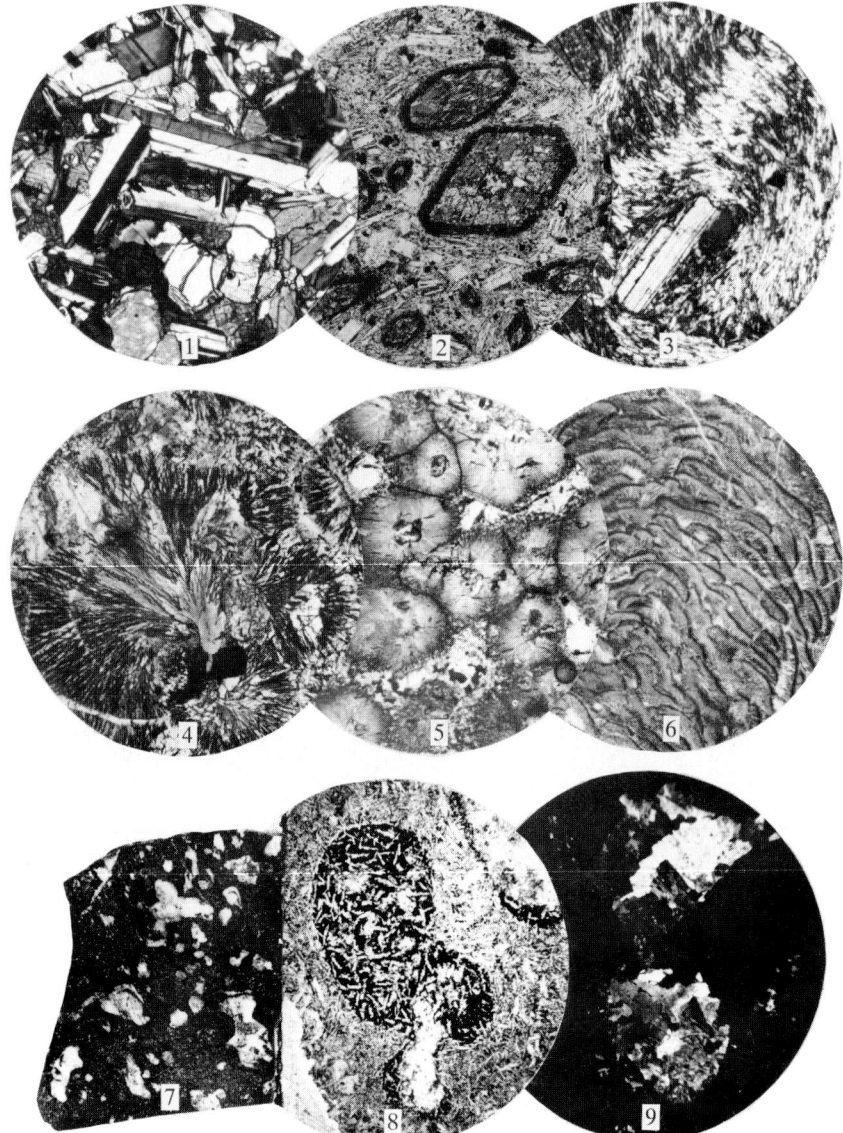

Plate XIV. 1, Augite–basalt, showing intergranular texture, Tertiary basalt, Wanquan, Hebei (+nicols, ×42); 2, Hornblende–andesite, showing opacitized border of the hornblende phenocrysts, Upper Jurassic volcanic formation, Nanjing, Jiangsu (//nicols, ×42); 3, Biotite-trachyte, showing trachytic texture and with alkaline feldspar phenocrysts in a groundmass of dominant alkaline feldspar microlites, Middle Cretaceous volcanic formation, Yao'an, Yunnan (+nicol, ×42); 4, Rhyolite, showing spherulitic texture, Middle Cretaceous volcanic formation, Doilungdeqen, Xizang (//nicol, ×42); 5, Alkaline rhyolite, showing microglobular texture, Upper Jurassic formation, Changbei Mountains, Jilin (//nicol, ×42); 6, Pearlite, showing vermicular texture, Upper Jurassic volcanic formation, Dongtou, Jinyun, Zhejiang (//nicol, ×42); 7, Amygdaloidal magnetite rock (iron ore of the ore magma type) containing 51.4% of Fe_2O_3 and 23.44% FeO, albitic amygdales in the groundmass composed mainly of minute magnetite grains and crystals with dispersed albite microlites. Upper Triassic volcanic formation, sample from Mengyang, Yunnan (2101–291, ×1); 8, Light colour spilite with spilitic iron droplets and albite amygdales, groundmass composed chiefly of albite and delessite with some magnetite, showing spilitic texture, the iron droplet composed of magnetite and albite microlites with scanty delessite showing a similar spilitic texture, lined with albite along the upper left margin; the albite amygdales (upper right and lower middle of the photograph) being partly lined with spilitic iron; geological age and locality same as 7 (2401–132.9; //nicol, ×17); 9, Microphoto of 7. The margin of the 'amygdales' is not even, with the albite partly protruding into the groundmass of dominant magnetite with scanty interstitial albite and delessite. (Photos 1, 2, 3, 5, and 6 by courtesy of Li Zhaonai and Fei Wenheng; 4 by courtesy of Wang Songchan; 7–9, by courtesy of Chen Yuchuan.)

Plate XV

Plate XV. 1, Layered olivine-gabbro (cumulate), Baigang, Xizang; 2, Zonal structure (longitudinal 'growth line') of the pillow lava, Qunrang, Xizang; 3, Pillow lavas, Qunrang, Xizang; 4, Sheeted sill (dyke) swarms (positive element within pillow lavas), Deji, Xizang; 5, Cretaceous–Tertiary porphyritic granite, showing the growth of clusters of unevenly distributed potash feldspar 'phenocrysts' of metasomatic origin, near the intensively assimilated relict patches of dark xenoliths, Xinfucun, along the Quxu–Lhasa highway, Xizang, similar scale to 6; 6, Dark xenoliths in Cretaceous–Tertiary granite, with potash feldspar of metasomatic origin, Quxu (Qushui), Xizang. (Photos 1–4, from *Tectonics of Yarlung Zangbo Suture Zone, Xizang—guide to geological Excursion*, Geol. Bureau and Geol. Soc. of Xizang, China, 1982; 5–6, by courtesy of Liu Guohui.)

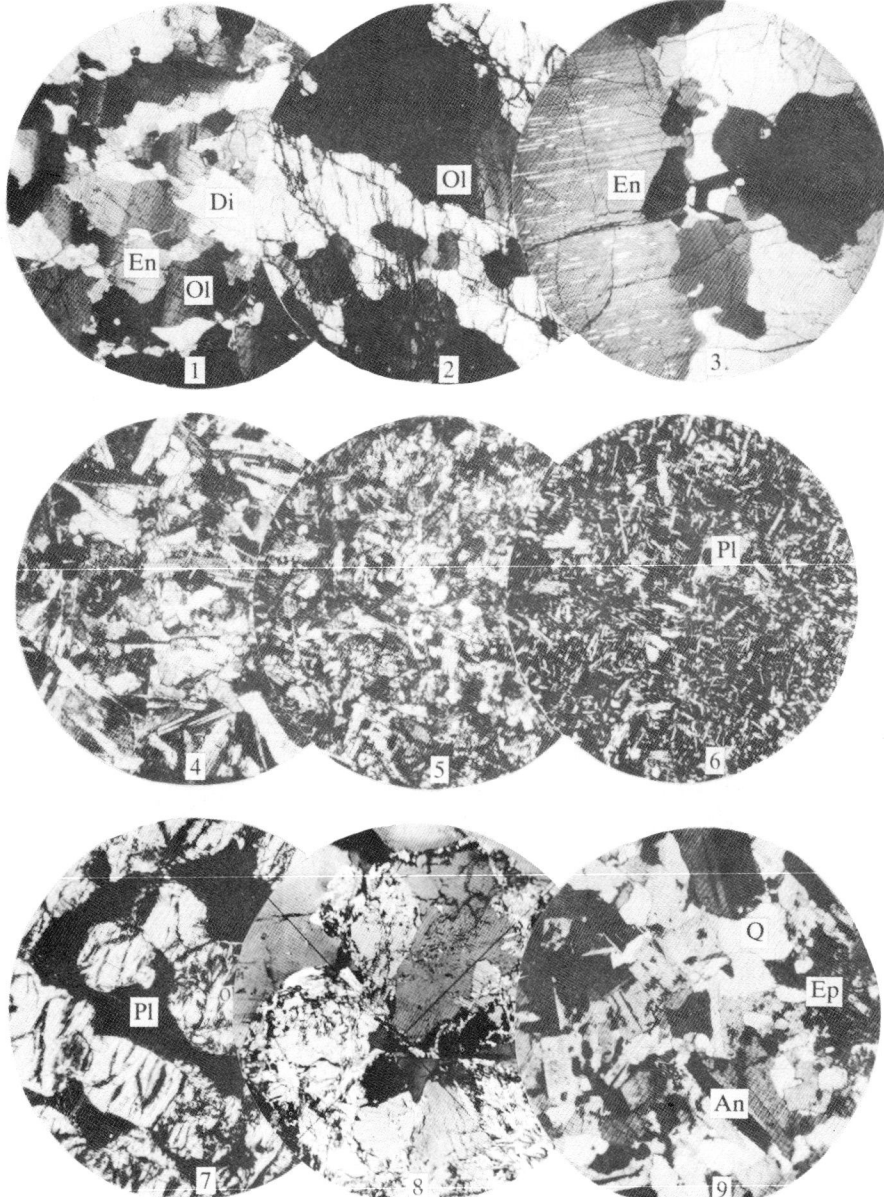

Plate XVI. 1, Diopside-bearing harzburgite, with granoblastic texture, consisting of olivine, enstatite and diopside; olivine with kink band structure and undulatory extinction, Dashuqu, Xizang (+nicols, ×12); 2, Dunite with inequigranular granoblastic texture, olivine occurring in inequigranular xenoblast and showing kink band and undulatory extinction, Dashuqu, Xizang (+nicols, ×12); 3, Harzburgite, enstatite with exsolution clinopyroxene (light-coloured narrow stripes or pegs), Luqu, south of Shigatse, Xizang (+nicols, +17); 4, Diabase, with diabasic texture, consisting of plagioclase, clinopyroxene and minor magnetite, Bazhuqu, Xizang (+nicols, ×12); 5, Dolerite, with holocrystalline intergranular texture, fine-grained clinopyroxene and minor magnetite being distributed in the interstices between plagioclase laths, Jiding, Xizang (+nicols, ×12); 6, Porphyritic basalt, with plagioclase phenocrysts in an interstitial to intergranular matrix composed of plagioclase, clinopyroxene and hyaline substance, locality and magnification same as 5; 7, Plagioclase-bearing dunite (cumulate), with accumulation texture, olivine (altered into serpentine) being panidiomorphic, with altered plagioclase filling the space between them, Dazhuqu, Xizang (+nicols, ×17); 8, Late Proterozoic cordierite-granite, with the periphery of cordierite (light grey) already sericitized. Jiuling, Jiangxi (+nicols, ×25). (By courtesy of Liu Guohui *et al.* and Xu Huifen); 9, Trondhjemite, consisting of oligoclase, xenomorphic quartz and minor epidote, Dazhuqu, Xizang (+nicols, ×17). (Photos 1–7, and 9, from *Tectonics of Yarlung Zangbo Suture Zone, Xizang—guide to geological Excursion*, Geol. Bureau and Geol. Soc. of Xizang, China, 1982.)

Plate XVII

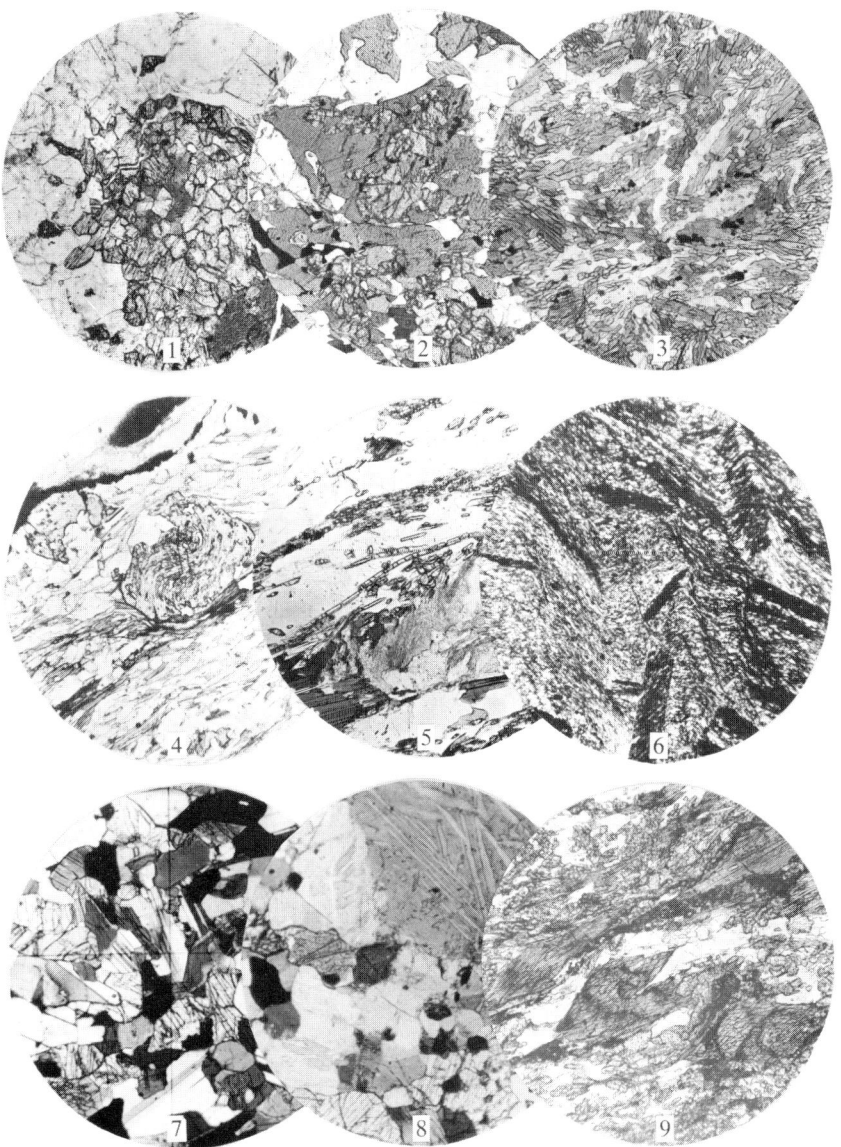

Plate XVII. 1, Diopside-trondhjemitic migmatitic granite in the Archaean Qianxi Group, showing pseudomorph of aggregates of later diopside granules after an earlier larger diopside, Gongdianzi, Qian'an, Hebei (G 76–475; //nicol, ×50). (By courtesy of Gao Jifeng); 2, Two-pyroxene-plagioclase-gneiss, showing the presence of relic diopside in a later hornblende, Archaean Qianxi Group. Qianxi, Hebei (G 027; //nicol, ×31). (By courtesy of Gao Jifeng); 3, Fine-grained plagioclase-amphibolite, with the pilotaxitic to intergranular texture of the protolithic basalt still partly preserved, the lath-shaped plagioclase microlites are now composed of andesine, but somewhat granulated, Upper part of the Archaean Yanlingguan Formation, Yanlingguan, Xintai, Shandong (Ay 1770 k; //nicol, ×25). (Cheng *et al.* 1977); 4, Mica–schist with snowball almandine, of dominant Himalayan metamorphic age, near Kangma, Xizang (//nicol, ×23). (By courtesy of Liu Guohui); 5, Sillimanite-biotite-quartz-schist, containing basic oligoclase, Nielamu Group of dominant Himalayan metamorphic age, Near Nielamu, Xizang (//nicol, ×23).(By courtesy of Liu Guohui); 6, Chloritoid-phyllite of dominant Himalayan metamorphic age, with puckered felts composed of sericite, chlorite, and quartz. The chloritoid porphyroblasts are formed after puckering or slightly later than the micro-folding, near Kangma, Xizang (+nicols, ×12). (By courtesy of Liu Guohui); 7, Granulite composed of hypersthene (pale colour), clinopyroxene (grey), hornblende and labradorite, Lower part of the Fuping Group, Fuping, Hebei (8098–4; +nicols, ×18). (By courtesy of Wu Jiashan); 8, Granulite composed of hypersthene (pale colour), clinopyroxene (grey), and antiperthitic plagioclase, Archaean Miyun Group, Huogezhuang, Miyun, Beijing (75043; +nicols, ×28). (By courtesy of Gao Boyu); 9, Glaucophane-schist composed of glaucophane (showing cleavage), epidote (granule), and quartz, Aksu Group of dominant Middle to Upper Proterozoic age, near Aksu, Xinjiang (//nicol, ×37). (By courtesy of Xiong Jibin.)

Plate XVIII

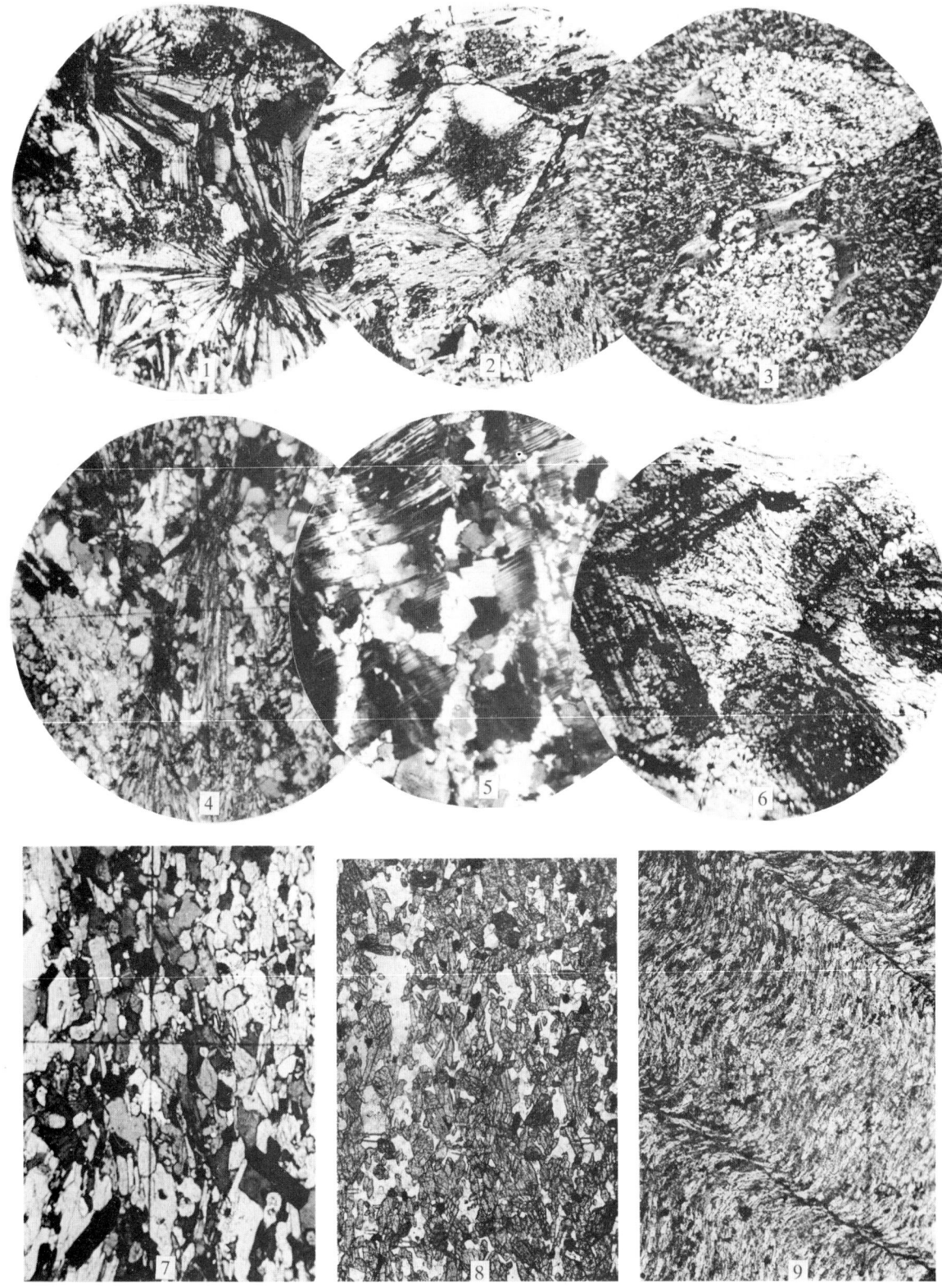

Plate XVIII. 1, Chloritoid rock composed of spherulitic aggregates of radiating chloritoid crystals, metamorphosed Early Jurassic basic volcanic formation, Fangshan, Beijing (+ nicols, × 32) (By coutesy of Liu Guohui and Wu Jiashan); 2, Andalusite-phyllite, with chiastolite porphyroblasts in a groundmass of sericite, quartz and carbon particles, metamorphosed Upper Carboniferous formation. Fangshan, Beijing (+ nicols, × 32) (By courtesy of Liu Guohui and Wu Jiashan); 3, Albite-chlorite-schist, with the amygdaloidal structure of the protolithic andesitic rocks still preserved, the amygdales are now composed chiefly of quartz and epidote granules surrounded by a chloritic rim, Hutuo Group, Wutai, Shanxi (+ nicols, × 30) (By courtesy of Wu Jiashan); 4, Hornfels with porphyroblasts of staurolite (light grey, on the left of the photo) and aluminium anthophyllite (dark grey sheaf-aggregates), Permian Hongmiaoling Formation, Fangshan, Beijing (+ nicols, × 32) (By courtesy of Liu Guohui and Wu Jiashan); 5, Crushed and sheared orthoclase-migmatitic granite probably of Archaean age, with crushed and strained orthoclase (dark grey to black) further resorbed and corroded by quartz. South-east of Gongchangling, Anshan, Liaoning (+ nicols, × 45) (Cheng Yuqi *et al.* 1963); 6, Chloritoid-phyllite, showing the chloritoid porphyroblasts formed slightly later than the alignment of sericite, chlorite and quartz in the groundmass, Fangshan, Beijing (62166D; + nicols, × 32) (By courtesy of Liu Guohui and Wu Jiashan); 7, Hornblende-schist composed chiefly of plagioclase (albite–oligoclase) and hornblende and subordinately of epidote and biotite, Wutai Group, Wutai, Shanxi (Ag 7–8. + nicols, × 30) (By courtesy of Wu Jiashan); 8, Quartz-bearing plagioclase-amphibolite, composed chiefly of plagioclase (andesine) and hornblende, with accessory quartz and magnetite, Cigou Formation, Anshan Group, Gongchangling, Anshan, Liaoning (AR377, S2520; //nicol, × 30) (Cheng Yuqi); 9, Sericite–phyllite. Liaohe Group, Anshan, Liaoning (As 304A; //nicol, × 30) (By courtesy of Wang Chaojun.)

Plate XIX. 1, Agmatite, with fragments of migmatized amphibolite embedded in a migmatitic granite, Songpeng, Xintai, Shandong (By courtesy of Shen Qihan and Wang Zejiu); 2, Biotite-bearing streaky migmatite with scattered potash feldspar 'augen', Hongjiang, Yichun, Jiangxi (Cheng Yuqi and Xu Huifeng); 3, Hornblendic nebulite with various proportions of hornblende and quartzo-feldspathic material in different parts of the rock, Huikou, Pingshan, Hebei (By courtesy of Liu Guohui); 4, Biotitic banded migmatite, strongly puckered, Archaean Badaohe Group, Yangshan, Lulong, Hebei (By courtesy of Shen Qihan *et al.*); 5, Biotitic streaky migmatite, notice the harmonious relationship between the coarse granitic 'veins' which contain relict biotite and the adjacent gneissic parts, portion of a specimen from Gongchangling, Anshan, Liaoning (× 38). (Cheng Yuqi *et al.*); 6, Plagioclase-amphibolite showing fusulina-shaped variolitic structure of the original subaqueous basic lava still partly preserved, but already drawn out and deformed; the variolites are considered as the fillings of the small vesicles, now composed of a dark green nucleus of minute plagioclase and hornblende with a white rim of plagioclase; uppermost part of the Archaean Yanlingguan Formation. Outcrop near Yanlingguan, Xintai, Shandong. About natural size (Cheng Yuqi *et al.*); 7, Partly gneissic biotite-quartz-dioritic contaminated rocks (grey, with white 'spots' of plagioclase of metasomatic origin) with relict bands of slightly transformed biotite–hornblende-granulitite ('enclaves') of variable shapes (dark grey to black), assimilation-contamination zone at the north-eastern periphery of the Fangshan granodiorite body, Beijing, this zone passes gradually outward into a zone of 'marginal migmatite', outcrop (By courtesy of Guo Huqi); 8, Laminated quartz–magnetite-taconite, composed of alternating layers of dominant quartz and dominant magnetite with very little actinolitic tremolite. Fe6, Archaean Cigou Formation, Gongchangling, Anshan, Liaoning (AO 390, (S 4435); //nicol, × 10): (Cheng Yuqi, 1956); 9, Quartz-diorite containing embryonic bluish green hornblende sieved with granules of quartz and plagioclase, probably formed by the further development of highly assimilated calcferromagnesian substance of the relics of the 'captured' country rock, border quartz-diorite zone (just inside the assimilation–contamination zone) of the Fangshan granodiorite body, Beijing (//nicol, × 27). (By courtesy of Guo Huqi).

(*see over*)

Plate XIX

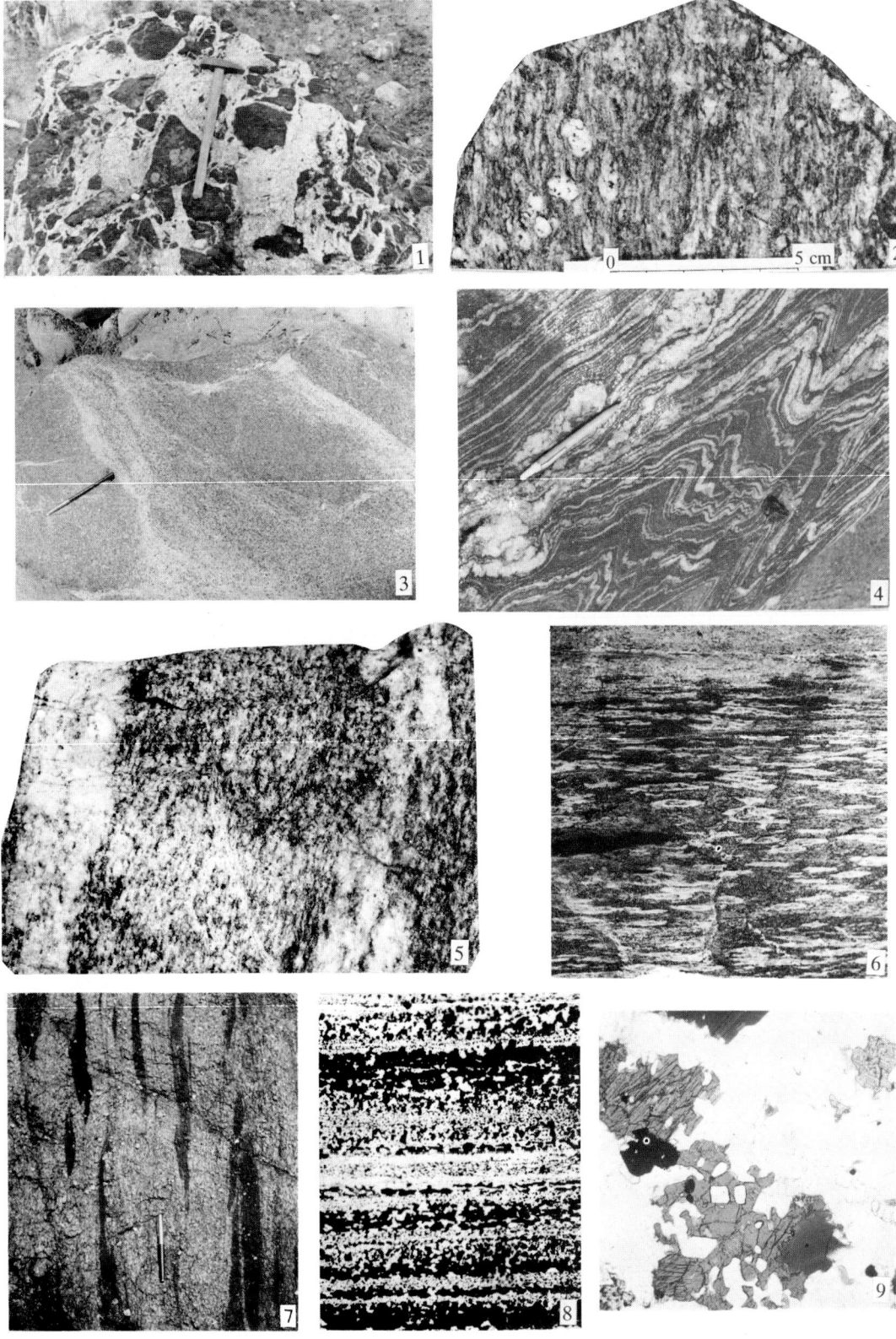

SELECTED REFERENCES

All-China Stratigraphic Committee (ed.) (1980). *Stratigraphic guide and explanations*.† Science Press, Beijing.

Anhui Regional Stratigraphic Tables Compilation Team (1978). *Regional stratigraphic tables of East China Region, Anhui volume*.† Geological Publishing House, Beijing.

Beijing Regional Stratigraphic Tables Compilation Team (1977). *Regional stratigraphic tables of Beijing*.† Geological Publishing House, Beijing.

Chinese Academy of Geological Sciences (ed.) (1982). *An outline of the stratigraphy of China*.* Vol. 1 of *Stratigraphy of China*. Geological Publishing House, Beijing.

Department of Geology, Nanjing University (1972). Granitic rocks of different geological periods of southeastern China and their genetic relations to certain metallic mineral deposits.* *Scientia Sinica*, Vol. 17.

—— (1980). Investigation on the time and spatial distribution of the granitic rocks of southeastern China, their petrographic evolution, petrogenetic types, and metallogenetic relations.* Special Issues on Geology, *J. Nanjing University*.

Fujian Regional Stratigraphic Tables Compilation Team (1979). *Regional stratigraphic tables of Fujian Province*.† Geological Publishing House, Beijing.

Gansu Regional Stratigraphic Tables Compilation Team (1980). *Regional stratigraphic Tables of Gansu Province*.† Geological Publishing House, Beijing.

Geological Bureau of Xizang and Geological Society of Xizang (compiled by Wan Ziyi et al.) (1982). *Tectonics of Yarlung Zangbo suture zone, Xizang (Tibet) — Guide to geological excursion*.

Geological Review Special Issue (1982). Marking the 60th anniversary of the founding of the Geological Society of China.† Vol. 28, No. 6.

Guiyang Institute of Geochemistry, Academia Sinica (1979). *Geochemistry of the granitoids of South China*.† Science Press, Beijing.

Guizhou Regional Stratigraphic Tables Compilation Team (1977). *Regional stratigraphic tables of Southwest region (Guizhou Province)*.† Geological Publishing House, Beijing.

Guizhou Stratigraphic and Palaeontologic Team (1978). *Palaeontologic atlas of southwestern China, Guizhou Volume*.† Geological Publishing House, Beijing.

Hebei and Tianjin Regional Stratigraphic Tables Compilation Team (1979). *Regional stratigraphic tables of Hebei and Tianjin*.† Geological Publishing House, Beijing.

Heilongjiang Regional Stratigraphic Team (1979). *Regional stratigraphic tables of Heilongjiang*.† Geological Publishing House, Beijing.

Hubei Bureau of Geology Gorges Stratigraphic Research Team (1978). *Sinian–Permian stratigraphy and palaeontology of the region east of the Yangzi Gorges*.† Geological Publishing House, Beijing.

Hubei Institute of Geological Sciences, et al. (1977). *Palaeontologic atlas of Central–South China*.† Geological Publishing House, Beijing.

Institute of Geology, Chinese Academy of Geological Sciences (1978). *Professional Papers of Devonian symposium of South China, 1974*.† Geological Publishing House, Beijing.

—— (1980). *Mesozoic stratigraphy and palaeontology of the Shaan-Gan-Nin Basin*.† Geological Publishing House, Beijing.

Institute of Geology, Ministry of Geology, P. R. China. *The Precambrian of China*† (ed. Cheng Yuqi and Wang Yuelen) Science Press, Beijing.

Institute of Vertebrate Palaeontology and Palaeoanthropology South China Red Beds Team (1977). Horizon and Faunas of Palaeocene mammalian fossils.* *Scientia Sinica* 3.

—— and Nanjing Institute of Geology and Palaeontology (1979). *Mesozoic and Cenozoic red beds of South China*.† Science Press, Beijing.

Iron Research Group of Middle–Lower Changjiang (Yangzi) Valley (1977). Porphyrite iron ore — a genetic model of a group of deposits in andesitic volcanic area.* *Acta Geol. Sinica*, 1977 vol. No. 1.

Jiangsu and Shanghai Regional Stratigraphic Tables Compilation (1978). *Regional stratigraphic tables of Jiangsu and Shanghai*.† Geological Publishing House, Beijing.

Jiangxi Regional Stratigraphic Tables Compilation

Team (1980). *Regional stratigraphic tables of Jiangxi.*† Geological Publishing House, Beijing.

Jilin Regional Stratigraphic Tables Compilation Team (1978). *Regional stratigraphic tables of Northeast China, Jilin volume.*† Geological Publishing House, Beijing.

Liaoning Regional Stratigraphic Tables Compilation Team (1979). *Regional stratigraphic tables of Liaoning.*† Geological Publishing House, Beijing.

Nanking (Nanjing) Institute of Geology and Palaeontology (ed.) (1982). *Stratigraphic Correlation Tables of China, with explanatory text.*† Science Press, Beijing.

—— (1976 1977). *Mesozoic fossils of Yunnan,* Vol. 1 (1976), Vol. 2 (1977).† Science Press, Beijing.

—— (1974). *Stratigraphic-palaeontologic handbook of south-western China.*† Science Press, Beijing.

—— (1983). *Studies on stratigraphic boundaries in China.*† Science Press, Beijing.

—— and Qinghai Institute of Geological Sciences (1979). *Palaeontologic atlas of northwestern China, Qinghai volume.*† Geological Publishing House, Beijing.

——, Yunnan Bureau of Geology, and Geologic Exploration Corporation of Yunnan Bureau of Metallurgy (1975). *Mesozoic Red Beds of Yunnan.*† Science Press, Beijing.

Nei Mongol Bureau of Geology and Northeast (Shengyang) Institute of Geological Sciences (1976). *Palaeontologic atlas, Nei Mongol volume.*† Geological Publishing House, Beijing.

Ningxia Regional Stratigraphical Tables Compilation Team (1980). *Regional stratigraphic tables of northwestern China, Ningxia volume.*† Geological Publishing House, Beijing.

Qinghai Regional Stratigraphic Tables Compilation Team (1980). *Regional stratigraphic tables of northwest China, Qinghai volume.*† Geological Publishing House, Beijing.

Scientific Expedition to Xizang (Tibet), Acad. Sinica (1974). *Scientific reports on the expedition to the Qomolangma area 1966–68, Geology.*† Science Press, Beijing.

—— (1975–1976). *Scientific reports on the expedition to the Qomolangma area 1966–68, Palaeontology.*† Vol. 1 (1975), Vol. 2 (1976), Vol. 3 (1976). Science Press, Beijing.

Shandong Regional Stratigraphic Tables Compilation Team (1978). *Regional stratigraphic tables of eastern China, Shandong volume.*† Geological Publishing House, Beijing.

Shanxi Regional Stratigraphic Tables Compilation Team (1979). *Regional stratigraphic tables of Shanxi.*† Geological Publishing House, Beijing.

Shengyang Institute of Geology and Mineral Resources (1980). *Palaeontologic atlas of northeastern China, Late Palaeozoic volume.*† Geological Publishing House, Beijing.

Sichuan Regional Stratigraphic Tables Compilation Team (1978). *Regional stratigraphic tables of southwestern China, Sichuan volume.*† Geological Publishing House, Beijing.

Southwest (Chengdu) Institute of Geology (1978). *Palaeontologic atlas of southwestern China, Sichuan volume.*† Geological Publishing House, Beijing.

Tianjin Institute of Geology and Mineral Resources (1980). *Research on Precambrian geology, Sinian Suberathem in China.** Tianjin Scientific Technical Press, Tianjin.

Xinjiang Bureau of Geology—Regional Geology Brigade, Xinjiang Institute of Geology, and Xinjiang Department of Geological surveying of Xinjiang Bureau of Petroleum (1981). *Palaeontologic atlas of northwestern China, Xinjiang volume.*† Geological Publishing House, Beijing.

Xinjiang Regional Stratigraphic Tables Compilation Team (1981). *Regional stratigraphic tables of northwestern China, Xinjiang volume.*† Geological Publishing House, Beijing.

Yichang Institute of Geological Sciences *et al.* (Central–South Regional Stratigraphic Tables Compilation Team) (1974). *Regional stratigraphic tables of Central–South China.*† Geological Publishing House, Beijing.

Yunnan Regional Stratigraphic Tables Compilation Team (1978). *Regional stratigraphic tables of southwestern China, Yunnan volume.*† Geological Publishing House, Beijing.

Zhejiang Regional Stratigraphic Tables Compilation Team (1979). *Regional stratigraphic tables of Zhejiang.*† Geological Publishing House, Beijing.

Bai Shunliang, Jin Shanyi, Ning Zhongshan (eds.) (1982). *Devonian biostratigraphy of Guanxi and adjacent area.*† Peking University Press, Beijing.

Cai Chongyang and Li Xingxue (1982). Subdivision and correlation of the Devonian continental strata in China.† In *Stratigraphic Correlation Chart with explanatory text.* Science Press, Beijing.

Cao Renguan, Wu Xiche, Ge Hongru, Luo Wanchen, and Liang Qizhong (1980). Sinian System of Wangjiawan section in Jinning county, Yunnan.* *The Sinian Suberathern.* Tianjin Scientific and Technical Press, Tianjin.

Cao Zhengyao, Li Baoxian, and Guo Shuangxing (1982). The Cretaceous plant-bearing strata in China.† In *Stratigraphic Correlation Chart with explanatory text.* Science Press, Beijing.

Chan Lipei and Li Li (1979). Distribution of Permian brachiopod faunas in China.† *Geological Papers Internal Ex.* 2, Stratigraphy and Paleontology.

Chang Chengfa and Pan Yushen (1981). A brief discussion on the tectonic evolution of Qinghai–Xizang plateau. *Proc. Symp. Qinghai–Xizang Plateau.* Vol. 1 Science Press, Beijing; Gordon and Breach Science Pub. Inc., New York.

—— and Zheng Xilan (1973). Tectonic features of the Jolmo Lungma in southern Tibet, China.* *Scientia Geologica Sinica,* 1973 vol., No. 1.

Chang (Zhang) Wentang, Yuan Kexing *et al.* (1979). *Carbonate biostratigraphy of Southwest China.*† Science Press, Beijing.

Chappell, B. W. and White, A. J. R. (1974). Two contrasting granite types. *Pacific Geol.* **8.**

Chen Bingwei (1983). Some new observations on the tectonic development of Sanjiang region, East Xizang (Tibet).* *Contr. Geol. Qinghai–Xizang (Tibet) Plateau,* **12,** Geological Publishing House, Beijing.

Chen Chuzhen (1978). The lower limit of Triassic of Southwest China.† *J. Stratig.* **2** (2).

Chen Chunyuan, Tsou Siping, Chen Tingen and Qi Dunluan (1979a). Late Cambrian cephalopods of North China—*Plectronocerida, Protactinocerida* (Order nov.) and *Yanhecerida* (Ord. nov.).* *Acta Palaeont. Sinica,* Vol. 18, No. 1.

—— (1979b). Late Cambrian *Ellesmerocerida* (cephalopods) of North China.* *Acta Palaeont. Sinica,* Vol. 18, No. 2.

Chen Jinbiao, Zhang Huimin, Xing Yusheng, and Ma Guogan (1981). On the Upper Precambrian (Sinian Sub-erathem) in China. In *Precambrian Research 15, Special Issue: Upper Precambrian correlations.*

Cheng, Y. C. (Cheng Yuqi) (1944). On successive zones of regional metamorphism in the vicinity of Tanpa, Sikang. *Science Record,* Vol. 1, Nos. 3–4, Academia Sinica.

——, Bai Jin, and Sun Dazhong (1982a). The lower and middle Precambrian of China.* In *An outline of the stratigraphy of China* (ed. Chinese Acad. Geol. Sciences (=CAGS)). Geological Publishing House, Beijing. (Also as a paper submitted to the Second All-China Conference of Stratigraphic Conference, 1979).

——, Chung Fudao, and Su Yunjun (1973). The Pre-Sinian of northern and northeastern china.† *Acta Geol. Sinica,* 1973 Vol. No. 1, Beijing.

——, Shen Qihan, Liu Guohui, and Wang Zejiu (1963). Some basic problems of metamorphic rocks and their related methods.† Industrial Press, Beijing.

——, *et al.* (1964). Geochronological study of magmatic and metamorphic rocks of the Taishan Group of the Sintai (Xintai) region, Shantung (Shandong).† *Geological Review (Dizhi Lunping),* Vol. 22, No. 3.

——, Shen Qihan and Wang Zejiu (1977). Preliminary study on the Taishan metamorphic group in the vicinity of Yanlingguan, Xintai, Shandong.† *Dizhi Kuangchan Yanjiu (Research on geology and mineral resources)* 1977 Vol., No. 3.

——, ——, —— (1982b). *Preliminary study of the metamorphosed basic volcano-sedimentary Yanlingguan Formation of the Taishan Group of Xintai, Shandong.** Geological Publishing House, Beijing.

—— and Zhang Shouguang (1982). Notes on the metamorphic series and metamorphic belts of various metamorphic epochs of China and related problems.* *Regional Geology of China,* 1982 Vol. No. 2.

——, Zhang Shouguang, and Zhang Mengyan. Metamorphic rocks and metamorphic belts of China.† In *Geology of China* (in press). Geological Publishing House, Beijing.

Chow Minchen and Rich, H. V. T. (1983). *Shuotheria dongi* n.g. and n.sp., therian with pseudotribosphenic molars from the Jurassic of Sichuan, China. *Aust. Mammals.* **5.**

Cowie, J. W. and Glaessner, M. F. (1975). The Precambrian–Cambrian boundary. A symposium. *Earth–Science Review,* Vol. II.

Daily, B. (1972). The base of the Cambrian and the first Cambrian faunas. Centre for Precambrian Research, *Univ. Adelaide Spec. Papers,* No. 1.

Fan Chengjun (1982). On geological characteristics of western Yunnan.† *Yunnan Dizhi (Geology of Yunnan)* Vol. 1, No. 4.

Fan Jiasong (1980). The marine features of marine Triassic sedimentary facies in southern China. *Riv. Ital. Paleont.* Vol. 85, No. 3–4.

—— and Sun (1984). On the Permian–Triassic boundary in southern China.* *Scientia Geol. Sinica,* No. 2.

Fu Yixian (1982). Preliminary analysis of Quaternary glacial climates in the mountainous regions of eastern China.† *Scientia Sinica,* Section B, 11.

Gao Jifeng (1981). The characteristics of metamorphism of ferruginous rock series in Qian'an and Luan Xian, eastern Hebei.* *Bull. Inst. Geol.,* Chinese Acad. Geol. Sciences, No. 3.

Gao Zhenjia, Peng Changwen, Li Yong'an, Qian Jianxin, Zhu Chengshun (1980). The Sinian System and its glacial deposits in Qurutagh, Xinjiang.* In *Sinian Suberathem in China.* Tianjin Scientific and Technical Press, Tianjin.

Gou Zhonghai and Li Xiaochi (1983). Preliminary study on the Jurassic bivalves of the Nyalam area,

Xizang.* *Contributions to the geology of the Qinghai–Xizang* (Tibet) *Plateau.* **12.** Geological Publishing House, Beijing.

Grabau, A. W. (1922). The Sinian System. *Bull. Geol. Soc. China.* Vol. 1, No. 4.

—— (1928). *Stratigraphy of China,* pt. 2, *Mesozoic,* Nat. Geol. Surv. China.

Gu Zhiwei (1982a). Distribution and development of non-marine Mesozoic bivalves and formations in China. *Scientia Sinica.* Section B, Vol. 25, No. 12.

—— (1982b). On the non-marine Jurassic-Cretaceous boundary and mid-Cretaceous events in China.† *J. Stratigraphy,* Vol. 6, No. 4.

—— (1982c). Correlation chart of the Jurassic in China with explanatory text.† In *Strat. Correlation Chart with explanatory text.* Science Press, Beijing.

—— (1982d). Correlation chart of the Cretaceous in China with explanatory text.† In *Strat. Correlation Chart with explanatory text.* Science Press, Beijing.

—— (1983). On the boundary of non-marine Jurassic and Cretaceous in China. *Studies on stratigraphic boundaries in China.*† Science Press, Beijing.

Guan Baode, Pan Zecheng, Geng Wuzhen, Rong Zhiquan, and Du Huiying (1980). Sinian Suberathem in the northern slope of the eastern Quinling ranges.* In *Sinian Suberathem in China.* Tianjin Scientific and Technical Press, Tianjin.

Guo Lingzhi, Shi Yangshen, Ma Ruishi (1980). The geotectonic frame of South China and crustal evolution.† *Sci. Papers Geol. Internal. Exch.* 1, Structural geology and geomechanics. Geological Publishing House, Beijing.

Hao Yichun, Su Deying, Yu Jingxian, Li Peixian, Li Yougui, Wang Naiwen, Qi Hua, Guan Shaozeng, Hu Huaguang, Luo Xun, Yang Wenda, Ye Liusheng, Shou Zhixi et al. (1982). The Cretaceous System of China.* In *An outline of the stratigraphy of China,* compiled by the Chinese Academy of Geological Sciences. Geol. Publishing House, Beijing.

Hao Ziwen, Yu Rulong et al. (1983). The Kunlun–Bayan Har Sea and its relation to evolution of Tethys.* *Contributions to the Geology of the Qinghai–Xizang* (Tibet) *Plateau,* **12.** Geological Publishing House, Beijing.

Harland, W. B., Cox, A. V., Llewellyn, P. G., Pickton, C. A. G., Smith, A. G. and Walters, R. (1982). *A geologic time scale.* Cambridge University Press, Cambridge.

Hedberg, H. D. (ed.) (1976). *International stratigraphic guide.* John Wiley, New York.

Ho Chunsun (1975). *An outline of the geology of Taiwan.*† Surv. Taiwan, China.

—— (1979). Explanation for the geological map of Taiwan.†

Hou Hongfei (1978). Devonian stratigraphy of South China. *Prof. Papers to the symposium on the Devonian System of South China.*† Geological Publishing House, Beijing.

——, Wang Shitao, Gao Lianda, Xian Siyuan, Bai Sunliang, Cao Xuanduo, P'an Kiang, Liao Weihua, et al. (1982a). The Devonian System of China.* In *An outline of the stratigraphy of China.* Geol. Pub. House, Beijing.

——, Wang Zengji, Wu Qianghe, Yang Shipu et al. (1982b). The Carboniferous System of China.* In *An outline of the stratigraphy of China.* Geological Publishing House, Beijing.

——, Xiang Liwen, Lai Caigeng, and Lin Baoyu (1979). Advances in the Paleozoic stratigraphy of Tienshan–Khingan region.† *Acta Strat. Sinica.*

Hou Shihchun, Li Sientzy, Song Chihkao, and Song Shuhe (1979). On some major types of igneous rocks of China and their metallogenetic characters.* In *Sci. Pap. Internat. Exch.* **3,** mineralogy, petrology and mineral deposits. Geological Publishing House, Beijing (English reprint published in Beijing in 1977).

Hou Hongfei and Xian Siyuan (1975). Lower and Middle Devonian brachiopods of Guangsi and Guizhou.† *Prof. Pap. Stratigr. Palaeontol.* No. 1.

Hou Youtang, He Junde, and Yang Hengren (1983). A tentative discussion on the boundary between continental Cretaceous and Tertiary in China.† In *Studies on stratigraphic boundaries in China.*† Science Press, Beijing.

——, Song Zhichen, Ho Junde, Huang Renjin, Li Manying, Zheng Yahui, He Yan, Tang Lingyu, Lan Xiu, Wang Huiji, Guo Shuangxing, Liu Jinling, Li Haomin, Hu Lanying, Yang Hengren, Huang Baoren, Zhang Yiyong, and Zhang Binggao (1982). Subdivision and correlation of the Cenozoic strata in China.† In *Stratigraphic Correlation Chart with explanatory text.* Science Press, Beijing.

Hsü Singwu C. (Xu Jie) (1983). *Selected works on graptolites* (English and Chinese). Geological Publishing House, Beijing.

Huang Benhong (1977). *Permian floras of southeastern Xiao Hingan Mountains.*† Geological Publishing House, Beijing.

Huang Jiqing (T. K. Kuang) (1962). A preliminary suggestion of the stratigraphic provinces of China.† *Sci. Reports, All-China Stratigraphic Conference 1959, general part.* Science Press, Beijing.

—— (1978). An outline of the geotectonic characteristics of China. *Eclog. geol. Helv.* Vol. 71, No. 3.

—— (1982). A brief account of the main achievements in geological science in China over the last 60 years and our tasks ahead.* *Geol. Rev.* **28** (4).

—— and Chen Bingwei (1980). On the formation of

Pliocene and Quaternary molasses in the Tethys-Himalayan tectonic domain and its relation with the Indian plate motion.† *Sci. Pap. Geol. Internat. Exch.* (26th IGC) **1**. Structural Geology and Geomechanics. Geological Publishing House, Beijing.

Ishihara, S. (1977). The magnetite-series and ilmenite-series granitic rocks. *Min. Geol.* Vol. 27, No. 5.

Jahn, Bor-ming (1974). Mesozoic thermal events in South-east China. *Nature*, Vol. 248, No. 5448.

—— and Zhang, Z. Q. (1984). In *Archean geochemistry* (eds A. Kröner, A. M. Goodwin, and G. N. Hason). Springer-Verlag, Berlin.

——, Chen, P. Y., and Yen, T. P. (1976). Rb–Sr ages of granitic rocks in southeastern China and their tectonic significance. *Bull. Geol. Soc. Amer.* Vol. 86, No. 5.

Juan, V. C. (1975). Tectonic evolution of Taiwan. *Tectonophysics*, Vol. 26.

Kobayashi, T. (1971). The Cambro-Ordovician faunal province and the interprovincial correlation discussed with special reference to the trilobites in eastern Asia. *J. Fac. Sci. Univ. Tokyo*, Sect. II, **16** (3).

Kozur, H. (1977). Die Lage der Perm/Trias-Grenze und die Anderung der Faunen und Floren im Perm/Trias-Grenzbereich, Teil 1, *Freiberger Forsch.* Leipzig.

Kummel, B. and Teichert, C. (1970). Stratigraphy and palaeontology of the Permian–Triassic boundary beds, Salt Range and Trans-Indus Ranges, West Pakistan. The stratigraphic boundary problems. In *Permian and Triassic of West Pakistan*. University of Kansas Press, Special Pub. 4.

Lai Caigen, Wang Xiaofeng, Fu Kun, An Taixiang, Yi Yongen, Lin Baoyu, Qiu Hongrong, Yang Jingzhi, Zhang Wentang, Liu Diyong, and Chen Tingen (1982*a*). The Ordovician System of China.* In *An outline of Stratigraphy of China*. Geological Publishing House, Beijing.

—— et al. (1982*b*). The Ordovician System of China.† Geological Publishing House, Beijing.

Lee, J. S. (Li Siguang) (1934*a*). Quaternary glaciation in the Yangtze valley. *Bull. Geol. Soc. China*, Vol. 13, No. 1.

—— (1934*b*). Data relating to the study of the problem of glaciation in the Lower Yangtze valley. *Ibid.* Vol. 13, No. 3.

—— (1939). *The geology of China*. Thomas Murby, London.

Lei Zuoqi (1982). The discovery of the microflora from Lancang Group and its significance.† *J. Stratigraphy* **6** (4).

Li Chuankui and Ding Suying (1983). Palaeogene mammals of China. *Bull. Carnegie Mus. Nat. Hist.*, Vol. 21.

——, Wu Wenyu, and Qiu Zhuding (in press). *Chinese Neogene—Subdivision and correlation*.

Li Chungyu (1980). A preliminary study of plate tectonics of China.† *Bull. Chinese Acad. Geol. Sciences*, Ser. 1, Vol. 2, No. 1 Geological Publishing House, Beijing.

——, Liu Yangwen, Zhu Baoqing, Feng Yimin, and Wu Hanquan (1978). History of tectonic development of Qinling and Qilian mountains.† *Sci. Pap. Geol. Internat. Exch.* 1. Structural Geology and Geomechanics. Geological Publishing House, Beijing.

Li Genkun, Zhang Jinhai, Chen Jiaomin and Li Changze (1983). Preliminary knowledge of the metamorphic belts in Fujian and Taiwan provinces.* *Geology of Fujian*, Vol. 2, No. 1.

Li Tong and Rao Jileng (1963). Average chemical composition of the igneous rocks of China.* *Acta Geol. Sinica*, Vol. 43, No. 3.

Li Xingxue and Cai Chongyang (1979). Devonian flora of China.† *J. Stratigraphy* **2** (3).

Li Yaoxi, Song Liseng, Chou Zhiqiang, and Yang Jingyao (1975). *Early Palaeozoic stratigraphy of western Da Bashan mountains*.† Geological Publishing House, Beijing.

Li Yuntong, Lei Yizhen, Wang Daning, Sun Mengrong, Sun Xiuyu, Wang Chongyou, Weng Shijie, et al. (1982). The Tertiary System of China.* In *An outline of stratigraphy of China*. Geological Publishing House, Beijing.

Li Zishun, Wang Sien, Yu Jingshan, Huang Huaizang, Zheng Shaolin, and Yu Xihan (1983). On the classification of the Upper Jurassic in North China and its bearing on the Jurassic–Cretaceous boundary.* *Acta Geol. Sinica*, **56**, 347–65.

Liang Dingyi, Nie Zetong, Guo Tieying, Zhang Yishi, Xu Baowen, and Wang Weipin (1983). Permo-Carboniferous Gondwana Tethys facies in southern Karakorum, Ali, Xizang.* *Earth Science (J. Wuhan Coll. Geol.)*.

Liao Weihua, Xu Hankui, Wang Chengyuan, Yuan Yeping, Cai Chongyang, Mu Daocheng, and Lu Lichang (1978). Subdivision and correlation of the Devonian stratigraphy of southwest China. *Prof. Papers to the Symposium on the Devonian System of South China*, 1974. Geological Publishing House, Beijing.

Liao Zhuoting (1979). Brachiopod assemblage zones of the Changxingian stage of South China and brachiopods of the Permo-Triassic mixed faunas.† *J. Stratigraphy* **3** (2).

Lin Baoyu (1979). The Silurian System of China.* *Acta Geol. Sinica*. Vol. 53, No. 3.

——, Guo Dianheng, Wang Xiaofeng, et al. (1982).

The Silurian System of China.* In *An outline of stratigraphy of China*. Geological Publishing House, Beijing.

Liou, J. G. (1981). Petrology of metamorphosed oceanic blocks in the central range of Taiwan. Abstracts of papers, *Seminar of plate tectonics and metamorphic geology*, Taipei.

Liu Beipei and Cui Xinsheng (1983). Discovery of Eurydesma-fauna from Rutog, northwestern Xizang (Tibet).* *Earth Science (J. Wuhan Coll. Geol.)*.

Liu Dongsheng et al. (1964). On the subdivision of Quaternary of China.† In *Problems of Quaternary Geology*. Science Press, Beijing.

—— and Ding Menglin (1983). Discussion on the age of 'Yuanmou Man'. *Acta Anthropologica Sinica* Vol. 2, No. 1.

Liu Guohui and Wu Jiashan (1977). On metamorphic zones and petrology of the country rocks of the Fangshan intrusive body, Beijing.† *Dizhi Kuangchan Yanjiu (Research on Geology and Mineral Resources)*, 1977 Vol., No. 3.

Liu Hongyun, Sha Qingan, and Hu Shiling (1973). The Sinian System of South China.† *Scientia Sinica*. 1973 (2).

Liu Hsienting and Wang Nienchung (1978). The upper Permian fish-fauna of Dzungaria Basin, Sinkiang. *Mem. Inst. Verteb. Paleont. and Palaeantrop.* Academia Sinica, No. 13.

Liu Xiaoliang (1981). Metazoan fossils from the Mashan Group near Jixi, Heilongjiang.* *Bull. Chin. Acad. Geol. Sci.* Vol. 3, No. 1.

Liu Yourui (1975). Lower Devonian agnathans of Yunnan and Sichuan.* *Verteb. Palasiatica.* 13 (4).

Liu Zengqian, Yu Xijing, Xu Xian, and Pan Guitang (1980). The basic geological characteristics of the Qinghai-Xizang Plateau.* *Bull. Chinese Acad. Geol. Sci.*, Ser. 1, Vol. 2, No. 1.

Lou Liangzhao, He Gaopin, Ye Huiwen, and Wei Yifang (1982). Metamorphism of the main Palaeozoic mobile belts of China.† In *Scientific Papers in commemoration of the 30th anniversary of the founding of the Changchun College of Geology*, Vol. 1.

Lu Yenhao (1962). *The Cambrian System of China.*† Science Press, Beijing.

—— (1975). Ordovician trilobite faunas of central and southwest China.* *Palaeont. Sinica*, New series B, No. 11. Science Press, Beijing.

—— (1983). Cambrian–Ordovician boundary in China.† In *Studies on stratigraphic boundaries in China*.

—— et al. (1974). Bio-environmental control hypothesis and its application to Cambrian biostratigraphy and palaeozoogeography.† *Mem. Nanjing Inst. Geol. Palaeont. Acad. Sinica* No. 5. Science Press, Beijing.

——, Zhu Zhaoling, Qian Yiyuan, Zhou Zhiyi, Chen Junyuan, Liu Gengwu, Yu Wen, Chen Xu, and Xu Hankui (1976). Ordovician biostratigraphy and palaeozoogeography of China.† *Mem. Nanjing Inst. Geol. and Palaeont. Acad. Sinica*, No. 7. Science Press, Beijing.

——, Zhu Zhaoling, Qian Yiyuan, Lin Huanling and Yuan Jinliang (1982). Correlation chart of Cambrian in China with explanatory text. In *Stratigraphic Correlation Chart with explanatory text*.† Science Press, Beijing.

Luo Zhili (1979). On the occurrence of Yangtze old plate and its influence on the evolution of the lithosphere in the southern part of China.* *Sci. Geol. Sin.* 1979, 2.

Ma Lifang (1985). Outcrop map of the Sinian of China. In: Wang Hongzhen (chief compiler) *Atlas of the Palaeogeography of China.*† The Cartographic Publishing House, Beijing.

Ma Xingyuan and Wu Zhengwen (1981). Early tectonic evolution of China.* *Precambrian Geology*, **14**.

Miyashiro, A. (1961). Evolution of metamorphic belts. *J. Petrology*, Vol. 2, No. 3.

—— (1972). Pressure and temperature conditions and tectonic significance of regional and ocean-floor metamorphism. *Tectonophysics*, Vol. 13.

Mo Zhusun, Ye Bodan, Pan Weizu, Wang Shaonian, Zhuang Jinliang, Gao Bingzhang, Liu Jinquan, and Liu Wenshang (1980). *Geology of the granites of the Nanling Ranges.*† Geological Publishing House, Beijing.

Mu Enzhi (1964). *The Silurian System of China.*† Science Press, Beijing.

—— (1983). Ordovician–Silurian boundary in China.† In *Studies on stratigraphic boundaries in China*. Science Press, Beijing.

——, Chen Xu, Ni Yunan, and Rong Jiayu (1982). Subdivision and correlation of the Silurian in China.† In *Stratigraphic Correlation Chart with explanatory text*. Science Press, Beijing.

Mu Daocheng (1978). On the Devonian tentaculite zones of south China. *Prof. Papers to the Symposium on the Devonian System of South China.*† Geological Publishing House, Beijing.

Nakazawa, K., Kapoor, H. M., Ishii, K., Bando, Y., Okimura, Y., and Tokuoka, T. (1975). The upper Permian and lower Triassic in Kashmir. *Mem. Fac. Sci. Kyoto Univ. Ser. Geol. and Mineral.* **42** (1).

Needham, J. (1959). *Science and civilization in China*. Vol. 3, *The Science of the Earth*. Cambridge University Press, Cambridge.

Norin, E. (1937). Geology of western Kuruktagh. *Pub. Sci. Exp. NW. Prov. China Dr. Sven Hedin*, **3**, Geology 1.

Nogami, Y. (1966). Kambrische conodonten von China. Teil 1. Conodonten aus den Oberkambrischen Kushan-schichten. *Kyoto Univ. Coll. Sci. Mem.* ser. B, **32**.

Palmer, A. R. (1973). Cambrian trilobites. In Hallam, A. (ed.) *Atlas of Palaeobiogeography.* New York, Elsevier.

—— (1977). Biostratigraphy of the Cambrian System—A progress report. *Ann. Rev. Earth Planet Sci.* **5**.

P'an Kiang (1981). Devonian antiarch biostratigraphy of China. *Geol. Mag.* **118**.

—— (1978). Devonian continental sedimentary formations of South China.† *Prof. Papers to the Symposium on the Devonian of South China, 1974.* Geological Publishing House, Beijing.

Pei Wenchung (1957). Discovery of the lower jaws of a giant ape in Kwangsi, South China. *Sci. Rec.* New Ser. Vol. 1, No. 3.

Peng Xingjie and Liu Wanling (1983). A preliminary identification of the paired metamorphic zones in the southern segment of Lancanjiang.* *Contributions to the Geology of the Qinghai-Xizang (Tibet) Plateau.* Vol. 13. Geological Publishing House, Beijing.

Pidgeon, R. T. (1980). 2480 Ma-old zircon from granulite facies rocks from East Hebei Prov. North China (also in Chinese) in *Geol. Review (Ti Chi Lun Ping)* Vol. 26, No. 3.

Pu Qingyu and Qian Fang (1977). Study on the fossil human strata—the Yuanmou Formation.* *Acta Geol. Sinica.* No. 1, 1977.

Qian Yi (1983). Sinian–Cambrian boundaries in China. In *Studies on stratigraphic boundaries in China.*† Science Press, Beijing.

Qiao Xiufu (1985). Outcrop map of the Middle and Upper Proterozoic of China. In: Wang Hongzhen (Chief compiler) *Atlas of the Palaeogeography of China.*† Cartographic Publishing House, Beijing.

——, Geng Shufang (1981). On late Precambrian plate tectonics of South China.* *Contr. Tect. China Adj. Reg. Inst. Geol. CAGS.* Geological Publishing House, Beijing.

Read, H. H. and Watson, J. (1962). *Introduction to geology,* Vol. 1. Principles. MacMillan, London.

Ren Jishun, Jiang Chunfa, Zhang Zhengkun, Qin Deyu (1980). *The geotectonic evolution of China.*† Science Press, Beijing.

Ren Mei'e, Yang Renzhang and Bao Haosheng (eds.) (1980). *An outline of physical geography of China.*† Commercial Press, Beijing.

Ridd, M. F. (1980). Possible Palaeozoic drift of Southeast Asia and Triassic collision with China. *J. Geol. Soc.* Vol. 137.

Robinson, R. A.; Rosova, A. V., Roweli, A. J.,

Fletcher, T. P. (1977). Carboniferous boundaries and divisions. *Lethaia*, **10** (3).

Rong Jiayu (1979). *Hirnantia* fauna of China and discussion on the Ordovician–Silurian boundary.† *J. Strat.* Vol. 3, 1.

——, Yang Xuechang (1981). Middle and Late Silurian brachiopod faunas in southwestern China.* *Mem. Nanjing Inst. Geol. Palaeont.* Acad. Sinica. No. 13.

Rozanov, A. Yu (1974). I.U.G.S. Precambrian/Cambrian boundary working Group in Siberia. *Geol. Mag.*, Vol. III, No. 3.

——, Missarzevski, B. B., Volkova, N. A. (1969). Tommotian stage and problem of lower boundary of the *Cambrian Mem. T.N.N.*

Russel, D. A. and Singh, C. (1978). The Cretaceous–Tertiary boundary problem. *Episodes* 1979 (4).

Schindewolf, O. H. (1970). Stratigraphic principles. *Newsl. Strat.* **1**, 1.

Sengor, A. M. C. (1981). The evolution of Palaeozoic Tethys in the Tibetan segment of the Alpides. *Proc. Symp. Qinghai–Xizang Plateau,* Vol. 1. Science Press, Beijing; Gordon and Breach, New York.

Shang Shanzhen, Yao Zhaoqi, Mo Zhuangguan, and Li Xingxue (1982). Subdivision and correlation of the Permian continental strata in China.† In *Stratigraphic Correlation Chart with explanatory text.* Science Press, Beijing.

Sheng Jinzhang (1962). *The Permian System of China.*† Science Press, Beijing.

—— and Li Xingxue (1965). Boundary between the Carboniferous and Permian of China.† *Papers on Carboniferous of China.* Science press, Beijing.

—— and —— (1974). Progress in the study of Permian biostratigraphy of China.† *Mem.* **5**. Science Press, Beijing.

——, Jin Yugan, Rui Lin, Zhang Linxin, Zhang Zhuoguan, Wang Yujing, Liao Zhuoting, and Zhao Jiamin (1980). Correlation Chart of the Permian in China with explanatory text. In *Stratigraphic Correlation Chart with explanatory text.*† Science Press, Beijing.

Sheng Xinfu (1964). Upper Ordovician trilobite fauna of Szechuan and Kweichow with special discussion on the classification and boundaries of the Upper Ordovician.* *Acta Palaeont. Sinica,* Vol. 12, No. 4.

—— (1974). *Subdivision and correlation of Ordovician System in China.*† Geological Publishing House, Beijing.

Shi Yafeng and Deng Yangxin (1982). Evidence of Quaternary solifluctional deposits on the Lushan piedmont—given on the example of Yangtoujiao at the north-western slope of Lushan.† *Kexue Tongbao* 1982, **20**. Science Press, Beijing.

Song Shuhe. Magmatic rocks of China.† In *Geology of China* (in press). Geological Publishing House, Beijing.

Song Zhichen, Liu Jinling and Tang Lingyu (1983). Discussion on the lower limit of continental Quaternary in China by means of palynologic materials.† *Studies on stratigraphic boundaries in China*. Science Press, Beijing.

Stockwell, C. H. (1968). Geochronology of stratified rocks in the Canadian Shield. *Canad. J. Earth Sci.* **5**, 3.

Sun Ailin (1978a). Two new genera of Dicynodontidae.* *Mem. Inst. Verteb. Palaeont. Palaeoanthrop.*, Acad. Sinica Sci. Press.

—— (1978b). On occurrence of *Parakannemeyeria* in Sinkiang.* *Mem. Inst. Verteb. Palaeont. Palaeanthrop.* Acad. Sinica Sci. Press.

Sun Dazhong and Wu Changhua (1981). The principal geological and geochemical characteristics of the Archaean greenstone–gneiss sequences in North China. In *Archaean Geology*, 2nd Internat. Symp., Perth, 1980. *Spec. Pub. Geol. Soc. Australia*, No. 7.

Sun Dianqing and Cui Shenqin (1980). On the major tectonic movements of China* *Sci. Pap. Geol. Internat. Exch.* 1. Structural Geology & Geomechanics.† Geological Publishing House, Beijing.

Sun Yunchu (1935). Upper Cambrian trilobites of North China.* *Pal. Sinica* B, **7**, 2.

—— (1961). Problems of classification of the Cambrian system in China.* *Scientia Sinica* **10** (6). Science Press, Beijing.

Tang Kedong and Su Yangzheng (1966). New data about the Palaeozoic formations and their significance in the northwestern Lesser Khingan.* *Acta Geol. Sinica* **46**, Science Press, Beijing.

——, Su Yangzhen and Wang Ying (1982). Some characteristic features of the geological development in the east of Ural-Mongolian folded region.* *Bull. Shenyang Inst. Geol. CAGS*, 3.

Tien, C. C. (1938). The Devonian brachiopoda of Hunan.* *Palaeont. Sinica*, New Series B, 4.

Tong Yongsheng (1978). Late Paleocene mammals of the Turfan Basin, Sinkiang.* *Mem. Inst. Vert. Palaeont. Palaeoanthrop.*, Acad. Sinica, No. 13.

Tozer, E. T. (1979). The significance of the ammonoids *Paratirolites* and *Otoceras* in correlating the Permian-Triassic boundary beds of Iran and the People's Republic of China. *Canadian J. Earth Sciences*, **16** (7).

Wang Chengyuan and Wang Zhihao (1978). Early and middle Devonian conodonts of Kwangsi and Yunnan.† Prof. Papers to the Symposium on the Devonian System of South China 1974. Geological Publishing House, Beijing.

——, Ruan Yiping, Yu Changmin and Wang Yu (1983). Subdivision of Devonian System in China.† In *Studies on stratigraphic boundaries in China*. Science Press, Beijing.

Wang Hongzhen (C. H. Wang) (1978). On the subdivision of the stratigraphic provinces of China.† *Acta Strat. Sinica*, Vol. 2, No. 2.

—— (1981). Geotectonic units of China from the viewpoint of mobilism.* *Earth Science* (J. Wuhan Coll. Geol.) 1981, 1.

—— (1982a). The main stages in the crustal development of China.* *Earth Science (J. Wuhan Coll. Geol.)* 1982, 3.

—— (1982b). On the use of 'Sinian' from the viewpoint of stratigraphic nomenclature and the chronostratigraphic classification of the Precambrian of China.† *J. Stratigraphy*, Vol. 6, No. 4.

—— (1983). On the problem of geotectonic units of Xizang.* *Earth Science (J. Wuhan Coll. Geol.)* 1983, 1.

——, Wang Ziqiang, Zhu Hong, Qiao Xiufu, and Yang Yuntong (1979). Palaeogeography of the Sinian System of China.† *Sci. Pap. Geol. Internat. Exch.* Geological Publishing House, Beijing.

——, Wang Ziqiang, and Zhu Hong (1980). Upper Proterozoic tectonopalaeogeography of China.* *Scientia Geol. Sinica* 1980, 2.

——, Wang Ziqiang, Zhu Hong, Qiao Liufu, and Yang Yuntong (1978). Stratigraphy and palaeogeography of the Changchengian, Jixianian and Qingbaikouan systems of North China.† *Sci. Pap. Geol. Internat. Exch.* Geological Publishing House, Beijing.

——, Xu Chengyan, and Zhou Zhengguo (1982). Tectonic development of the continental margins on both sides of the Qinling marine realm.* *Acta Geol. Sinica*, Vol. 62, 3.

——, Yang Sennan, and Li Sitian (1983). Mesozoic and Kainozoic basin formation and development of continental margin in eastern China and adjacent regions.* *Acta Geol. Sinica*, Vol. 63, 2.

Wang Liankui, Zhu Weifeng and Zhang Shaoli (1982). On the evolution of two petrographic and mineralized series of granites in South China.* *Geochemica*. 1982 Vol. No. 4.

Wang Naiwen et al. (1983). The Jurassic System of China.* In *Outline of stratigraphy of China*. Geological Publishing House, Beijing.

Wang Qichao, Chen Boyan, Wu Tieshan, Xu Chaolei and Wu Zhenshan (1980). Stratigraphy of Sinian Suberathem in Taihang and Wutai Ranges and discussion on its relationship with the Hutuo Supergroup.* *Sinian Suberathem in China*. Tianjin Scientific and Technical Press, Tianjin.

Wang Rimin and Chen Zhenzhen (1982). A discussion

of the temperature–pressure condition of Early Archaean metamorphism in Qian'an, eastern Hebei.* *Acta Petrologia Mineralogica et Analytica*, Vol. 1, No. 4.

Wang Sien, Cheng Zhengwu, Wang Naiwen, Zhang Zhicheng, Xu Fuxiang, Sun Donli, Zhang Renjie, Yan Yonkui *et al.* (1982). The Jurassic System of China.* In *An outline of the Stratigraphy of China*. Geological Publishing house, Beijing.

Wang Xiaofeng (1977). The discovery of the latest Silurian and early Devonian monograptids from Qinzhou, Guangxi and its significance.* *Acta Geol. Sinica*, 1977, Vol. 2.

—— (1980). The Ordovician System in China.* *Acta Geol. Sinica*, **54** (1).

Wang Yangeng, Chen Yulin, Wang Ruigang, Chen Xianwei and Wei Xushou (1980). Stratigraphic types and characteristics of the Sinian in Hunan, Guizhou and Guangxi.* In *Sinian Suberathem in China*. Tianjin Scientific and Technical Press, Tianjin.

Wang Yikang (Yigang) (1978). Latest early Triassic ammonoids of Ziyun, Guizhou—with notes on the relationship between Early and Middle Triassic ammonoids.* *Acta Palaeont. Sinica*. Vol. 17, No. 2.

—— *et al.* (1974). Advances in the Devonian biostratigraphy of South China.† *Mem. Nanjing Inst. Geol. Palaeont.* No. 6.

Wang Yu, Yu Changmin, Liao Weihua, Shi Congguang, Hu Zhaoxun, Deng Zhanqui, Rong Jiayu, Ni Yunan, Wang Chengyuan, Ruan Yiping, Cai Chongyang, Wang Shangqi, Mu Daocheng, Xia Fengcheng, and Wang Zhihao (1982). Correlation chart of the Devonian in China with explanatory text.† In *Stratigraphic Correlation Chart in China with explanatory text*. Science Press, Beijing.

——, Lu Zongbin, Xing Yusheng, Gao Zhenjia, Lin Weixing, Ma Guogan, Zhang Luyi, and Lu Songnian (1980). Subdivision and correlation of the Upper Precambrian in China.* *Sinian Suberathem in China*.

Wang Yunshan, Zhuang Qingxing, Shi Congyan, Liu Tigang, and Zheng Liangzhi (1980). Quanji Group along northern border of Qaidam basin.* *Sinian Suberathem in China*.

Weidenreich, F. (1943). The skull of *Sinanthropus pekinensis*. A comparative study on a primitive hominid skull.* *Pal. Sinica*, New series D, No. 10.

Wen Shixuan (1981). Palaeobiogeography of Qinghai–Xizang plateau, evidence of continental drift. *Proc. Symp. Qinghai–Xizang Plateau*, Vol. 1, Science Press, Beijing.

Windley, B. F. (1977). *The evolving continents*. Wiley, London.

Woo Jukang, Wu Xinzhi and Wang Cunyi (1959). New reconstruction of physiognomy of *Sinanthropus* woman.* *Verteb. Palasiatica*, Vol. 3, No. 3.

Wu Hanquan (1980). The glaucophane-schists of eastern Qinling and northern Qilian Mountains, China.* *Acta Geol. Sinica*, Vol. 54, No. 3.

—— (1982). Petrology and mineralogy of high pressure metamorphic zones in northern Qilian Mountains, China.* *Bull. Xi'an Inst. Geol. Min. Res., CAGS*, No. 4.

Wu Haorou (1981). Geological outline and geological history of the Yarlungzangbo suture line. *Proc. Symp. Qinghai–Xizang Plateau*. Vol. 1. Science Press, Beijing; Gordon and Breach, New York.

Wu Liren, Qi Jinying, Wang Tingdu, Zhang Xinqi, and Xu Yonshang (1982). Mesozoic rocks in the eastern part of China.* *Acta Geol. Sinica*, Vol. 56, No. 3.

Wu Shunqing and Wu Xianwu (1982). The Triassic plant-bearing strata in China.† In *Stratigraphic Correlation Chart in China with explanatory text*. Science Press, Beijing.

Wu Xiuyuan and Zhao Xiuhu (1982). Subdivision and correlation of the Carboniferous continental strata in China.† In *Stratigraphic Correlation Chart in China with explanatory text*. Science Press, Beijing.

Xia Xiangyong, Li Zhongjun, and Wang Genyuan (1980). *History of ancient Chinese mining industry*.† Geological Publishing House, Beijing.

Xiang Liwen *et al.* (1981). *The Cambrian System of China, Stratigraphy of China* (No. 4).† Geological Publishing House, Beijing.

——, Li Shanji, Nan Runshan, Guo Zhenming, Yang Jialu, Zhou Guoqiang, An Taixiang, Yuan Kexing, and Qian Yi (1982). The Cambrian System of China.* In *An outline of the stratigraphy of China*. Geological Publishing House, Beijing.

Xiao Xuchang and Gao Yanlin (1982). New finds of the high pressure and low temperature greenschist facies from Yarlung Zangpo (Tsangpo) suture zone, Xizang (Tibet).* *Bull. Inst. Geol., CAGS*, No. 5.

——, Chen Guoming, and Zhu Zhizhi (1978). A preliminary study on the tectonics of ancient ophiolites in the Qilian Mountains, North-west China.* *Acta Geol. Sinica*, Vol. 52, No. 4.

Xing Yusheng (1976). The Sinian System of China. *Inst. Geol. Min. Res. CAGS*. Beijing.†

Xu Keqin, Hu Shouxi, Sun Mingzhi and Ye Jun (1982). On the two genetic series of granites in southeastern China and their metallogenetic characteristics.* *Mineral Deposits*, Vol. 1, No. 2.

Xu Yunpeng (1982). New knowledge of the age of the Kongling Group in western Hubei.† *J. Stratigraphy* **6**, No. 4.

Xue Chunting, Su Yangzheng, Zhang Hairi, and Cui Ge (1980). Upper Silurian and Lower Devonian of the northwestern Xiao Hinganling (Lesser Khingan Mountains.† *J. Stratigraphy,* **4** (1).

Yang Chungchien (C. C. Young) (1978a). A complete skeleton of *Chasmatosaurus yuani* from Sinkiang.* *Mem. Inst. Verteb. Palaeont. Palaeoanthrop.* Academia Sinica. No. 13.

—— (1978b). A late Triassic vertebrate fauna from Fukang, Sinkiang.* *Mem. Inst. Verteb. Palaeont. Palaeoanthrop.* Academia Sinica. No. 13.

Yang Jingzhi and Wang Chengyuan (1983). Devonian-Carboniferous boundary in China. *Studies in stratigraphic boundaries in China.* Science Press, Beijing.

Yang Jingshi, Wu Wangshi, Chang Lingxin, Wang Keliang, Lu Linhuang et al. (1982). On the subdivision and correlation of the Carboniferous System of China.† In *Correlation charts with explanations of the various systems of China.* Science Press, Beijing.

Yang Qinghe, Zhang Youli, Zheng Wenwu, and Xu Xuesi (1980). *Correlation of Sinian Suberathem in northern Jiangsu and Anhui.** Tianjin Scientific and Technical Press, Tianjin.

Yang Shipu, Pan Jiang (P'an Kiang), and Hou Hungfei (1979). The Devonian System of China.* *Acta Geol. Sinica,* Vol. 53 (3). Also in *Geol. Mag.* Vol. 118, No. 2, 1981.

Yang Wanrong and Jiang Nayan (1981). On the depositional characters and microfossils of the Changhsing Formation of Permian–Triassic boundary in Changxing, Zhejiang.* *Bull. Nanjing Inst. Geol. Palaeont. Acad. Sinica.* No. 2.

Yang Zigeng and Mou Yunzhi (1979). Some fundamental problems of Quaternary geology of eastern Hebei plain.* *Acta Geol. Sin.* Vol. 53 (4).

—— (1983). Latest conceptions of the Late Cenozoic strata in Zhoukoudian (Choukoutien). *Kexue Tongbao (A monthly of Sci.)* Vol. 27, No. 12.

Yang Zunyi (Yang Tsun-yi, T. I. Young) and Xu Guirong (1966). Triassic brachiopods of central Guizhou (Kweichou) province, China.* Chinese Industrial Press, Beijing.

——, Li Zishun (1980). Chronostratigraphic classification of the marine Triassic in China. *Pal. Riv. Ital.* Vol. 85, 3–4.

——, Li Zishun, Qu Lifan, Lu Chongming, Zhou Huiqin, Zhou Tongshun, and Liu Guifang et al. (1982). The Triassic System of China.* In *An outline of the stratigraphy of China.* Geological Publishing House, Beijing.

——, Li Zishun, Qu Lifan, Lu Chongming, Zhou Huiqin, Zhou Tongshun, Liu Guifong, Liu Beipei and Wu Ruitang (1982). The Triassic System of China.* *Acta Geol. Sinica,* Vol. 56, No. 1.

——, Wu Shunbao, Yang Fengqing (1981). Permo-Triassic boundary in the marine regime of South China. In *Gondwana Five* (ed. M. M. Cresswell and P. Vella). A. A. Balkema, Rotterdam. Also in Chinese with English abst. in *Earth Science (J. Wuhan Coll. Geol.)* 1981, Vol. 1 (1).

—— et al. (1983). *Triassic of the South Qilian Mountains.** Geological Publishing House, Beijing.

Yao Zhaoqi, Xu Juntao, Cheng Zhuoguan, Zhao Xiugu, and Mo Zhuangguan (1980). Late Permian biostratigraphy and the boundary between Permian and Triassic of western Guizhou and eastern Yunnan. In *Late Permian coal-bearing beds and biotas of western Guizhou and eastern Yunnan.*† Science Press, Beijing.

Ye Meina and Li Baoxin (1982). Subdivision and correlation of the Jurassic plant-bearing strata in China.† In *Stratigraphic Correlation Chart in China with explanatory text.* Science Press, Beijing.

Yen, T. P. (1981). Review of metamorphic geology of the central range of Taiwan. *Abstract of papers, Seminar on plate tectonics and metamorphic geology.* Taipei.

Yin Hongfu (1962). Biostratigraphic problems on the Triassic of Kueichou province, China.* *Acta Geol. Sinica,* Vol. 42, No. 2.

Yin Zanxun, Zhang Shouxin and Xie Cuihua (1978). *On folding phase.*† Science Press, Beijing.

You Zhengdong and Han Youqing (1981). Regional metamorphism of the Dengfeng Group in Songshan area, Henan prov. *Earth Sci. (J. Wuhan Coll. Geol.)* 1981, 2.

Young C. C. (1964). The pseudosuchians in China.* *Pal. Sinica.* New Series, C. No. 19.

Young T. I. (Yang Zunyi) (1935). *Bibliography of Chinese geology up to 1934.* National Academy of Peiping.

Yu Guanming, Chang Qihua, Gou Zhonghai, Lan Belong, Wang Chengshang, Xu Yulin, Wang Guorong, Li Xiaochi, Wang Xiaoqiao and Huang Yaping (1983). Subdivision and correlation of Jurassic System in the Nyalam area, Xizang (Tibet).* *Contr. to Geol. Qinghai–Xizang (Tibet) Plateau (11).* Geological Publishing House, Beijing.

Yu Wen and Chang Xianqiu (1982). Late Cretaceous and early Tertiary non-marine gastropods from Sanshui Basin, Guangdong.† *Mem. Nanjing Inst. Geol. Palaeont. Acad. Sinica,* No. 17.

Zhai Renjie, Zheng Jiajian and Tong Yongsheng (1978). Stratigraphy of the mammal-bearing Tertiary of the Turfan basin, Sinkiang.* *Mem. Inst. Verteb. Palaeont. Palaeoanth.* Acad. Sinica, No. 13.

Zhan Lipei, Chen Yuling, Li Li, Chen Binguei et al. (1982). The Permian System of China.* In *An outline* *of the stratigraphy of China.* Geological Publishing House, Beijing.

Zhang Qi and Li Zhaohua (1981). Metamorphic belts and metamorphism of Xizang.† In *Magmatism and metamorphism in Xizang.* Science Press, Beijing.

Zhang Qihua, Huang Yaping (1983). Jurassic and lower Cretaceous ammonites from the Nyalam county, Xizang (Tibet).* *Contr. Geol. Qinghai–Xizang (Tibet) Plateau* (11).

Zhang Qiusheng, Zhu Gaoling, Zhu Yongzheng and Yan Yongliang (1977). The polymetamorphic and deformational history of east Qinling Mountains.† *J. Changchun Geol. Institute,* 1977 vol., No. 3.

Zhang Ruyuan, Cong Bolin, and Ying Yupu (1982). Study on quartz-almandine eulitite from the Louzi Mountain, Qian'an, eastern Hebei province, China.* *Acta Geol. Sinica,* Vol. 56, No. 1.

Zhang Shangzhen, Yao Zhaoqi, Mo Zhuangguan, and Li Xingxue (1982). Subdivision and correlation of the Permian continental strata in China. In *Stratigraphic Correlation Chart in China with explanatory text.*† Science Press, Beijing.

Zhang Wentang (W. T. Chang) (1979). On the northern rift of the Africo-Arabian and Indian plates. *Proc. Roy. Soc. S. Victoria,* Vol. 22.

——, Li Jijin, Ge Meiyu, and Cheng Junyuan (1982). Classification and correlation of the Ordovician in China.† In *Stratigraphic Correlation Chart in China with explanatory text.* Science Press, Beijing.

Zhang Zongqing and Jiahn Borming (1982). Certain recent progress on the geochronology and geochemistry of Archean granulitic gneisses from eastern Hebei.† *Geological Review,* Vol. 28, No. 5.

Zhao Jinke, Chen Chuzhen, Wang Yigang, He Guoxiong and Chen Jinhua (1982). Problems on subdivision and correlation of the marine Triassic strata in China with explanatory text.† In *Stratigraphic Correlation Chart in China with explanatory text.* Science Press, Beijing.

——, Liang Xiluo and Zheng Zhuoguan (1978). Late Permian cephalopods of South China. *Pal. Sinica,* New series B, 12. Science press, Beijing.

——, Shen Jinzhang, Yao Zhaoqi, Liang Xiluo, Chen Chuzhen, Rui Lin and Liao Zhuoting (1981). The Changhsingian and Permian-Triassic boundary of South China.* *Bull. Nanjing Inst. Geol. Palaeont., Acad. Sinica,* No. 2.

——, Sheng Jinzhang, Yao Zhaoqi, Liang Xiluo, Chen Chuzhen, Rui Lin, and Liao Zhuoting (1982). Studies on stratigraphic boundaries in China.† Science Press, Beijing.

Zhao Rongli (1982). About the Upper Triassic Batang Group in Yushu district, Qinghai.* *Contr. Geol. Qinghai-Xizang (Tibet) Plateau,* 10.

Zhao Zijiang, Xing Yusheng, Ma Guogan, Yu Wen, and Wang Ziqiang (1980). The Sinian System of eastern Yangzi Gorges.* In *Sinian Suberathem in China.* Tianjin Scientific and Technical Press, Tianjin.

Zhong Fudao (1977). On the Sinian time scale as based on the isotopic ages of the Sinian strata in the Yanshan region.† *Scientia Sinica* 1977, No. 2.

Zhou Mingzhen (Chow Minchen) *et al.* (1977). Mammalian fauna from the Paleocene of Nanxiong basin, Guangdong.* *Pal. Sinica,* New series C, No. 20.

Zhou Mulin, Liu Yourui *et al.* (1982). The Quaternary System of China.* In *An outline of the stratigraphy of China.* Geological Publishing House, Beijing.

Zhou Tingru (1982). *Palaeogeography.*† Beijing Teachers University Press, Beijing.

Zhou Yunsheng, Zhang Qi, Cui Chengwei, and Deng Wanming (1981). The migration and evolution of magmatism and metamorphism in Xizang since the Cretaceous and their relation with the Indian plate motion, a possible model for the uplift of the Qinghai–Xizang Plateau. *Proc. Symp. Qinghai-Xizang Plateau.* Vol. 1, Science Press, Beijing; Gordon and Breach, New York.

Zhu Xia (1983). Notes on ancient global tectonics and Palaeozoic petroliferous basins.* *Oil and Gas Geology,* Vol. 4, No. 1.

—— and Chen Huanjiang (1980). Tectonic evolution of Chinese petroleum basins. *Colloques C2 Energy Resources.* 26th IGC, Mem. BRGM.

Zhu Shixing, Cao Ruiji, Zhao Wenjie, and Liang Yuzuo (1978). An outline of the studies on stromatolites from the stratotype section of the Sinian Suberathem in Chihsien County, North China.* *Acta Geologica Sinica* **52**, No. 2.

Zhu Zhiwen, Zhu Xiangyuan, Zhang Yiming (1981). Palaeomagnetic observation in Xizang and continental drift.* *Acta Geophysics Sinica,* Vol. 24, No. 1.

Zhuravleva, I. T. and Meshkova, N. P. (1979). Biostratigraphy and palaeontology of Lower Cambrian of Siberia (in Russian). *Izvestiya 'Nauka'* No. 406.

*in Chinese with English abstract or summary.
†in Chinese only.

STRATIGRAPHIC INDEX

Aijiashan (Aichiashan) Fm　艾家山组
Ailaoshan Gr　哀牢山群
Ailiankate Gr　艾连卡特群
Airjigan Gr　爱尔基干群
Akalehe Fm　阿恰勒河群
Aikuan (Yanguan) Fm　岩关组
Aksu Fm　阿克苏组
Alashan Complex　阿拉善杂岩
Altai Fm　阿尔太组
Altungol Gr　阿勒通沟群
Amushan Fm　阿木山组
Anding Fm　安定组
Angzanggou Fm　肮脏沟组
Anshan Gr　鞍山群
Aodi Fm　澳底组
Aoqu Fm　鳌曲组
Asizha Gr　阿斯札群
Atoush Gr　阿图什群
Bacun Gr　八村群
Badaohe Fm　八道河组
Badatu Fm　巴达图组
Badaowan Fm　八道湾组
Bagongshan Fm　八公山组
Bahe Fm　坝河组
Baicaoping Fm　白草坪组
Baihu Gr　白湖群
Bailiuping Fm　白柳坪组
Bailongjiang Gr　白龙江群
Baishanian (Peishanian) St　白沙阶
Baitianba Fm　白田坝组
Baiyantang Fm　白岩塘组
Baiyaya Fm　白崖垭组
Baiyixi (Baiyisi) Fm　贝义西组
Baizhu Fm　白竹组
Baishushan Gr　柏树山群
Balikelike Fm　巴立克立克组
Banqiao Fm　板桥组
Banqiuguan Fm　斑鸠关组
Banxi (Panchi) Gr　板溪群
Baode Fm　保德组
Baoshi Fm　宝石组
Baoqing Member　保菁段
Baota (Pagoda) Fm　宝塔组
Baota'an St　宝塔阶
Bashili Xiaohe Fm　八十里小河组
Bayan Obo Gr　白云俄博群
Bayinhe Fm　巴音河组
Bazulu Fm　坝注路组
Beianzhuang Fm　北庵庄组
Beidajian Fm　北大尖组
Beijiangjun Fm (Pt₃)　北将军组
Beikuang Fm　北矿组

Beimenxia Fm　北门峡组
Benbatu Fm　本巴图组
Benxi (Penchi) Fm　本溪组
Bikou Gr　碧口群
Bingdou Fm　兵斗组
Binggou Gr　冰沟群
Binyang Fm　宾阳组
Bocigou Fm　菠茨沟组
Bolila Fm　波里拉组
Boluoan Mvt　博罗运动
Boqu Gr　波曲群
Bulong Fm　布龙组
Buxin Fm　布心组
Cangfanggou Gr　沧房沟群
Canglangpu (Tsanglangpu) Fm　沧浪铺组
Canglangpuan (Tsanglangpuan) St　沧浪铺阶
Cangshan Fm　苍山组
Cangshan St　苍山阶
Caodi Gr　草地群
Caotangou Gr　草滩沟群
Chaishiling Fm　采石岭组
Chaltai Gr　渣尔太群
Chang'an Fm　长安组
Chaganbulage Fm　查干布拉格组
Changcheng Gr　长城群
Changcheng System　长城系
Changhe Fm　张河群
Changkengshui Fm　长坑水组
Changping Fm　昌平组
Changshan Fm　长山组
Changshanian St　长山阶
Changxingian (Changhsingian) St　长兴阶
Changzhougou Fm　常州沟组
Chasang Fm　札桑组
Chejiang Fm　车江组
Chengjiang (Chengkiang) Fm　澄江组
Chengjiangian Mvt　澄江运动
Chengqiangou Fm　城墙沟组
Chengzitan Fm　城子瞳组
Chigu Fm　齐古组
Chisian (Qixian) St　栖霞阶
Chijiachuan Fm　祁家川组
Chijiang Fm　池江组
Chongshan Gr　崇山群
Chongyu Fm　崇右组
Chuandong Fm　穿洞组
Chuanlinggou (Chuanllingkou) Fm　串岭沟组
Chuanshan Fm　船山组

Chujiang Fm　珠江组
Chujiang Fm　衢江组
Cigou Fm　茨沟组
Ciying Fm　四营组
Congla Fm　丛拉组
Coalhill Memb　煤山段
Cuifeng Fm　翠峰组
Cuijiatun Fm　崔家屯组
Cuizhuang Fm　崔庄组
Dabie Gr　大别群
Dabasu Fm　达巴苏组
Dabuka Fm　大步勘组
Dagangou Fm　大干沟组
Dahongshan Gr　大红山群
Dahongyu (Tahungyu) Fm　大红峪组
Dahuichang Fm　大灰厂组
Dahuoluoshan Gr　大豁落山群
Daijiaping Gr　戴家坪群
Dakdaban (Dakendaban) Gr　达肯大板群
Daklakbulak Gr　达克拉克布拉克群
Dalangshan Gr　大郎山群
Daliugou Gr　大柳沟群
Dalo Fm　大陆组
Dalong Fm　大隆组
Dalongkou Fm　大垅口组
Dameigou Fm　大煤沟组
Dameguaihe Fm　大磨拐河组
Daminshan Fm　大民山组
Dananao Gr　大南澳群
Dananao schist　大南澳片岩
Danghe Gr　党河群
Danshanshi Gr　担山石群
Dangchong Fm　当冲组
Dangzehe Fm　荡泽河组
Dantazi Gr　单塔子群
Danzhou Gr　丹洲群
Dashalong Fm　达沙隆组
Dashigou Fm　大石沟组
Dashiqiao Fm　大石桥组
Datangian St　大塘阶
Datong Fm　大同组
Datongshan Fm　大桶山组
Dawangshan Fm　大王山组
Dawanian St　大湾阶
Daye (Tayeh) Fm　大冶组
Dayingpan Fm　大营盘组
Dege Gr　德格群
Dengfeng Gr　登封群
Dengjia Fm　邓家组
Denglouku Fm　登娄库组
Dengying (Tengying) Fm　灯影组

Dengyingian (Tengyinging) St 灯影阶
Derirong Fm 德日荣组
Dewuian St 德坞阶
Diaoyutai Fm 钓鱼台组
Dingjiashan Fm 丁家山组
Doba Gr 多巴群
Dongchagou Fm 东叉沟组
Dongganglingian (Tungkanglingian) St 东岗岭阶
Donghe Gr 洞河群
Dongjia Fm 董家组
Dongjiao Gr 东焦群
Dongjing Fm 东井组
Dongjun Fm 洞均组
Donglingtai Fm 东岭台组
Dongmen Fm 峒门组
Dongtang Fm 东塘组
Dongtujinghe Fm 东图津河组
Dongwuling Fm 东屋岭组
Dongxuanguan Fm 东宣关组
Dongye (Tungyeh) Subgr 东冶亚群
Dongziyan Fm 童子岩组
Doucun (Toutsun) Subgr 豆村亚群
Doumu Fm 痘母组
Douposi Fm 陡坡寺组
Doushanguan Fm 陡山关组
Doushantuo (Toushantuo) Fm 陡山沱组
Dunhuang Gr 敦煌群
Duobaoshan Gr 多宝山群
Duojiu Fm 多久组
Dushanian St 独山阶
Dushanzi Fm 独山子组
Erdaogou Gr 二道沟群
Erdowa Gr 二道洼群
Ergunahe Gr 爱尔古那河群
Ermaying Fm 二马营组
Falang Fm 法朗组
Fangcheng Gr 防城群
Fanghu Fm 方壶组
Feixianguan (Feihsien Kuan) Fm 飞仙关组
Fengfanghe Fm 芬芳河组
Fengfeng Fm 峰峰组
Fengjiahe Fm 冯家河组
Fengjiawan Fm 冯家湾组
Fenglaizhen Fm 蓬莱镇组
Fengningian St 丰宁阶
Fengshan Fm 凤山组
Fengshanian St 凤山阶
Fengxiang Fm 分乡组
Fengyang Gr 凤阳群
Fenzishan Gr 粉子山群
Fujunshan Fm 府君山组
Fulu Fm 富禄组
Fuping Gr 阜平群
Fupingian Mvt 阜平运动
Fuxian Fm 富县组

Fuxian Fm 复县组
Fuxin (Fuhsin) Gr 阜新群
Fuyang Fm 富阳组
Gahai Fm 尕海组
Gaixian Fm 盖县组
Gangou Fm 乾沟组
Ganjingzi Fm 甘井子组
Ganquan Fm 干泉组
Gantaohe Gr 甘陶河群
Ganxi Fm 甘溪组
Gaofan Subgr 高凡亚群
Gaojiayu Fm 高家峪组
Gaoshanhe Fm 高山河组
Gaotai (Kaotai) Fm 高台组
Gaotaian (Kaotaian) St 高台阶
Gaoyuzhuang (Kaoyuchuang) Fm 高于庄组
Gaozhiping Gr 高芝坪组
Gedicun Fm 格底村组
Gelaohe Gr 革老河组
Gongdong Fm 拱洞组
Gongyanghe Gr 公养河群
Gouhou Fm 沟后组
Guandian (Kuantian) St 关底阶
Guangou Fm 官沟组
Guanling Fm 关岭组
Guanshan Fm 观山组
Guanwushan Fm 观雾山组
Guanyinqiao (Kuanyinchiao) Fm 观音桥组
Guanyinya Fm 观音垭组
Gufeng Fm 孤峰组
Guijiatun Memb 桂家屯段
Gulangian Mvt 古浪运动
Gulanhe Fm 古浪河组
Guniutan Fm 牯牛潭组
Guniuntanian St 牯牛潭阶
Guojiazhai Gr 郭家寨群
Gushan Fm 崮山组
Gushan Fm 姑山组
Gushanian St 崮山阶
Gyirong fauna 吉荣动物群
Haikou Fm 海口组
Haitaoping Fm 核桃坪组
Haitong Fm 海通组
Handaqi Fm 罕达气组
Hangelchaok Fm 汉克尔乔克组
Hannan Complex 汉南杂岩
Hanxia Gr 旱峡群
Haujiagou Fm 郝家沟组
Heilongjiang Gr 黑龙江组
Heitupo Fm 黑土波组
Heishantou Fm 黑山头组
Heijiagou Fm 黑甲沟组
Hekou Gr 河口群
Kelixi Fm 河沥溪组
Hengtang Fm 亨堂组
Hepingxian Fm 和平乡组

Hepu Gr 合浦群
Heshanggou Fm 和尚沟组
Hetong Fm 合桐组
Himalayan Mvt 喜马拉雅运动
Homoshan Fm 荷磨山组
Hongan Gr 红安组
Hongchicun Fm 虹赤村组
Hongchunping Fm 洪春坪组
Hongguleleng Fm 洪古勒楞组
Honghuayuan Fm 红花园组
Honghuayuanian St 红花园阶
Hongliugou Fm 红柳沟组
Hongliuxia Fm 红柳峡组
Hongshantou Fm 红山头组
Hongshuigou Fm 红水沟组
Hongshuizhuang (Hungshuichuang) Fm 红水庄组
Hongtiegou Fm 红铁沟组
Hongtupo Fm 红土坡组
Hongweikeng Fm 红卫坑组
Hongzaoshan Fm 红藻山组
Houcheng Fm 后城组
Houershan Fm 猴儿山组
Huainan Gr 淮南群
Huaitoutala Fm 环头他拉组
Huakaizuo Fm 花开左组
Hualong Gr 化龙群
Huangben Fm 黄垄组
Huangboling Fm 黄柏岭组
Huangcaopo Fm 黄草坡组
Huanghuagou Fm 黄花沟组
Huangjiadeng Fm 黄家凳组
Huangkeng Fm 黄坑组
Huanglian Fm 黄连组
Huanglianduo Fm 黄连垛组
Huanglishu Fm 黄栗树组
Huangnigang Fm 黄泥岗组
Huangshanjie Fm 黄山街组
Huangshuihe Gr 黄水河群
Huangyuan Gr 湟源群
Huaqiao Fm 花桥组
Huashibanian St 滑石板阶
Huashishan Gr 花石山群
Huixiangping Fm 迥香坪组
Hujiersite Fm 呼吉尔斯特组
Hulan Gr 呼兰组
Huluo Fm 胡乐组
Huobaoshan Fm 火包山组
Huodiya Gr 火地垭群
Houjiashan Fm 猴家山组
Huili Gr 会理群
Hunjiang Gr 浑江群
Huoqiu Gr 霍丘群
Hutianian St 壶天阶
Hutuo Gr 滹沱群
Indosinian Mvt 印支运动
Jenheqiao Fm 人和桥组
Jiageda Gr 嘉格达群

Jialingjiang Fm　嘉陵江组
Jialu Fm　甲路组
Jiamlecu Fm　姜姆勒曲组
Ji'an Gr　辑安群
Jianchang Fm　碱厂组
Jiande Gr　建德群
Jiangge Fm　姜格组
Jiangkou Gr　江口群
Jiangmu Fm　江木组
Jiangxian Gr　绛县群
Jian'ou Gr　建瓯群
Jianping Fm　建平群
Jiaodong Gr　胶东群
Jiapeila Fm　甲不拉组
Jiayuan Fm　贾园组
Jidula Fm　基堵拉组
Jiangzi Gr　江子群
Jiayuqiao Gr　嘉玉桥群
Jiehekou Gr　界河口群
Jiezha Gr　结札群
Jijicao Gr　芨芨草群
Jijitaizi Fm　芨芨台子组
Jilong Fm　基龙组
Jingle Fm　静乐组
Jingeryu (Chingeryu) Fm　景儿峪组
Jingtieshan Gr　镜铁山群
Jingxing Gr　景星群
Jining Gr　集宁群
Jinji Fm　金鸡组
Jinningian Mvt　晋宁运动
Jinshanzhai Fm　金山寨组
Jinshuikou Gr　金水口群
Jinxian Gr　金县群
Jishishan Gr　积石山群
Jiucaiyuan Fm　韭菜园组
Jiudingshan Fm　九顶山组
Jiuling Gr　九岭群
Jiuliqiao Fm　九里桥组
Jiulongshan Fm　九龙山组
Jiusi Fm　旧司组
Jixi Gr　鸡西群
Jixian System　蓟县系
Junshao Fm　军哨组
Kabo Gr　嘎波群
Kadong Fm　卡东组
Kalpin Tag Fm　柯坪塔克组
Kalundal Fm　卡仑达尔组
Kangding Gr (Complex)　康定群（杂岩）
Kangkelin Fm　康克林组
Kangshare Fm　康沙热组
Kangzhugou Fm　康主沟组
Kapsalian Fm　卡普沙良群
Karakax Gr　喀拉喀什群
Karamay Fm　克拉玛依组
Karaza Fm　喀拉札组
Karton Fm　卡东组
Kawabulak Gr　卡瓦布拉克群
Keguqinshan Gr　科古琴山群

Kekehiongdukuke Fm　克克雄都库克组
Keluke Fm　克鲁克组
Kezilesu Gr　克孜勒苏群
Kizilenur Fm　克孜勒努尔组
Kongkang Fm　公康组
Konsu Fm　康苏组
Kuanping Gr　宽坪群
Kueilin Fm　桂林组
Kufeng Fm　孤峰组
Kukebai Fm　库克拜组
Kunyang Gr　昆阳群
Kushuixia Fm　苦水峡组
Laibuxi Fm　赖布西组
Lancang Gr　澜沧群
Langshan Gr　狼山群
Langzishan Fm　浪子山组
Lanhe Gr　岚河群
Lanongla Fm　拉弄拉组
Lantang Gr　烂塘群
Lantian Fm　兰田组
Laobao Fm　老堡组
Laocun Fm　劳村组
Laohutai Fm　老虎台组
Laoling Gr　老岭群
Leigongwu Fm　雷公坞组
Leijiahe Fm　雷家河组
Leikoupo Fm　雷口坡组
Lengjiaxi Gr　冷家溪群
Lengshuigou Fm　冷水沟组
Lengwu Fm　冷坞组
Liangfengpo Fm　凉风坡组
Liangjiashan (Liangchiashan) Fm　亮甲山组
Liangquan Fm　凉泉组
Liangshan Memb　梁山段
Lianhuashan Fm　莲花山组
Lianhuashanian St　莲花山阶
Liankang Fm　连坎组
Liantan (Lientan) Fm　连滩组
Liantuo Fm　莲沱组
Liaohe Gr　辽河群
Licha Fm　里查组
Lichaiba Fm　梨柴坝组
Lieguliu Fm　列古六组
Lieryu Fm　里尔峪组
Lincang Gr　临沧群
Linjiang Fm　临江组
Linshan Gr　林山群
Linxiangian St　临湘阶
Lishan Fm　梨山组
Liubatang Fm　柳坝塘组
Liujiatai Fm　刘家台组
Liulaobei Fm　刘老碑组
Liuling gr　刘岭群
Liumei Fm　留眉组
Lixian Coal Series　礼贤煤系
Lizigou Fm　梨子沟组
Longgang Gr　龙岗群

Longjiagou Fm　龙家沟组
Longli Fm　隆里组
Longmaxian (Lungmachian) St　龙马溪阶
Longmaxi (Lungmachi) Fm　龙马溪组
Longquanguan Gr　龙泉关群
Longshan Fm　龙山组
Longshoushan Gr　龙首山群
Longtanian St　龙潭阶
Longtouzhai Fm　龙头寨组
Longwangmiaonian (Lungwangmiaonian) St　龙王庙阶
Longwangshan Cycle　龙王庙（火山）旋迴
Longyin Fm　龙吟组
Longzhuagou Gr　龙爪沟群
Lonongla Fm　弄弄格拉组
Lopingian St　乐平阶
Loreping Fm　罗惹坪组
Louziba Gr　楼子坝群
Luanxian Gr　滦县群
Lueyang Gr　略阳组
Lufeng Fm　禄丰组
Luliang Gr　吕梁群
Luliangian Mvt　吕梁运动
Luochangxia Fm　乐昌峡群
Luojiamen Fm　骆家门组
Luokedong Fm　落可栋组
Luoquan Fm　罗圈组
Luotuoshan Fm　骆驼山组
Luoyang Fm　洛阳组
Luoxie Fm　落雪组
Luoyu Gr　洛峪群
Luoyukou Fm　洛峪口组
Lushan Fm　庐山组
Lutiaoshan Fm　绿条山组
Luzhijiang Fm　绿什江组
Maantang Fm　马鞍塘组
Macaoyuan Fm　马槽园组
Machala Fm　乌查拉组
Madiyi Fm　马底驿组
Mahuanggou Fm　蚂蟥沟组
Majiadian Gr　马家店群
Majiagou Fm　马家沟组
Majian Fm　马涧组
Majiatun Fm　马家屯组
Majin Fm　马金组
Mangang Fm　曼岗组
Mangcuo Fm　莽错组
Mangeer Fm　曼格组
Mangkelu Fm　芒格鲁组
Mankuanhe Fm　曼宽河组
Manshan Fm　漫山组
Mantou Fm　馒头组
Maoba Fm　茅坝组
Maokouan St　茅口阶
Maotian Fm　毛田组
Maozhuang (Maochuang) Fm　毛庄组

Maozhuangian St　毛庄阶
Maping Ls　马平灰岩
Mapingian St　马平阶
Mashan Gr　麻山群
Mati Fm　马蹄组
Meidang Fm　美党组
Meidiping Fm　麦地坪组
Meishucun Fm　梅树村组
Meishucunian St　梅树村阶
Meitan Fm　眉潭组
Melukahe Fm　美路卡河组
Menbu Fm　门布组
Menggongao Fm　孟公坳组
Mentougou Fm　门头沟组
Miandiancun Fm　面店村组
Miaogao (Miaokao) Fm　妙高组
Miaogaoian (Miaokao'an) St　妙高阶
Miaopo Fm　庙坡组
Miaopoan St　庙坡阶
Mingshan Gr　名山群
Mingshui Fm　明水组
Minhe Fm　民和组
Minshan Fm　民山组
Minxian Gr　勉县群
Miyun Gr　密云群
Moshigou Fm　磨石沟组
Muchang Fm　木厂组
Mukeng Fm　木坑组
Muzhart Gr　木札尔特群
Nabiao Fm　纳标组
Nadu Fm　那读组
Nagaoling (Nakaoling) Fm　那高岭组
Nagaolingian (Nakaolingian) St　那高岭阶
Naj Tai Gr　纳赤台群
Nancheng Fm　南城组
Nandaling Fm　南大岭组
Nanfen Fm　南坟组
Nanguanling Fm　南关岭组
Nanjinguan Fm　南津关群
Nankou Gr　南口群
Nantuo Fm　南沱组
Nanyan Fm　南岩组
Nanying Fm　南营组
Naomugen Fm　脑木根组
Naxing Fm　纳兴组
Nenjiang Fm　嫩江组
Niangniangshan Fm　娘娘山组
Nielamu (Nyalam) Gr　聂拉木群
Nieniexiongla Fm　聂聂雄拉组
Nihewan Fm　泥河湾组
Ningguo Fm　宁国组
Ningundang Fm　牛滚档组
Niujiaohe Fm　牛角河组
Niushang Fm　牛上组
Niutoushan Fm　牛头山组
Niuwu Fm　牛屋组
Nouyinhe Fm　诺音河组

Nyanqentanglha Gr　念青唐古拉群
Nyalam (Nielamu) Gr　聂拉木群
Obian Gr　峨边群
Oheblake Gr　俄霍布拉克群
Omeishan Basalt Fm　峨眉山玄武岩组
Otouchang Fm　鹅头厂组
Panzhao Fm　番召组
Paoquanchang Fm　跑泉厂组
Pargangtag Gr　帕尔冈塔格群
Penjiayuan Fm　彭家园组
Pengkuang Fm　蓬夼组
Penglai Gr　蓬莱群
Pingbian Gr　屏边群
Pingjing Fm　平井组
Pingtoushan Gr　平头山群
Pingyipu Fm　平驿铺组
Piyuancun Fm　皮园村组
Puli Mvt　普里运动
Puling Fm　铺岺组
Pupiao Fm　蒲缥组
Pupuga Fm　普普嘎组
Pushang Fm　铺上组
Pushuiqiao Fm　普水桥组
Qakmaklik Fm　恰克马克里克组
Qianfuyan Fm　千佛崖组
Qianxi Gr　迁西群
Qiaoenbulak Fm　巧恩布拉克组
Qiaotou Fm　桥头组
Qihulin Fm　奇虎林组
Qikbulak Fm　奇格布拉克组
Qiktai Fm　七克台组
Qilianian (Guangxian) Mvt　祁连 (广西) 运动
Qingbaikou System　青白口系
Qinling (Tsinling) Gr　秦岭群
Qinglonghe Gr　青龙河群
Qingshan Fm　青山组
Qingshankou Fm　青山口组
Qingshipo Fm　青石坡组
Qingshishan Fm　青石山组
Qingshuijiang Fm　清水江组
Qingxi Fm　清溪组
Qingyan Fm　青岩组
Qingyuangang Fm　青元岗组
Qiongzhusi (Chiungchussu) Fm　筇竹寺组
Qiongzhusian (Chiungchussuan) St　筇竹寺阶
Qiyunshan Fm　齐云山组
Qomolangma (Jolmo Lungma) Gr　珠穆朗玛群
Quanaogou Fm　泉脑沟群
Quanji Gr　全吉群
Quanjiagou Fm　全家沟组
Quantou Fm　泉头组
Quberga Fm　曲布日嘎组
Qubu Fm　曲布组
Qujing (Chutsing) Fm　曲靖组

Qulonggongba Fm　曲龙共巴组
Rehe (Jehol) Fm　热河组
Renjiazhuang Fm　任家庄组
Ridang Fm　日当组
Rimova Fm　日莫瓦组
Ruifang Gr　瑞芳群
Saerbuer Fm　沙尔布尔组
Saerburte Mountain Fm　沙尔布尔特山组
Saiwa Fm　赛瓦组
Sandouping Gr　三斗坪群
Sanggan Gr　桑干群
Sangonghe Fm　三公河组
Sangxiu Fm　桑秀组
Sangzhutag Gr　桑珠塔格群
Sanheming Gr　三合明群
Sanjiaotang Fm　三教堂组
Sanliting Fm　三里亭组
Sanmen Fm　三门组
Sanmentan Fm　三门滩组
Sanqiutian Fm　三丘田组
Sanshuanghe Fm　三双河组
Sawozi Fm　沙窝子组
Sanxia Gr　三峡群
Sanying Fm　三营组
Sanyoudong gr　三游洞群
Serongsi Fm　色容寺组
Shadui Gr　沙堆群
Shagou Fm　沙沟组
Shahegou Fm　沙河沟组
Shahezi Fm　沙河子组
Shajingzi Fm　沙井子组
Shalitashi Fm　莎里塔什组
Shancaoyu Fm　山草峪组
Shangdamingshan Fm　三大明山组
Shanghu Fm　上湖组
Shangshi Gr　上施群
Shangshu Gr　上墅群
Shangsi Fm　上司组
Shangxi Gr　上溪群
Shanwang Fm　山旺组
Shanxi Fm　山西组
Shaodong Fm　韶东组
Shaofanggou Fm　烧房沟组
Shaping Fm　沙平组
Shawan Fm　沙湾组
Shaximiao Fm　沙溪庙组
Shedian Fm　蛇店组
Shennongjia Gr　神农架群
Shenshan Gr　神山群
Shenwangshan Fm　神王山组
Shetianchiaonian St　佘田桥阶
Shidian Fm　施甸组
Shihhotse (Shihezi) Fm　石盒子组
Shihuiba Fm　石灰坝组
Shijia Fm　史家组
Shilengshui Fm　石冷水组
Shimen Fm　石门组

STRATIGRAPHIC INDEX

Shinagan Fm　什那干组
Shiniulan Fm　石牛栏组
Shiniulanian St　石牛栏阶
Shipanshan Fm　石板山组
Shiqianfeng (Shichianfeng) Fm　石千峰组
Shiqipo Gr　石器坡群
Shisanjianfang Fm　十三间房组
Shisanlitai Fm　十三里台组
Shixiagou Fm　石峡沟组
Shiyingliang Fm　石英梁组
Shizhongshan Fm　石钟山组
Shizikou Fm　狮子口组
Shizong Fm　师宗组
Shizui (Shihtsui) Fm　石嘴组
Shouxian Fm　寿县组
Shuangqiaoshan Gr　双桥山群
Shuangshanzi Gr　双山子群
Shuangxiwu Gr　双溪坞群
Shuicheng Gr　水城群
Shuidigou Fm　水底沟组
Shuiquan Fm　水泉组
Sibao Gr　四堡群
Sibaoan Mvt　四堡运动
Sidaohe Fm　四道河组
Sidaolazi Fm　四道砬子组
Sidingshan Fm　四顶山组
Sifangtai Fm　四方台组
Sijiaoyanggou Fm　四角羊沟组
Siling Fm　四棱组
Sipai Fm　四排组
Sipainian St　四排阶
Songhuajiang Gr　松花江群
Songshan Gr　嵩山群
Subashi Fm　苏巴什组
Suchiawan Fm　苏家湾组
Sugaitbulak Fm　苏盖特布拉组
Suhuhe Gr　苏呼河群
Suining Fm　遂宁组
Sujiazhuang Fm　索家庄组
Sunjiagou Fm　孙家沟组
Susong Gr　宿松群
Suxiong Fm　苏雄组
Taihua Gr　太华群
Taihuai Fm　台杯组
Taiping Gr　太平群
Taipingding Fm　太平顶组
Taishan Complex (Gr)　泰山杂岩（群）
Taiyuan Fm　太原组
Taizicun Fm　台子村组
Talejihe Fm　塔勒吉河组
Talga Fm　塔尔尕组
Taliqike Fm　塔里奇克组
Tangbagou Fm　汤巴沟组
Tangding Fm　塘丁组
Tangdingian St　塘丁阶
Tangjiawu Gr　唐家坞群
Taoshuyuanzi Fm　桃树园子组
Taowan Gr　陶湾群

Tashihe Fm　塔西河组
Taye (Daye) Fm　大冶组
Tayean St　大冶阶
Tayugou Fm　大峪沟组
Tereecken Fm　特列艾肯组
Tianhechi Fm　天河池组
Tianjingshan Fm　天井山组
Tianshanian Mvt　天山运动
Tianzhushan Mb　天柱山段
Tiaojishan Fm　髻髻山组
Tiaosu Fm　缔敖苏组
Tiedonggou Gr　铁洞沟群
Tieling Fm　铁岭组
Tiemulik Fm　铁木里克组
Tieshanling Fm　铁山岭组
Tieyegou Fm　铁野沟组
Tingzhonglong Fm　丁宗隆组
Toba Fm　妥垅组
Tongchuan Fm　铜川组
Tongmuliang Gr　通木梁群
Tongshicun Fm　通什村组
Tongtianhe Gr　通天河群
Tongziyan Fm　童子岩组
Toudunghe Fm　头顿河组
Tuanpokou Fm　团泊口组
Tuanshanzi Fm　团山子组
Tungur Fm　通古组
Tuoba Fm　妥垅组
Tuodian Fm　妥甸组
Tuolainanshan Gr　托赖南山群
Tuolie Gr　陀烈群
Tuqiaozi Fm　土桥子组
Turpan Gr　土尔番群
Tuyilok fm　吐依洛克组
Tyokar Fm　交嘎组
Unuoer Fm　乌奴尔组
Urumchi Fm　乌鲁木齐组
Utubulake Fm　乌图克拉克组
Wanghudun Fm　望虎墩组
Wangjiawan Fm　王家湾组
Wangshan Fm　望山组
Wangshi Gr　王氏群
Wangyu Fm　王佑组
Wanshanzhuang Fm　万山庄组
Wanyiaoshu Fm　弯腰树组
Wanzhangzi Fm　湾丈子组
Watasi Fm　瓦塔寺组
Weiji Fm　魏集组
Weimei Fm　魏美组
Weiningian St　威宁阶
Wenbishan Fm　文笔山组
Wenbiyan Mb　文笔岩段
Wendu'ermiao (Wendurmiao) Gr　温都尔庙群
Wenquan Fm　温泉组
Wentong Fm　文通组
Wentoushan Fm　文头山组
Woduhe Fm　沃都河组

Wudang Gr　武当群
Wufengian St　五峰阶
Wuhangshan Gr　五行山群
Wuhe Gr　五河群
Wujiahe Fm　吴家河组
Wujiapingian (Wuchiapingian) St　吴家坪阶
Wulang Fm　乌郎组
Wulanbulage Fm　乌兰布拉格组
Wulashan Gr　乌拉山群
Wulonggou Fm　五龙沟组
Wumishan Fm　雾迷山组
Wuqiangxi Fm　五强溪组
Wuqingna Fm　乌箐纳组
Wushan Fm　伍山组
Wutai Gr　五台群
Wutai'an Mvt　五台运动
Wutonggou Fm　梧桐沟组
Wutung Fm　五通组
Wuye Fm　乌叶组
Wuyitake Fm　乌依塔克组
Xiacaowan Fm　下草弯组
Siadamingshan Fm　下大明山组
Siafang Fm　下坊组
Xiamaling (Hsiamaling) Fm　下马岭组
Xiangshan Gr　象山群
Xiangxi Fm　香溪组
Xianshuihe Fm　咸水河组
Xiaochaka Fm　小茶卡组
Xiaogoubei Fm　小沟背组
Xiaogaolu Fm　小高炉组
Xiaojiahe Fm　肖家河组
Xiaolongtan Fm　小龙潭组
Xiaomeigou Fm　小煤沟组
Xiaoping Fm　小坪组
Xiaoshihugou Fm　小石胡沟组
Xiaoshui Fm　小水组
Xiaotangxi Fm　小塘子组
Xiaoyan Fm　小岩组
Xiaoputonggou Fm　小普通沟组
Xiare Fm　蔗拉组
Xiashan Fm　小山组
Xiatigu Fm (= Zhuwo Gr)　下底谷组（ = 侏倭群）
Xiaxiang Fm　下巷组
Xicun Fm　西村组
Xiejia Fm　谢家组
Xiejiawan Fm　谢家湾组
Xifengsi Fm　西峰寺组
Xihanshui Gr　西汉水群
Xihe (Hsiho) Gr　细河群
Xikang Gr　西康群
Xikuangshan Fm　锡矿山组
Xikuangshanian St　锡矿山阶
Ximen Gr　西门群
Xinchang Fm　新厂组
Xincun Fm　新村组
Xindianzi Fm　新店子组

Xinditag Gr 兴地塔格群
Xinduqiao Fm 新都桥组
Xineitai Fm 洗奶台组
Xinghuadukou Gr 新华渡口群
Xingkai'an Mvt 兴凯运动
Xingshikou Fm 杏石口组
Xingxingjia Gr 星星峡群
Xinhe Fm 新河组
Xinji Fm 辛集组
Xinmincun Fm 新民村组
Xionger (Hsiunger) Gr 熊耳群
Xishanbulak Fm 西山布拉克组
Xishancun Memb 西山村段
Xitun Memb 西屯段
Xiumo Fm 休莫组
Xiuning (Hsiouning) Fm 休宁组
Xiushan Fm 秀山组
Xiushanian St 秀山阶
Xiyanghe Gr 西洋河群
Xiyangshan Fm 西阳山组
Xiyu Gravels 西域砾石层
Xuantasi Fm 双塔寺组
Xuanwei Fm 宣威组
Xuehuagou Fm 雪花沟组
Xueshan Fm 雪山组
Xuexianggang Fm 雪乡岗组
Xuhuai Gr 徐淮群
Xujiachong Memb 许家冲段
Xujiahe Fm 须家河组
Xuzhuang Fm 徐庄组
Xuzhuangian (Hsuchuangian St 徐庄阶
Yali Fm 亚里组
Yan'an Fm 延安组
Yanbian Gr 盐边群
Yanchang Fm 延长组
Yangba Fm 阳坝组
Yangchun Fm 杨村组
Yangfang Mvt 杨坊运动
Yangjiadian Fm 杨家店组
Yangjiang Fm 漾江组
Yangjibulak Gr 杨吉布拉克群
Yangmaba Fm 养马坝组
Yangqiao fm 堰桥组
Yanguanian (Aikuanian) St 岩关阶
Yangxia Fm 阳霞组
Yangye Fm 杨叶组
Yangzhuang (Yangchuang) Fm 杨庄组

Yankou Fm 堰口组
Yanlingguan Fm 雁翎关组
Yanshanian Mvt 燕山运动
Yanshiping Gr 雁石坪群
Yaochicun Fm 瑶池村组
Yaojia Fm 姚家组
Yaojie Fm 窑街组
Yaolinghe Gr 耀岭河群
Yejishan Gr 野鸡山群
Yeli Fm 冶里组
Yejisa (Yegisar) Gr 英吉沙群
Yeliu Fm 野柳组
Yeljiang Gr 叶尔羌群
Yemashan Gr 野马山群
Yenchang (Yanchang) Fm 延长组
Yichangian St 宜昌阶
Yichang Fm 宜昌组
Yigeziya Fm 依格孜牙组
Yikebulge Fm 伊克布拉格组
Yimen Fm 益门组
Yindaoyuan Fm 樱桃园组
Yindonggou Fm 银洞沟组
Yingchengzi Fm 营城子组
Yiningian Mvt 伊宁运动
Yinjiajian Fm 殷家涧组
Yinmin Fm 因民组
Yintang Fm 应堂组
Yintangian St 应堂阶
Yinwashan Fm 砚瓦山组
Yinzhubu Fm 印渚埠组
Yipinglan Gr 一平浪群
Yiwagou Fm 依瓦沟组
Yiyuan Fm 倪园组
Yongdingzhuang Fm 永定庄组
Yongkang Gr 永康群
Yongning Gr 永宁群
Yongningzhen Fm 永宁镇组
Youganwo Fm 油干锅组
Yuhe Fm 孟河组
Yujiang Fm 郁江组
Yujiangian St 郁江阶
Yukengou Fm 育肯沟组
Yulongshan Fm 玉龙山组
Yulongsi Fm 玉龙寺组
Yungan Fm 云岗组
Yunlong Fm 云龙组
Yunmengshan Fm 云梦山组
Yunshan Fm 云山组
Yuqian Fm 于潜组

Yurtus Fm 玉尔吐斯组
Yushanjian Fm 渔山尖组
Yushe Fm 榆社组
Yushulazi Fm 榆树砬子组
Yushuwan Fm 榆树湾组
Yuxi Fm 鱼西组
Zagunao Gr 杂谷脑群
Zaige Fm 宰格组
Zalainor Gr 札赉诺尔群
Zamnazu Fm 札姆那曲组
Zanhuang Gr 赞皇群
Zaoshi Gr 枣市群
Zaoshang Fm 造上组
Zewan Fm 者王组
Zhamkti Fm 查谟克提组
Zhalagongga Fm 札拉贡嘎组
Zhamure Fm 札木热组
Zhangbaling Fm 张八岭组
Zhanggou Fm 张沟组
Zhangqian Fm 漳前组
Zhangqu Fm 张渠组
Zhangxia (Changhsia) Fm 张夏组
Zhangxian (Changhsian) St 张夏阶
Zhangwu Fm 彰武组
Zhanjin Fm 展金组
Zhaobishan Fm 照壁山组
Zhaowei Fm 赵圩组
Zhawangzi Fm 札旺子组
Zhegui Fm 者贵组
Zhengmuguan Fm 正目观组
Zhenjiang Fm 镇江组
Zhesi (Jesu) Fm 哲斯组
Zhidan Gr 志丹群
Zhilo Fm 直罗组
Zhitang Fm 志堂组
Zhongning Fm 中令组
Zhongpeng Fm 中棚组
Zhongpur Gr 宗浦群
Zhongshan Fm 宗山组
Zhongtiao Gr 中条群
Zhongtiao'an Mvt 中条运动
Zhongyu'an Mvt 中岳运动
Zhougang Fm 周岗组
Zhoujieshan Fm 皱节山组
Zhuoguodong Fm 卓戈洞组
Zhulongguan Gr 珠龙关群
Ziliujing Fm 自流井组
Zongji Fm 宗集组

INDEX

Acanthopteris-Ruffordia flora 155, 160
accretional crustal consumption zones 239, 242, 252, 267, 274
acritarch 34, 37, 39–40, 46–7, 50, 52, 55, 57, 59, 61
Asperatopsophoshaera 34, 37
agnostids 69–70
Ailiankate Group 35
Ailaoshan Group 47, 61, 214
   polymetamorphic belt 214, 217, 231
Airjigan Group 41
Akalehe Formation 110
Aksu Group 35, 56
algae 39, 40, 56, 60, 65, 68–9, 114, 141, 159
Altai Formation 243
   Palaeozoic fold zone 243
Altungol Formation 55–6
aluminium 102
ammonoids 93, 96, 98–9, 102–3, 107, 109–10, 112, 116–20, 127–9, 132, 140–2, 145, 157, 269
amphibians 131
amphibolite facies 18–19, 24, 212–16, 218, 220, 223, 225, 227
Ampyx zone 76
Anabarites 65
   –*Circotheca* assemblage 70
Anagymnotoceras sp. 134
Angaran flora 14, 119, 242–4, 247, 267
angiosperms 156–60
Angzanggou Formation 86
Anisian Stage 127, 129, 132–4
Anluolin Formation 46
Anshan Group 17–18, 25, 27, 213, 232, 244
   Movement 17
Aoqu Formation 109
apatite 19, 189, 210
Archaean A Movement 190–1
archaeocyathids 65, 68
Arumberia 61
Atlantic faunas 78
atmosphere 259, 261
aulacogen (rift zones) 54, 62, 239, 244, 246–7, 249, 254–5, 261, 263, 275

Bacun Group 54
Badaohe Group 22, 27, 212, 232
Bagong Formation 132
Bagongshan Formation 41
Bahe Formation 173
Baicalia sp. 40–3
   –*Chihsienella* assemblage 39
Baicaoping Formation 40

Baihu Group 41
Bailiuping Formation 98
Bailongjiang Group 87
Baisha'an Stage 83
Baishan Formation 110
Baitanba Formation 144–5
Baiyanding Formation 45
Baiyaya Formation 87
Baizhu Formation 45
Baizushan Group 142
Bajocian Stage 141–2
Balatonites sp. 132
Balikelike Formation 119
banded iron formations 18, 21, 27, 31
Banqiao Formation 46
Banqiuguan Formation 87
Banxi Group 44–5, 54, 192, 204, 214, 249
Baode Formation (Hipparion Red Clay) 168, 173
Baota Formation 76, 78
Bashili Xiaohe Formation 86
Batang Group 198
Bathonian Stage 140, 142
bauxite 114
Bayan Obo Group 34, 213, 245
Bayisi Formation 55–6
Baytik Hercynian fold zone 243
Beianzhuang Formation 77
Beidajian Formation 40
Beikuang Formation 98
belemnites 140–1, 157
Bendong granite 45
Benxi Formation 108
Birmanites zone 76
bivalves 68, 87, 109–10, 116–17, 119, 127–9, 131–3, 140–3, 145, 153, 156–9, 160, 170
Blackwelderia zone 68, 70
Bocigou Formation 132
Bolila Formation 132
Boqu Group 97
Boreal faunas 14, 119–21
boron 34
brachiopods 65, 68–9, 76, 83, 92–3, 96–9, 102–3, 107–12, 114–21, 127, 133, 141, 157, 267
bryozoans 114, 117, 119–21, 133
Bulong Formation 86, 170
Burmesia sp. 133

Caledonian or Qilianian (Guangxian) Movement 91, 190, 192, 195, 197, 199, 204–5, 214–16, 219, 221, 230, 240, 242–3, 247, 249, 251, 259, 263

fold zones 215, 245
Callipteris zeilleri 14, 267
Callovian Stage 140–1
Calvinella-Metacalvinella zone 65, 68
   –*Mictasaukia* zone 81
camarotoechids 133
Cambrian 192, 196
Cangfanggou Group 120, 131
Canglangpuan Stage 65, 68
Cangshan Group 47, 61, 214
Carboniferous 242–3
Carnian Stage 127–9, 132
Cathaysian flora 119–21, 244–5, 247, 251–2, 267, 269, 270
cephalopods 83–6, 127, 133, 158
   zones 76, 79
Chafangzi limestone 39
Chaltai Group 34
Chang'an Formation 54
Changcheng Group 37
Changchengian System 13, 33, 37–8, 192, 195
Changhe Formation 145
Changkengshui Formation 78
Changlingzi Formation 57–8
Changping Formation 68
Changshanian Stage 68
Changxian Stage 65, 68
Changxinian Stage 114, 116–18, 120–1
Changzhougou Formation 37, 40
Chara 145–6, 158, 172
Charnia 52, 61
charophytes 155–60
Chejiang Formation 158
Chelonia 155, 159
Chengqiangou Formation 109
Chengzitan Formation 17
Chenjiang Formation 52
   Movement 50, 190, 214, 248
Chihsian (Qixian) Stage 114, 117–18
Chilotherium xizangensis 170
Chlamys sp. 142
Chuandong Formation 96
Chuanlinggou Formation 37, 39
Chuaria 41, 60
Chuifengshan Formation 96
Cigou Formation 17, 27
Circotheca sp. 65, 69
Cishan granite 229
Cladophlebis sp. 132, 142, 145–6, 158
Claraia sp. 132
Clathrophyopsis sp. 132
coal 3, 69, 102, 108–9, 114, 116–18, 120, 127–30, 132, 140, 142, 144–6, 159–60, 169, 269

conchostracans 65, 68–9, 131, 159
Congla Formation 132
conodonts 76, 78, 85, 87, 92–3, 98–9, 103, 112, 117–18, 127, 133
Conophyton sp. 37, 41, 43, 59, 60
  lituum–Jacutophyton assemblage 39
convergent crustal consumption zone 239–40, 424–3, 246, 248, 267–8, 273–4
copper 3, 18, 33, 114, 191, 208, 210
corals 83–8, 96, 98–9, 102, 107–12, 114–15, 117–21, 141, 245, 251, 267, 273
  compound 120
  hexacorals 132
  rugose 92–3, 133
  tabulate 103
Coreanoceras–Manchuroceras zone 76
Cornuicardia sp. 132
crinoid 129, 132–3
crocodilia 155–6
Cuipingshan Formation 118
Cuccoceras sp. 132
Cuizhuang Formation 40
Cupressinocladus sp. 146
Cypridea sp. 154, 156–7, 159–60

Dabie Group 20, 25, 27, 35, 213, 249
Dabie uplift 197
Dabsu Formation 157
Dagangou Formation 111
Dahongshan Group 35, 213, 249, 262
Dahongyu Formation 37, 39, 40
Dahuichang Formation 144
Dahuoloshan Group 42
Daijiaping Formation 158
Dakendaban Group 35, 42, 57, 214
Daklakbulak Group 25, 34
Dala'an Stage 103, 107, 109
Daliugou Group 42, 214
Dalong Formation 118, 146
Dalongkou Formation 43
Dameguaihe Formation 160
Damesella zone 68, 70
Danghe Group 42, 214
Danshanshi Group 33
Dangzehe Formation 24, 27
Dantazi Group 21–2, 27
Danyingpan Formation 43
Daonella sp. 132
Darwinula sp. 129–31, 142, 145, 154
Dashalong Formation 129
Dashigou Formation 142
Dashiqiao Formation 34
Datangian Stage 103, 108–11
Dawangshan Cycle 203
Dawan Formation 76–8
Daye Formation 128–9
Denfeng Group 24–5, 212, 221, 232
Dengjia Formation 46
Denglouka Formation 159

Dengying Formation 52, 61, 65
Dengyingian 69
Derirong Formation 129
Development stages 237, 256, 273
Dewuian Stage 103, 109–11
Diaoyutai Formation 41
Dicellograptids 76, 79, 88
Dictyophyllum sp. 133
Dicynodon sp. 134
Dingwuling Formation 55
dinosaurs 155–7, 159
Diplograptids 76, 79
Discocerida 76
Discophyllia sp. 132
Doba Group 158
Dongganglingian Stage 93, 96–8
Dongjia Group 33, 57
Dongjing Formation 158
Dongjun Formation 172
Donglingtai Formation 144
Dongpo Formation 57
Dongtang Formation 158
Dongtujinghe Formation 110
Dongye Subgroup 32–3
Doucun Subgroup 32
Doushanguan Formation 87
Doushantuo Formation 52, 59
Drepanura zone 68, 70
Drepanuroides 65
Duobaoshan Group 243
Doujiu Formation 157
Duoroner Group 42
Dushanzi Formation 171

Edentates 172
Ediacara fauna 50, 59, 61, 243, 265
Ellesmerocerids 76
Ellipsophysa 41, 60
Emeishan Basalt Formation 117
Endocerids 76
Eocene 169–72
Eoredlichia 65
Ephemeropsis trisetalis 154
Equisetites sp. 130–1
Erdaowa Group 33, 213
Ergunahe Group 215
Ermaying Formation 130
estherids 131, 143, 145–6, 153–4, 156, 158–60

Falang Formation 134
Fangcheng Group 87
Fanjingshan Group 44, 189, 249
Fanzhao Formation 44
Feidong Group 47
Feixianguan Formation 127
Fengfeng Formation 78
Fengningian Series 102, 109, 111
Fengshanian Stage 68
Fengxiang Formation 76
Fengyang conglomerate 60

Fenwei Graben 15, 272
Fenzishan Group 34, 213
fishes 3, 92–3, 96–9, 117, 146, 154–6, 159–60
foraminifera 112, 117, 119, 120, 127, 133, 141, 169–70
Fujanshan Formation 39
Fuping Group 22–3, 25, 27, 213, 232
Fuping Movement 25, 190, 223, 259
fusulinids 102–3, 107–11, 114–21, 133
Fuxian Formation 134, 142, 161
Fuyang Formation 19

Gahai Formation 108
Gaixian Formation 34
Gandise-Nyainqentanglha volcanic belt 199, 201, 231
Ganjingzi Formation 58
Ganquan Formation 110
Gantaohe Group 33
Ganxi Formation 98, 129
Gaofan Subgroup 32
Gaojiayu Formation 34
Gaoligong metamorphic belt 217
Gaoshanhe Formation 40
Gaoyuzhuang Formation 37, 39
Gaozhiping Group 145
Garantiana sp. 140
Gastrioceratidae 102
gastropods 65, 68, 110, 119, 131, 133, 158–60, 170
Gedicum Formation 132
Gelaohe Formation 103, 109, 112
glaciation 62, 174, 262
  glacial deposits 50, 52, 55–7
Glaessnerina (*Charnia*) 61
Globigerina sp. 158
Glossopteris flora 113, 119, 129
Glytograptus zone 76, 78–9, 81, 86–8
gold 18
Gondolella sp. 133
Gondong Formation 45, 54
Gondwana continent 14, 49, 63, 252, 262, 270, 275
  flora and fauna 15, 111, 113, 154, 251
  suture zone 199
Gongyanghe Group 61
goniatites 92
Gouhou Formation 60
Grammatodon sp. 141
granulite facies 18, 24, 27, 34, 210, 223
graphite 55
graptolites 76–9, 81, 83–8, 92, 97
greenschist facies 18, 24, 213–16, 218, 222, 225
greenstone belts (Archaean) 18, 20, 25
Griesbachian Stage 127
Grunneria sp. 41
  –Xiayingella assemblage 37
Guandian Stage 83, 85

INDEX

Guanwushan Formation 98
Guanyinqiao Formation 76–7, 81
Gulangian Movement 247, 257
Gulanhe Formation 86
Guniutan Formation 76–8, 81
Guojiazhai Subgroup 32, 34
Gushan Cycle 203
Gushanian Stage 68, 70
Gyirong fauna 170
Gymnosolen 32, 41, 42, 59, 60
gymnosperm 158–9
Gymnotoceras sp. 134
gypsum 65, 68, 102, 109, 153, 158, 170–1

hadrosaurs 157
Haikou Formation 96
Haitaoping Formation 61
Haitong Formation 99
Halobia sp. 132
Halovella sp. 132
Hannan Complex 189
Hanxia Group 87
Hebaoshan Formation 98
Heichashan Group 33
Heilongjiang Group 47, 213
Heishantou Formation 43
Heitugou Formation 56–7
Hekou Group 35, 160, 249, 262
Helixi Formation 87
Hepu Group 87
Hercynian Movement 190–3, 195, 197–8, 204–5, 215–16, 219, 231, 240, 243, 246, 248, 259
Heshanggou Formation 130, 134
Hetong Formation 45
Hettangian Stage 140, 142, 146
Heyao Formation 40
Himalayan Movement 168, 190, 246, 272
  biota 121
  magmatism 199
  metamorphism 217–18, 231
'Hipparion red clay' *see* Baode Formation
Hipparion xizangensis 170–1
Hirnantia fauna 81, 83, 88
Hoffetella–Redlichia zone 65, 69
Hollanites sp. 134
Homo erectus pekinensis 174
Hongan Group 20, 213
Hongchicun Formation 47
Hongguleleng Formation 98
Honghuayuan Formation 76–8
Hongliujia Formation 242
Hongshuizhuang Formation 39
Hongtupo Formation 23
Hongweikeng Formation 129
Hongzaoshan Formation 56–7
Hongzixi Formation 44
Houchang Formation 144

Huaitontala Formation 109
Hualong Group 35, 213
Huangben Formation 129
Huangcaopo Group 243
Huanghuagou Formation 86
Huangkeng Formation 78
Huanglian Formation 55
Huanglianduo Formation 57
Huangling anticlinorium 35
Huangnigan Formation 79
Huangshanjie Formation 131
Huangshuihe Group 44
Huangyuan Group 42, 214
Huashibanian Stage 103, 108, 110
Huashishan Group 42
Huili Group 35, 43–4, 192, 249, 262
Huixiangping Formation 44
Hulan Group 215, 245
Huluo Formation 79
Huoqiu Group 24
Huoshiling Formation 159
Hutianian Series 102, 108
Huto Group 24
Hutuo Group 32, 212–13, 220, 223, 225
Hutuoan 13, 31–4, 191, 244, 261
hydrosphere 259
hyolithids 52, 56, 65, 68–9

Indosinian Movement 15, 126, 128, 140, 190–2, 195, 197, 204–5, 216–17, 231, 240, 244–5, 248, 253, 255, 259, 269, 273
Induan Stage 126, 129, 132
inland basin deposits 15, 142, 153
inland basin development 140, 250–1
Inoceramus 140
insects 131, 145–6, 154
Inzeria 41, 43, 59
  –*Linella* assemblage 39
iron 3, 17, 19, 24, 31, 68–9, 102, 114, 191, 208, 210
island arcs 20, 237, 247, 249, 252–3, 255, 259, 262–3, 267, 269
isotopic dating
  Anshan Group 213
  Archaean 18–21, 23–4, 189
  Caledonian 199, 204
  Cretaceous 203
  Dakendaban Group 214
  Devonian 191
  Fuping Group 213
  Hercynian 204
  Himalayan 199, 218
  Indosinian 198, 204
  Jining Group 213
  Jinningian 204
  Jurassic 191
  Kunyang Group 214
  Mashan Group 214

Proterozoic 33–5, 38–9, 43, 45–9, 189
Qianxi Group 213
Sibaoan 204
Sinian 52, 57, 60
Taihua Group 213
Taishan Group 213
Yanshanian 191, 199, 204, 218

Jenheqiao Formation 88
Jesu Formation 243
Jiageda Group 47, 60
Jialingjiang Formation 127, 133
Jialu Formation 44
Jiamlechu Formation 142
Jian Group 34
Jiande Group 153
Jiangge Formation 99
Jiangkou Group 54–5, 62
Jiangnan uplift 54
Jiangxian Group 32, 213
Jian'ou Group 55
Jianping Group 27
Jiaodong Group 18, 27, 213
Jiapeila Formation 132
Jiehekou Group 22, 27
Jijicao Group 119
Jijitaizi Formation 110
Jilong Formation 110, 119
Jingeryu Formation 39, 41
Jingtieshan Group 42
Jingxing Formation 160
Jining Group 21, 27, 212–13
Jinji Formation 142
Jinningian Movement 44, 50, 190, 204–5, 214, 219, 248, 250, 259, 261–2
Jinshuikou Group 35, 42
Jinxian Group 58
Jiucaiyuan Formation 131, 134
Jiulianshan intrusion 204
Jiuliqiao Formation 60
Jiulongshan Formation 144
Jiusi Formation 103, 108
Jiuxiao Formation 45
Jixi Group 146–7
Jixian section 13, 36
Jixianian System 13, 38–40, 42, 192, 195
Jongkang Group 153
Jurusania 41–3, 59

Kadong Formation 157
Kaixinling fold zone 252
Kalundal Formation 119
Kangshare Formation 129
Kanzhugou Formation 132
Kaotaia–Kunmingaspis zone 65
Karamay Formation 131, 134
Kashi Group 170
Kawabulak Group 41, 214

INDEX

Keguqinshan Group 110
Kekexiongkuduk Formation 86, 243
Keluke Formation 109
Kezilesu Group 158
Kham-Yunnan uplift 62
Kimmeridgian Stage 140
Kuanping Group 40, 214, 221
Kueiling Formation 158
Kukebai Formation 158
Kunyang Group 35, 43, 192, 214, 249, 262
Kussiella sp. 33, 41–3
Kuximchik Group 41

Ladinian Stage 127, 129, 132–3
Laibuxi Formation 129
Laminarites antiquissimus 52, 57
Lancang Group 47, 61
Langcanjiang metamorphic belt 214–15, 217, 231
Langshan Group 213
Langzishan Formation 34
Lanhe Group 33
Lanongla Formation 141
Lantang Formation 145
Lanzhi granite 32
Laocun Formation 146
Laoling Group 34
Laurasian flora 129
Leigongwu Formation 54
Leikoupo Formation 127, 133
Leizhou peninsula 15
Lenotropites sp. 134
Liangfengpo Formation 117
Liangjiashan Formation 77
Liangquan Formation 97
Liangshan Formation 117
Lianhuashanian Stage 92–3, 98
Liantan Formation 87
Liantuo Formation 51–2
Liaohe Group 17, 34, 213, 261
Lichaiba Formation 88
Lieryu Formation 34
Linshan Group 24
Linxiang Formation 76, 78
Liubatang Formation 43
Liujiagou Formation 118, 130
Liulaobei Formation 41
Liumei Group 54
Lixian Coal series 116
Longli sandstone 45
Longmaxian Stage 83, 87
Longquanguan Group 22–3, 25, 223
Longshan Formation 39, 60
Longshoushan Group 25, 213, 261
Longtan Formation 117–18
Longtanian Stage 116, 120–1
Longwangmiaoian Stage 65, 68
Longwangshan Cycle 203
Longyin Formation 114, 121
Longzhuagou Group 142, 146

Lophosphaeridium 59
Lopingian Series 114, 118
Louziba Group 55
Luanxian Group 212
Lueyang Formation 108
Lufeng Formation 144–5
Luliang Group 32–3, 213
Luliangian Movement (Zhongtianoan) 31, 190, 212–13, 244, 250, 259, 261–2
Luochangjia Group 54
Luojiamen Formation 47
Luokedong Formation 46
Luoquan tillites 57
Luotuoshan Formation 98
Luoxue Formation 43
Luoyu Group 40, 57
Luoyukou Formation 40, 57
Lutiaoshan Formation 110
Luzhijiang Formation 43
Lycoptera 154 6, 159
Lystrosaurus 131, 134

Maantang Formation 127
Macaoyuan Group 43
Machala Formation 109
Macrocephalites sp. 140
Madiyi Formation 44–5
magnesite 34
magnetite 55
Majiadian Group 33, 213
Majiagou Formation 78
Majian Formation 146
Majiatun Formation 60
mammals 146, 171–3
manganese 102, 114
Mangang Formation 160
Mangcuo Formation 120
Mangeer Formation 97
Mangkelu Formation 97
Mankuanhe Formation 160
Manshan Formation 23
Mantou Formation 68
Maoba Formation 98
Maokouan Stage 114, 116–18, 121
Maotian Formation 65
Maozhuangian Stage 65, 68, 70
Mapingian Stage 102, 107–11
marble 22–4, 33–4, 47, 110
marginal seas 54, 62, 237, 239, 262, 270
Mashan Group 47, 61, 213, 215
fauna 61
Matuyama palaeo-magnetic reversal 174
medusoids 58
Megastage 256–9
Meishucunian Stage 65, 70
Melukahe Formation 110
Menbu Formation 141
Mengtongou Formation 144

Mesozoic folding 240, 243
metazoans 50, 58–60
Miandiancun Formation 96
Miaogao'an Stage 83, 85
Miaopo Formation 76–8, 81
Micristylus–Pseudogymnosolen assemblage 39
migmatization 190–1, 205, 208, 229–30, 232–3
mineral deposits 3, 16, 64, 68–9, 102, 114, 190–1, 198, 203, 207–10
Mingshui Formation 159
Minhe Formation 108, 160
Minshan Formation 19
Min-Yue fold zone 253
Miocene 169–72
Miyun Group 212
Mobilism 237, 273
Montlivaltia sp. 132
Moshshan migmatite 230
Movement 256
Muchang Formation 22, 27
Mukeng Formation 46
Multisiphonia 60
Muzat gneiss 35

Nabiaonian Stage 93, 98–9
Nadanhada fold zone 254–5
Nagaolingian Stage 92–3, 98
Nammalian Stage 127
Nandaling Formation 144
Nanfen Formation 41–57
Nanguanling Formation 58
Nanjinguan Formation 76
Nankou Group 37
  section 36
Nanling series 208
Nantuo Formation 51–2, 56
Nanyan Formation 55
Nanying Formation 23
Naomugen Formation 172
nautiloids 76, 78, 98, 265
Naxing Formation 109
Nei Mongol fold zone 194
Nenjiang fault 243
  fold zone 194
  Formation 159
Neocalamites sp. 130–1, 142, 144
Neogene 170
Niangniangshan Cycle 203
nickel 18
Nielamu Group 219
Nieniexiongla Formation 141
Ningguo Formation 78
Niujiaohe Group 55
Niushang Formation 78
Niuwu Formation 46
Nongshan Formation 172
Norian Stage 127–9, 132
Northern biotas 132, 134
North Himalayan geosyncline 160

Notoungulates 172
Nujiang fracture 199
Nummulites sp. 170
   limestone 273
Nyainquentanglha Group 214, 253
Nyalam Group 49, 214

Obian Group 44
Oheblake Group 131
oil and gas 4, 15, 114, 130, 168, 172, 270
Olenekian Stage 126, 129, 132–3
Oligocene 169, 171–2
Oncocerida 76
Ophiceras–Claraia wangi assemblage 133
ophiolite suites 132, 193, 199, 201, 239, 242–3, 246, 249, 251, 255, 263, 265, 267, 269, 273
Ordos massif 14
Ordovician 192, 196
ostracods 114, 119, 127–31, 142–3, 145–6, 154, 156–9, 171–2
Otouchang Formation 43
Oxfordian Stage 141–2

Pacific fauna 65, 78, 128, 250, 255
Palaeocene 171–2
Palaeogene 274
Palaeolenus 65, 68
palynological assemblages 157
Pangaea 259, 262, 275
Paoquanzhang Formation 23
Paraboxonia 32
Pargontagh Group 41, 214
Pengkuang Formation 19, 27
Penglai Group 41, 213
phosphate 34, 55, 65, 68–9, 102, 114
Pingjing Formation 65
Pingtoushan Group 42
Pingyipu Formation 98
plants 92, 96–8, 109, 111, 116–20, 128–32, 134, 144–6, 153, 155–6, 158–60, 242
Pleistocene 169, 173
Pleuromeia sp. 129–30, 134
Pleurotomaria sp. 141
Pliensbachian Stage 142
Pliocene 169–70, 173
Polyedryxium 52
polymetamorphism 221
Posidonia sp. 132
potash 37
Praesolenopora 60
primates 172–3
Procolophonidae fugusuchus 134
productids 133
Progonoceratites sp. 134
Protospongia 55, 69
Pseudohydria 157, 159
Psiloceras sp. 140, 145

pteridophytes 155, 158–9
pterosaur 155
ptychoparids 70
Puli Movement 255
Puling Formation 46
Pushang Formation 31–2
Pushuiqiao Formation 132

Qakmaklik Formation 61
Qianxi Group 21–2, 25, 27, 210–13, 221, 232, 244
Qiaotou Formation 57, 59
Qilian fold belt 263
   geosyncline 14, 34–5, 190, 193, 196, 214
Qilianian Movement 257
Qilian–Qinling–Dabie metamorphic belt 231, 247
Qingbaikouan System 13, 39–41, 45, 62, 192, 195, 262
Qinglonghe Group 33, 213
Qingshan Formation 159
Qingshankou Formation 159
Qingshuijiang Formation 44
Qingyan Formation 134
Qiongzhusian Stage 65
Qixia Formation 117–18
Qixian (Chihsian) Stage 114
Qiyunshan Formation 158
Quanji Group 56
Quannaogou Group 86
Quantou Formation 159
Quaternary cave deposits 173
Quberga Formation 119
Qubu Formation 119
Qulonggongba Formation 129

radiolarians 140
Ramapithecus 173
rare earths 69
Redlichia 68, 70
Red River fault zone 35, 198, 214
regression
   Indosinian 140
   Ladinian 127
Rehe fauna 146, 161
   Group 153
Renjiazhuang Formation 229–30
reptiles 129–31, 145–6, 155–60
Rhaetian Stage 127–8
rhinoceroids 171
Ridang Formation 140, 146
Rimova Formation 157
Rudistae 158
Ruifang Group 169
Ruyang Group 40

Sabellidites 58, 60
Saerburte Formation 86
Saitula Group 61
Saiwa Formation 142

salt deposits 3, 52, 65, 68, 153
Sandaogou Formation 18
Sandouping Group 35
Sanggan Group 20–1
Sangxiu Formation 157
Sanheming Group 33, 213
Sanshuanghe Formation 96
Sanxia Group 169
Sawozi Formation 98
scolecodonts 56–7
Serongsi Formation 132
Serratognathus zone 76
Shaanganning Basin 15
Shadui Group 157
Shaerbur Formation 86
Shahezi Formation 159
Shajingzi Formation 119
Shancaoyu Formation 19, 227, 229–30
Shangdamingshan Formation 98
Shanghu Formation 172
Shangshi Group 55
Shangshu Group 47, 54
Shangsi Formation 103
Shangxi Group 46
Shanmentan Formation 97
Shantungaspis zone 68, 70
Shanwang Formation 173
Shanxi Formation 118
Shaodong Formation 103, 112
Shaofanggou Formation 131, 134
Shawan Formation 171
shelly facies 65, 78, 83, 87, 265
Shennongjia Group 35, 42, 192
Shenshan Group 55
Shenwangshan Formation 158
Shetianqiaonian Stage 93, 96–8
Shihezi Formation 118
Shihtienfenia sp. 134
Shihugou Formation 86
Shihuiba Formation 173
Shijia Formation 60
Shilengshui Formation 65
Shinagan limestone 33, 39
Shiniulanian Stage 83, 87
Shipanshan Formation 110
Shiqianfeng Formation 118, 130, 134
Shiqipo Group 86, 97
Shixiagou Formation 97
Shiyingliang Formation 56–7
Shizui Formation 32
Shouxian Formation 41
Shuangjiaoshan Group 46
Shuangshanzi Group 33, 213
Shuangxiwu Group 46
Shuicheng Formation 160
Shuidigou Formation 24
Sibao Group 45, 192, 214, 249
   Movement 45, 190, 204, 259, 262
Siberian fauna 267
Sidaohe Formation 23
Sidaolazi Formation 18

Sidingshan Formation 60
Sifangtai Formation 159
Sijiaoyanggou Formation 111
Siling Formation 169
Silurian 192, 196
Sinamia 154–6
Sinian System 36, 192
Sinograptids 76
Sinokannemeyeria sp. 130, 134
Sivapithecus 173
Siwalik Formation 170
  fauna 170
Songshan Group 33, 213
Spathian Stage 127
Sphaerium 157, 159
sponge spicules 69
spores and pollen 93, 128, 131, 134, 142, 146, 155, 157–9, 171–2
Stage 256
Stepanoviella 113
stromatolites 32–4, 37, 39–43, 50, 56, 59–60, 261
stromatoporoids 103
subduction zones 239, 243, 250, 259, 263, 265, 267, 269, 272
Suhuhe Group 215, 230
Sujiazhuang Formation 23
Susong Group 35, 213
syntexis series granites 208

Taeniatum sp. 56–7, 61
Taibai–Danfeng subduction zone 248
Taihua Group 24, 27, 32, 34, 189, 213, 232
Taiping Group 87
Taipingding Formation 19
Taishan Group 19, 27, 213, 232, 244
Taiwan fold zone 255
Taiyuan Formation 109
Taizicun Formation 171–2
Taliqike Formation 131
Tancheng–Lujian fault zone 18–20, 34
Tangbagou Formation 103, 108–9
Tangdingian Stage 93, 98–9
Tangjiawu Group 87
Taoshuyuanzi Formation 171
Taowan Group 40, 214, 221, 246
Tarphycerida 76
Tayagou Formation 17
tentaculitids 92–3, 99
terebratulids 133
Tereeken Formation 55–6
Tethyan fauna and flora 14, 111, 120, 132–4, 245, 247, 250, 267,
Tethys ocean 63, 91, 113, 153, 169, 247, 250, 263, 267, 269–70, 275
Tianjingshan Formation 127
Tianshan–Nei Mongol belt 230
Tianshan fold belt 191, 193, 195, 242, 246, 259, 265

Tianshan trough 263
Tiaojishan Formation 144
Tiaosu Formation 111
Tiedonggou Formation 34
Tieling Formation 39
Tieshanling Formation 24
tin 191, 208
Tingzhonglong Formation 99
tintinnids 93
Tirolites sp. 133
Tithonian Stage 140–2
Toba Formation 120
Todites sp. 144–6
Tongchuan flora 130, 134
Tongshicun Formation 17
Tongziyan Formation 118
Trachyceras sp. 132
Trachysphaeridium 59, 61
transcurrent fracture zones 239, 255, 269, 272
transformation series granites 208
transgressions
  Cambrian 245
  Carboniferous 102, 111
  Devonian 91
  Palaeozoic 250
  Permian 113
  Sinian 63
  Triassic 250
Trematosphaeridium sp. 40, 43, 52, 56
Triassic subduction zone 198
trilobites 65, 68–9, 76, 83, 85, 87, 92–3, 97, 99, 133, 265
  zones 81
Tuanpokou Formation 23
Tuanshanzi Formation 37, 39
tungsten 191, 208
Tungur Formation 173
Tuolainanshan Group 42
Tuqiaozi Formation 98
Tutuella sp. 145–6
Tuvaella fauna 86, 88, 242–3, 267
Tuyilok Formation 158
Tyokar (Jiaoga) Formation 120

Utubulake Formation 97

vanadium 69
vertebrates 120, 130–1, 134–5, 145–6, 172, 174

Wangchang Formation 27
Wangshan Formation 60
Wangshi Group 153, 157, 159
Wanshanzhuang Formation 19
Wanzhangzi Formation 27
Watasi Formation 132
Weichselia 155–6
Weimei Formation 140
Weiningian Stage 102, 109–10
Wenbishan Formation 118

Wendurmiao Group 215, 230, 245
Wenquan Formation 142
Wentong Formation 45
Wentoushan Formation 87
Woduhe Formation 86
worm 52
Wudang Group 35, 213
Wufeng Formation 76, 78
Wuhangshan Group 57
Wujiahe Formation 87
Wujiapingian Stage 114
Wulanbulage Formation 172
Wulashan Group 21, 27
Wulonggou Formation 110
Wumishan Formation 38
Wuqiangxi Formation 44
Wuqingna Formation 109
Wutai Group 22, 24, 27, 31–2, 212, 220, 223, 225
Wutaiian Stage 13, 31, 33–4, 190–1, 244, 259, 261
Wuye Formation 44
Wuyitake Formation 158

Xenodiscus sp. 133
Xiafang Formation 55
Xiamaling Formation 39
Xianggui fold zone 253
Xiaojiahe Formation 44
Xiaoshui Formation 129
Xia Putonggou Formation 98–9
Xiare Formation 140
Xiashan Formation 97
Xiatigu Formation 132
Xiaxiang Formation 87
Xicun Formation 169
Xiejia Formation 173
Xiejiawan Formation 98
Xifengsi Formation 54
Xihe Group 34, 41, 57–8
Xikuangshanian Stage 93, 96–8, 103
Ximeng Group 61
Xinchang Formation 78
Xindianzi Formation 98
Xinduqiao Formation 132
Xinfushan granite 229–30
Xingditag Group 25, 34, 41, 213
Xinghuadukou Group 47, 60, 213
Xingkaian orogeny 61, 194, 215, 240, 243
Xingshikou Formation 144
Xingxingjia Group 35
Xinhe Formation 142–3
Xinmin Formation 46
Xiumo Formation 141
Xiushanian Stage 83, 85, 87
Xiyanghe Group 33–4, 40, 192
Xuehuagou Formation 24
Xueshan Formation 142
Xujiahe Formation 128
Xuzhaungian Stage 65, 68

INDEX

Yali Formation 109
Yanbian Group 44, 249
Yanchang Formation 130, 134, 142
 flora 131
Yangba Formation 61
Yangfang Movement 44
Yangjiadan Formation 18
Yangjibulak Group 41, 214
Yangqiao Formation 118
Yangmaba Formation 98
Yangsinian Series 114
Yanguanian Stage 103, 107, 109–11
Yangzhuang Formation 38
Yangzi Gorge section 36, 50–1
Yangzi Series granitoids 208
Yanlingguan Formation 19–20, 27, 225–7
Yanshanian Movement 190–5, 197–9, 202, 204–5, 218, 221–3, 231–3, 246, 251, 255, 270
Yanshiping Group 142
Yaochicun Formation 32
Yaojia Formation 159
Yarlung Zanbo fracture zone 199, 201
 –West Yunnan belt 231, 273
Yejishan Group 33
Yeki Formation 77
Yeliu Group 169
Yemashan Group 42
Yiaotangxi Formation 128
Yichangian 76–8

Yigeziya Formation 158
Yikebulge Formation 172
Yiliangella 65
Yimin Formation 160
Yindaoyuan Formation 17
Yindonggou Formation 43
Yingcheng Formation 159
Yiningian or Boluoan orogeny 242, 247, 257, 259, 265
Yinjisa Group 158
Yinmin Formation 43
Yinwashan Formation 79
Yinzhubu Formation 78
Yiwagou Formation 108
Yongning Group 41
Youganwo Formation 172
Youjiang fold zone 254
Yujiangian Stage 93, 98
Yukengol Group 56
Yulangbaijia Group 157
Yuli metamorphic belt 231
Yunmengshan Formation 40
Yunnanotheca–Pupoella assemblage 65
Yunshan Formation 97
Yuqian Formation 79
Yushanjian Formation 146
Yushulazi Group 34
Yushuwan Formation 23
Yuxi Formation 45

Zagunao Group 132
Zalainor Group 160
Zamnazu Formation 142
Zanhuang Group 22
Zewan Formation 103
Zhalagongga Formation 120
Zhamure Formation 129
Zhangbaling Group 47
Zhanggou Formation 24
Zhangguangcailing fold zone 194
Zhangping Group 145
Zhangqian Formation 46
Zhaobishan Formation 55
Zhawangzi Formation 157
Zhenmuguan Formation 57
Zhepure Formation 170
Zhidan Group 153, 161
Zhitang Formation
Zhongning Formation 97
Zhongpeng Formation 97
Zhongpur Group 161
Zhongtiao Group 33, 213
Zhongyuean Movement 257
Zhouji Formation 24
Zhoujieshan Formation 56–7
Zhulongguan Group 41, 192, 214
Zhulumite Formation 97
Zhuoguodong Formation 99
Zongpu Formation 169